A HANDBOOK OF ENVIRONMENTAL MANAGEMENT

A Handbook of Environmental Management

Edited by

Jon C. Lovett

Professor of Sustainable Development in a North South Perspective, University of Twente, The Netherlands and former Director, Centre for Ecology, Law and Policy, University of York, UK

David G. Ockwell

Lecturer in Geography and Fellow, Sussex Energy Group and Tyndall Centre for Climate Change Research, University of Sussex, UK

Edward Elgar
Cheltenham, UK • Northampton, MA, USA

© Jon C. Lovett and David G. Ockwell 2010

All rights reserved. No part of this publication may be reproduced, stored in a retrieval system or transmitted in any form or by any means, electronic, mechanical or photocopying, recording, or otherwise without the prior permission of the publisher.

Published by
Edward Elgar Publishing Limited
The Lypiatts
15 Lansdown Road
Cheltenham
Glos GL50 2JA
UK

Edward Elgar Publishing, Inc.
William Pratt House
9 Dewey Court
Northampton
Massachusetts 01060
USA

A catalogue record for this book
is available from the British Library

Library of Congress Control Number: 2009940631

Mixed Sources
Product group from well-managed forests and other controlled sources
www.fsc.org Cert no. SA-COC-1565
© 1996 Forest Stewardship Council

FSC

ISBN 978 1 84064 207 0 (cased)

Printed and bound by MPG Books Group, UK

Contents

List of contributors		vii
1	Introduction *Jon C. Lovett and David G. Ockwell*	1
2	Global biodiversity conservation priorities: an expanded review *Thomas M. Brooks, Russell A. Mittermeier, Gustavo A.B. da Fonseca, Justin Gerlach, Mike Hoffmann, John F. Lamoreux, Cristina G. Mittermeier, John D. Pilgrim and Ana S.L. Rodrigues*	8
3	Integrated conservation and development projects: a positive role for forest conservation in tropical Africa? *Neil Burgess, David Thomas, Shakim Mhagama, Thomas Lehmberg, Jenny Springer and Jonathan Barnard*	30
4	Biodiversity conservation in managed landscapes *Tom M. van Rensburg and Greig A. Mill*	75
5	How do institutions affect the management of environmental resources? *Bhim Adhikari*	119
6	Analysing dominant policy perspectives – the role of discourse analysis *David G. Ockwell and Yvonne Rydin*	168
7	Theoretical perspectives on international environmental regime effectiveness: a case study of the Mediterranean Action Plan *Sofia Frantzi*	198
8	The price of fish and the value of seagrass beds: socioeconomic aspects of the seagrass fishery on Quirimba Island, Mozambique *Fiona R. Gell*	241
9	The link between ecological and social paradigms and the sustainability of environmental management: a case study of semi-arid Tanzania *Claire H. Quinn and David G. Ockwell*	282
10	Exploring game theory as a tool for mapping strategic interactions in common pool resource scenarios *Vanessa Pérez-Cirera*	309

11	Economic valuation of different forms of land-use in semi-arid Tanzania *Deborah Kirby*	331
12	Economic growth and the environment *Dalia El-Demellawy*	359
13	Biodiesel as the potential alternative vehicle fuel: European policy and global environmental concern *Mahesh Poudyal and Jon C. Lovett*	408

Index 431

Contributors

Bhim Adhikari, Institute for Water, Environment and Health, United Nations University, Canada

Jonathan Barnard, BirdLife International, UK

Thomas M. Brooks, NatureServe, USA; World Agroforestry Centre (ICRAF), University of the Philippines; and School of Geography and Environmental Studies, University of Tasmania, Australia

Neil Burgess, Institute of Biology, University of Copenhagen, Denmark

Dalia El-Demellawy, Environment Department, University of York, UK

Gustavo A.B. da Fonseca, Global Environment Facility, USA and Departamento de Zoologia, Instituto de Ciencias Biologicas, Universidade Federal de Minas Gerais, Brazil

Sofia Frantzi, Institute for Environmental Studies (IVM), Vrije Universiteit Amsterdam, The Netherlands

Fiona R. Gell, Wildlife and Conservation Division, Department of Agriculture, Fisheries and Forestry, Isle of Man Government, British Isles

Justin Gerlach, Nature Protection Trust of Seychelles, UK

Mike Hoffmann, IUCN Species Survival Commission, c/o United Nations Environment Programme – World Conservation Monitoring Centre, UK

Deborah Kirby, Kirby Consulting, Nairobi, Kenya

John F. Lamoreux, Randolph, New Hampshire, USA

Thomas Lehmberg, Danish Ornithological Society, Denmark

Jon C. Lovett, University of Twente, The Netherlands

Shakim Mhagama, Wildlife Conservation Society of Tanzania

Greig A. Mill, De Montfort University, UK

Cristina G. Mittermeier, International League of Conservation Photographers, USA

Russell A. Mittermeier, Conservation International, USA

David G. Ockwell, Department of Geography, University of Sussex, UK

Vanessa Pérez-Cirera, WWF Mexico

John D. Pilgrim, BirdLife International in Indochina, Vietnam and Global Biodiversity Network, Vietnam

Mahesh Poudyal, University of York, UK

Claire H. Quinn, School of Earth and Environment, University of Leeds, UK

Ana S.L. Rodrigues, Department of Zoology, University of Cambridge, UK and Centre d'Ecologie Fonctionnelle et Evolutive, CNRS–UMR 5175, France

Yvonne Rydin, Bartlett School of Planning, University College London, UK

Jenny Springer, WWF US, USA

David Thomas, BirdLife International, UK

Tom M. van Rensburg, Department of Economics, National University of Ireland, Galway, Republic of Ireland

1. Introduction
Jon C. Lovett and David G. Ockwell

Environmental management has risen from being the task of technical natural resource specialists to being the concern of everyone on the planet. This has led to a rapid expansion in the range of jobs dealing with environmental issues. Not only are ecologists, conservationists, hunters, farmers and fishers involved; we now also have professionals in social science fields such as environmental economics, law and politics. Previously a topic that was dominated by the application and interpretation of technical measures such as species diversity and population growth rates, environmental management is now being debated in terms such as property rights and market trading. Sometimes the technical and social aspects make uneasy bedfellows: for example, ecologically minded conservationists can find themselves at loggerheads with human rights lawyers seeking equitable access to protected areas for indigenous peoples. In this book we aim to provide overviews and specific examples of case studies and techniques that are used in environmental management from the local level to international environmental regimes.

The recognition of a division between technical and social fields of study is not new. In 1959 the scientist, administrator and novelist C.P. Snow gave a lecture in Cambridge entitled 'The Two Cultures and the Scientific Revolution'. This focused on the idea that the 'intellectual life of the whole of western society is increasingly split into two groups', literary intellectuals and scientists (Snow, 1998). The 'Two Cultures' theme was taken up again nine years later in another famous paper, 'The Tragedy of the Commons', written by the biologist Garrett Hardin (Hardin, 1968). In this paper, Hardin addressed the classic academic divide between social and natural sciences. He described the gulf thus:

> An implicit and almost universal assumption of discussions published in professional and semi-popular scientific journals is that the problem under discussion has a technical solution. A technical solution may be defined as one that requires a change only in the techniques of the natural sciences, demanding little or nothing in the way of change in human values or ideas of morality. The class of 'no technical solution problems' has members ... They think that farming the seas or developing new strains of wheat will solve the problem – technologically. I try to show here that the solution they seek cannot be found. (Hardin, 1968, p.1243)

2   *A handbook of environmental management*

The point of Hardin's paper is that technical solutions alone cannot solve environmental management problems. The environment does not exist in isolation from human society or the economic systems that operate within society. The environment both defines, and is shaped by, the activities of human beings. In the past, however, there has been a tendency amongst environmental managers to try to implement technical natural-science-based solutions to environmental problems without any attempt to understand the socioeconomic dynamics that underlie the context within which such technical solutions are applied. This has often resulted in misaligned management objectives and ultimately management failure.

Leach and Mearns' (1996) study of the fuel wood shortage in Africa in the 1980s provides a good example of this (Ockwell and Rydin, 2006). The fuel wood shortage was perceived by most environmental managers as the result of a wood supply gap. In other words, demand for fuel wood exceeded supply and the technical solution was simple – plant more trees. Unfortunately, this tree planting approach failed to address the problem of the fuel wood shortage for the many African people whose livelihoods were affected. It emerged later that the problem was not a simple issue of a lack of supply, but a far more complex problem related to the nature of ownership and use of trees as a source of fuel in Africa. Following the broad-scale failure of the tree planting policy to address the fuel wood shortage, social scientists working together with natural scientists later demonstrated that the basic assumptions that define the idea of a supply gap ignore more subtle issues such as the fact that most fuel wood comes from clearing wood for agriculture or from lopping branches valued for fruit and shade. From the perspective of people affected by the fuel wood shortage there was not one simple problem of a lack of supply, but many more complex socioeconomic problems associated with command over trees and their products to meet a wide range of basic needs. This highlights the need to attend to a range of socioeconomic issues in environmental management, such as the nature of property rights regimes, local cultural practices and the subjective, often conflicting, understanding of different resource users (Ockwell, 2008).

The constructive outcomes of natural and social scientists working together to solve environmental management problems have led to an increasing awareness of the need for interdisciplinary approaches to environmental management. This requires managers to combine insights from both the natural and social sciences in order to ensure sustainable outcomes. In an attempt to help current and future managers to understand how they might complement their natural science approaches with insights from the social sciences, this *Handbook of Environmental Management* contains a range of case studies that demonstrate the complementary application of different social science techniques in combination with ecological management thinking.

Tom Brooks et al. (Chapter 2) highlight the importance of an awareness of how conservation funding is spent. Allocation of money essentially represents a key way in which human beings interact with the environment. The approach taken to prioritizing which areas benefit from the $6bn spent annually on conservation has obvious consequences for global biodiversity. In their chapter, a shorter version of which was originally published in the journal *Science*, Brooks et al. present a comprehensive review of the concepts, methods, results, impacts and challenges of approaches to nine templates of global biodiversity priorities that have been proposed by biodiversity conservation organizations over the last decade. Their review is rooted within the theoretical irreplaceability/vulnerability framework of systematic conservation planning. This chapter makes an important contribution to improving understanding of these prioritization approaches, which in turn makes it possible to orient more efficient allocations of geographically flexible conservation funding.

Neil Burgess et al. (Chapter 3) discuss people versus environment and people *and* environment policies in the context of wildlife conservation as a divide between those promoting 'fortress conservation' and those promoting 'people-focused conservation'. In their chapter they argue that for environmental managers involved with implementing conservation projects on the ground in the developing world, these polarized views often represent impractical extremes. Furthermore, for people living in rural areas of developing countries, the divide between 'development' and 'conservation' is also often quite artificial. Burgess et al. highlight a third approach to environmental management that falls between the two extremes. Projects that take this middle ground approach are known as Integrated Conservation and Development Projects (ICDPs). The authors present a detailed analysis of the successes and failures of ICDPs over the years and develop some practical ecological, social and economic criteria by which ICDPs might be assessed. They then provide a practical example of how to apply such criteria by using them to analyse the successes and failures of two ICDPs with which they have had personal involvement. By combining ecological criteria with social and economic criteria, the authors' analysis enables them to make a series of practical management recommendations for making the ICDP model more effective in achieving conservation at the same time as sustaining and improving the lives of the people that live in these areas.

Management efforts to maintain biodiversity do not need to be focused solely on protected areas. Economically productive landscapes contain many of the world's species, and many protected areas now regarded as wilderness were formerly managed for livestock, hunting or extensive agriculture. Tom van Rensburg and Greig Mill (Chapter 4) take a functional approach to the ecology of biodiversity conservation with an emphasis on disturbance. With rising human population leading to ever-increasing demand for food and other natural products, policies that offer incentives for combining biodiversity conservation with other productive management objectives will become ever more important in the future.

The move away from protection to production has resulted in new laws that shift the focus of management from central government control of natural resources such as forests, towards community involvement in management with corresponding changes in access and utilization. Bhim Adhikari (Chapter 5) demonstrates the role that social institutions play in the management of common pool resources (CPRs – natural resources that are communally owned and managed). Drawing on the new institutional economics literature, Adhikari shows how an understanding of the nature of social institutions is vital if environmental managers are to be successful in intervening in the management of CPRs. This includes a need to understand the property rights determining the nature of resource ownership as well as any unwritten social 'contracts' that permit members of the community to access and use the resource. When CPRs begin to be degraded, it is often as a result of external pressures that erode the social institutions that have traditionally governed resource management regimes. Management interventions that fail to understand these traditional institutions and the way in which they have been disrupted are unlikely to be successful in restoring natural resource use to a sustainable pattern.

David Ockwell and Yvonne Rydin (Chapter 6) explore the idea of policy discourses in theoretical and methodological terms. They provide a practical example of how environmental managers might formally approach the analysis of the hidden assumptions, values and beliefs that often underpin dominant framings of environmental problems (for example, the fuel wood supply gap mentioned above) and expose them to more critical scrutiny. These dominant framings often prevent more sustainable, alternative policy solutions from gaining policy influence. Exposing them to critical scrutiny is one way in which to demonstrate the policy relevance of alternative knowledge. Ockwell and Rydin focus on the now well-established field of 'discourse analysis'. They introduce some of the core theoretical principles behind different approaches to discourse analysis before demonstrating the methodological and practical implications of these different approaches via their application to a case study of fire management in Cape York, northern Australia. Their chapter provides a practical example of 'how to do discourse analysis'. At the same time it clearly highlights the insights that environmental managers might derive by using discourse analysis to better understand the hidden assumptions that lie behind different management options.

Differing discourses are not only found in approaches to the management of specific resources or areas of land, but also in much higher-level policy. Sofia Frantzi (Chapter 7) reviews different perspectives on international environmental regime effectiveness using the Mediterranean Action Plan (MAP) as an example. Originally conceived by an 'epistemic community' of scientists as a means of combating pollution in the Mediterranean Sea, the MAP can also be regarded as a tool that enables Mediterranean countries to come together for negotiation, with the technical management goal of pollution reduction being subsidiary to the objective of more general political cooperation. This insight is fundamental to understanding why scientists often become frustrated with policy-makers when the science findings they are trying to promote take a back seat to considerations of trade and security, which are the main drivers of national interests.

Staying in the marine environment, but with a local rather than international focus, Fiona Gell (Chapter 8) provides a detailed demonstration of how understanding the economics behind natural resource use can lead to a better understanding of how to ensure that resource use is sustainable. Gell looks at the economics of a seagrass fishery in the Quirimba Archipelago, northern Mozambique. Through an in-depth analysis of the socioeconomic dynamics of the people who rely on the fishery for their livelihoods, Gell makes an informed set of management recommendations for the long-term sustainable management of the fishery.

Claire Quinn and David Ockwell (Chapter 9) highlight an issue common to many of the chapters in this handbook – that environmental management for sustainable development needs to protect the environment at the same time as protecting and developing the livelihoods of those people who depend on it. This is particularly important for some of the world's poorest people whose livelihoods are often most dependent on natural resources. Using the case study of semi-arid Tanzania, Quinn and Ockwell highlight the reciprocal relationship that exists between the environment and society. They then demonstrate how the paradigms that define environmental managers' and policy-makers' conceptions of ecological and social problems are integral to defining the policy discourses that shape the choice of management solutions. The authors provide a clear example of how the traditional emphasis on the ecologically centred 'equilibrium theory' led to a view of pastoralism as responsible for environmental degradation in semi-arid regions of sub-Saharan Africa. In contrast, the alternative, more recently emerging ecological paradigm based on 'non-equilibrium theory' lends itself to a new perspective that sees pastoralists and other indigenous populations as being an integral part of the environment. Most importantly, Quinn and Ockwell show that these different ecological and social paradigms have fundamental implications for the policy discourses that are adopted and that define appropriate management strategies. Equilibrium theory was fundamental to the policy discourse of people versus environment that has traditionally defined colonial-influenced approaches to environmental management in Africa. Non-equilibrium theory, on the other hand, supports a policy discourse of people *and* environment, which sees the appropriate management response to be inclusive of indigenous people and their knowledge. Quinn and Ockwell's central argument is that if the ecological paradigms that underpin policy discourses fail to recognize the reciprocal link between the environment and society, the resulting management solutions can only protect the environment at the expense of the livelihoods of poor people, thus failing to achieve sustainable development.

The two following chapters by Vanessa Pérez-Cirera (Chapter 10) and Deborah Kirby (Chapter 11) provide differing methodological perspectives on the semi-arid environment discussed by Quinn and Ockwell. Pérez-Cirera explores the application of game theory whereas Kirby illustrates the use of production function economics. Together, these chapters provide detailed examples of the potential for different approaches to environmental decision-making.

The scale of environmental management changes to the macroeconomic considerations of economic growth and the environment in Chapter 12 by Dalia El-Demellawy. Whilst environmental riches are associated with either a complete absence of economic activity in pristine wilderness, or low-level hunter-gatherer economies, it is perhaps wealthier countries that can afford to have both the technological benefits of development and sustainable environmental policies. Intermediate economies are characterized by natural resource exploitation and pollution. This observation is formalized in the 'environmental Kuznet's curve', which suggests that there is an inverted 'U'-shaped curve of environmental degradation associated with development. If this is the case, then the macroeconomic environmental management solution is to enhance economic development to the point where the whole planet is enjoying environmental sustainability. However, as explained by El-Demellawy, reality is a bit more complicated.

The final chapter by Mahesh Poudyal and Jon Lovett (Chapter 13) deals with a controversial issue that has created an environmental management conundrum. Scientists are agreed that the release of greenhouse gases by modern economic activity, in particular the burning of fossil fuels, has resulted in global warming. The environmental effects of global warming are predicted to be enormous, changing the whole ecology of the planet. A technical solution is to replace fossil fuel with renewable resources such as biofuel derived from agricultural crops. Policy-makers in Europe can see a wide range of benefits from this move: meeting commitments under the Kyoto Protocol, improved security of fuel supplies, enhanced European integration through agricultural subsidies to new European partners and a strong market for biofuels from developing countries, which can help meet the Millennium Development Goals. Set against these potential benefits are environmentally negative changes in land use such as the destruction of tropical rainforest for biofuel crops and the introduction of large-scale mono-cultures. The future will reveal if 'second generation' biofuels from wood products are the answer, or if biofuels offer a false dawn for maintaining our fuel dependence in light of global warming.

Each of the chapters in this handbook provides practical examples of the ways in which insights from the social sciences can complement knowledge from the natural sciences to make environmental management more effective. Sustainable development presents the dual challenge of maintaining environmental quality whilst improving the livelihoods of the people who rely on natural resources. In the past, implementation of environmental management has been hampered by a tendency to rely on technical solutions without understanding the socioeconomic context within which these technical solutions were applied. The complementary application of different social science techniques in combination with ecology-based management thinking, as demonstrated in this handbook, provides practical solutions to overcoming this problem. Such an interdisciplinary approach to environmental management, working across the social and natural sciences, is integral to developing effective management solutions and achieving sustainable development.

References

Hardin, G. (1968), 'The tragedy of the commons', *Science*, **162**(3859), 1243–8.

Leach, M. and R. Mearns (1996), *The Lie of the Land. Challenging Received Wisdom on the African Environment*, Oxford, UK: James Currey.

Ockwell, D.G. (2008), '"Opening up" policy to reflexive appraisal: a role for Q Methodology? A case study of fire management in Cape York, Australia', *Policy Sciences*, **41**(4), 263–92.

Ockwell, D. and Y. Rydin (2006), 'Conflicting discourses of knowledge: understanding the policy adoption of pro-burning knowledge claims in Cape York Peninsula, Australia', *Environmental Politics*, **15**(3), 379–98.

Snow, C.P. (1998), *The Two Cultures*, Cambridge, UK: Cambridge University Press.

2. Global biodiversity conservation priorities: an expanded review[1]
Thomas M. Brooks, Russell A. Mittermeier, Gustavo A.B. da Fonseca, Justin Gerlach, Mike Hoffmann, John F. Lamoreux, Cristina G. Mittermeier, John D. Pilgrim and Ana S.L. Rodrigues[2]

Human actions are causing a biodiversity crisis, with species extinction rates now up to 1000 times higher than the background rate (Pimm et al., 1995). Moreover, the processes driving extinction are eroding the environmental services on which humanity depends (Millennium Ecosystem Assessment, 2005). People care most about what is close to them, so most responses to this crisis will be local or national (Hunter and Hutchinson, 1994). Thus, approximately 90 per cent of $6bn annual conservation funding originates in, and is spent within, economically rich countries (James et al., 1999). However, this still leaves globally flexible funding of hundreds of millions of dollars annually from multilateral agencies (for example, Global Environment Facility), bilateral aid and private sources including environmentally focused corporations, foundations and individuals (Balmford and Whitten, 2003). These resources are frequently the only ones available where conservation is most needed, because biodiversity is unevenly distributed and the most biodiverse places are often the most threatened and poorest economically (Balmford and Long, 1994; Balmford et al., 2003; Baillie et al., 2004). Accordingly, geographically flexible resources exert disproportionate influence on conservation worldwide, and have a key role in the recently agreed intergovernmental 2010 target to reduce significantly the rate of biodiversity loss (Balmford et al., 2005).

Since the pioneering work of Myers (1988) on how to best allocate flexible conservation resources, no less than nine major institutional templates of global biodiversity conservation prioritization have been published, each with involvement from non-governmental organizations (Figure 2.1). These strategies have attracted considerable attention (Figure 2.2), resulting in much progress as well as controversy. The diversity of approaches has raised criticisms of duplication of efforts and lack of clarity (Mace et

↑ High irreplaceability ↓	High vulnerability	**Crisis ecoregions** (Hoekstra et al., 2005) 305 ecoregions with ≥20% habitat conversion and within which the percentage conversion is ≥2 times the percentage protected area coverage	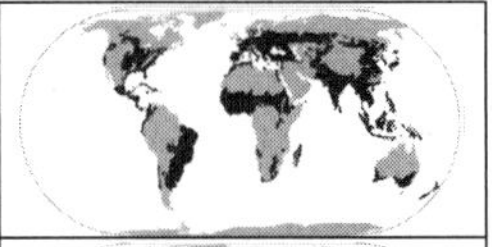
		Biodiversity hotspots (Myers et al., 2000) 34 biogeographically similar aggregations of ecoregions holding ≥0.5% of the world's plants as endemics, and with ≥70% of primary habitat already lost [Myers 1990, 1991; Mittermeier et al., 1998, 1999, 2004]	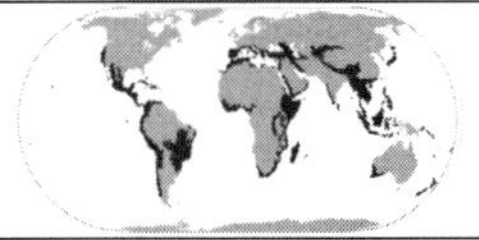
		Endemic bird areas (Stattersfield et al., 1998) 218 regions holding ≥2 bird species with global ranges of <50 000 km², and with more of these endemic than are shared with adjacent regions [ICBP 1992; Crosby 1994; Long et al., 1996]	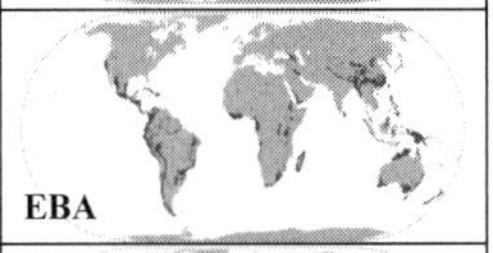 EBA
		Centres of plant diversity (WWF and IUCN 1994–97) 234 mainland sites holding >1000 plant species, of which ≥10% are endemic either to the site or the region; or islands containing ≥50 endemic species or ≥10% of flora endemic	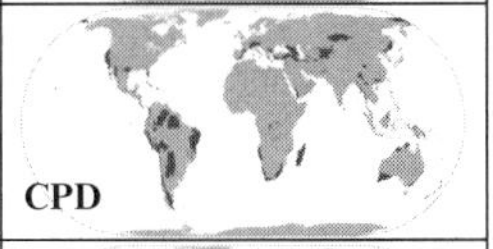 CPD
		Megadiversity countries (Mittermeier et al., 1997) Countries holding ≥1% of the world's plants as endemics [Mittermeier, 1988]	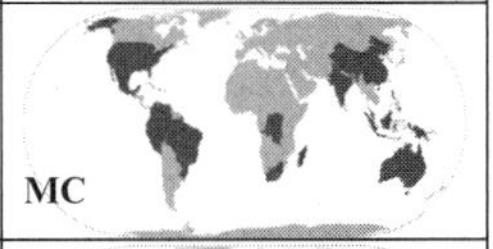 MC
		Global 200 ecoregions (Olson and Dinerstein, 1998) 142 terrestrial ecoregions within biomes characterized by high species richness, endemism, taxonomic uniqueness, unusual phenomena, or global rarity of major habitat type [Olson and Dinerstein, 2002]	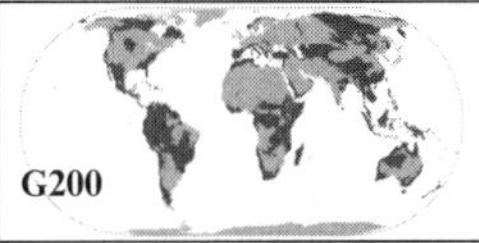 G200
		High-biodiversity wilderness areas (Mittermeier et al., 2003b) Five biogeographically similar aggregations of ecoregions with ≥0.5% of the world's plants as endemics, and with ≥70% of primary habitat remaining and ≥5 people per km² [Mittermeier et al., 1998, 2002]	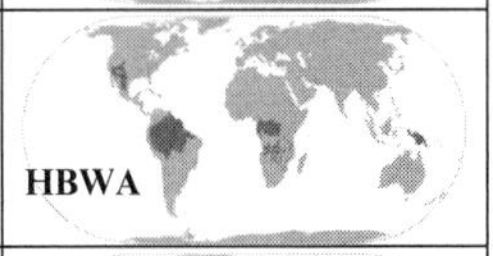 HBWA
	Low vulnerability	**Frontier forests** (Bryant et al., 1997) Forested regions large enough to support viable populations of all native species, dominated by native tree species, and with structure and composition driven by natural events	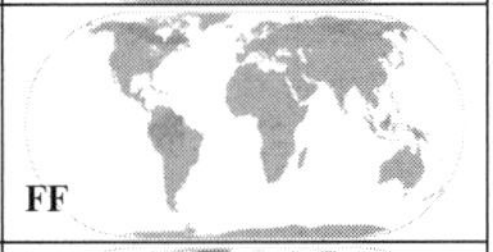 FF
		Last of the wild (Sanderson et al., 2002) 10% wildest 1-km² grid cells in each biome, with wildness measured with an aggregate index of human density, land transformation, access and infrastructure	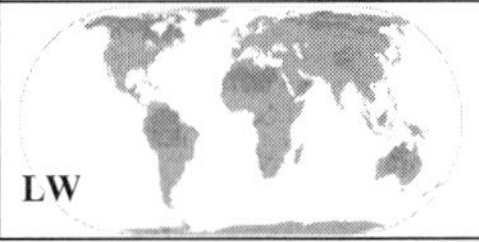 LW

Note: For each proposal, we note primary references, definitions paraphrased from the primary references, maps from the primary references except for Global 200 ecoregions (Olson and Dinerstein, 2002) and biodiversity hotspots (Mittermeier et al., 2004) for which more recent maps are available, and secondary references in square brackets. The Global 200 ecoregions also include 53 freshwater and 43 marine ecoregions, not mapped.

Figure 2.1 Global biodiversity conservation prioritizations

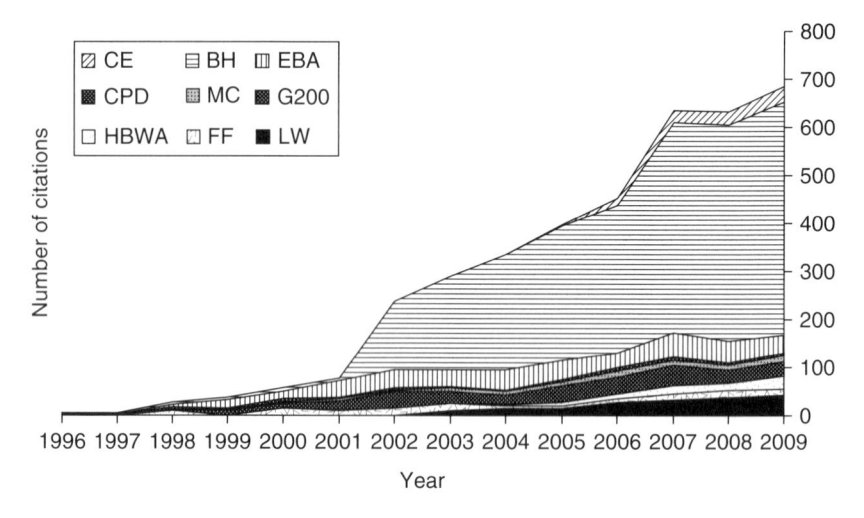

10   *A handbook of environmental management*

Note: Codes as in Figure 2.1. Growth measured by number of citations (in parentheses) of the primary (most cited) reference for each global conservation prioritization recorded in the Web of Science (accessed 23 December 2009, searching comprehensively using 'Cited Reference Search').

Figure 2.2   Growth of attention to global biodiversity conservation prioritization

al., 2000). Although attempts have been made to summarize conservation planning strategies by scale (Redford et al., 2003), no one has considered them within the framework of conservation planning generally (Margules and Pressey, 2000). We review the published concepts and methods behind global biodiversity conservation prioritization, assess the remaining challenges and highlight how this synthesis can already inform allocation of globally flexible resources.

Global prioritization in context
A framework of 'irreplaceability' relative to 'vulnerability' is central to conservation planning theory (Margules and Pressey, 2000). Conservation of all components within such a framework – 'representation' of biodiversity – is not antagonistic to prioritization, as some have argued (Schmidt, 1996). Rather, the latter is a subset of the former; representation identifies everything that biodiversity conservation aims to preserve, whereas prioritization identifies what it aims to preserve first (Ginsberg, 1999). Importantly, in the conservation context, prioritization is quite distinct from, and should not be confused with, triage. Prioritization

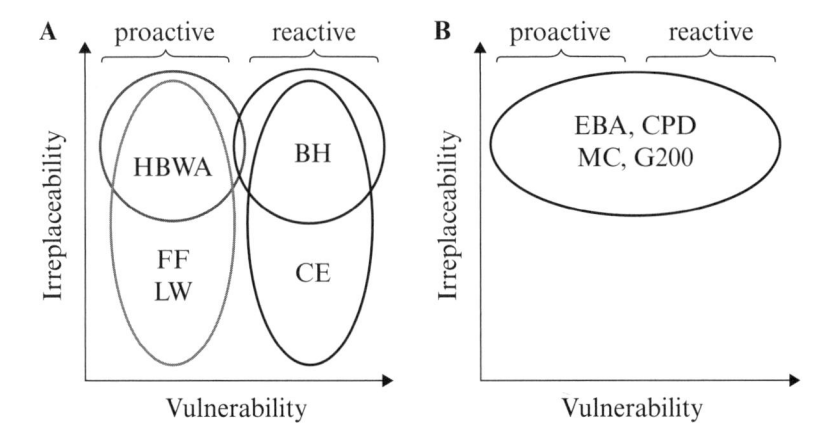

Note: Codes as in Figure 2.1. (**A**) Purely reactive (prioritizing high vulnerability) and purely proactive (prioritizing low vulnerability) approaches. (**B**) Approaches that do not incorporate vulnerability as a criterion (all prioritize high irreplaceability).

Source: Revised version of figure originally published in Brooks et al. (2006).

Figure 2.3 Global biodiversity conservation priority templates placed within the conceptual framework of irreplaceability and vulnerability

provides a means of scheduling responses within such an overall framework (Mittermeier et al., 2003a). Triage, by contrast, has been interpreted as writing threatened biodiversity off the conservation agenda as beyond hope (Pimm, 2000) – discounting the high vulnerability components of the framework.

Conceptually all nine templates of global biodiversity conservation priority fit within the framework of conservation planning theory (Figure 2.3). Importantly, though, they map onto different portions of the framework – while most of the templates prioritize high irreplaceability, some prioritize high and others low vulnerability. These differences are key to understanding how, and why, the nine prioritizations differ, yielding priority maps that cover from less than one-tenth to more than one-third of Earth's land surface.

Measures of irreplaceability
Six of the nine templates of global conservation priority incorporate irreplaceability – measures of spatial conservation options (Margules and Pressey 2000; Pressey and Taffs, 2001). The most common measure of irreplaceability is plant (WWF and IUCN, 1994–97; Mittermeier et al., 1997,

2003b; Myers et al., 2000) or bird (Stattersfield et al., 1998) endemism, often supported by terrestrial vertebrate endemism overall (Mittermeier et al., 1997, 2003b; Myers et al., 2000). The logic for this is that the more endemics a region holds, the more biodiversity is lost if that region is lost (even if anywhere holding even one endemic is irreplaceable in a strict sense). In addition to numbers of endemic species, other aspects of irreplaceability have been proposed including taxonomic uniqueness, unusual phenomena and global rarity of major habitat types (Olson and Dinerstein, 1998), but these remain difficult to quantify. Despite the fact that species richness within a given area is sometimes assumed to be important in prioritization (Prendergast et al., 1993), none of the approaches rely on species richness alone. This is because species richness is driven by common, widespread species, thus strategies focused on species richness tend to miss exactly those biodiversity features most in need of conservation (Orme et al., 2005; Possingham and Wilson, 2005; Lamoreux et al., 2006). Three approaches do not incorporate irreplaceability (Bryant et al., 1997; Sanderson et al., 2002; Hoekstra et al., 2005).

The choice of measures of irreplaceability is to some degree subjective, in that data limitations currently preclude the measurement of biodiversity wholesale. Further, these same data constraints have meant that, with the exception of endemic bird areas (Stattersfield et al., 1998), the measures of irreplaceability used in global conservation prioritization have necessarily been derived from specialist opinion. Subsequent tests of plant endemism estimates (Krupnick and Kress, 2003) have proven this expert opinion to be quite accurate. However, reliance on specialist opinion means that results cannot be replicated, raising questions concerning the transparency of the approaches (Humphries, 2000; Mace et al., 2000). It also prevents formal measurement of irreplaceability, which requires the identities of individual biodiversity features, such as species names, rather than just estimates of their magnitude expressed as a number (Balmford, et al., 2000; Humphries, 2000; Mace et al., 2000; Brummitt and Lughadha, 2003).

Measures of vulnerability
Five of the templates of global conservation priority incorporate vulnerability – measures of temporal conservation options (Margules and Pressey, 2000; Pressey and Taffs, 2001). A recent classification of vulnerability (Wilson et al., 2005) recognizes four types of measures based on: environmental and spatial variables; land tenure; threatened species; and expert opinion. Of these, environmental and spatial variables have been used most frequently in global conservation prioritization, measured as proportionate habitat loss (Myers et al., 2000; Sanderson et al., 2002;

Mittermeier et al., 2003b; Hoekstra et al., 2005). Species–area relationships provide justification that habitat loss translates into biodiversity loss (Pimm et al., 1995; Brooks et al., 2002). However, use of habitat loss as a measure of vulnerability has several problems, in that it is difficult to assess using remote sensing for xeric and aquatic systems, does not incorporate threats such as invasive species and hunting pressure and is retrospective rather than predictive (Wilson et al., 2005). The frontier forests approach (Bryant et al., 1997) uses absolute forest cover as a measure, although this has been criticized as not reflective of vulnerability (Innes and Er, 2002).

Beyond habitat loss, land tenure, measured as protected area coverage, has also been incorporated into two approaches (Olson and Dinerstein, 1998; Hoekstra et al., 2005). Other possible surrogates not classified by Wilson et al. (2005) include human population growth and density, which are widely thought to be relevant (Sisk et al., 1994; Cincotta et al., 2000; O'Connor et al., 2003; Shi et al., 2005; Veech, 2003; Balmford et al., 2001), and were integral to two of the systems (Sanderson et al., 2002; Mittermeier et al., 2003b). None of the global conservation prioritization templates used threatened species or expert opinion as measures of vulnerability. Household dynamics (Liu et al., 2003) and political and institutional capacity and governance (O'Connor et al., 2003; Smith et al., 2003) affect biodiversity indirectly, but have not been incorporated to date. This is true for climate change as well, which is worrying given that its impact is likely to be severe (Parmesan and Yohe, 2003; Thomas et al., 2004; McClean et al., 2006). Finally, while costs of conservation generally increase with threat, no proposals for global biodiversity conservation priority have yet incorporated costs directly, despite the availability of techniques to do this at regional scales (Moran et al., 1997; Wilson et al., 2006). Two of the templates of global conservation prioritization do not incorporate vulnerability (WWF and IUCN, 1994–97; Mittermeier et al., 1997) and the remaining two only incorporate it peripherally (Olson and Dinerstein, 1998; Stattersfield et al., 1998).

Spatial units
The spatial units most commonly used in systematic conservation planning are equal-area grids. However, data limitations have precluded their use in the development of actual templates of global biodiversity conservation priority to date. Instead, all proposals, with the exception of megadiversity countries (Mittermeier et al., 1997), are based on biogeographic units. Typically, these units are defined a priori by specialist perception of the distribution of biodiversity. For example, 'ecoregions', one of the most commonly used such classifications, are 'relatively large units of land containing a distinct assemblage of natural communities and species'

(Olson et al., 2001, p.933). Only in the endemic bird areas approach are biogeographic units defined a posteriori by the distributions of the species concerned (Stattersfield et al., 1998). Relative to equal-area grids, biogeographic units bring advantages of ecological relevance, while megadiversity countries (Mittermeier et al., 1997) bring political relevance.

However, reliance on biogeographic spatial units raises several complications. Various competing bioregional classifications are in use (Jepson and Whittaker, 2002), with the choice of system having considerable repercussions for resulting conservation priorities (Pressey and Logan, 1994). Further, when unequally sized units are employed, priority may be biased towards large areas as a consequence of species–area relationships. Assessment of global conservation priorities should, therefore, factor out area, either by taking residuals about a best-fit line to a plot of species against area (Balmford and Long, 1995; Brooks et al., 2002; Werner and Buszko, 2005; Lamoreux et al., 2006) or by rescaling numbers of endemics using a power function directly (Veech, 2000; Brummitt and Lughadha, 2003; Hobohm, 2003; Ovadia, 2003). Nevertheless, use of a priori bioregional units for global conservation prioritization will be essential until data of sufficient resolution become available to enable the use of grids.

Spatial patterns
In Figure 2.4, we map the overlay of the global biodiversity conservation priority systems into geographic space from the conceptual framework of Figure 2.3. Figure 2.4A illustrates the large degree of overlap between templates that prioritize highly vulnerable regions of high irreplaceability: tropical islands and mountains (including montane Mesoamerica, the Andes, the Brazilian Atlantic forest, Madagascar, montane Africa, the Western Ghats of India, Malaysia, Indonesia, the Philippines and Hawaii), Mediterranean-type systems (including California, central Chile, coastal South Africa, south-west Australia and the Mediterranean itself), and a few temperate forests (the Caucasus, the central Asian mountains, the Himalaya and south-west China). Highly vulnerable regions of lower irreplaceability (generally, the rest of the northern temperate regions) are prioritized by fewer approaches. Figure 2.4B shows a large amount of overlap between templates for regions of low vulnerability but high irreplaceability, in particular the three major tropical rainforests of Amazonia, the Congo and New Guinea. Regions of simultaneously lower vulnerability and irreplaceability, such as the boreal forests of Canada and Russia, and the deserts of western USA and central Asia are prioritized less often.

Two general observations are apparent. First, most land (79 per cent) is highlighted by at least one of the prioritization systems. Second,

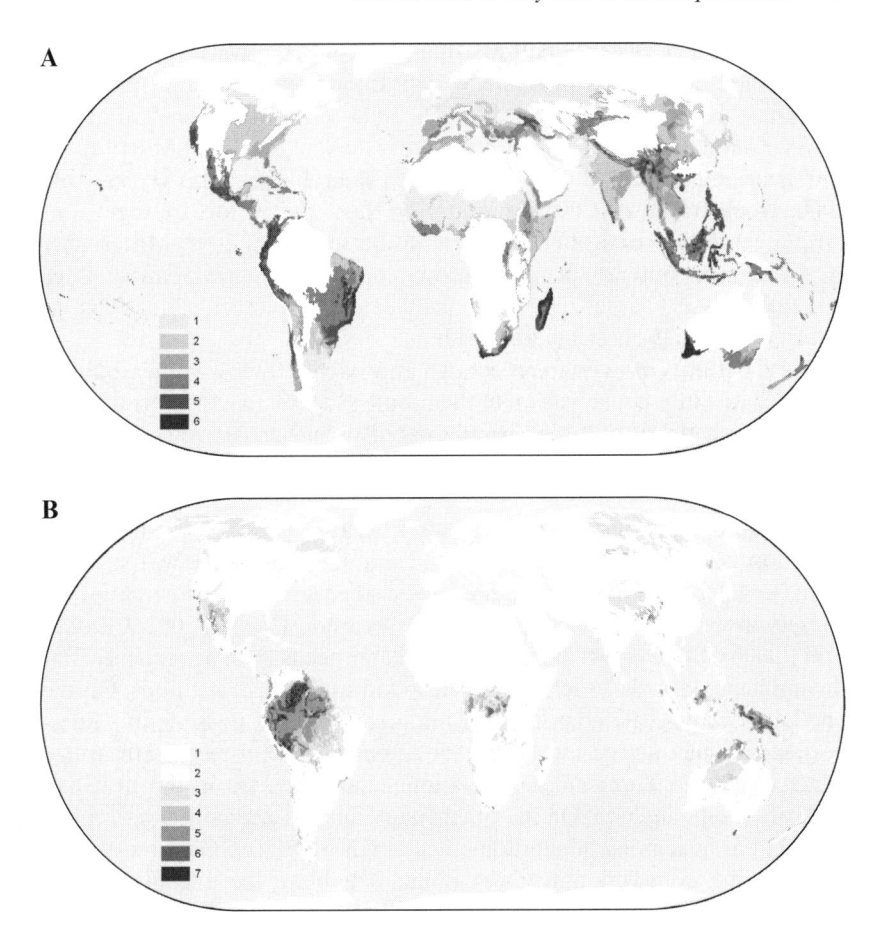

Global biodiversity conservation priorities 15

A

B

Note: (**A**) Reactive approaches, corresponding to the right-hand side of Figure 2.3A, that prioritize regions of high threat and those that do not incorporate vulnerability as a criterion (Figure 2.3B); the latter are only mapped where they overlap with the former. (**B**) Proactive approaches, corresponding to the left-hand side of Figure 2.3A, that prioritize regions of low threat and those that do not incorporate vulnerability as a criterion (Figure 2.3B); again, the latter are only mapped where they overlap with the former. Shading denotes number of global biodiversity conservation prioritization templates, in both cases.

Source: Revised version of figure originally published in Brooks et al. (2006).

Figure 2.4 Mapping the overlay of approaches prioritizing reactive and proactive conservation

despite this, a noticeable pattern emerges from the overlay of different approaches. There is significant overlap among templates that prioritize irreplaceable regions (WWF and IUCN, 1994–97; Mittermeier et al., 1997, 2003b; Olson and Dinerstein, 1998; Stattersfield et al., 1998; Myers et al., 2000), among those that prioritize highly vulnerable regions (Myers et al., 2000; Hoekstra et al., 2005), and among those that prioritize regions of low vulnerability (Bryant et al., 1997; Sanderson et al., 2002; Mittermeier et al., 2003b), but not between approaches across each of these three general classes (Table 2.1). This provides useful cross-verification of priority regions (Fonseca et al., 2000).

These patterns of overlap reflect two approaches to how vulnerability is incorporated into conservation in the broadest sense: reactive (prioritizing areas of high threat and high irreplaceability) and proactive (prioritizing areas of low threat but high irreplaceability). The former are considered the most urgent priorities in conservation planning theory (Margules and Pressey, 2000; Pressey and Taffs, 2001) because unless immediate conservation action is taken within them, unique biodiversity will soon be lost. The latter are often de facto priorities, because the opportunities for conservation in these are considerable (Norris and Harper, 2003; Cardillo et al., 2006). Biodiversity conservation clearly needs both approaches, but the implementation of each may correspond to different methods. On the one hand, large-scale conservation initiatives may be possible in wilderness areas, such as the establishment of enormous protected areas, like the 3 800 000 ha Tumucumaque National Park, created in the Brazilian state of Amapá in 2003. On the other hand, finely tuned conservation will be essential in regions of simultaneously high irreplaceability and threat, where losing even tiny patches of remnant habitat, like the sites identified by the Alliance for Zero Extinction (Ricketts et al., 2005), would be tragic.

Impact of global prioritization
The appropriate measure of impact is the success of prioritization in achieving its main goal – influencing globally flexible donors to invest in regions where these funds can contribute most to conservation. Precise data are unavailable for all of the approaches (Halpern et al., 2006), but hotspots alone have mobilized at least $750m of funding for conservation in these regions (Myers, 2003). More specifically, conservation funding mechanisms have been established for several of the approaches, such as the $100m, ten-year Global Conservation Fund focused on high-biodiversity wilderness areas and hotspots, and the $125m, five-year Critical Ecosystem Partnership Fund, aimed exclusively at hotspots (Dalton, 2000). The Global Environment Facility, the largest financial

mechanism addressing biodiversity conservation, is currently exploring a resource allocation framework that builds on existing templates. Both civil society and government organizations often utilize the recognition given to regions as global conservation priorities as justification when applying for geographically flexible funding. In addition, the global prioritization systems must have had sizeable effects in the cancellation, relocation or mitigation of environmentally harmful activities, even in the absence of specific legislation (Kunich, 2001). Unfortunately, resources still fall an order of magnitude short of required conservation funding (James et al., 1999). Nevertheless, the dollar amounts are impressive, and represent dramatic increases in conservation investment in these regions.

Challenges facing global prioritization
Limitations of data have thus far generally restricted global conservation prioritization to specialist estimates of irreplaceability, to habitat loss as a measure of vulnerability, and to coarse geographic units defined a priori. Over the last five years, spatial datasets have been compiled with the potential to reduce these constraints (Baillie et al., 2004), particularly for mammals (Ceballos et al., 2005), birds (Orme et al., 2005), and amphibians (Stuart et al., 2004). When these maps are combined with assessment of conservation status, they enable the development of threat metrics based on threatened species directly (Sisk et al., 1994; Rodrigues et al., 2004a). So far, the main advances to global prioritization enabled by these new data are validation tests of existing templates (Fonseca et al., 2000; Burgess et al., 2002). Encouragingly, global gap analysis of priorities for the representation of terrestrial vertebrate species in protected areas (Rodrigues et al., 2004a, b) and initial regional assessment of plants (Küper et al., 2004) yield results similar to existing approaches (Figure 2.5).

A few have argued that global conservation priorities should be driven solely by those vertebrates known and loved by society (Jepson and Canney, 2001). However, invertebrates represent the bulk of eukaryotic diversity on Earth with over a million known species (Baillie et al., 2004) and many more yet to be described (Novotny et al., 2002). The conservation status of only ~3500 arthropods has been assessed (Baillie et al., 2004), and so even setting aside microbes as near-irrelevant to conservation (Nee, 2004), global conservation priority is still far from being able to incorporate megadiverse invertebrate taxa (Mace et al., 2000; Brummitt and Lughadha, 2003). While regional data can show little overlap between priority areas for arthropods and those for plant and terrestrial vertebrate taxa (Dobson et al., 1997), there are strong correlations between phytophagous insects and plant species richness (Kelly and Southwood, 1999), and preliminary global data for groups like tiger beetles and termites suggest

Table 2.1 Spatial overlap between the nine proposed global biodiversity conservation prioritizations

		% land area	Crisis ecoregions	Biodiversity hotspots	Endemic bird areas	Centres of plant diversity	Megadiversity countries	Global 200 ecoregions	High-biodiversity wilderness areas	Frontier forests	Last of the wild
			30	16	10	9	35	37	8	9	24
High irreplaceability	High vulnerability	Crisis ecoregions		33*	14	10	44	36	2§	1§	4§
		Biodiversity hotspots	61*		33*	21*	46*	78*	0§	5	6§
		Endemic bird areas	43*	50*		24*	68*	70*	7	11	11§
		Centres of plant diversity	34	40*	28*		48*	66*	18*	14	21
		Megadiversity countries	38	21	19*	12		53*	18*	11	24
		Global 200 ecoregions	28	33*	19*	15*	49*		16*	16*	28
	Low vulnerability	High-biodiversity wilderness areas	6§	0§	15	19*	79*	72*		41*	53*
		Frontier forests	4§	8§	11	13	39	64*	35*		73*
		Last of the wild	5§	4§	4	7	34	43	17*	28*	

18

Note: Cells show the percentage overlap of the approach in a given row by the approach in the corresponding column. The three blocks of cells outlined in bold correspond to overlaps between approaches within three categories, indicated in the bars on the left: reactive approaches, those prioritizing high irreplaceability and proactive approaches. Significance was tested against overlap expected based on the percentage of the land area covered by each approach, using a chi-squared test (P = 0.05; * significantly greater overlap than expected; § significantly less overlap than expected). There are many significant overlaps between approaches within each of the three categories: those prioritizing highly vulnerable regions (2/2 overlaps significantly more than expected), those prioritizing irreplaceable regions (24/30), and those prioritizing regions of low vulnerability (6/6). The opposite is true when comparing approaches prioritizing vulnerable regions with those prioritizing regions of low vulnerability (11/12 overlaps significantly less than expected). The approaches that incorporate irreplaceability almost all show significantly more overlap than expected with and by at least four other systems. In contrast, those that do not incorporate irreplaceability only overlap significantly more than expected with or by three other approaches at the most.

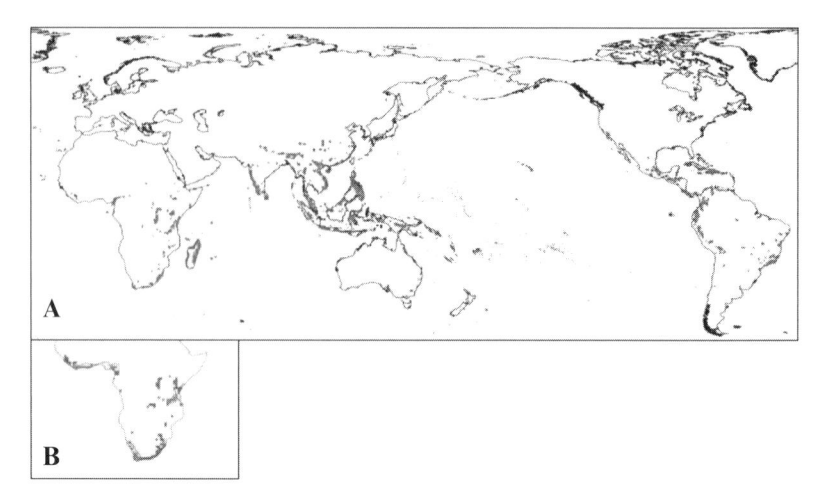

Note: Global conservation prioritization templates have been based almost exclusively on bioregional classification and specialist opinion, rather than primary biodiversity data. Such primary datasets have recently started to become available under the umbrella of the IUCN Species Survival Commission (Baillie et al., 2004), and they allow progressive testing and refinement of templates. (**A**) Global gap analysis of coverage of 11 633 mammal, bird, turtle and amphibian species (~40% of terrestrial vertebrates) in protected areas (Rodrigues et al., 2004b). It shows unprotected half-degree grid cells characterized simultaneously by irreplaceability values of at least 0.9 on a scale of 0–1, and of the top 5% of values of an extinction risk indicator based on the presence of globally threatened species (Rodrigues et al., 2004a). (**B**) Priorities for the conservation of 6269 African plant species (~2% of vascular plants) across a 1-degree grid (Küper et al., 2004). These are the 125 grid cells with the highest product of range-size rarity (a surrogate for irreplaceability) of plant species distributions and mean human footprint (Sanderson et al., 2002). Comparison of these two maps, and between them and Figure 2.4, reveals a striking similarity among conservation priorities identified using independent datasets. The difference in taxonomic and geographic coverage between **A** and **B** also highlights the challenge facing the botanical community to compile comprehensive primary data on plant conservation in order to inform global conservation prioritization (Callmander et al., 2005). Rectifying this is part of the Global Strategy for Plant Conservation of the Convention on Biological Diversity.

Source: Revised version of figure originally published in Brooks et al. (2006).

Figure 2.5   Incorporating primary biodiversity data in global conservation priority-setting

much higher levels of congruence (Mittermeier et al., 2004). Similarly, pioneering techniques to model wholesale irreplaceability by combining point data for megadiverse taxa with environmental datasets produce results commensurate with existing conservation priorities (Ferrier et al., 2004). These findings, while encouraging, in no way preclude the need to use primary invertebrate data in global conservation prioritization as they become available.

Aquatic systems feature poorly in existing conservation templates, although there is evidence that marine biodiversity may be at least as threatened as (Dulvy et al., 2003) and freshwater biodiversity even more threatened (McAllister et al., 1997) than that on land (Baillie et al., 2004). Only one conservation prioritization explicitly incorporates aquatic systems (Olson and Dinerstein, 1998). The irreplaceability dimension has been particularly overlooked in the seas, with the traditional emphasis of marine conservation being on species richness (Briggs, 2002) despite little correspondence between marine species richness and endemism (Hughes et al., 2002; Price, 2002). Nevertheless, the most comprehensive study yet, albeit restricted to tropical coral reef ecosystems, identified ten priority regions based on endemism and threat (Roberts et al., 2002). Eight of these regions lie adjacent to priority regions highlighted in Figure 2.4, raising the possibility of correspondence between marine and terrestrial priorities despite the expectation that surrogacy of conservation priorities will be low between different environments (Reid, 1998). Efforts to identify freshwater priorities lag further behind, although initial studies reveal a highly uneven distribution of freshwater fish endemism at regional (Darwall et al., 2005) and global (Mittermeier et al., 2004) scales.

Most measurement of irreplaceability is species-based, raising the concern that phylogenetic diversity may slip through the net of global conservation priorities (Mace et al., 2000; Jepson and Canney, 2001; Brummitt and Lughadha, 2003; Kareiva and Marvier, 2003). However, analyses for mammals (Sechrest et al., 2002) and birds (Brooks et al., 2005) find that priority regions represent higher taxa and phylogenetic diversity better than would be predicted by the degree to which they represent species. Islands such as Madagascar and the Caribbean hold especially high concentrations of endemic genera and families (Mittermeier et al., 2004). A heterodox perspective argues that the terminal tips of phylogenetic trees should be higher priorities than deep lineages (Erwin, 1991). In any case, the balance of work implies that even if phylogenetic diversity is not explicitly targeted for conservation, global prioritization based on species provides a solid surrogate for evolutionary history.

That global conservation priority regions capture phylogenetic history does not necessarily mean they represent evolutionary process (Mace et al., 2000; Myers and Knoll, 2001; Smith et al., 2001). For example, transition zones or 'biogeographic crossroads', frequently overlooked by conservation prioritization, could be of particular importance in driving speciation (Smith et al., 1997; Spector, 2002). On the other hand, the scale of global conservation priorities is often coarse enough to capture the transitions necessary in facilitating evolutionary processes, such as resilience to climate change (Midgley et al., 2001). There is also evidence that areas of greatest importance in generating biodiversity are those of long-term climatic stability, especially where they occur in tropical mountains (Fjeldså and Lovett, 1997), which are incorporated in most approaches to global conservation prioritization. A related hypothesis states that regions with particularly high percentage endemism – tropical mountains and islands – necessarily capture evolutionary process (Fa et al., 2004). The development of metrics for the maintenance of evolutionary process is in its infancy, and represents an emerging research front (Araújo, 2002).

A final dimension that will prove important to assess in the context of global conservation prioritization concerns ecosystem services (Jepson and Canney, 2001; Kareiva and Marvier, 2003; Odling-Smee, 2005). Although the processes threatening biodiversity and ecosystem services are likely similar, the relationship between biodiversity per se and ecosystem services remains unresolved (Loreau et al., 2001). Thus, while it is important to establish distinct goals for these conservation objectives (Sarkar, 1999), identification of synergies between them is strategically vital. This research avenue has barely been explored and questions of how global biodiversity conservation priorities overlap with priority regions for carbon sequestration, climate stabilization, maintenance of water quality, minimization of outbreaks of pests and diseases, and fisheries, for example, remain unanswered. However, the correspondence between conservation priorities and human populations (Cincotta et al., 2000; Balmford et al., 2001; Baillie et al., 2004) and poverty (Balmford et al., 2003; Baillie et al., 2004) is an indication that the conservation of areas of high biodiversity priority will deliver high local ecosystem service benefits.

From global to local priorities
The establishment of global conservation priorities has been extremely influential in directing resources towards broad regions. However, a number of authors have pointed out that global conservation prioritization has had little success in informing actual conservation implementation (Dinerstein and Wikramanayake, 1993; Mace et al., 2000; Jepson, 2001; Brummitt and Lughadha, 2003). Separate processes are necessary to identify actual conservation targets and priorities at much finer scales (Supriatna, 2001), because even within a region as uniformly important as (say) Madagascar, biodiversity and threats are not evenly distributed. Bottom-up processes of identification of priorities are therefore essential to ensure the implementation of area-based conservation (Whittaker et al., 2005).

Indeed, numerous efforts are underway to identify targets for conservation implementation. Many focus on the site scale, drawing on two decades of work across nearly 170 countries in the designation of important bird areas (BirdLife International, 2004). There is an obvious need to expand such work to incorporate other taxa (Eken et al., 2004), and to prioritize the most threatened and irreplaceable sites (Ricketts et al., 2005). Such initiatives have recently gained strong political support under the Convention on Biological Diversity, through the development of the Global Strategy for Plant Conservation and the Programme of Work on Protected Areas. Both mechanisms call for the identification, recognition and safeguarding of sites of biodiversity conservation significance. Meanwhile, considerable attention is also targeted at the scale of landscapes and seascapes, to ensure not just the representation of biodiversity but also of the connectivity, spatial structure and processes that allow its persistence (Cowling et al., 2003).

Global conservation planning is key for strategic allocation of flexible resources. Despite divergence in methods between the different schemes, an overall picture is emerging in which a few regions, particularly in the tropics and in Mediterranean-type environments, are consistently emphasized as priorities for biodiversity conservation. It is crucial that the global donor community channels sufficient resources to these regions, at the very minimum. This focus will continue to improve if the rigour and breadth of biodiversity and threat data continue to be consolidated, especially important given the increased accountability demanded from global donors. However, it is through the conservation of actual sites that biodiversity will ultimately be preserved, or lost, and thus drawing the lessons of global conservation prioritization down to a much finer scale is now the primary concern for conservation planning.

Notes

1. This chapter is an expanded version of Brooks et al. (2006), 'Global biodiversity conservation priorities', *Science*, **313**(5783), 58–61, originally published by AAAS.
2. We thank G. Fabregas, D. Knox, T. Lacher, P. Langhammer, N. Myers, A. Sugden and W. Turner for help with the manuscript, anonymous peer reviewers for comments, the Gordon and Betty Moore Foundation for funding, D. Ockwell and J. Lovett for the invitation for this contribution and for editorial help, and AAAS for permission to reprint this expanded version of our original review paper.

References

Araújo, M.B. (2002), 'Biodiversity hotspots and zones of ecological transition', *Conservation Biology*, **16**(6), 1662–3.
Baillie, J.E.M., L.A. Bennun, T.M. Brooks, S.H.M. Butchart, J.S. Chanson, Z. Cokeliss, C. Hilton-Taylor, M. Hoffmann, G.M. Mace, S.A. Mainka, C.M. Pollock, A.S.L. Rodrigues, A.J. Stattersfield and S.N. Stuart (2004), *A Global Species Assessment*, Gland, Switzerland: IUCN – International Union for Conservation of Nature.
Balmford, A. and A. Long (1994), 'Avian endemism and forest loss', *Nature*, **372**(6507), 623–4.
Balmford, A. and A. Long (1995), 'Across-country analyses of biodiversity congruence and current conservation effort in the tropics', *Conservation Biology*, **9**(16), 1539–47.

Balmford, A. and T. Whitten (2003), 'Who should pay for tropical conservation, and how could the costs be met?', *Oryx*, **37**(2), 238–50.

Balmford, A., K.J. Gaston, A.S.L. Rodrigues and A. James (2000), 'Integrating costs of conservation into international priority setting', *Conservation Biology*, **14**(3), 597–605.

Balmford, A., J. Moore, T. Brooks, N. Burgess, L.A. Hansen, P. Williams and C. Rahbek (2001), 'Conservation conflicts across Africa', *Science*, **291**(5513), 2616–19.

Balmford, A., G.J. Gaston, S. Blyth, A. James and V. Kapos (2003), 'Global variation in conservation costs, conservation benefits, and unmet conservation needs', *Proceedings of the National Academy of Sciences of the U.S.A.*, **100**(3), 1046–50.

Balmford, A., L. Bennun, B. ten Brink, D. Cooper, I. Côté, P. Crane, A. Dobson, N. Dudley, I. Dutton, R.E. Green, R.D. Gregory, J. Harrison, E.T. Kennedy, C. Kremen, N. Leader-Williams, T.E. Lovejoy, G. Mace, R. May, P. Mayaux, P. Morling, J. Phillips, K. Redford, T.H. Ricketts, J.P. Rodriguez, M. Sanjayan, P.J. Schei, A.S. van Jaarsveld and B.A. Walther (2005), 'The Convention on Biological Diversity's 2010 target', *Science*, **307**(5707), 212–13.

BirdLife International (2004), *State of the World's Birds 2004*, Cambridge, UK: BirdLife International.

Briggs, J.C. (2002), 'Coral reef biodiversity and conservation', *Science*, **296**(5570), 1027.

Brooks, T.M., R.A. Mittermeier, C.G. Mittermeier, G.A.B. da Fonseca, A.B. Rylands, W.R. Konstant, P. Flick, J. Pilgrim, S. Oldfield, G. Magin and C. Hilton-Taylor (2002), 'Habitat loss and extinction in the hotspots of biodiversity', *Conservation Biology*, **16**(4), 909–23.

Brooks, T.M., J.D. Pilgrim, A.S.L. Rodrigues and G.A.B. da Fonseca (2005), 'Conservation status and geographic distribution of avian evolutionary history', in A. Purvis, J.L. Gittleman and T.M. Brooks (eds), *Phylogeny and Conservation*, Cambridge, UK: Cambridge University Press, pp. 267–94.

Brooks, T.M., R.A. Mittermeier, G.A.B. da Fonseca, J. Gerlach, M. Hoffmann, J.F. Lamoreux, C.G. Mittermeier, J.D. Pilgrim and A.S.L. Rodrigues (2006), 'Global biodiversity conservation priorities', *Science*, **313**(5783), 58–61.

Brummitt, N. and E.N. Lughadha (2003), 'Biodiversity: where's hot and where's not?', *Conservation Biology*, **17**(5), 1442–8.

Bryant, D., D. Nielsen and L. Tangley (1997), *Last Frontier Forests: Ecosystems and Economies on the Edge*, Washington DC: World Resources Institute.

Burgess, N.D., C. Rahbek, F.W. Larsen, P. Williams and A. Balmford (2002), 'How much of the vertebrate diversity of sub-Saharan Africa is catered for by recent conservation proposals?', *Biological Conservation*, **107**(33), 327–39.

Callmander, M.W., G.E. Schatz and P.P. Lowry II (2005), 'IUCN Red List assessment and the Global Strategy for Plant Conservation: taxonomists must act *now*', *Taxon*, **54**(4), 1047–50.

Cardillo, M., G.M. Mace, J.L. Gittleman and A. Purvis (2006), 'Latent extinction risk and the future battlegrounds of mammal conservation', *Proceedings of the National Academy of Sciences of the U.S.A.*, **103**(11), 4157–61.

Ceballos, G., P.R. Ehrlich, J. Soberón, I. Salazar and J.P. Fay (2005), 'Global mammal conservation: what must we manage?', *Science*, **309**(5734), 603–7.

Cincotta, R.P., J. Wisnewski and R. Engelman (2000), 'Human population in the biodiversity hotspots', *Nature*, **404**(6781), 990–92.

Cowling, R.M., R.L. Pressey, M. Rouget and A.T. Lombard (2003), 'A conservation plan for a global biodiversity hotspot – the Cape Floristic Region, South Africa', *Biological Conservation*, **112**(1–2), 191–216.

Crosby, M.J. (1994), 'Mapping the distributions of restricted-range birds to identify global conservation priorities', in R.I. Miller (ed.), *Mapping the Diversity of Nature*, London: Chapman and Hall, pp. 145–54.

Dalton, R. (2000), 'Biodiversity cash aimed at hotspots', *Nature*, **406**(6798), 818.

Darwall, W., K. Smith, T. Lowe and J.-C. Vié (2005), *The Status and Distribution of Freshwater Biodiversity in Eastern Africa*, Gland, Switzerland: IUCN.

Dinerstein, E. and E.D. Wikramanayake (1993), 'Beyond "hotspots": how to prioritize investments to conserve biodiversity in the Indo-Pacific region', *Conservation Biology*, **7**(1), 53–65.
Dobson, A.P., J.P. Rodriguez, W.M. Roberts and D.S. Wilcove (1997), 'Geographic distribution of endangered species in the United Space', *Science*, **275**(5299), 550–53.
Dulvy, N.K., Y. Sadovy and J.D. Reynolds (2003), 'Extinction vulnerability in marine populations', *Fish and Fisheries*, **4**(1), 25–64.
Eken, G., L. Bennun, T.M. Brooks, W. Darwall, M. Foster, D. Knox, P. Langhammer, P. Matiku, E. Radford, P. Salaman, W. Sechrest, M.L. Smith, S. Spector and J. Tordoff (2004), 'Key biodiversity areas as site conservation targets', *BioScience*, **54**(12), 1110–18.
Erwin, T.L. (1991), 'An evolutionary basis for conservation strategies', *Science*, **253**(5021), 750–52.
Fa, J.E., R.W. Burn, M.E. Stanley-Price and F.M. Underwood (2004), 'Identifying important endemic areas using ecoregions: birds and mammals in the Indo-Pacific', *Oryx*, **38**(1), 91–101.
Ferrier, S., G.V.N. Powell, K.S. Richardson, G. Manion, J.M. Overton, T.F. Allnutt, S.E. Cameron, K. Mantle, N.D. Burgess, D.P. Faith, J.F. Lamoreux, G. Kier, R.J. Hijmans, V.A. Funk, G.A. Cassis, B.L. Fisher, P. Flemons, D. Lees, J.C. Lovett and R.S.A.R. Van Rompaey (2004), 'Mapping more of terrestrial biodiversity for global conservation assessment', *BioScience*, **54**(12), 1101–9.
Fjeldså, J. and J.C. Lovett (1997), 'Geographical patterns of old and young species in African forest biota: the significance of specific montane areas as evolutionary centres', *Biodiversity and Conservation*, **6**(3), 325–46.
Fonseca, G.A.B. da, A. Balmford, C. Bibby, L. Boitani, F. Corsi, T. Brooks, C. Gascon, S. Olivieri, R. Mittermeier, N. Burgess, E. Dinerstein, D. Olson, L. Hannah, J. Lovett, D. Moyer, C. Rahbek, S. Stuart and P. Williams (2000), 'Following Africa's lead in setting priorities', *Nature*, **405**(6785), 393–4.
Ginsberg, J. (1999), 'Global conservation priorities', *Conservation Biology*, **13**(1), 5.
Halpern, B.S., C.R. Pyke, H.E. Fox, J.C. Haney, M.A. Schlaepfer and P. Zaradic (2006), 'Gaps and mismatches between global conservation priorities and spending', *Conservation Biology*, **20**(1), 56–64.
Hobohm, C. (2003), 'Characterization and ranking of biodiversity hotspots: centers of species richness and endemism', *Biodiversity and Conservation*, **12**(2), 279–87.
Hoekstra, J.M., T.M. Boucher, T.H. Ricketts and C. Roberts (2005), 'Confronting a biome crisis: global disparities of habitat loss and protection', *Ecology Letters*, **8**(1), 23–9.
Hughes, T.P., D.R. Bellwood and S.R. Connolly (2002), 'Biodiversity hotspots, centres of endemicity, and the conservation of coral reefs', *Ecology Letters*, **5**(6), 775–84.
Humphries, C.J. (2000), 'Hotspots: going off the boil?', *Diversity and Distributions*, **7**(1–2), 103–4.
Hunter, M.L., Jr. and A. Hutchinson (1994), 'The virtues and shortcomings of parochialism: conserving species that are locally rare, but globally common', *Conservation Biology*, **8**(4), 1163–5.
ICBP (1992), *Putting Biodiversity on the Map*, Cambridge, UK: International Council for Bird Protection.
Innes, J.L. and K.B.H. Er (2002), 'The questionable utility of the frontier forest concept', *BioScience*, **52**(12), 1095–1109.
James, A., K. Gaston and A. Balmford (1999), 'Balancing the earth's accounts', *Nature*, **401**(6751), 323–4.
Jepson, P. (2001), 'Global biodiversity plan needs to convince local policy-makers', *Nature*, **409**(6816), 12.
Jepson, P. and S. Canney (2001), 'Biodiversity hotspots: hot for what?', *Global Ecology and Biogeography*, **10**(13), 225–7.
Jepson, P. and R.J. Whittaker (2002), 'Ecoregions in context: a critique with special reference to Indonesia', *Conservation Biology*, **16**(1), 42–57.
Kareiva, P. and M. Marvier (2003), 'Conserving biodiversity coldspots', *American Scientist*, **91**(14), 344–51.

Kelly, C.K. and T.R.E. Southwood (1999), 'Species richness and resource availability: a phylogenetic analysis of insects associated with trees', *Proceedings of the National Academy of Sciences of the U.S.A.*, **96**(14), 8013–16.
Krupnick, G.A. and W.J. Kress (2003), 'Hotspots and ecoregions: a test of conservation priorities using taxonomic data', *Biodiversity and Conservation*, **12**(11), 2237–53.
Kunich, J.C. (2001), 'Preserving the womb of the unknown species with hotspots legislation', *Hastings Law Journal*, **52**(16), 1149–1253.
Küper, W., J.H. Sommer, J.C. Lovett, J. Mutke, H.P. Linder, H.J. Beentje, R.S.A.R. Van Rompaey, C. Chatelain, M. Sosef and W. Barthlott (2004), 'Africa's hotspots of biodiversity redefined', *Annals of the Missouri Botanical Garden*, **91**(4), 525–35.
Lamoreux, J.F., J.C. Morrison, T.H. Ricketts, D.M. Olson, E. Dinerstein, M.W. McKnight and H.H. Shugart (2006), 'Global tests of biodiversity concordance and the importance of endemism', *Nature*, **440**(7082), 212–14.
Liu, J., D.C. Daily, P.R. Ehrlich and G.W. Luck (2003), 'Effects of household dynamics on resource consumption and biodiversity', *Nature*, **421**(6922), 530–33.
Long, A.J., M.J. Crosby, A.J. Stattersfield and D.C. Wege (1996), 'Towards a global map of biodiversity: patterns in the distribution of restricted-range birds', *Global Ecology and Biogeography Letters*, **5**(4–5), 281–304.
Loreau, M., S. Naeem, P. Inchausti, J. Bengtsson, J.P. Grime, A. Hector, D.U. Hooper, M.A. Huston, D. Raffaelli, B. Schmid, D. Tilman and D.A. Wardle (2001), 'Biodiversity and ecosystem functioning: current knowledge and future challenges', *Science*, **294**(5543), 804–8.
Mace, G.M., A. Balmford, L. Boitani, G. Cowlishaw, A.P. Dobson, D.P. Faith, K.J. Gaston, C.J. Humphries, R.I. Vane-Wright, P.H. Williams, J.H. Lawton, C.R. Margules, R.M. May, A.O. Nicholls, H.P. Possingham, C. Rahbek and A.S. van Jaarsveld (2000), 'It's time to work together and stop duplicating conservation efforts. . .', *Nature*, **405**(6785), 393.
Margules, C.R. and R.L. Pressey (2000), 'Systematic conservation planning', *Nature*, **405**(6783), 243–53.
McAllister, D.E., A.L. Hamilton and B. Harvey (1997), 'Global freshwater biodiversity: striving for the integrity of freshwater ecosystems', *Sea Wind*, **11**(2), 1–106.
McClean, C.J., N. Doswald, W. Küper, J.H. Sommer, P. Barnard and J.C. Lovett (2006), 'Potential impacts of climate change on sub-Saharan Africa plant priority area selection', *Diversity and Distributions*, **12**(16), 645–55.
Midgley, G.F., L. Hannah, R. Roberts, D.J. MacDonald and J. Allsopp (2001), 'Have Pleistocene climatic cycles influenced species richness patterns in the Greater Cape Mediterranean Region?', *Journal of Mediterranean Ecology*, **2**(2), 137–44.
Millennium Ecosystem Assessment (2005), *Ecosystems and Human Well-being: Biodiversity Synthesis*, Washington DC: World Resources Institute.
Mittermeier, R.A. (1988), 'Primate diversity and the tropical forest', in E.O. Wilson (ed.), *BioDiversity*, Washington DC: National Academy Press, pp. 145–54.
Mittermeier, R.A., P. Robles Gil and C.G. Mittermeier (1997), *Megadiversity*, Mexico: CEMEX.
Mittermeier, R.A., N. Myers, J.B. Thomsen, G.A.B. da Fonseca and S. Olivieri (1998), 'Biodiversity hotspots and major tropical wilderness areas: approaches to setting conservation priorities', *Conservation Biology*, **12**(3), 516–20.
Mittermeier, R.A., N. Myers, P. Robles Gil and C.G. Mittermeier (1999), *Hotspots*, Mexico: CEMEX.
Mittermeier, R.A., C.G. Mittermeier, P. Robles Gil, J.D. Pilgrim, W.R. Konstant, G.A.B. da Fonseca and T.M. Brooks (2002), *Wilderness: Earth's Last Wild Places*, Mexico: CEMEX.
Mittermeier, R.A., G.A.B. da Fonseca, T. Brooks, J. Pilgrim and A. Rodrigues (2003a), 'Hotspots and coldspots', *American Scientist*, **91**(5), 384.
Mittermeier, R.A., C.G. Mittermeier, T.M. Brooks, J.D. Pilgrim, W.R. Konstant, G.A.B. da Fonseca and C. Kormos (2003b), 'Wilderness and biodiversity conservation', *Proceedings of the National Academy of Sciences of the U.S.A.*, **100**(18), 10309–10313.

Mittermeier, R.A., P. Robles Gil, M. Hoffmann, J. Pilgrim, T. Brooks, C.G. Mittermeier, J. Lamoreux and G.A.B. da Fonseca (2004), *Hotspots: Revisited*, Mexico: CEMEX.
Moran, D., D. Pearce and A. Wendelaar (1997), 'Investing in biodiversity: an economic perspective on global priority setting', *Biodiversity and Conservation*, **6**(9), 1219–43.
Myers, N. (1988), 'Threatened biotas: "hot spots" in tropical forests', *The Environmentalist*, **8**(3), 187–208.
Myers, N. (1990), 'The biodiversity challenge: expanded hotspots analysis', *The Environmentalist*, **10**(4), 243–56.
Myers, N. (1991), 'Extinction "hot spots"', *Science*, **254**(5034), 919.
Myers, N. (2003), 'Biodiversity hotspots revisited', *BioScience*, **53**(10), 916–17.
Myers, N. and A.H. Knoll (2001), 'The biotic crisis and the future of evolution', *Proceedings of the National Academy of Sciences of the U.S.A.*, **98**(10), 5389–92.
Myers, N., R.A. Mittermeier, C.G. Mittermeier, G.A.B. da Fonseca and J. Kent (2000), 'Biodiversity hotspots for conservation priorities', *Nature*, **403**(6772), 853–8.
Nee, S. (2004), 'More than meets the eye', *Nature*, **429**(6994), 804–5.
Norris, K. and N. Harper (2003), 'Extinction processes in hot spots of avian biodiversity and the targeting of pre-emptive conservation action', *Proceedings of the Royal Society of London*, B **271**(1535), 123–30.
Novotny, V., Y. Basset, S.E. Miller, G.D. Weiblen, B. Bremer, L. Cizek and P. Drozd (2002), 'Low host specificity of herbivorous insects in a tropical forest', *Nature*, **416**(6883), 841–4.
O'Connor, C., M. Marvier and P. Kareiva (2003), 'Biological vs. social, economic and political priority-setting in conservation', *Ecology Letters*, **6**(8), 706–11.
Odling-Smee, L. (2005), 'Dollars and sense', *Nature*, **437**(7059), 614–16.
Olson, D.M. and E. Dinerstein (1998), 'The Global 200: a representation approach to conserving the earth's most biologically valuable ecoregions', *Conservation Biology*, **12**(3), 502–15.
Olson, D.M. and E. Dinerstein (2002), 'The Global 200: priority ecoregions for global conservation', *Annals of the Missouri Botanical Garden*, **89**(12), 199–224.
Olson, D.M., E. Dinerstein, E.D. Wikramanayake, N.D. Burgess, G.V.N. Powell, E.C. Underwood, J.A. D'Amico, I. Itoua, H.E. Strand, J.C. Morrison, C.J. Loucks, T.F. Allnutt, T.H. Ricketts, Y. Kura, J.F. Lamoreux, W.W. Wettengel, P. Hedao and K.R. Kassem (2001), 'Terrestrial ecoregions of the world: a new map of life on earth', *BioScience*, **51**(11), 933–8.
Orme, C.D.L., R.G. Davies, M. Burgess, F. Eigenbrod, N. Pickup, V.A. Olson, A.J. Webster, T.-S. Ding, P.C. Rasmussen, R.S. Ridgely, A.J. Stattersfield, P.M. Bennett, T.M. Blackburn, K.J. Gaston and I.P.F. Owens (2005), 'Global hotspots of species richness are not congruent with endemism or threat', *Nature*, **436**(7053), 1016–19.
Ovadia, O. (2003), 'Ranking hotspots of varying sizes: a lesson from the nonlinearity of the species–area relationship', *Conservation Biology*, **17**(5), 1440–41.
Parmesan, C. and G. Yohe (2003), 'A globally coherent fingerprint of climate change impacts across natural systems', *Nature*, **421**(6918), 37–42.
Pimm, S.L. (2000), 'Against triage', *Science*, **289**(5488), 2289.
Pimm, S.L., G.J. Russell, J.L. Gittleman and T.M. Brooks (1995), 'The future of biodiversity', *Science*, **269**(5222), 347–50.
Possingham, H.P. and K.A. Wilson (2005), 'Turning up the heat on hotspots', *Nature*, **436**(7053), 919–20.
Prendergast, J.R., R.M. Quinn, J.H. Lawton, B.C. Eversham and D.W. Gibbons (1993), 'Rare species, the coincidence of diversity hotspots and conservation strategies', *Nature*, **365**(6444), 335–7.
Pressey, R.L. and V.S. Logan (1994), 'Level of geographic subdivision and its effects on assessments of reserve coverage: a review of regional studies', *Conservation Biology*, **8**(4), 1037–46.
Pressey, R.L. and K.H. Taffs (2001), 'Scheduling conservation action in production landscapes: priority areas in western New South Wales defined by irreplaceability and vulnerability to vegetation loss', *Biological Conservation*, **100**(33), 355–76.

Price, A.R.G. (2002), 'Simultaneous "hotspots" and "coldspots" of marine biodiversity and implications for global conservation', *Marine Ecology Progress Series*, **241**, 23–7.
Redford, K.H., P. Coppolillo, E.W. Sanderson, G.A.B. da Fonseca, E. Dinerstein, C. Groves, G. Mace, S. Maginnis, R.A. Mittermeier, R. Noss, D. Olson, J.G. Robinson, A. Vedder and M. Wright (2003), 'Mapping the conservation landscape', *Conservation Biology*, **17**(1), 116–31.
Reid, W.V. (1998), 'Biodiversity hotspots', *Trends in Ecology and Evolution*, **13**(7), 275–80.
Ricketts, T.H., E. Dinerstein, T. Boucher, T.M. Brooks, S.H.M. Butchart, M. Hoffmann, J.F. Lamoreux, J. Morrison, M. Parr, J.D. Pilgrim, A.S.L. Rodrigues, W. Sechrest, G.E. Wallace, K. Berlin, J. Bielby, N.D. Burgess, D.R. Church, N. Cox, D. Knox, C. Loucks, G.W. Luck, L.L. Master, R. Moore, R. Naidoo, R. Ridgely, G.E. Schatz, G. Shire, H. Strand, W. Wettengel and E. Wikramanayake (2005), 'Pinpointing and preventing imminent extinctions', *Proceedings of the National Academy of Sciences of the U.S.A.*, **102**(51), 18497–18501.
Roberts, C.M., C.J. McClean, J.E.N. Veron, J.P. Hawkins, G.R. Allen, D.E. McAllister, C.G. Mittermeier, F.W. Schueler, M. Spalding, F. Wells, C. Vynne and T.B. Werner (2002), 'Marine biodiversity hotspots and conservation priorities for tropical reefs', *Science*, **295**(5558), 1280–84.
Rodrigues, A.S.L., H.R. Akçakaya, S.J. Andelman, M.I. Bakarr, L. Boitani, T.M. Brooks, J.S. Chanson, L.D.C. Fishpool, G.A.B. da Fonseca, K.J. Gaston, M. Hoffmann, P.A. Marquet, J.D. Pilgrim, R.L. Pressey, J. Schipper, W. Sechrest, S.N. Stuart, L.G. Underhill, R.W. Waller, M.E.J. Watts and X. Yan (2004a), 'Global gap analysis – priority regions for expanding the global protected area network', *BioScience*, **54**(12), 1092–1100.
Rodrigues, A.S.L., S.J. Andelman, M.I. Bakarr, L. Boitani, T.M. Brooks, R.M. Cowling, L.D.C. Fishpool, G.A.B. da Fonseca, K.J. Gaston, M. Hoffmann, J.S. Long, P.A. Marquet, J.D. Pilgrim, R.L. Pressey, J. Schipper, W. Sechrest, S.N. Stuart, L.G. Underhill, R.W. Waller, M.E.J. Watts and X. Yan (2004b), 'Effectiveness of the global protected area network in representing species diversity', *Nature*, **428**(6983), 640–43.
Sanderson, E.W., M. Jaiteh, M.A. Levy, K.H. Redford, A.V. Wannebo and G. Woolmer (2002), 'The human footprint and the last of the wild', *BioScience*, **52**, 891–904.
Sarkar, S. (1999), 'Wilderness preservation and biodiversity conservation – keeping divergent goals distinct', *BioScience*, **49**(5), 405–12.
Schmidt, K. (1996), 'Rare habitats vie for protection', *Science*, **274**(5289), 916–18.
Sechrest, W., T.M. Brooks, G.A.B. da Fonseca, W.R. Konstant, R.A. Mittermeier, A. Purvis, A.B. Rylands and J.L. Gittleman (2002), 'Hotspots and the conservation of evolutionary history', *Proceedings of the National Academy of Sciences of the U.S.A.*, **99**(4), 2067–71.
Shi, H., A. Singh, S. Kant, Z. Zhu and E. Waller (2005), 'Integrating habitat status, human population pressure, and protection status into biodiversity conservation priority setting', *Conservation Biology*, **19**(4), 1273–85.
Sisk, T.D., A.E. Launer, K.R. Switky and P.R. Ehrlich (1994), 'Identifying extinction threats', *BioScience*, **44**(9), 592–604.
Smith, R.J., R.D.J. Muir, M.J. Walpole, A. Balmford and N. Leader-Williams (2003), 'Governance and the loss of biodiversity', *Nature*, **426**(6962), 67–70.
Smith, T.B., R.K. Wayne, D.J. Girman and M.W. Bruford (1997), 'A role for ecotones in generating rainforest biodiversity', *Science*, **276**(5320), 1855–7.
Smith, T.B., S. Kark, C.J. Schneider, R.K. Wayne and C. Moritz (2001), 'Biodiversity hotspots and beyond: the need for preserving environmental transitions', *Trends in Ecology and Evolution*, **16**(8), 431.
Spector, S. (2002), 'Biogeographic crossroads as priority areas for biodiversity conservation', *Conservation Biology*, **16**(6), 1480–87.
Stattersfield, A.J., M.J. Crosby, A.J. Long and D.C. Wege (1998), *Endemic Bird Areas of the World: Priorities for Biodiversity Conservation*, Cambridge, UK: BirdLife International.
Stuart, S.N., J.S. Chanson, N.A. Cox, B.E. Young, A.S.L. Rodrigues, D.L. Fischman and R.W. Waller (2004), 'Status and trends of amphibian declines and extinctions worldwide', *Science*, **306**(5695), 1783–6.

Supriatna, J. (2001), 'All sectors of society must work together to save biodiversity', *Nature*, **410**(6824), 14.
Thomas, C.D., A. Cameron, R.E. Green, M. Bakkenes, L.J. Beaumont, Y.C. Collingham, B.F.N. Erasmus, M. Ferreira de Siqueira, A. Grainger, L. Hannah, B. Huntley, A.S. van Jaarsveld, G.F. Midgley, L. Miles, M.A. Ortega-Huerta, A.T. Peterson, O.L. Phillips and S.E. Williams (2004), 'Extinction risk from climate change', *Nature*, **427**(6970), 145–8.
Veech, J.A. (2000), 'Choice of species–area function affects identification of hotspots', *Conservation Biology*, **14**(1), 140–47.
Veech, J.A. (2003), 'Incorporating socioeconomic factors into the analysis of biodiversity hotspots', *Applied Geography*, **23**(1), 73–88.
Werner, U. and J. Buszko (2005), 'Detecting hotspots using species–area and endemics–area relationships: the case of butterflies', *Biodiversity and Conservation*, **14**(8), 1977–88.
Whittaker, R.J., M.B. Araújo, P. Jepson, R.J. Ladle, J.E.M. Watson and K.J. Willis (2005), 'Conservation biogeography: assessment and prospect', *Diversity and Distributions*, **11**(1), 3–23.
Wilson, K.A., R.L. Pressey, A. Newton, M. Burgman, H. Possingham and C. Weston (2005), 'Measuring and incorporating vulnerability into conservation planning', *Environmental Management*, **35**(5), 527–34.
Wilson, K.A., M. McBride, M. Bode and H.P. Possingham (2006), 'Prioritizing global conservation efforts', *Nature*, **440**(7082), 337–340.
WWF and IUCN (1994–97), *Centres of Plant Diversity: A Guide and Strategy for their Conservation*, Gland, Switzerland: WWF and IUCN.

3. Integrated conservation and development projects: a positive role for forest conservation in tropical Africa?[1]

*Neil Burgess, David Thomas,
Shakim Mhagama, Thomas Lehmberg,
Jenny Springer and Jonathan Barnard*

Background

A debate has been going on for a number of years on the best ways to achieve conservation in Africa (and elsewhere). Two elements of the debate involve those espousing 'fortress conservation' and those promoting 'people-focused conservation'. In some circles this debate has become highly polarized, with a considerable divide on the best ways to achieve conservation opening between biologists (Spinage, 1996, 1998; Kramer et al., 1997; Oates, 1999; Attwell and Cotterill, 2000; Bruner et al., 2000) and social scientists (Grove, 1995; Neumann, 1996, 1998; Borrini-Feyerabend and Buchan, 1997; Ghimire and Pimbert, 1997; Hackel, 1999; Leach et al., 2002). However, for those involved with implementing conservation projects on the ground in the developing world, the polarized views often represent impractical extremes. Moreover, for the people living in the rural areas of developing countries, the divide between 'development' and 'conservation' is often quite artificial. The third element of the debate involves attempts to merge human development and wildlife conservation issues within a single integrated programme, ideally where all sides benefit, the basis of Integrated Conservation and Development Projects (ICDPs). These kinds of projects can be considered to fall between 'fortress conservation' and 'sustainable resource use for rural development'. The ICDP has become one of the dominant approaches to field implementation of conservation in the developing world over the past 30 years. In this chapter we look at where ICDPs have come from, what their successes and failures have been and where they are heading.

Where have ICDPs come from?

In the earlier part of the twentieth century, national governments in African countries (often the colonial power, but in some cases African royalty) tended to take a preservationist approach to the conservation of wildlife and forestry resources (Neumann, 1996; Schrijver, 1997). Areas of land were set aside for colonial and royal hunting and local uses were prohibited. In other cases, areas of mountain forests were reserved to ensure a water supply for people over a broader area (see Rodgers, 1993).

In the later part of the twentieth century the preservationist models came under significant criticism from social scientists working primarily with human development issues (Anderson and Grove, 1987; IIED, 1994; Pimbert and Pretty, 1995; Alpert, 1996; Chambers, 1997). In an attempt to create a more socially equitable model for conservation in poor developing countries, conservation agencies borrowed ideas from development practitioners and created a single integrated model for conservation and development – the ICDP. The model proved extremely popular, not least amongst the development assistance agencies from developed countries, which were charged with assisting developing countries to solve their human development and environmental problems. This popularity provided significant new funding opportunities, and the ICDP model was rapidly adopted by many of the larger conservation NGOs working in developing countries. The first operational projects were established in the middle 1980s (Hannah, 1992; Stocking and Perkin, 1992; Sanjayan et al., 1997; Larsen et al., 1998; MacKinnon, 2001; Jeanrenaud, 2002a, b; Franks and Blomley, 2004; Wells and McShane, 2004). Since that time, hundreds of ICDPs have operated across the world, especially in Africa where people live side by side with biological resources (Brandon and Wells, 1992; Brown and Wyckoff-Baird, 1993; Barrett and Arcese, 1995; Fisher, 1995; Caldecott, 1998; Margolius and Salafsky, 1998; Wainwright and Wehrmeyer, 1998; Newmark and Hough, 2000; Adams and Hulme, 2001; Hughes and Flintan, 2001).

The ICDP differs from protectionist approaches to conservation in that the local people surrounding the areas of high natural resource value also form a focus for project attention (Franks, 2001). In the protected area management systems of the past, people were often excluded by force and regarded as 'poachers' who should be punished or even shot if they ventured into the protected area without government permit. If they suffered costs from living close to the protected area these were ignored and it was often suggested that they should move elsewhere. This caused much antagonism between protected area staff and local residents. Most protected area management agencies have softened their stance in recent years and many agencies share some of their revenues, assist with local development projects, or allow local populations to take some resources from the reserve. Such changes have improved relations between protected areas and people in many places, and such changes bring the park closer to the ICDP model.

The ICDP differs from community-based natural resource management because it always contains a core area protected for conservation reasons (either government or community-managed), whereas community-focused approaches may or may not contain core conservation areas (Dubois and Lowore, 2000; Songorwa et al., 2000; Roe and Jack, 2001). However, in some of the developing models of community-based forest or community-based wildlife management the villagers are taking over the management of former government forest or wildlife reserves. Hence, there is a blending of the ICDP model and that of community-based natural resource management – which is perhaps not surprising as many of the management and implementation issues are the same.

The advantages of ICDPs
There are a number of ways in which ICDPs have been seen as advantageous implementation models in the context of poor African societies. A major advantage to the poor is that these projects aim to be socially just (Carney, 1999; Koziell, 2000; Franks, 2001). They do not aim to work to further the interests of elites, but instead are trying to achieve a long-term solution to poverty alleviation and natural resource management with the participation of local communities.

The fact that ICDPs have aimed to address the needs of the rural poor has also been a major advantage in terms of their acceptability to agencies engaged in poverty alleviation (Wells and Brandon, 1992). The attention of ICDPs to development issues has allowed the approach to be mainstreamed within the portfolio of development assistance agencies, which provide by far the largest source of external funding to poor developing countries (see, for example, Wells et al., 1999). Stricter conservation approaches can be funded using private foundation money, or by money collected by NGOs from wealthy individuals in northern countries, but this represents a significantly smaller pool of resources.

The declarations from the World Summit on Sustainable Development in South Africa in September 2002 focused heavily on the alleviation of poverty and the benefits of good environmental management to achieve this aim (WSSD, 2002). This was regarded as especially important in Africa where there remains the greatest levels of global poverty (UNDP, 2001). The potential of the ICDP approach to tackle both poverty and natural resource management issues provides significant advantages over other conservation approaches.

The problem with ICDPs
Over the past ten years there have been increasing critiques of ICDPs, from both the development and the conservation communities. These critiques arise from a perception that ICDPs have not performed as well as

expected in delivering either conservation or development results. ICDPs have also been criticized for their focus on defining problems and undertaking solutions at local levels, when many local problems are driven by powerful social, economic and political forces external to project sites (Sanjayan et al., 1997; Larson et al., 1998). Critics have also noted the high level of funding required for ICDPs, and difficulties in achieving long-term sustainability. The fact that ICDPs place natural resource conservation as a central goal is often seen as privileging an international conservation agenda while the local communities bear the local costs of that conservation (Leach et al., 2002). For those involved with human development work, projects should focus exclusively on the needs of poor local people, rather than find ways in which development activities can be devised to maintain the status quo of established protected areas (for wildlife or forest conservation).

On the other side, some conservationists have advocated returning to the 'core values' of conservation – that of protecting wild habitat in official conservation areas (generally government owned in Africa), that will ensure the survival of species and habitats valued at a national or international level. It is stated that ICDPs put too much effort into the development interests of local communities, to the detriment of the conservation work (Oates, 1995, 1999).

Among economists there have also been critiques of the ICDP approach as not being sufficiently direct to create incentives for conservation (Ferraro and Kiss, 2002). It is argued that direct payments to local communities for the costs they incur from conserving natural resources would create a direct link between the receipt of money and the conservation of a resource value. ICDPs do not make such a direct link as the natural resource benefits are a more indirect product of multiple programme interventions, including human development.

One difficulty with much of the debate on the value of the ICDP approach is that it reflects the political and social perspectives of those framing the debate, what their goals are, and where in the world they live (Adams and Hulme, 2001). Management approaches for conservation interventions exist along a gradient from strict protection through to community management (see matrix on p. 34). Strict Nature Reserves and most National Parks would come on the left, and community-based management approaches to the right. As ICDPs have a philosophy of integrating their management approach, they fall somewhere near the centre of this gradation. For those people favouring a protectionist approach to conservation, ICDPs are too 'social' in their approach, and for those focusing on community management, ICDPs represent a means to support the established government and 'outsider' elites.

CATEGORIES OF INTERVENTION

Protectionist	Integrated	Communal Management
	EXAMPLES	
Strict Nature Reserve	Joint Forest Management	Community-based Forest Management
National Park		Community-based Wildlife Management
Forest Wilderness Area		

Improving the ICDP model

We believe that recent experiences with the ICDP model, despite all the harsh criticism, do not argue for the cessation of the approach. The blending of poverty alleviation with conservation goals (or conservation with poverty alleviation) is in line with many international conservation agreements, and is evidently morally correct in poor regions of the world. ICDPs are not aiming for conservation at the expense of lives and livelihoods – but instead are trying to find a solution to this *dual* goal through a multidisciplinary approach. This is not only an issue of relevance to conservation staff, as it has been shown that the environment does matter to poor people in Africa (Posey, 1999; Songorwa, 1999), and that maintaining a strict divide between 'development' and 'conservation' is often quite artificial in the minds of the people living in the sites where conservation projects are implemented.

Instead of abandoning the overall ICDP approach, we believe that the limitations identified in ICDP design and performance need to be addressed, new interventions tested and performance of these interventions evaluated in a rigorous way. Failure to build upon the ICDP model and to truly create the conditions where livelihood enhancement is linked carefully with natural resource conservation will almost certainly result in a significant decline of funding for conservation-related activities. This becomes more likely as funding from northern governments is targeted to the improvement of livelihoods and the reduction of poverty across Africa.

Through our experience, and from discussions with colleagues and a review of the literature, we have identified seven issues that need more attention during the design and the implementation phases of future ICDPs in Africa.

Scale is appropriate
Recent discussion has focused on whether the ICDP model only applies to small sites of isolated habitat, or whether it can be used as a model

BOX 3.1 DEFINING GOALS, FOCAL BIOLOGICAL ELEMENTS AND TARGETS

- A *goal* is essentially the long-term conservation outcome you wish to achieve. For example, in a forested landscape of south-west Cameroon, the goal might be, 'Conserve sufficient connected habitat to maintain viable long-term populations of lowland gorilla, forest elephant, and forest buffalo'.
- *Focal elements* refer to the set of biological characteristics that make an area significant for conservation. Focal elements include species (such as elephants, endemic birds and plant species), habitats (cloud forest, wetlands, miombo woodlands) and processes (colonial nesting sites for birds, elephant migration routes). For example, in the Cameroon Highlands the endemic birds would be focal species; the montane forests would be focal habitat; and the hydrological function of water catchment to lowland areas would be a focal ecological process.
- A *target* is the amount, type, and configuration of the land needed to conserve the focal elements. It might also contain, in the case of species, a determination of viable population levels. Following the example above, a target for, for example, mountain gorillas might be, 'Conserve minimum of 500 km^2 of interconnected forests ranging in altitude from 1400 metres to 3000 metres, and access to water'. A full set of targets should maintain the focal elements of an area.

for interventions across larger landscapes (Franks, 2001; Burgess et al., 2002b). In recent years, a landscape approach to the design and implementation of ICDPs has been in favour, for example in the Participatory Environment Management Programme in Uganda and Tanzania.[2] The design of ICDPs at the landscape scale is perhaps even more challenging than at smaller scales. Landscape-level design can use biological features (forest cover, area-dependent species, ecological processes) to derive the conservation targets (Box 3.1). Alternatively, the targets for the ICDP can be derived from analyses of the threats (or pressures) to the natural resource values (Box 3.2), or can use human socioeconomic indicators to design the conservation interventions. Whether using biological features,

> **BOX 3.2 POTENTIAL SOCIOECONOMIC ISSUES THAT CAN FORM THREATS TO BIOLOGICAL DIVERSITY**
>
> - Patterns of land and resource use:
>   damaging land and resource use (including forest, water, wildlife);
>   damaging development plans (roads, dams, and so on) or projected changes in land use;
>   lack of existing zoning regulations;
>   lack of protected areas.
> - Governance and land/resource ownership and management:
>   political boundaries (provinces, districts);
>   insecure or conflicting land tenure (private, public, ancestral/communal areas);
>   conflicting responsibilities for management (for example, Forestry vs. Agriculture Departments).
> - Population data:
>   high population density and growth rates, population centres;
>   unhelpful human migration patterns (in- and out-migration);
>   high levels of poverty.

threats, or a combination of attributes including human development needs will make a more targeted ICDP able to function across landscapes is not yet known, but this thinking may have helped to clarify some of the structural issues of ICDP design.

There is real local ownership
Experience indicates that ICDPs are more likely to succeed if they have two levels of support. First, in African countries, it is important that the government authorities support the project, as it will be impossible to implement if their support is lacking. Once operational, the issue of ownership moves to the people who live in close proximity to the natural resource values the ICDP is seeking to conserve. In general the involvement of many different groups of people (stakeholders) during the design and implementation stages of a project is fundamental to making the project work and for it to become sustainable. However, this is often difficult to

achieve as different groups of local people can have highly conflicting points of view about the best use of the natural resources (for example, government forestry/wildlife staff, pitsawyers, commercial agriculturalists, traditional healers, hunters, rich people/poor people). Making all of these groups feel that they are involved in the ICDP and that it is working on their behalf is extremely difficult, and some trade-offs are inevitable.

Protection components are included
It is important for a conservation project to define what it is trying to conserve and for whose benefit. In most cases ICDPs have listed the species that the core habitat area is important for, and perhaps some from the surrounding lands as a part of the project justification. Many projects proceed no further, and do not know if the habitat is large enough, suitable enough, or in the right spatial configuration to maintain the biological values. In recent years attempts have been made to structure these questions, so that an ICDP could establish targets for species, habitats and important ecological processes as a part of its planning phase. Box 3.1 outlines simple approaches for defining focal biological elements and setting targets, as core parts of the project design. These targets could be built as indicators into a logical framework for an ICDP, typically at the level of overall project goal or objectives.

The management of any form of natural resource implies a need for rules, regulations, boundaries and enforcement mechanisms. It is therefore not surprising that supporting protection and regulation is a common part of ICDPs. Indeed, it has been concluded that protection of the core natural resource values is essential for all ICDPs (Hannah, 1992; Rodgers, 1993; Abbot et al., 2001). In many African ICDP locations, there is a government-managed protected area at the core of the ICDP. In such cases, the government officers traditionally undertake the protection component. This system works well in some places and poorly in others. Reasons for a poor performance range from a lack of staff or equipment to allow them to do the job, through to their personal exploitation of the resource for their own profit, or for the profit of the government agency managing the area.

Poor governance by the official managers of the core area (or perceptions of it) can result in negative perspectives of the local people towards the official managers, which makes ICDP models difficult to operate without considerable work on building trust and cooperation. In other ICDPs a locally controlled and managed area of natural habitat may be found instead of a government reserve. In many places these local systems are long-standing and can provide good protection. If there is no governmental protected area, and no traditional management systems, some

ICDPs have attempted to create management systems that take on their own institutional form over the long term.

A related issue is whether there is an appropriate and functioning legal framework for the ICDP, and for actions related to conservation and development. If there are no legal frameworks (either from the national government, or from traditional leaders) then it is difficult to design project interventions that require the 'control' of use of natural resources. In many parts of Africa, projects start by working with the government-authorized village leadership, and gradually discover that there may be an equally (or sometimes more) powerful system of traditional leadership. Working with the right people from the outset of an ICDP has obvious benefits, but is often difficult to achieve.

Project targets are clearly formulated
Importantly, within the context of an ICDP there needs to be a clear and strong linkage between the development activities of the project and the conservation objectives of the project. Often the ICDP development activities are executed separately to the conservation activities and have no obvious linkage to the overall conservation goals, for example the improvement of roads, purchase of sewing machines, or improvements in local health care facilities. In such cases it is difficult to link the development interventions to the conservation, which may cause the project to fail. However, this strategy may work if the linkages have been developed carefully over a long period of time, or people know that they are receiving these benefits as a form of 'payment' for not damaging species, habitats or ecological processes.

Most ICDPs are developed by teams of consultants, generally from outside the area. There is a tendency for these consultant teams to build assumptions about the state of the local community and the natural resource management systems into the targets for the ICDP, without having the data to validate or refute these assumptions. They also bring their own biases on the 'best' ways to achieve conservation in a particular situation. One kind of assumption that is often inadequately tested is the source of threats to natural resources. Over-simplistic analyses are often made of the threats to the system, and to the natural resource value that the ICDP is targeting. The main threat to a mountain forest may come from agricultural expansion, but this may not be aimed at providing more food to the local populations, but rather a commercial crop for export to townspeople. The clearance of natural habitats to expand the commercial agriculture may be organized by a powerful person based well away from the site of the ICDP and who is simply paying local people to work on his/her behalf.

> **BOX 3.3 DEVELOPMENT ACTIVITIES THAT ARE WHOLLY CONSISTENT WITH FOREST CONSERVATION OBJECTIVES**
>
> | Activity | Benefit |
> |---|---|
> | Tree planting | Reduces dependence on extraction from natural forest |
> | Community management | Empowers local populations to protect forest (generally outside reserves) |
> | Agricultural improvement | Reduces shifting cultivation, controls soil erosion and reduces loss of natural forest |
> | Ecotourism | Provides a value to the natural forest resource |
> | Sustainable extraction | Provides a value to the natural forest resource |
> | Small animal husbandry | Can reduce hunting pressure in natural forest (but care has to be taken to maintain these links) |
> | Alternative incomes | Can take pressure from forest (but care has to be taken to maintain the beneficial links between the utilization of the resource and the conservation of the habitat and species of concern |

Assumptions about threats need to be validated before the project starts, or at least early in its lifespan. Targeted studies using appropriate specialists can explore different assumptions, and in cases where assumptions are shown to be false, this can be a painful experience. Failure to understand assumptions is another major cause of project failure. Linked to understanding assumptions is the idea of undertaking better analyses of the threats to the natural resource value in question. Methodologies exist to analyse both the surface and 'root-cause' threats that are impacting on a system.

A number of development actions are entirely consistent with the objectives of, for example, forest conservation (Box 3.3). These and activities like them form suitable interventions for future ICDPs.

One of the best ways to ensure that project targets are well structured and logically arranged is to use the Logical Framework Approach,[3] a methodology that is popular in both government development agencies and conservation NGOs. This approach starts with an analysis of threats to develop a problem tree for an area. By reversing the problem tree a set of solutions is developed, which forms the basis for the project interventions in the area. The log frame structures the project in terms of broad goals, narrower objectives and the activities that need to be done to achieve the objectives (and ultimately the goal). Although some criticize the framework as too restrictive and not permitting adaptive management approaches, the 'log frame' provides a very effective way to organize a project. This method has the additional advantage that it can be designed using participatory methods in workshops of the relevant stakeholders. It also takes consideration of the assumptions made when proposing actions to solve threats, and can be used to develop a framework to measure actual conservation success (a monitoring and evaluation framework) (Caldecott, 1998; Margolius and Salafsky, 1998).

During the design phase of an ICDP, and throughout its lifespan, people coming from a natural resource management background (biologists and foresters and so on) and those coming from a human development background (social scientists and development scientists) need to pool their ideas and approaches to achieve the conservation of an area of high importance for biological diversity. Compromise over the idealized scenarios of a number of different stakeholders is an essential component of the development and long-term management of such projects. The theme of compromise and negotiation continues into the implementation phase, and without it the project becomes difficult to manage. The lack of compromise and a failure to appreciate the points of view of others involved in ICDP design and management is, we believe, the fundamental cause of the conflict that has arisen around these projects in recent years. But it is also the nature of the compromises that need to be made that have prevented ICDPs achieving as much conservation or development as was originally envisaged – in most cases a neat win-win situation is not possible and success is much vaguer and difficult to measure.

Specialized computer software is now available that allows biological and social data to be integrated easily and cheaply and that might provide some assistance to the spatial design of ICDP interventions and to achieving 'consensus' over different possible conservation scenarios (Box 3.4). This process is possible at a remote computer lab, but can also be done using participatory methods in the field. Stakeholder negotiations can form a part of this form of 'land use planning' and may assist in resolving some of the conflicting stakeholder interests, or at least exposing them clearly, as the ICDP is designed and interventions for different geographical areas are proposed.

Integrated conservation and development projects 41

BOX 3.4 USE OF SPATIAL DECISION SUPPORT SYSTEM (SDSS) SOFTWARE TO INTEGRATE CONSERVATION TARGETS AND THREATS

What is an SDSS?
A decision support software (DSS) can be generally defined as an interactive computer-based software to help decision-makers utilize data and models to solve problems. For our purposes, we further define a DSS to include a spatial (mappable) component, commonly provided through a GIS. Therefore a spatial decision support system (SDSS) is a software program that uses a variety of spatial data and analytical and statistical modelling capabilities to answer problems in a map or graphic display. Most SDSSs can be adapted to meet decision-makers' needs to solve problems, formulate alternative displays, interpret and select appropriate implementation options and modify or include new data.

Why use an SDSS?
SDSS allows users to input a variety of spatial data (that is, species ranges, future land use plans, cost of land parcels, development zones) into a computer system and specify a set of requirements (for example, select areas of forests that maximize tiger habitat but minimize cost and distance from roads). These requirements can be modified or changed and therefore are useful in developing a series of conservation planning options. A common output of an SDSS is a map or series of maps that indicate various land use configuration options that will meet conservation, development, or combined goals. The outputs of an SDSS can provide compelling visualization of the conservation landscape that can aid in communicating the plan with broad stakeholder groups. However, as with all computer-generated solutions, your answer is only as good as the data used in the program.

Currently, a range of SDSS software is being used by some conservation organizations, including SITES (Marxan and Spexan), C-PLAN, IDRISI, NatureServe DSS, TAMARIN and GeoNetWeaver. All of these programs can accept both conservation-orientated and development-orientated information.

Clearly identify livelihoods opportunities (productive potential of different habitats is taken into account)
In Africa the ICDP approach has perhaps been most successful in savanna woodland regions with large mammal populations that generate tourist revenues, and/or can be used for hunting and meat production for local people. There are numerous examples of successful or nearly successful savanna woodland ICDPs from southern and eastern Africa (Leader-Williams et al., 1996; Hulme and Murphree, 1999). In the dense forest habitats there is a lower density of large mammals, a lower rate of biomass production that can be hunted for food and a lower potential for tourists to visit the area (see Lukumbuzya, 2000; Wily and Mbaya, 2001). Most direct forest values have been realized by logging timber trees, or hunting animals as bush-meat, and conservationists argue that these uses are unsustainable everywhere they have been attempted in Africa. The difficulties of designing sustainable management approaches with poverty-stricken people whose short-term survival is likely to override long-term management opportunities are well known (Hackel, 1999). Far more of the forest values are indirect and realized by people away from the forest edge, for example clean water supply, reliable water flows to downstream users (both people and industry), carbon sequestration, genetic resource conservation and so on. Exceptions are rare, but include some of the montane forests of the Albertine Rift in Central Africa; here the presence of mountain gorillas makes the conservation of the forests financially viable. There is a large direct economic benefit to the local people from tourists and a large international interest in the conservation of the gorilla as a species.

A sustainable end point is defined
All ICDPs have the ultimate aim of solving resource management problems and leaving a sustainable system in place that can carry on the work of the project at the local level, forever. One of the first problems is deciding how to measure sustainability. Ecological sustainability is often different from agricultural sustainability, or economic sustainability. Most experiences so far indicate that in poor developing countries the problems are not solved entirely, even after a decade of ICDP interventions, and creating the funding and institutional mechanisms to sustainably manage the resources is complex. Typically, an ICDP will remain in an area working to solve these issues for as long as funding is in place. Once the funding ends, or is set to end, then attempts are made to leave as sustainable a system as possible. It is a common perception that interventions quickly disappear as soon as the project leaves the area and the situation returns to how it was before the project was operational.

Although hard to achieve, it is perhaps best to try to define a sustainable end point for the ICDP from the outset, and work towards this end point. If the natural resource values are in a government reserve then building the strength of that agency to achieve its mandate in an equitable manner should assist long-term management sustainability. If the resources are found in community-controlled areas, then building the capacity and management authority of locally constituted bodies (village committees or similar) to take on the role of managing the resource may be the best way to achieve sustainability.

If the natural resource requires funding to manage, then working to achieve financial sustainability is critical. Most sources of sustainable funding for long-term conservation interventions in Africa come from tourism (for example, ecotourism, see Ashley and Roe, 1998), or use the natural resource to generate funds. One other source of long-term funding is the trust fund, where a capital sum is used to generate some interest to manage the resource in perpetuity. Another mechanism is the national or local taxation system that might – for example – generate funds from water users and then provide these funds to the managers of the relevant watersheds – in particular in some relatively dry countries like Kenya, Zimbabwe, Tanzania and Malawi. Considerable discussion now takes place around the issue of whether schemes that entail 'Payments for Environmental Services' can provide a real solution to the matter of sustainably financing conservation around ICDP sites. The most popular schemes – and those where there has been the greatest amount of work – are those related to carbon and water payments.[4]

A number of other issues can make achieving a sustainable end point to project interventions challenging. Changing politics and changes in the local/global economies can alter an apparently 'sustainable' outcome to one that is unsustainable. There may also be local differences in interpreting a sustainable end point. Local farmers may accept a balance between habitat conservation and agricultural development, but a powerful local politician may have different aims for use of these resources. Finally, valued biodiversity sites may need a subsidy (in perpetuity) to have a sustainable conservation outcome, that is, if in economic terms the benefits from conversion greatly outweigh the benefits from conservation.

Monitoring and evaluation confirms project success
Most ICDPs have failed to measure their conservation impact (Kremen et al., 1994; Salafsky and Margolius, 1999; Klieman et al., 2000; Danielsen et al., 2001; Wells and McShane, 2004). Without such measures it has been relatively easy for them to be heavily criticized (for example, Oates, 1999; Leach et al., 2002).

A major challenge for ICDPs is to devise systems to collect data systematically that show that they have delivered conservation better than doing nothing (the null hypothesis), or when compared with other conservation (or development) approaches. ICDPs, especially those set up using development assistance funds, were in the past often only required to measure their activities, such as numbers of meetings held, numbers of people trained, newsletters produced, study tours completed, accounts produced on time and audited correctly and so on. Such measurements illustrate that the project is working according to its schedule and that its activities are being done. However, this level of measurement does not illustrate real conservation impact in terms of reducing the threats to the natural resources within the ICDP area, and does not measure the biodiversity and societal state of the region before, during and after the project intervention. These fundamental problems are harder to solve. Successful changes would generally be regarded as a successful 'outcome' of the project, and much effort is now placed on devising monitoring systems to look at project impact and the achievement of the relevant outcomes.

In addition to assisting project design, the Logical Framework Approach can also be used to assist in the development of a scheme for measuring the effectiveness of an ICDP. At all levels of the log frame (but especially at the higher levels of the goal and objectives) the development of measurable indicators of pressure (threat), state (biodiversity and people) and response (biological and people) would provide a real mechanism to assess ICDP success. Monitoring of activities and budgets and reports produced on time and so on, would continue to be important as well, but if conservation success were measured across a wide suite of ICDPs this would provide the data to answer many currently unanswerable questions.

In addition to ICDPs themselves seeking to measure their conservation success by looking at how effectively they address threats, preserve or change the biodiversity state, there is also a clear need for multiple ICDPs to roll up their results into a more formal assessment of the success of the approach in general. Currently most of the statements of project success or failure are opinion- or case-study-driven. We are not aware of any study that uses quantitative data to test the success of ICDPs, when compared with any other conservation mechanism, either in monetary terms, or in terms of conservation delivery or livelihood improvements. This lack of hard data allows considerable unproductive debate to occur.

We believe that monitoring and evaluation schemes that look at conservation impact need to be built into every ICDP. For species, surveys of endemic, rare or focal species can provide the evidence that these elements of value are maintained, or increased. Satellite, aerial photographic or ground surveys of habitats (and reserve boundaries) provide an additional measure whereby the success of the project at conserving biological elements of value can be assessed. Rates of extraction, levels of disturbance and habitat 'condition' provide a further tranche of data whereby conservation impact can be determined. Measuring the effect the project has, such as reduction of threatening activities, increased protection, fewer purchases of illegally hunted animals, and so forth links the response of society to the state of the resources. Surveys of the knowledge, attitudes and practices of the local population provide an indirect measure of the success of a project at changing practices and perceptions – which can be repeated to assess change over time.

Measuring development impact is also possible, for example using measures such as numbers of people within the project area who are short of food, do not send their children to school, or have very poor levels of agricultural productivity. As income levels increase, then the number of people making a living from the sustainable use of the forest resource, or forest-related activities (for example, tourism) can also be used as a measure of development impact. Measuring the combination effect of conservation and development interventions within an ICDP is the most challenging issue of all. We are not aware of any attempt to measure the synergistic effects of different interventions.

Building success in ICDPs in the forests of Africa

Here we look at two case-study ICDPs from African forests and assess how well they have addressed, or are addressing the issues we outline above as fundamental concerns in improving the performance of ICDPs in the region. These projects are ones with which we have had a long personal involvement, hence our comments are certainly not unbiased. However, due to this involvement we also know some of the failures that the projects have experienced, and have seen in some cases how these have later been solved.

The Uluguru Mountains of Tanzania

The Uluguru Mountains are located in eastern Tanzania (Figure 3.1) and comprise a forest-capped area covering 1500 sq km of highlands, with forested areas found discontinuously from 150 m up to 2630 m above sea level. The main ridge runs almost north–south, and there are a number of smaller outlying hills that are broadly regarded as a part of the Uluguru Mountains landscape. Administratively the area falls within two Tanzanian rural districts and a part of one urban municipality. More than 50 villages have a border with the larger forests of the area, which are found in 22 government Forest Reserves (Figure 3.2a). Over 100 villages cover the entire Uluguru Mountains landscape, and these often have their

46   *A handbook of environmental management*

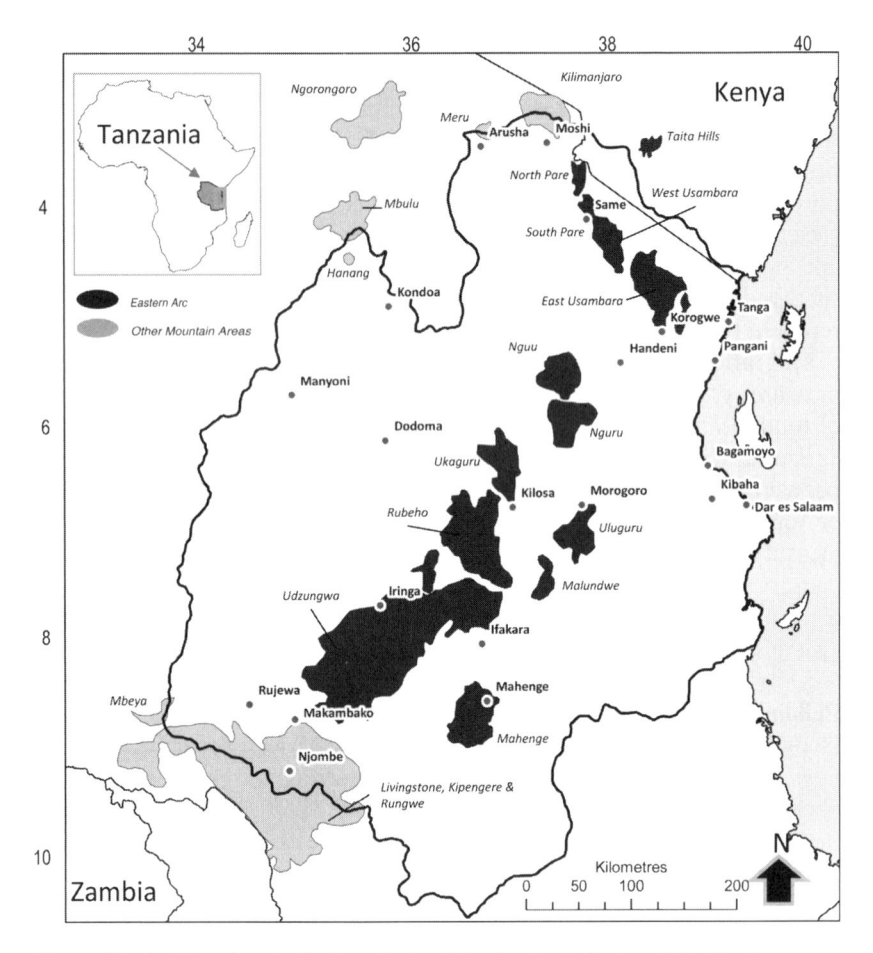

Note: The dark line denotes the boundaries of the river basin that are linked to the Eastern Arc.

Source: Valuing the Arc Programme, Cambridge University, UK.

Figure 3.1   The location of the Uluguru Mountains in eastern Tanzania in relation to other mountain forest blocks in the 'Eastern Arc' biogeographical region of Africa

own smaller patches of forest, often for traditional spiritual purposes, and 50 of these villages touch the forest margins (Figure 3.2b). Population density is somewhat variable, but typically there are high densities of people up to the borders of the Forest Reserves and in some parts of the area human populations increase at higher altitudes where the farming

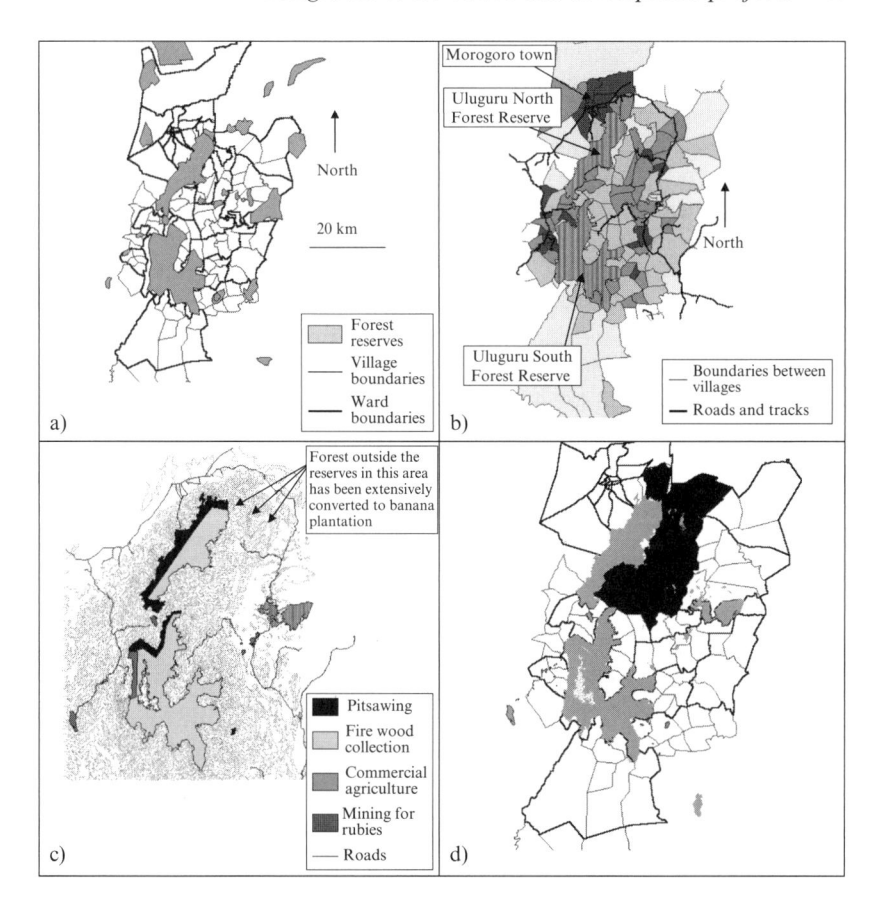

Integrated conservation and development projects 47

Note: a) location of government-owned Forest Reserves, b) human population density in the wards around the Ulugurus in 1988, c) main threats facing the forests of the area, and d) main area of project intervention during Phase II DANIDA support.

Source: WCST Uluguru Mountains Biodiversity Project.

Figure 3.2 The Uluguru Mountains landscape, showing geographical features of importance when designing an ICDP

potential is better (Figure 3.2c). Most people living on the mountains are poor (Hartley and Kaare, 2001). For an outsider it is quite a daunting region and a car can only reach a few higher regions of the mountain. However, for the people of the Ulugurus, paths link nearby villages and there are paths crossing the mountains to facilitate longer-distance travelling. It is often quicker to walk (or run) over the mountain than it is to drive around it. Distances are measured in hours of walking.

The biological importance of the Uluguru Mountains is well known, with more than 135 species of strictly endemic plants and 16 species of strictly endemic vertebrates (Lovett and Pócs, 1993; Lovett and Wasser, 1993; Svendsen and Hansen, 1995; Burgess et al., 1998, 2002a; Burgess and Clarke, 2000; Doggart et al., 2005). The mountains are also important hydrologically as they provide the source for the main water supply to the largest city in Tanzania, Dar es Salaam (see for example, Pócs, 1974, 1976), grow considerable quantities of food for export to towns and are home to over 100 000 people – primarily from the Luguru tribe. The Ulugurus therefore have high international values for biodiversity conservation, high national values for water supply to the national and regional capital cities, and high local importance as a living place for the Luguru people.

Externally funded ICDP activities have been undertaken in the Ulugurus for over a decade (see for example, Bhatia and Buckley, 1998; Burgess et al., 2002b), but have tended to cover small geographical areas (Figure 3.2d). Three project phases are recognized here:

- Phase 1. The planning phase where socioeconomic surveys were undertaken to try to understand the interventions that were appropriate and needed in the area (Bhatia and Ringia, 1996; Bhatia and Buckley, 1998). This phase was funded by the European Union and the Royal Society for the Protection of Birds and worked with Tanzanian NGO and government partners.
- Phase 2. This phase (after a gap of three years) started to implement the findings of the preparatory studies. Danish International Development Agency (DANIDA) funded this part of the work, which involved Danish (Dansk Ornitologisk Forening – DOF) and Tanzanian (Wildlife Conservation Society of Tanzania – WCST) NGOs and Tanzanian government staff.[5]
- Phase 3. This phase represented an expansion of Phase 2, where the existing Danish government-supported programme was joined by a second project funded by GEF and managed by the development agency, CARE. The DANIDA-supported programme started in 2002 and the GEF one started in 2003. Both ran for five years.

The focal area for intervention by these projects is the region of highest forest biodiversity and greatest threats to the forests. Until 2002, this caused project actions to concentrate in the forest remnants outside the official government Forest Reserves on the northern end of the range (Figure 3.2d). The area was heavily forested in 1955 (from aerial

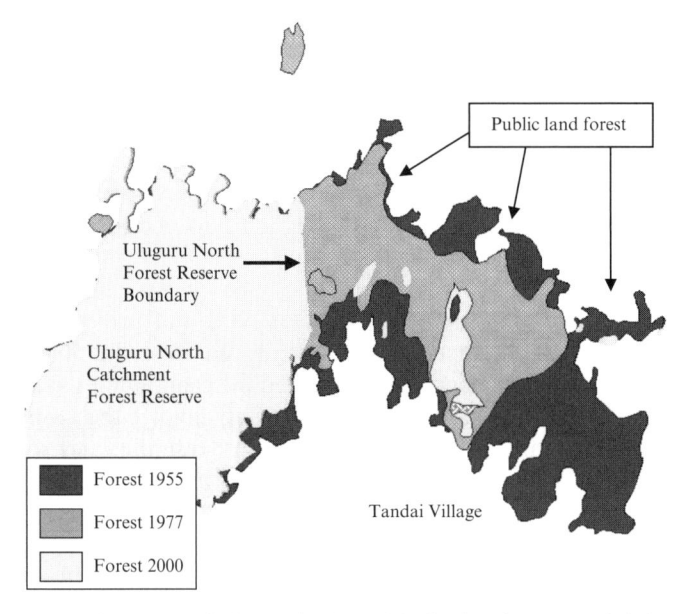

Integrated conservation and development projects 49

Note: This area is closely co-incident with that of the focal project area as it is here where most conflicts between biodiversity conservation and human use of the land occur.

Source: WCST Uluguru Mountains Biodiversity Conservation Project.

Figure 3.3   Loss of the non-reserved forest outside the Uluguru North Catchment Forest Reserve, 1955–2000

photographs), but had been largely deforested by 2000, apart from some remnants (Burgess et al., 2002a; Figure 3.3). The project attempted to preserve these fragments by building them into the village land use pattern as 'Village Forest Reserves'. In the Phase 3 project there will be a greater spread of interventions around the mountain, both north (DOF/WCST) and south (GEF/CARE), and more funding for the conservation of the Forest Reserves that contain the highest levels of biodiversity.

Scale is appropriate   The Ulugurus are a large and topographically complex range of mountains and are occupied mainly by one tribe, the Luguru, although an influx of other tribes is noted. Biologically there is a considerable similarity in the flora and fauna of the forests across the entire mountain range, with the major differences being due to altitude. Due to these factors it makes sense to consider a large-scale ICDP design across the entire Uluguru landscape. During Phase 2 support there was a mismatch between the scale of the project planning and monitoring (covering the entire landscape) and the scale of project interventions (Figure 3.2d). Project interventions were focused in a smaller part of the mountain that had the greatest problems in terms of forest and biodiversity loss (Figure 3.3), and where there was the highest rate of commercial agriculture and impact from outside interests. During the design of the third phase of activities on the mountain, efforts were made to expand activities to cover the entire landscape. However, funding remains insufficient to fully tackle the whole area.

There is real local ownership The earliest phases of project input into the Ulugurus were designed and implemented in collaboration with regional government partners, who managed the programme as their own. However, these earlier phases did not adequately involve the people living around the mountains and apart from studies, training, and some tree planting efforts, no significant impacts of project interventions were felt in the villages on the mountain.

The second and third phases of this ICDP engaged with district government officers, village governments, farmers' groups, tree planting groups, women's groups and government forestry officers based in the mountains. By working at these levels the project has found greater resonance with local people, but has also experienced some problems of 'partnership', especially within the elements being coordinated by CARE. By working at the local level, the communities agreed to set aside patches of both natural forest and human-made woodlots as Village Forest Reserves. Towards the end of the second phase, the project started to work closely with the traditional chiefs as well as with a broader range of government-appointed staff in the villages (Figure 3.4). This has further improved the perception and ownership of project interventions among the local communities, and has also raised awareness of the project and the conservation issues of these mountains at the national level. Changes to a more people-focused approach to forest conservation in the Ulugurus have been facilitated by the publication of a new *Forest Policy* (GOT, 1998), *Guidelines for Community-based Forest Management* (GOT, 2001), and a new Forest Act (GOT, 2002). These legal changes provide the mechanisms to empower local populations to manage forest resources, either in collaboration with the central government, or alone.

Project targets are clearly formulated The initial project, starting in 1993, was designed using the traditional approaches – teams from outside working with government authorities in the nearby towns developed project proposals based on their best-available understanding of the situation on the ground, and some targeted case studies (Lyamuya et al., 1994).

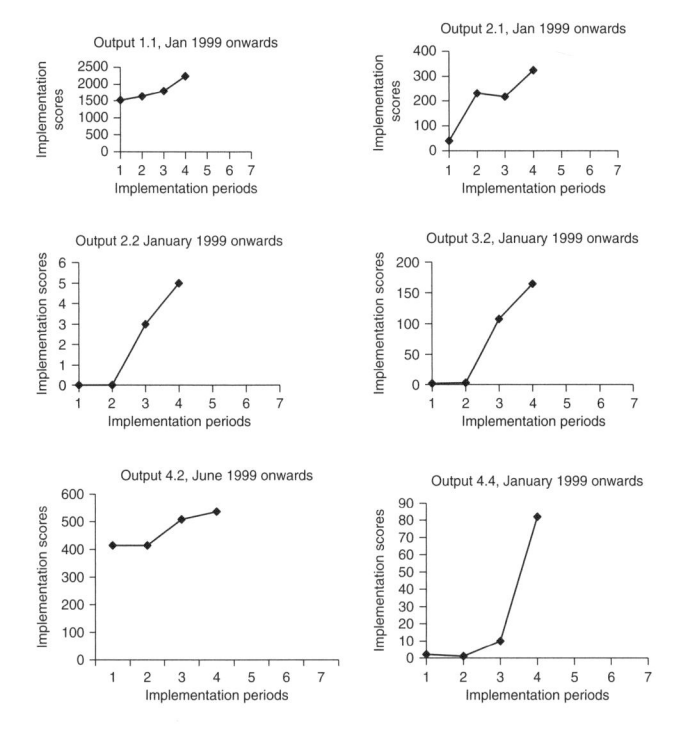

Integrated conservation and development projects 51

Note: Implementation periods refer to six month reporting periods to DANIDA, with the first reporting period (1) being the baseline when the project started. Scores relate to a variety of measures easily gathered by the project management, often with several measures added to produce this index of change. The scheme was not continued for the full three years of the project phase due to changes in staff and because political changes in government priorities in Denmark resulted in much additional work for staff in Tanzania.

Explanation of outputs:
Output 1.1   Administration of WCST and skills of its staff enhanced.
Output 2.1   Activities provided to involve Tanzanian members and volunteers in the work of the Society.
Output 2.2   Training provided to develop skills of members and volunteers as a resource to promote the work of the Society.
Output 3.2   Uluguru project assists the formulation of agreements between District Forestry and local people on sustainable uses of the Uluguru Public Land.
Output 4.2   Uluguru project supports sustainable agricultural practices in villages adjacent to Uluguru Forest Reserves.
Output 4.4   Uluguru project produces written materials to promote its work within Tanzania and internationally.

Source: WCST Uluguru Mountains Biodiversity Conservation Project.

Figure 3.4   Simple measures of success in the implementation of Logical Framework Outputs during part of the second phase of the Uluguru ICDP (with DANIDA funding)

Between 1993 and 1995 extensive further studies were undertaken to feed into the development of further ICDP activities in the Ulugurus (Bhatia and Ringia, 1996; Bhatia and Buckley, 1998). The results of these studies were used to design the Phase 2 DANIDA support on this mountain. At this point a logical framework approach was used to capture the links between the inputs and outputs of the project, and the ways to measure conservation achievement. Planning of the Phase 3 project involved Tanzanian and foreign development experts, social scientists and a series of consultative meetings with relevant local people living on the mountain to further refine the series of project interventions (Hartley and Kaare, 2001). A further detailed logical framework was developed, but despite the increased stakeholder involvement it was quite similar to that from the second phase. Designing the third phase did, however, highlight the importance of the traditional leadership on the mountain. The village government that was part of the Tanzanian government structure held some power, but it was not in charge of land allocation, which passed through the female line of different clans of the Luguru people. By looking closely at different stakeholders, the Phase 3 proposals were able to capture the facts that:

- Almost all government officers (national, regional, district, division, ward and village) wished to retain the reserves in their present form to ensure the maintenance of national (water flow) and local (better climate for farming) values. As the reserves had been under government control for a long time (almost 100 years) most of the local population also accepted their existence as a part of the local pattern of land use.
- The traditional chiefs wanted to retain the Forest Reserves (and other non-reserved forest areas) as places where their ancestors could remain undisturbed, and also because of some sacred sites and so animals (in particular a mythical giant snake) could find refuge.
- Some groups of people within the villages also wanted to retain the Forest Reserves as sources of clean water, medicinal herbs (traditional healers), for hunting, to provide fuel and building wood (which was often not available in farmland areas), or for spiritual reasons (a place for the ancestors to live).

However, other stakeholder groups wished to obtain greater freedom of access to the forests to pursue a number of activities:

- For some people, growing bananas has become a lucrative activity (Hymas, 1999). This crop grows best on recently cleared forest soil,

and has resulted in the loss of over 20 sq km of non-reserved forest on the Ulugurus over the past 20 years (Burgess et al., 2002a, 2002b). Some of those people involved in this activity would like to expand their production into the reserved forest, which would provide even greater financial rewards.
- For most people farming on the Ulugurus, their crop yields are very low, typically only one to three bags of maize per acre of cultivated land (Hymas, 2000, 2001). Much better yields can be obtained from newly cleared areas of forest – at least for a few years. Hence many farmers have a desire to obtain 'fresh' land from the forest, and some manage to obtain 'permissions' to cultivate on the edge of the Forest Reserve from village authorities.
- Pitsawyers regarded their extraction of high-value timber from the forest as beneficial as it generated money, did not destroy the forest and was a better option for using the forest than converting it to farmland. In areas being cleared for banana farming the pitsawyers were actively sawing up felled trees, and expressed regret at the loss of their future work opportunities through the clearance. However, they also wished for greater legal access to the reserve so that they could extract more trees and sell them – activities that were currently illegal.
- Other villagers sought the right to access the forest to collect firewood and other woody products that they need (and already obtain illegally) from the forest (Hymas, 2000). Even the issue of crossing the mountain through the forest was a vexed one – the paths across the forests are essential lifelines for villagers, saving days of travelling in some cases, but they do not have the official right to use the paths, even though they are centuries old in some cases.

The biggest compromise during the design process from the side of those wishing to retain the reserves was to agree to explore methods of collaborative management of the forest resources with forest-adjacent villages. Devolution of power and authority from district and national government level to village and traditional authorities would be required for this, and its effective implementation will need a period of negotiation and formalization through signed agreements (bye-laws). Currently ICDPs on the Ulugurus have only been gathering experience on the potential ways to achieve this goal by working in remaining patches of non-reserved forests to establish Village Forest Reserves. The experience provides some models that can be applied to the village communities living around the official government Forest Reserves.

Protection components are included The Uluguru ICDP is working at three levels to ensure that forest protection components are included. Most of the remaining natural forest and the highest biodiversity values are found within large Forest Reserves on the top of the mountain managed by the Catchment Forest Project, a branch of the Forestry Department that is under central government control. These reserves are maintained for their water catchment functions. Smaller patches of forest are also found in a number of Local Authority Forest Reserves managed by the district, generally on the slopes of the mountains. These smaller forest patches are of lower biological importance when compared with the Catchment Forest Reserves. Finally, there are traditionally protected areas of forest on the Ulugurus, which are under the authority of the traditional chief (one large area) or individual villages or clans (smaller patches). The current Uluguru ICDP is working with Catchment Forestry, with the district Forest Officer and with the traditional chief and the relevant village authorities to try to enhance the protection of these different categories of forest.

However, the funding allocation within the Uluguru ICDP for protection activities is modest. Most funding has been provided for developing arrangements between the local village authorities and the government agencies over some kinds of collaborative management of the forest on the Ulugurus. The extent that nationally important Catchment Forest Reserves *should* be managed collaboratively with communities, and what the community benefits will be, remains a major issue of debate. The current view is that collaborative management should go ahead, but that much more work is required to identify the benefits that communities can get. Another reality is that with only three Forest Officers stationed on the mountain, but with more than 50 villages and several tens of thousands of people, there remains a strong need to involve villagers in the management of these reserves. Since 2005, it has also been proposed that the Uluguru Forest Reserves be upgraded to the status of Nature Reserve. This status would not prevent collaborative management agreements, but might bolster the government management capacity.[6]

Productive potential of different habitats is taken into account Resource extraction agreements have not been formulated for the Uluguru Mountains, as legally the majority of the forest (within reserves) is protected and utilization is not allowed. However, the local villagers do utilize the forests as a source of firewood, building materials and medicines. There is also some pitsawying of valuable timber species. If and when agreements are formulated between the local villages and the catchment forestry managers, they will need to consider the potential of the forest to supply materials sustainably. Hunting pressure has already removed large mammals and reduced populations of smaller ones, and some timber species are no longer of harvestable size due to past logging. The true sustainable level of resource utilization on the mountain has not been assessed, and will be difficult to achieve because of the centuries of exploitation that have already occurred.

A sustainable end point is defined   In the Ulugurus the maintenance of all remaining forest cover, the reconnection of separated forest patches, and full local participation in forest management could represent a sustainable end point. Having the funding in place to manage these areas forever would also represent a sustainable end point. But such sustainability has not been defined by the existing projects and given the scale of the area and the problems that it faces in terms of population growth and resource needs, reaching a sustainable conclusion is highly problematic. In simple financial terms, the various government offices with a responsibility for the management of the reserves on these mountains lack the funds to effectively undertake their jobs. Moreover, the local populations are poor and the forests do not provide them with much in terms of cash revenues. It's much more profitable for them to convert the forests to banana plantations or other forms of agriculture, at least in the short term. But the forests have huge national cash value as a source of water to the capital city and much of the industry of the country. They also have an international value as the home of hundreds of unique species including species of commercially valuable genera (African violet, busy lizzie, begonia, coffee and so on). If these indirect values could be captured using tax on water users, or through other kinds of monetary system then the management of the Ulugurus over the longer term could be ensured. The World Bank has recognized the indirect monetary values to the Tanzanian economy and started a trust fund mechanism for the region during 2002, with initial capitalization of $7m. In its initial phase the trust fund will not fund projects in the Uluguru Mountains, but may in future years. Current projects through **WWF-CARE-IIED** are investigating whether the establishment of water payment schemes (Payments for Water Environmental Services) is a viable conservation funding mechanism for the Ulugurus. Such a mechanism could provide the required management funds, while at the same time ensuring that the natural values required for Tanzanian development (water supply) continue to be provided.

In many ways these forests epitomize the problems of ending an ICDP programme – there is no end point because management interventions will be needed forever to assist agriculture to improve, to resolve disputes over land use, to maintain the existing reserves and so on. Working to develop sustainable funding mechanisms is perhaps the only way to achieve the sustainability of project interventions and thus end the ICDP cycle of donor support.

Monitoring and evaluation confirms project success   The Uluguru ICDP has undertaken various forms of biological, social, habitat, disturbance and attitude surveys since 1993.[7] Most of these studies are baselines against which future trends can be measured. A few studies have also tried to look backwards in time to assess what has changed over time, in terms of habitat and species values (see for example, Burgess et al., 2002b). The WWF-World Bank Management Effectiveness Tracking Tool has been recently used to gauge the impact of project activities on the Uluguru Forest Reserves, showing an improvement over time. The other system that might have been used to measure the impact of conservation interventions is the 'Threat Reduction Assessment' methodology, which measures the extent to which the projects have reduced pressures/threats in the area (Salafsky and Margolius, 1999).

Conservation and development in the Bamenda Highlands, Cameroon
The Bamenda Highlands support remnant areas of montane forest of global biodiversity importance for endemic species, within a landscape dominated by farmland and high human population densities. People are poor and derive most of their livelihoods from farming and exploiting natural resources. Most of the natural forests remaining in this area are managed either by traditional authorities, or by the more remote government forest department.

The Kilum-Ijim forest (Figure 3.5) forms the core area of the community-based conservation in the Bamenda Highlands Programme. This forest extends over about 17 300 hectares on the slopes of Mount Oku (3011 metres) and the adjoining Ijim Ridge and is the last significant remnant of Afro-montane forest in West Africa. Mount Oku lies in the Bamenda Highlands, part of the Cameroon Mountain chain. The forest is a globally important centre of endemism: 15 bird species endemic to the Cameroon Mountains can be found at Kilum-Ijim, of which Bannerman's turaco (*Tauraco bannermani*) and the banded wattle-eye (*Platysteira laticincta*) are restricted to the Kilum-Ijim forest and a few other forest remnants within the Bamenda Highlands. Both species are threatened and the forest represents the only possibility of conserving viable populations of these two species. Surveys of other taxa also demonstrate very high endemicity, not only of the Bamenda Highlands, but of the Kilum-Ijim forest.

The alternative forest conservation strategies suggested at the start of the project in the 1980s were either to advocate the establishment of a protected area (an option that had been proposed and attempted, by

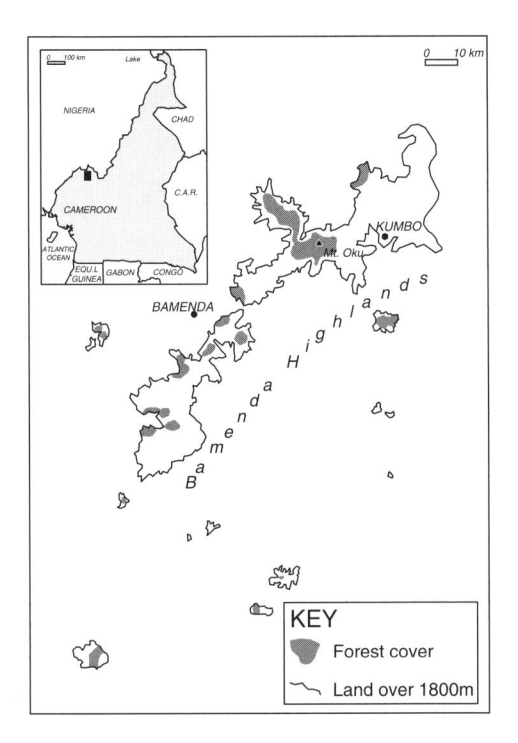

Source: BirdLife International – Cameroon Highlands Project.

Figure 3.5 Location of forest patches that form part of the Bamenda Highlands ICDP network, within the geographical extent of the Cameroon Highlands

government, for several decades) or to support conservation without an official protected area that was based instead on community agreements and sustainable use of forest resources. When the project started in 1987, the first step was to agree and demarcate a forest boundary. This was urgently needed to avoid further loss to encroaching farmland, and the project worked closely with community leaders to agree a boundary that was then marked with *Prunus africana* trees. In these earlier phases, the project focused on education and awareness, sustainable use of forest products (especially honey) and a 'livelihoods' programme that aimed to assist farmers to improve production on their farmlands bordering the forest. Following the passing of a new Forestry Law in 1994, the Kilum-Ijim Forest Project started a major new phase, in which the focus became support to decentralized forest management by adjacent communities (under management plans drawn up by the community and approved by the Ministry of Forests and Wildlife (MINFOF)).[8]

Scale is appropriate   Initially the project focused only on the Kilum-Ijim forests. But over time the approach was requested by more and more surrounding villages with their own patches of similar forest and similar development issues to be solved. Therefore, whilst the Kilum-Ijim forest is still the top conservation priority, the project was expanded to a regional scale of approach, and the scale of operations became the wider Bamenda Highlands. Under this enlarged programme, communities in the region were invited to approach the project for assistance – but support came with conservation-related conditions attached. This aimed to ensure genuine commitment on behalf of the village before any work commenced. The project also comprised components focused on building a local constituency in support of forest conservation (a diverse education and awareness programme) and components focused on capacity-building for supporting community forest management within government (MINFOF) and local NGOs. Indications are that this approach had significant success in some components – especially the education and awareness programme.[9] Possibly as a result of the success of this awareness campaign, many more communities approached the project for support for forest management than had been anticipated (the project planned to work with 30 communities, and focus on supporting the eight highest priorities, but received over 60 applications for support). The project was reluctant to turn communities away and so initially attempted to respond to all inquiries. However, it was clear that limited resources required activities to be more focused, and the project subsequently prioritized ten key communities who received high levels of direct support, 15 communities who received limited direct support, and supported the remaining communities indirectly through local NGOs.

Within the context of national and local constraints (for example, a poorly resourced and motivated civil service, political interference, inadequate levels of funding), the regional, landscape approach was judged a success. It allowed cost-effective use of resources and materials (trained staff, methodologies, awareness materials), encouraged engagement of a wider constituency both in government and civil society (since Bamenda, the Provincial capital, lies at the heart of the project area) and has the potential, through the institutional structures built through the project, to justify a mechanism for the long-term management of the forests (see below). The attention to the wider Bamenda Highlands landscape has also revealed the biodiversity importance of many of the smaller forest fragments, which had previously not been well surveyed, and helped to put Kilum-Ijim into a wider context. This approach would seem to be suited to similar landscapes, where habitats of high biodiversity value are scattered in an agricultural landscape, but of course solutions need to be situation specific.

There is real local ownership   When BirdLife International (then ICBP) began working at the Kilum-Ijim forest in 1987 one of the first steps was to assist local communities in the demarcation of a forest boundary beyond which no further clearance for agriculture would take place. Nineteen years on that boundary remains largely intact, and traditional authorities have dealt with the few infringements that have taken place promptly and effectively. Each of the communities surrounding the forest has a working and legally recognized Forest Management Institution (FMI) and 11 communities have produced forest management plans and now hold legal title to the resources of their forest (with the management responsibilities that implies). The project has supported communities in the process of developing the necessary institutional capacity, and with the highly complex legal process of forest registration and management plan approval. This has been combined with a development programme that has demonstrated methods of improving productivity and sustainability of land use outside the forest, as well as enhanced production and value of products harvested from inside the forest (such as honey).

There is real overlap at Kilum-Ijim between long-term community interests from forest conservation (especially watershed protection but also non-timber forest products like honey, medicinal plants, fuel wood and bamboo) and biodiversity conservation. Most of the remaining forest is on very steep slopes and conversion to agriculture is almost certainly not sustainable (as landslides in adjacent areas testify) and would lead to loss of watershed protection functions (again, as demonstrated by many adjacent areas where springs have dried up). However, whilst most people recognize the convergence of biodiversity conservation interests with their own social and economic concerns, and therefore support the project, the process continues to be undermined by individuals or groups with no immediate interest in forest conservation – especially graziers (most of whom are wealthy individuals who are no longer living in the area) and *Prunus africana* exploiters (external, and very often armed). Regrettably these interest groups have been able to influence the judicial process, and their illegal activities (which are also unwanted by the community) have been extremely difficult to control. What this shows is that the interests and wishes of the majority, who have organized themselves into democratic community institutions, and who have legally backed rights, can still be frustrated by powerful or stubborn individuals in situations where capacity (or commitment) in state authorities is lacking.

60   *A handbook of environmental management*

Project targets are clearly formulated   Over the 19 years that BirdLife International has been working in the Bamenda Highlands, the planning mechanisms have been constantly developed. Project interventions were based on a detailed logical framework that was itself created from a detailed problem tree developed with project staff and villagers around the forests.

Detailed logframes and workplans have proved critical tools in the regular cycle of planning, implementation, monitoring and evaluation, and have been used in a flexible way, with managers prepared to modify objectives (with the agreement of donors) if circumstances made that appropriate. However, what the full problem tree demonstrates is an immense web of problems that can only partly be solved by a project approach of fixed duration (and budget). The Bamenda Highlands project aimed to put in place a sustainable outcome within the constraints of politics and national economics (the context – national debt and an underpaid civil service). These factors are not within a single project's ability to influence and yet clearly they significantly affect what can be achieved – and the strategies that must be put in place to try to accomplish a project's objectives.

Protection components are included   A 1000 hectare plantlife sanctuary forms a core protection area in the centre of the Kilum-Ijim forests, and the people-centred approach at Kilum-Ijim also takes place within a legal context (provided by a Provincial Decree) that imposes (state) controls on land use (banning inappropriate activities such as burning and grazing). The project additionally chose to support a traditional protected area in the form of an ancestor living place (forest) under the authority of the traditional chief (*fon*). This traditional protected area has been formalized through the creation of local forest management institutions around the forest who ensure that the boundaries are respected and that resource utilization agreements are adhered to. Effective implementation of the management plans they produce is in turn monitored and enforced by MINFOF. Although different from the model of a government-managed Forest Reserve or National Park, the proximity of the people to the resource and their acceptance of traditional authority make this model work. Although no detailed comparisons have been made it is believed that the level of protection of the forests under the Bamenda Highlands programme is stronger than those under government protection. The forest is highly valued culturally, and many taboos and regulations exist that are enforced by the *fon* and his ruling council (*kwifon*). De jure government 'control' over forests had eroded this traditional authority and management system, without replacing it with an effective alternative (MINFOF Forest Officers were too few, poorly resourced and often not motivated to manage the forest). The project has helped to rebuild the traditional authority of the *fon* and *kwifon* within a parallel and integrated system of management and control by community-based institutions of users (FMIs). However, experience shows that government support is still essential as a back-up and endorsement of FMIs and traditional authority rules, especially where powerful outside (and internal) interests threaten to destabilize the situation. For example, the valuable bark of *Prunus africana* is illegally exploited by outside gangs that are often armed, and illegal livestock graziers in the forest enter with the backing of wealthy and influential livestock owners. This experience suggests that a three-pronged approach (FMIs–traditional authorities–government) may be the most effective combination, and that each group has its role to play in an effective overall protection system.

Productive potential of different habitats is taken into account The project has made efforts to assess the sustainable levels of offtake for key forest products – those studies examined harvesting and production of fuel wood, honey and small mammals (mainly rodents), and future work will attempt to assess the sustainable offtake for *Prunus africana* bark. These studies suggest that the forest is already exploited at levels equal to or exceeding its productive potential. For resources such as mammals this is self-evident – it is very rare now to see duikers, monkeys or other mammals in the forest, and most hunters have resorted to trapping a range of small rodent species. Although the reproductive potential of these species is probably very high, their detailed biology in this respect is not known and in most cases studies have not been carried out. Since rodents are likely to be key seed dispersers for some tree species the long-term impact of over-harvesting on ecosystem functioning could be severe. The project's approach of integrating conservation and development aimed to address this issue, by finding alternatives (supplements) to harvest of products from the forest.

A sustainable end point is defined The main projects in the Bamenda Highlands lasted for 17 years and cost a total of over US$3m with the objective of conserving montane forests with a total area of about 30 000 ha. In its very early stages the project was 'fire-fighting' – its priority was to try to halt the rapid encroachment of farmland into the forest. The strategy then shifted to education and awareness and building sustainability of local livelihoods through support to agricultural development and income-generating activities linked to the forest. However, it is now recognized that the conservation of the forest is a long-term enterprise, and that to be sustainable key elements are going to be strong institutions to carry the process forward (first and foremost) possibly combined with some form of 'subsidy' to enable national government and local communities to conserve a resource of global significance into the future. As a result the present project strategy for sustainability is based on achieving sustainable institutions for the long-term management of the forest. This will work at three levels – that of community institutions (FMIs), the traditional authorities and the government. As suggested above in relation to protection activities, each of these levels needs to work effectively for forest management to succeed. It was envisaged that these three institutional elements would be supported financially in the long term through a trust fund (CAMCOF – CAmeroon Mountains COnservation Foundation) to support sustainable management and conservation of forests throughout the Cameroon Highlands, with an initial focus on the Kilum-Ijim Forest and Mount Cameroon. However, despite considerable investment from the Global Environment Facility and the UK's Department for International Development towards project development, CAMCOF failed to get the necessary support and the Foundation has been closed. Sustainable financing for conservation and sustainable use in the region remain unresolved issues.

Monitoring and evaluation confirms project success For much of the past 19 years there has been no formal monitoring scheme in place to measure the conservation success of project interventions in the Bamenda Highlands. An initial survey of the extent of the montane forest carried out in 1983/84 noted the rapid rates at which montane forest in the region was being cleared for cultivation, and concluded that without rapid intervention the remaining Kilum-Ijim forest would be destroyed by the year 2000. The fact that the forest remains is an important measure of success, especially considering the continued loss and degradation of other montane forests, including Forest Reserves, over the same period. Monitoring of forest extent using satellite images and aerial photography has shown the impact of the project's interventions. They show that between 1958 and 1988 (that is, pre-project) more than 50 per cent of the forest was lost. This was followed by a period of forest regeneration and in the period 1988–2001 the forest extended by 7.8 per cent of the 1988 area. This extension of forest cover has taken place largely through forest regeneration on cleared areas within the boundary that was demarcated in 1988.

In 1994 a conscious change in approach was made to ensure that the success of the project in terms of conserving the key bird species in the forest (a key conservation target) could be reported. A thorough analysis of the conservation objectives helped to define what the project hoped to achieve in terms of key species (especially endemics and threatened species of plant and animal), habitats and ecosystem functions (Maisels, 1998). The monitoring system then evolved to comprise a comprehensive monitoring system designed to operate on a number of scales (Maisels and Forboseh, 1999) and combining point counts of birds, permanent vegetation quadrats and occasional sightings of mammals with fixed point photography and satellite image analysis. It aimed to combine species monitoring with 'functional' monitoring (monitoring of keystone species, such as pollinators (sunbirds, bees) and seed dispersers (frugivorous birds and mammals). However, this level of monitoring is very labour intensive (and therefore costly) and proved not to be sustainable when major-donor funding of the project came to an end in 2004. However, initial results suggested some stability in forest condition (although the effect of 'lag' is not known). For example, survey data from the census of Bannerman's turaco from 1991, 1995–97 and 1999–2001 failed to detect significant changes in the population of this 'flagship' species (Kilum-Ijim Forest Project, 2002).

It was recognized by the project that measurement of 'biodiversity' rather than 'resources' is mainly of interest to conservationists (be it MINFOF, or national or international NGOs) rather than local communities. In order to address this the project also put in place a programme of institutional monitoring and natural resource monitoring designed to be implemented by FMIs. Based on discussions with forest users, these participatory measures allow FMIs to measure the state of their institution (by asking questions on certain indicators such as 'What percentage of the population are members of the FMI', 'What proportion of FMI members participate in collective forest management activities such as firetracing?' and 'How representative is the FMI (number of women, men, youth, elders, Fulani and so on)?)'. Another set of indicators are designed to allow FMIs to assess and monitor the condition of the forest resources, and the threats it faces, using measures that are relevant to them and their livelihoods. This uses indicators such as 'number of streams flowing in the dry season', 'annual yield from bee hives' and 'number of seedlings of selected species (mainly species of economic importance such as food trees and those whose timber is used for carving) that are over knee height'. This community-based monitoring is at an early stage, and its effectiveness in directing management by FMIs has not yet been assessed.

The project has also undertaken specific (one-off) exercises in order to measure both the uptake of technologies introduced by the project and the attitudes of people towards the forest and its conservation. An uptake survey in 1999 interviewed 950 farmers in villages around the forest to determine their uptake of technologies promoted by the project through its 'livelihoods' programme (such as contour ridging, tree production, live fencing, erosion control and improved livestock feeding). This demonstrated levels of outreach (that is, percentage of respondents who had received information on the technology) ranging from 24 per cent (for training in raising fruit trees) to 93 per cent (for training in methods of controlled burning and alternative land preparation methods) and uptake levels (percentage of farmers who had received training who then adopted the technology) of between 49 per cent and 98 per cent (for uptake of controlled burning) (Tsongwain, 1999; Kilum-Ijim Forest Project, 2002).

Assessment of changes in behaviour have been complicated by the lack of any baseline statistics. However, using qualitative, participatory methods and a comparative approach (comparing communities that had little contact with the project, with those that had had high levels of contact and uptake) has demonstrated the significant positive impact of the project on people's attitudes to the forest and its conservation (Abbot et al., 1999, 2001).

Discussion
Looking at our two ICDP examples from the forests of Africa, a number of issues emerge. First, it is clear that neither of these ICDPs is perfect, even to those who have been involved with them over a period of years. In this regard we accept some of the criticisms that the ICDP approach does not deliver as much conservation or development as some originally proposed it would. But, both of these projects are working in some of the least developed regions of the world, where billions of dollars of development assistance and the best economists in the world have largely failed to improve national economic fortunes (UNDP, 2001). We therefore believe that any conservation successes need to be set in the context of greater economic failures.

Scale and ownership/benefits
In both of our examples, the projects operate over relatively small parts of the landscape of biological importance, and focus on maintaining critical patches of forest cover for their biodiversity values – principally endemic species with no known economic importance. The values of the biodiversity are uncosted. The local values for villagers living adjacent to the forest habitats, in terms of direct cash generated from these forests or their endemic species, are small. However, in both of our case examples the local perception of the value of the forest in terms of providing a water supply through the dry season is high. Numerous villagers in both regions tell stories of the drying of their local stream when its catchment area was deforested. The cultural value of the remaining forest is also high in

both of these areas, as places for the ancestors to live in. Moreover, in the Uluguru Mountains of Tanzania the national cash values of the forests in terms of maintaining water flow to the capital city and most of the nation's industry are huge. The appreciation of these values and the development of systems that capture some of this value and use it for long-term management, offers one of the best ways to achieve sustainable management of the forest resources and some assistance to the development of local human populations.

Ownership

Both of our two example projects illustrate a gradual shift in emphasis from forest ownership by the national government to a situation where local people have a role in the management of government-owned reserves (for example, in the Ulugurus), or more promisingly perhaps where they establish and manage Village Forest Reserves on their own land and where the government, or a local NGO, is a facilitating and problem-solving agency. The legal mechanisms that allow these community approaches to forest ownership and conservation may provide a boost to forest conservation efforts more widely across Africa in coming decades. An evolving issue in development thinking involves attempts to ensure that issues of equity and rights are considered fully in development projects. This novel thinking, especially on rights, is likely to become an important aspect of ICDP design and implementation, especially given the global re-emphasis on solving some of the fundamental issues of poverty and inequality across the developing world, especially in Africa. One of the risks of this approach is that the choices that local people make might not be ones that 'conservationists' want, but equally there is a potential for people to reject centralized government models (for example, large-scale concessions, clearance for industrial agriculture – oil palms and so on) and choose their own pathway.

Planning

In our opinion one of the most helpful mechanisms for designing and managing an ICDP to ensure that it works towards its goals, and has the chance to periodically review and alter these goals, has been the Logical Framework Approach (or related planning systems). By building a set of interventions around solving the environmental (and human development) problems in an area, and monitoring these carefully over time, best practice conservation management can result. However, even in cases where log frames have been developed, many are poorly formulated and do not attempt to measure conservation impact – instead they monitor whether activities are progressing and funding is being used. The majority of ICDPs in operation have this flaw, which is one reason why it has been easy for them to be criticized – most cannot provide data to answer what their conservation achievements have been. In this regard we would endorse exploring ways to measure the pressure/threat to an area and how it has changed, the state of the system (both biodiversity/habitat and socioeconomic) and what responses occur (reserves declaration, attitude changes, activity changes, degradation changes, governance changes and so forth). If logical frameworks could be used to develop and help maintain monitoring programmes that capture information on these broad areas, this would be a significant step forward. If a sufficiently large sample of ICDPs were operating similar monitoring schemes then some statistically analysed conclusions about the success or failure of the ICDP model might be developed.

Sustainability
Long-term sustainability has rarely been achieved by ICDPs, even those that have operated for a decade or more. It may be that this model of project cannot achieve a sustainable end point, especially when working in developing countries where the economic situation is stagnant and there are few funds from government or local leaders to support conservation activities. Solving the macroeconomic problems of many African nations would enhance the chances of long-term sustainability for projects operating in poor rural areas. In the absence of economic improvements, working to establish mechanisms that ensure the supply of modest funding over the long term, and the availability of reliable and trained staff, may offer the best chances to reach a sustainable end point. If there are no funds then good staff will move to other areas, or will start to engage in other activities to support themselves. If there is too much funding for a known short period, then people may seek to capitalize on it while it is there.

Neither of our two example ICDPs are close to being economically sustainable. The models used by these projects have involved significant sums of money, expatriate support, purchase of transport and fuel, international travel and some foreign management costs. The local economy of these areas involves small amounts of money, local decision-making, a lack of mechanized transport, few income-generating opportunities and traditional beliefs and norms. In order for the ICDP approach to become more sustainable it has to become financially, technically and operationally closer to the normality for the areas concerned, which is a serious challenge using current project implementation models.

Protection components
We endorse the criticism that ICDPs do not expend as much effort as required on natural resource conservation. Our experience with ICDPs

funded by development agencies shows how their focus on poverty alleviation, while providing valuable funds, can force the ICDP to expend nearly all its effort on development with only weak and insufficient components targeting the protection of natural resource values. Part of the current criticism that ICDPs are trying to achieve conservation using indirect mechanisms comes from this mismatch of agendas – on the one hand the funding agency wants to see conservation by development, whereas the conservation agency wants to see conservation gain and will try to design development interventions to achieve this goal. However, there need not always be this dichotomy, and there are many situations where conservation and development agendas directly overlap, especially in poor rural communities in Africa.

Implementation constraints
Our two example ICDPs, and especially the Uluguru Mountains ICDP managed by CARE have been dogged by implementation constraints.[10] This has affected the efficiency of the projects, and has resulted in some significant time delays and failures to achieve some of the planned targets. This issue has already been explored in some detail in the review by Wells and McShane (2004) where they note that even a well-designed project can fail because of problems with its implementation strategy – including problems of obtaining suitable staff, and so on.

Learning components
One general observation is that ICDPs would benefit from a more scientific approach, including designing interventions to test hypotheses about different kinds of interventions. The learning component of many ICDPs is generally small and often given a low priority. However, this means that ICDPs will continue to fail to show what they have achieved, or indicate what interventions give the best value for money.

Local people can collect much of the data required to monitor conservation impact. This can both simplify the methods, allow local understanding of what is required and what the project is trying to achieve, and also keep costs low (Danielsen et al., 2001). However, as with many other elements of ICDPs, the data collected by local investigators will need periodic field validation to ensure that standards are maintained. These locally based methods of monitoring can tell a lot about resource use and management actions, but they might not satisfy the requirements of conservation biologists interested in the number of particular species or details of ecological services. In these cases more professionally based monitoring may be required, involving trained scientists.

In addition to measuring biological and threat-related measures within an ICDP, measuring perceptions and attitudes of the local population over time can also indicate conservation trends. Local populations are often reported to have a negative perception of protected areas. However, our Bamenda Highlands case study provides evidence that ICDPs can change these negative perceptions over time. Such changes go a long way to making the interventions sustainable in the long term.

Other management options
Given the long-term funding input to our two example ICDPs, and the relatively low level of impact, it is relevant to consider whether other forms of conservation might have been more successful. In many parts of the world National Parks or Strict Nature Reserves are regarded as the most secure means for conserving wild areas. Would these have worked in the case of the Ulugurus or the Bamenda Highlands? The possible scenarios for these two sites are highly different. For the Ulugurus, most of the forest remains in two large central government Forest Reserves and the boundaries are fairly well marked. As the national government already controls the forest area, then changing the Forest Reserves into a National Park would only involve a transfer of control within the government, from the Forest Department to Tanzania National Parks. An alternative that is now being pursued in Tanzania is to upgrade the status of these Forest Reserves to a Nature Forest Reserve, which would make the reserves more secure but keep them under the authority of the same part of government. For Kilum-Ijim in the Bamenda Highlands, any protected area would be strongly resisted by local people – for example, previous attempts to designate a Forest Reserve failed. The creation of a National Park or other form of government reserve would be highly problematic, strongly resisted by the local people, and difficult to implement in a meaningful way.

Other models of conservation management that have been advocated in recent years include direct land purchase, purchase of concessions for conservation purposes and direct payments to communities. Direct land purchase and concession purchase models cannot easily work in these two project areas, although legal changes are making this more possible (see Wily and Mbaya, 2001 for Tanzania). Current ideas about making direct payments to communities in return for specific conservation targets (area of forest, population of a species, number of annual fires and so on) may help to focus attention on making conservation interventions as direct as possible (Ferraro and Kiss, 2002). A relevant question is whether if all the funds placed in ICDPs had instead been used for direct payments, would the communities have been better off? The total funds used in the Uluguru Mountain ICDPs since 1993 are around US$2.5m. If all that funding had been available in Tanzania at the project site it would represent less than US$2 per locally based person (assuming around 100000 people on the mountain) per annum over eight years. Although this is not a major financial inducement, it is a more tangible sum than most received from the project activities. However, we predict that the direct payments model would be difficult to implement in poor rural communities. The mechanism of delivering funds would need to be carefully worked out and monitoring this would require considerable effort – and would therefore rapidly become expensive. Funds would also need to be available over the long term if this was the only mechanism employed to conserve the forests. The direct payment approaches may work best linked to a trust fund mechanism that provides small funding to local communities over long periods.

In conclusion, the ICDP model has certainly been challenged in recent years but we believe that the jury is still out and nothing is yet in place/proposed that can realistically (proven to) replace it. Few would argue that it makes sense (moral, practical) to conserve areas that are rich in biodiversity without involving the people that are living in the area. Falling back totally on the strict 'protection' approach with the exclusion and marginalization it implies is not a direction most conservationists in Africa want to take. The integration of conservation at high biodiversity sites with alleviation of social and economic problems is a goal that has popular support, but the challenge is to find ways of making it work, to learn from the successes (and failures) of existing projects and not to dismiss the approach without proposing viable alternatives backed by detailed scientific analysis.

Notes

1. BirdLife International managed the Kilum-Ijim Forest Project and Bamenda Highlands Forest Project in collaboration with the Cameroon Ministry of the Environment and Forestry and forest-adjacent communities. The projects received funding from the Global Environment Facility (GEF/UNDP), the UK Community Fund (National Lottery Charities Board), the British Department for International Development (DFID) through the Civil Society Challenge Fund, the Joint Funding Scheme (DFID), the Global Environment Facility (GEF-World Bank) through the Cameroon Biodiversity Conservation and Management Programme, the Dutch Ministry of Agriculture, Nature Management and Fisheries through the Programme International Nature Management (PIN) and WWF Netherlands. Current work at Kilum-Ijim is funded by the JJ Charitable Trust and the Dutch Ministry of Foreign Affairs.

    Conservation Projects in the Uluguru Mountains landscape have been funded by the European Union, the Royal Society for the Protection of Birds (RSPB), the Government of Tanzania, Danish International Development Agency (DANIDA), Dansk Ornitologisk Forening (DOF) and the Wildlife Conservation Society of Tanzania (WCST), Christian Aid, the Spanish NGO INTERMON and will soon receive funds from the GEF. We would like to the thank the following Tanzanian staff from the various parts of government and projects working around the Ulugurus for their support and assistance: Lameck Noah, Cheyo Mayuma, Elisha Mazengo, Elisa

Pallangyo, Sia Aroko, Phanuel Bwimba, John Mejissa, Eliakim Enos, Ernest Moshi, Prof. Amon Mattee, Athman Mgumia. The following also collected data in the field in Tanzania, some of which is used here: Andrew Perkin, Tom Romdal, Marcel Rahner, Anders Tøttrup, Klaus Tølbøl Sørensen and Olivier Hymas.
2. See http://www.care.org/careswork/projects/TZAO38.asp, accessed 11 August 2009.
3. See http://www.wikipedia.org/wiki/logical_framework_approach, accessed 11 August 2009.
4. See www.ecosystemmarketplace.com, accessed 12 August 2009.
5. See www.africanconservation.com/uluguru, accessed 12 August 2009.
6. See www.easternarc.or.tz for details of the Nature Reserve Process, accessed 12 August 2009.
7. See www.easternarc.or.tz, accessed 12 August 2009.
8. Formerly the Ministry of Environment and Forestry – MINEF.
9. A recent (2003) independent mid-term review of the project wrote that 'Regarding general awareness and support for natural resource conservation the work of the past has paid off. The Northwest Province is without doubt the most conservation minded in the entire land, and past efforts by the Kilum-Ijim Forest Project and more recently the Bamenda Highlands Forest Project have contributed immensely to this achievement through its hard work in the field of Environmental Education. This has been so successful, that the output may be considered achieved at this point in time. An impressive list of Environmental Education material has been developed and disseminated, ranging from e.g. 22000 posters, courses for 60 biology teachers, to 32 newspaper articles, 25 radio programmes, etc. The conservation constituency is so overwhelmingly present at all levels of North-western society, that it would not be irresponsible for the Project to consider winding down its activities in this domain, unless in direct support of achievement of other outputs, and reallocate the available resources in time, money and staff to the implementation of other activities' (Bamenda Highlands Forest Project, 2003, independent mid-term review, BHFP, Cameroon).
10. See progress reports on www.easternarc.or.tz, accessed 13 August 2009.

References

Abbot, J., S.E. Neba and M.W. Khen (1999), 'Turning our eyes from the forest: the role of the livelihoods programme at Kilum-Ijim Forest Project, Cameroon in changing attitudes and behaviour towards forest use and conservation', Cambridge, UK: unpublished report to BirdLife International.

Abbot, J.I.O., D.H.L. Thomas, A.A. Gardner, S.A. Neba and M.W. Khen (2001), 'Understanding the links between conservation and development in the Bamenda Highlands, Cameroon', *World Development*, **29**(7), 1115–36.

Adams, W.M. and D. Hulme (2001), 'If community conservation is the answer in Africa, what is the question?', *Oryx*, **35**(3), 193–200.

Alpert, P. (1996), 'Integrated Conservation and Development Projects: examples from Africa', *BioScience*, **46**(11), 845–55.

Anderson, D. and R. Grove (eds) (1987), *Conservation in Africa: Peoples, Policies and Practice*, Cambridge, UK: Cambridge University Press.

Ashley, C. and D. Roe (1998), *Enhancing Community Involvement in Wildlife Tourism: Issues and Challenges*, London: IIED Wildlife and Development Series No. 11.

Attwell, C.A.M. and F.P.D. Cotterill (2000), 'Postmodernism and African conservation science', *Biodiversity and Conservation*, **9**(5), 559–77.

Barrett, C.B. and P. Arcese (1995), 'Are Integrated Conservation-Development Projects (ICDPs) sustainable? On the conservation of large mammals in sub-Saharan Africa', *World Development*, **23**(7), 1073–84.

Bhatia, Z. and P. Buckley (1998), 'The Uluguru Slopes Planning Project: promoting community involvement in biodiversity conservation', *Journal of East African Natural History Society*, **87**(1), 339–49.

Bhatia, Z. and O. Ringia (1996), *Socio-economic Survey of Selected Villages in the Uluguru Mountains, Tanzania. Uluguru Slopes Planning Project Report No. 3*, Sandy, UK: Royal Society for the Protection of Birds.
Borrini-Feyerabend, G. and D. Buchan (1997), *Beyond Fences. Seeking Social Sustainability in Conservation. Volume 1: A Process Companion*, Gland, Switzerland: IUCN.
Brandon, K. and M. Wells (1992), 'Planning for people and parks: design dilemmas', *World Development*, **20**(4), 557–70.
Brown, M. and B. Wyckoff-Baird (1993), *Designing Integrated Conservation and Development Projects*, Washington DC: Biodiversity Support Program.
Bruner, A., R.E. Gullison, R.E. Rice and G.A.B. da Fonseca (2000), 'Effectiveness of parks in protecting tropical biodiversity', *Science*, **291**(5501), 125–8.
Burgess, N.D. and G.P. Clarke (eds) (2000), *The Coastal Forests of Eastern Africa*, Gland, Switzerland and Cambridge, UK: IUCN Forest Conservation Programme.
Burgess, N.D., M. Nummelin, J. Fjeldså, K.M. Howell, K. Lukumbuzya, L. Mhando, P. Phillipson and E. Vanden Berghe (eds) (1998), 'Biodiversity and conservation of the Eastern Arc Mountains of Tanzania and Kenya', *Special Issue: Journal of the East African Natural History Society*, **87**(1–2), 1–367.
Burgess, N.D., N. Doggart and J.C. Lovett (2002a), 'The Uluguru Mountains of eastern Tanzania: the effect of forest loss on biodiversity', *Oryx*, **36**(2), 140–52.
Burgess, N.D., T. Lehmberg, C. Loucks, J. D'Amico and J. Morrison (2002b), 'Some aspects of designing conservation landscapes: with an example from eastern Africa', in P. Schultz and D. Noppen (eds), *Integrating Conservation and Development with a Landscape Approach*, Copenhagen, Denmark: Environment and Development Network.
Caldecott, J. (1998), *Designing Conservation Projects*, Cambridge, UK: Cambridge University Press.
Carney, D. (1999), *ODI Poverty Brief: Approaches to Sustainable Livelihoods*, London: Overseas Development Institute.
Chambers, R. (1997), *Whose Reality Counts? Putting the First Last*, London: Intermediate Technology Publications.
Danielsen, F., D.S. Balete, M.K. Poulsen, M. Enghoff, C.M. Nozawa and A.E. Jensen (2001), 'A simple system for monitoring biodiversity in protected areas of a developing country', in H. Agersnap and M. Funder (eds), *Conservation and Development: New Insights and Lessons Learnt*, Copenhagen: Environment and Development Network, pp. 231–72.
Doggart, N., J. Lovett, B. Mhoro, J. Kiure and N.D. Burgess (2005), 'Biodiversity surveys in eleven Forest Reserves in the vicinity of the Uluguru Mountains, eastern Tanzania', Dar es Salaam, Tanzania: Wildlife Conservation Society of Tanzania, and Tanzania Forest Conservation Group, www.easternarc.or.tz, accessed 12 August 2009.
Dubois, O. and J. Lowore (2000), *The Journey towards Collaborative Forest Management in Africa: Lessons Learned and Some Navigational Aids*, London: IIED, Forestry and Land Use Series No. 15.
Ferraro, P.J. and A. Kiss (2002), 'Direct payments to conserve biodiversity', *Science*, **298**(5599), 1718.
Fisher, R.J. (1995), *Collaborative Management of Forests for Conservation and Development. Issues in Forest Conservation Series*. Gland, Switzerland: IUCN and WWF.
Franks, P. (2001), 'Poverty and environmental degradation in the context of Integrated Conservation and Development Projects: what makes an ICDP integrated?', in H. Agersnap and M. Funder (eds), *Conservation and Development: New Insights and Lessons Learnt*, Copenhagen: Environment and Development Network, pp. 95–104.
Franks, P. and T. Blomley (2004), 'Fitting ICD into a project framework: A CARE perspective', in T.O. McShane and M.P. Wells (eds), *Getting Biodiversity Projects to Work: Towards More Effective Conservation and Development*, New York: Columbia University, pp. 77–97.
Ghimire, K.B. and M. Pimbert (eds) (1997), *Social Change and Conservation*, London: Earthscan.

Govt. of Tanzania (GOT) (1998), *Tanzanian Forest Policy*, Dar es Salaam, Tanzania: Division of Forestry and Beekeeping.
Govt. of Tanzania (GOT) (2001), *Community-based Forest Management Guidelines*, Dar es Salaam, Tanzania: Division of Forestry and Beekeeping.
Govt. of Tanzania (GOT) (2002), *Forest Act*, Dar es Salaam, Tanzania: Division of Forestry and Beekeeping.
Grove, R. (1995), *Green Imperialism*, Cambridge, UK: Cambridge University Press.
Hackel, J.D. (1999), 'Community conservation and the future of Africa's wildlife', *Conservation Biology*, **13**(4), 726–34.
Hannah, L. (1992), *African People, African Parks: An Evaluation of Development Initiatives as a Means of Improving Protected Area Conservation in Africa*, Washington DC: Conservation International.
Hartley, D. and S. Kaare (2001), 'Institutional, policy and livelihoods analysis of communities adjacent to the Uluguru Mountains Catchment Reserves, Eastern Arc Mountains', Dar es Salaam, Tanzania: CARE Tanzania, www.africanconservation.com/uluguru, accessed 12 August 2009.
Hughes, R. and F. Flintan (2001), *Integrated Conservation and Development Experience: A Review and Bibliography of the ICDP Literature*, London: IIED, Biodiversity and Livelihoods Issues Series No. 3.
Hulme, D. and M. Murphree (1999), 'Communities, wildlife and the "new conservation" in Africa', *Journal of International Development*, **11**(2), 277–86.
Hymas, O. (1999), 'Bananas in the hills', MSc dissertation, University College London, www.africanconservation.com/uluguru, accessed 12 August 2009.
Hymas, O. (2000), 'Assessment of the remaining forests on the Uluguru Mountains and the pressures they face, Morogoro, Tanzania: Uluguru Mountains Biodiversity Conservation Project', www.africanconservation.com/uluguru, accessed 12 August 2009.
Hymas, O. (2001), 'People on the hills: an assessment of forest disturbance of the Ulugurus in relation to access, Morogoro, Tanzania: Uluguru Mountains Biodiversity Conservation Project', www.africanconservation.com/uluguru, accessed 12 August 2009.
IIED (1994), *Whose Eden? An Overview of Community Approaches to Wildlife Management*, Nottingham, UK: Russell Press.
Jeanrenaud, S. (2002a), 'Changing people/nature representations in international conservation projects', *IDCS Bulletin*, **33**(11), 111–22.
Jeanrenaud, S. (2002b), *People-orientated Approaches in Global Conservation: Is the Leopard Changing its Spots?*, London: IIED, Institutionalizing Participation Series.
Kilum-Ijim Forest Project (2002), *Report from the Ecological Monitoring Programme, July 2001–October 2002*, Cameroon: Kilum-Ijim Forest Project, BirdLife International/MINEF.
Kleiman, D.G., R.P. Reading, B.J. Miller, T.W. Clark, J.M. Scott, J. Robinson, M.L. Wallace, R.J. Cabin and F. Felleman (2000), 'Improving the evaluation of conservation programmes', *Conservation Biology*, **14**(2), 356–65.
Koziell, I. (2000), *Diversity not Adversity: Sustaining Livelihoods with Biodiversity*, London: IIED.
Kramer, R., C. van Schaik and J. Johnson (eds) (1997), *Last Stand. Protected Areas and Defense of Tropical Biodiversity*, New York and Oxford, UK: Oxford University Press.
Kremen, C., A.M. Merenlender and D.D. Murphy (1994), 'Ecological monitoring: a vital need for Integrated Conservation and Development Programs in the tropics', *Conservation Biology*, **8**(2), 388–97.
Larson, P.S., M. Freudenberger and B. Wyckoff-Baird (1998), *WWF Integrated Conservation and Development Project: Ten Lessons from the Field 1985–1996*, Washington DC: WWF.
Leach, M., J. Fairhead and K. Amanor (eds) (2002), 'Science and the policy process: perspectives from the forest', *IDS Bulletin*, **33**(1), 1–126.
Leader-Williams, N., J.A. Kayera and G.L. Overton (1996), *Community-based Conservation in Tanzania*, Gland, Switzerland and Cambridge, UK: IUCN.
Lovett, J.C. and T. Pócs (1993), *Assessment of the Condition of the Catchment Forest*

Reserves, a Botanical Appraisal, Dar es Salaam, Tanzania: Catchment Forestry Report No. 93.3, Forest and Beekeeping Division, Ministry of Tourism, Natural Resources and Environment.

Lovett, J.C. and S.K. Wasser (eds) (1993), *Biogeography and Ecology of the Rain Forests of Eastern Africa*, Cambridge, UK: Cambridge University Press.

Lukumbuzya, K. (2000), 'The effects of different management regimes on the diversity and regeneration of woody vegetation under a Joint Forest Management model in Tanzania', in P. Virtanen and M. Nummelin (eds), *Forests, Chiefs and Peasants in Africa: local management of natural resources in Tanzania, Zimbabwe and Mozambique*, Silva Carelica, 34, University of Joensuu, Finland, pp. 159–77.

Lyamuya, V.E., L.G. Noah, E.J., Kilasara and N.D. Burgess (1994), *Socio-economic and Land Use Factors Affecting the Degradation of the Uluguru Mountains Catchment in Morogoro Region, Tanzania*, Morogoro, Tanzania: Regional Natural Resources Office.

MacKinnon, K. (2001), 'Integrated Conservation and Development Projects – can they work?', *Parks*, **11**(2), 1–4.

Maisels, F. (ed.) (1998), *Conservation Objectives of the Kilum-Ijim Forest*, Ecological Monitoring Programme, Kilum-Ijim Forest Project, BirdLife International/MINEF.

Maisels, F. and P. Forboseh (1999), *Methods Manual and Biodiversity List. First Draft*, Cameroon: Kilum-Ijim Forest Project, BirdLife International/MINEF.

Margolius, R. and N. Salafsky (1998), *Measures of Success: Designing, Managing and Monitoring Conservation and Development Projects*, Washington DC: Island Press.

Neumann, R.P. (1996), 'Dukes, earls, and ersatz edens: aristocratic nature preservationists in colonial Africa', *Environment and Planning*, **14**(1), 79–98.

Neumann, R.P. (1998), *Imposing Wilderness: Struggles over Livelihood and Nature Preservation in Africa*, Berkeley, CA: University of California Press.

Newmark, W.D. and J.L. Hough (2000), 'Conserving wildlife in Africa: Integrated Conservation and Development Projects and beyond', *BioScience*, **50**(7), 585–92.

Oates, J. (1995), 'Challenges to nature conservation with community development – a case study from the forests of Nigeria', *Oryx*, **29**(2), 115–22.

Oates, J.F. (1999), *Myth and Reality in the Rain Forest: How Conservation Strategies are Failing in West Africa*, Berkeley, USA: University of California Press.

Pimbert, M.C. and J.M. Pretty (1995), 'Parks, people and professionals: putting "participation" into protected area management', Geneva: UNRISD Discussion Paper No. DP57.

Pócs, T. (1974), 'Bioclimatic studies in the Uluguru Mountains (Tanzania, East Africa) I', *Acta Botanica Academiae Scientarium Hungaricae*, **20**(1–2), 115–35.

Pócs, T. (1976), 'Bioclimatic studies in the Uluguru Mountains (Tanzania, East Africa) II. Correlations between orography, climate and vegetation', *Acta Botanica Academiae Scientarium Hungaricae*, **22**(1–2), 163–83.

Posey, D.A. (1999), *Cultural and Spiritual Values of Biodiversity*, London: UNEP and Intermediate Technology Publications.

Rodgers, W.A. (1993), 'The conservation of the forest resources of eastern Africa: past influences, present practices and future needs', in J.C. Lovett and S. Wasser (eds), *Biogeography and Ecology of the Rain Forests of Eastern Africa*, Cambridge, UK: Cambridge University Press, pp. 283–331.

Roe, D. and M. Jack (2001), *Stories from Eden: Case-studies of Community-based Wildlife Management*, Evaluating Eden Series No. 9, London: IEED.

Salafsky, N. and R. Margolius (1999), 'Threat reduction assessment: a practical and cost-effective approach to evaluating conservation and development projects', *Conservation Biology*, **13**(4), 830–41.

Sanjayan, M.A., S. Shen and M. Jansen (1997), 'Experiences with conservation-development projects in Asia', Washington, DC: World Bank Technical Paper No. 388, The World Bank.

Schrijver, N. (1997), *Sovereignty over Natural Resources: Balancing Rights and Duties*, Cambridge, UK: Cambridge University Press.

Songorwa, A. (1999), 'Community-based wildlife management (CWM) in Tanzania: are the communities interested?', *World Development*, **27**(12), 2061–79.
Songorwa, A.N., T. Buhrs and K.F.D. Hughey (2000), 'Community-based wildlife management in Africa: a critical assessment of the literature', *Natural Resources Journal*, **40**(3), 603–43.
Spinage, C.A. (1996), 'The rule of law and African game – a review of some recent trends and concerns', *Oryx*, **30**(3), 178–86.
Spinage, C.A. (1998), 'Social change and conservation misrepresentation in Africa', *Oryx*, **32**(4), 265–76.
Stocking, M. and S. Perkin (1992), 'Conservation-with-development; an application of the concept in the Usambara Mountains, Tanzania', *Transactions of the Institute of British Geographers*, **17**(3), 337–49.
Svendsen, J.O. and L.A. Hansen (eds) (1995), *Report on the Uluguru Biodiversity Survey 1993. Parts A and B*, Morogoro, Tanzania: Tanzania Forestry Research Institute.
Tsongwain, D.V. (1999), 'Conservation and development: agricultural intensification and the uptake of project initiated alternative livelihood activities around the Kilum-Ijim Forest (Cameroon)', unpublished MSc dissertation, Department of Geography, University of Cambridge, UK.
UNDP (2001), *World Development Report*, New York: UNDP.
Wainwright, C. and W. Wehrmeyer (1998), 'Success in integrating conservation and development? A study from Zambia', *World Development*, **26**(6), 933–44.
Wells, M. and K. Brandon (1992), 'People and parks – linking protected area management with local communities', Washington, DC: The World Bank, WWF, USAID.
Wells, M. and T.O. McShane (2004), 'Integrating protected area management with local needs and aspirations', *Ambio*, **33**(8), 513–519.
Wells, M., S. Guggenheim, A. Khan, W. Wardojo and P. Jepson (1999), *Investing in Biodiversity: A Review of Indonesia's Integrated Conservation and Development Projects*, Washington, DC: The World Bank.
Wily, L.A. and S. Mbaya (2001), *Land, People and Forests in Eastern and Southern Africa at the Beginning of the 21st Century: The Impact of Land Relations on the Role of Communities in Forest Future*, Nairobi, Kenya: IUCN Eastern Africa Programme Office.
World Summit on Sustainable Development (WSSD) (2002), www.un.org/jsummit/html basic_info/basicinfo.html, accessed 11 August 2009.

4. Biodiversity conservation in managed landscapes
Tom M. van Rensburg and Greig A. Mill

Introduction

Biodiversity conservation has emerged as one of the most important and controversial global environmental issues in recent years (UNEP, 1995). First, it has been suggested that we are on the verge of mass extinctions, the like of which have not been observed in the fossil record (Wilson, 1985). Second, it is argued that biodiversity loss matters because it is of fundamental importance to human society. It provides food, shelter, fuel, supports recreation and tourism and is thought to play an important part in global life support and in the functioning of ecosystems (Lindberg, 1991; Raven et al., 1992; Brown et al., 1994).

A decline in habitat is thought to be one of the most significant causes of the loss in terrestrial biodiversity (Wilson, 1985). A large proportion of the earth's fertile land has been converted into managed agricultural, forest and urban landscape. Recent estimates by the FAO (2004) indicate that some 38 per cent of land globally is now utilized for agriculture. One solution proposed by ecologists is to expand reserves and protected areas. However, there are a number of problems with this approach: protected areas cover a limited area – approximately 11 per cent of the earth's surface (WRI, 2005); protected areas generally exclude economic activities and they can impose costs on land managers and prevent future economic opportunities from taking place. Consequently, it is unlikely that the proportion of land allocated to protected areas will be sufficient to maintain all biodiversity.

In recent years a number of studies indicate that biodiversity conservation must focus on managed human-dominated ecosystems (Miller, 1996; Reid, 1996; Daily et al., 2001; Rosenzweig, 2003; Polasky et al., 2005). Economically valuable managed landscapes do not necessarily have to exclude biodiversity conservation goals. A wide range of species occur in the presence of human activities and much of the world's biodiversity is found in human-dominated ecosystems (Pimmental et al., 1992). Instead of threatening biodiversity, many managed systems may actually enhance biodiversity because of, rather than in spite of, the day-to-day management activities carried out by land managers. Indeed, the phenomenon of land abandonment has become a subject of major concern in many countries because it results in the loss of biodiversity. It is also thought that land managers may conserve biodiversity because it supports the productivity and resilience of the ecosystems they manage and there is now a significant literature on the functional and ecosystem service values associated with biodiversity (Ellis and Fisher, 1987; Daily, 1997; Barbier, 2000; Daily et al., 2000).

However, the intensity and mode of disturbance clearly play an important part in the management of ecosystems. Heavily disturbed agricultural and forest ecosystems in many parts of the world are threatened by human intervention that has resulted in a loss of biodiversity and resilience. Clearly it is important to distinguish between managed landscapes that undergo disturbance in which biodiversity appears to be thriving and those in which it is threatened. In this chapter we explore the relationship between human-induced disturbance and biodiversity. We also consider the relationship between disturbance regimes and the properties of a managed ecosystem – its productivity, resilience and stability.

All too frequently the application of 'good science' is not in itself sufficient to guarantee desirable environmental outcomes with respect to biodiversity conservation. This also requires an understanding of socioeconomic and policy considerations: how markets allocate scarce resources, how they influence decisions taken by land managers and why they frequently fail to protect biodiversity and other non-market values. To a degree, most land managers are affected by the market. Thus, the chapter also explores the economic linkages between markets and natural resources, and the possibilities of exploiting the powerful creative forces of the market in a manner that conserves biodiversity whilst also providing useful marketable goods and services. Finally the chapter concludes with some policy recommendations on making markets account for the benefits of biodiversity conservation in managed landscapes. We begin by distinguishing between genetic and functional diversity.

Genetic and functional diversity in managed landscapes
Several definitions have been proposed to capture the multifaceted nature of biodiversity (ecosystems, species and genes). This is acknowledged in the definition developed in the Convention on Biodiversity as follows: '"Biological diversity" means the variability among living organisms from all sources, including inter alia, terrestrial, marine and other aquatic ecosystems and the ecological complexes of which they are part; this includes diversity within species, between species and among ecosystems' (UNEP, 1995. p. 8).[1]

A distinction is made in the literature between genetic diversity and functional diversity. Genetic diversity usually refers to the genetic variation that exists within a species (the gene pool). Genes are the fundamental unit of biodiversity and the ultimate source of all variation among all animal and plant species (Dobzhansky, 1970; Soulé and Wilcox, 1980). Genetic diversity has been proposed as the basis on which to make conservation decisions using the evolutionary distinctiveness of taxa when assigning them priorities for preservation (Vane-Wright et al., 1991; Crozier, 1992; Solow et al., 1993; Weitzman, 1998). Here, the relative ecological value is based on how far away species are from one another genetically and an objective value is assigned to the taxonomic distinctiveness or degree of 'independent evolutionary history' (IEH) that is vested in a species (Vane-Wright et al., 1991).

Conservation organizations frequently employ descriptions and measures of ecosystem diversity based on genetic diversity and they tend to place great emphasis on species and their populations (IUCN, 1988). Although considerable sums have been allocated towards species preservation, there is frequently a bias towards charismatic species, large birds or mammals that are very familiar to the public. Conservation groups and professional conservationists often exploit certain species and ecosystems to further their own conservation goals. Conservationists have called the charismatic species that win the hearts of the general public 'flagship species'. This may be at the expense of less well-known species that may be critical for the functioning of ecosystems (Metrick and Weitzman, 1994).

Understanding the value of biodiversity to land managers requires a different perspective that is linked to the functional value of biodiversity. How then does functional diversity differ from genetic diversity and if so, why do these differences matter to managers? First, genes are after all just chemicals that have no value in and of themselves. Instead, genes have value in what they do – control the structure and function of life – instead of in what they are. Measures of genetic distance may not capture the relative values of species such as the complex functional relationships embodied in ecosystems. Two species might be very similar with respect to genetic distance but they may perform very different functions within the ecosystem. One might be a keystone species that is vital to the well-being of the managed ecosystem whilst the other is 'functionally redundant'. Species diversity is relevant to land managers because some species appear to play a more important functional role than others. An individual who is evaluating a species in terms of its functional role would be more sensitive to a change in the ecosystem's productivity than would a person focusing on biodiversity. A person assessing an ecosystem from an ecosystem function perspective would be more likely to focus on key species and processes and might overlook the disappearance of a rare species. There is greater

78 *A handbook of environmental management*

emphasis on the biological integrity of the system than simply ensuring that all the biotic elements are present.

Functional diversity refers to the characteristics of ecosystems and includes ecosystem complexity at different levels of organization such as trophic levels (Cousins, 1991). This approach uses trophic-level analysis to relate species diversity to functional ecosystem parameters such as food web structure or the transfer of energy, water and chemicals between different trophic levels. Functional diversity can be interpreted as the number of species required for a given ecological process.

In managed ecosystems, although every species may have a particular role, it does not follow that these roles are of equal importance. Ecologists acknowledge that some species have a greater ecological impact than one might expect from their abundance or biomass and these have been referred to as keystone species (Power et al., 1996). Some ecosystem studies indicate that only a small number of the numerous species found in ecosystems perform key functions or so-called keystone roles and that most species perform a perfunctory role (Holling, 1992). For example, beavers have been shown to have a profound impact on streams, forests and wetlands through dam construction.

Many species may play keystone roles that involve interdependencies with other species (Daily et al., 1993). The elimination of any single component of an ecosystem could lead to an unanticipated unravelling of community structure and to local extinctions of some species. Bird dispersers such as the blackcap (*Sylvia atricapilla*) interact with shrub species to influence floristic diversity, spatial patterns of vegetation development and plant dynamics in patchy Mediterranean vegetation in southern France (Debussche and Isenmann, 1994). Seeds deposited by blackcaps underneath pioneer shrubs (which have positive 'nurse' effects on other plants) were more likely to survive than in the open field and birds may actually trigger dynamic successional processes initiated by pioneer woody perennials in Mediterranean grasslands and shrublands.

The 'keystone role' of a species may also depend on whether a number of other species can assume its functional role within the ecosystem (Schindler et al., 1989). Functional redundancy is known to occur if other species can perform similar roles (Hutchinson, 1961: Walker, 1992). Although there is evidence that the deletion of some species has very little effect on ecosystem functioning, in many diverse tropical forests there are so many rare species that collectively they may have an important impact on the ecosystem. Here, top predators in ecosystems are relatively scarce because they are large in size and at the top of the food chain but may nevertheless be important in terms of ecosystem structure.

Most research has focused on which species are important here and now.

However, millions of species have not even been identified let alone evaluated for their potential values to humans. There are difficulties in predicting which species will be important in the future since the present functions performed by a species may provide no clues as to its role when environmental conditions change (Main, 1982; Lovejoy, 1988). Tree species that colonize gaps in tropical forests and species that require fire to enhance germination provide such examples. The population of *Cactoblastis cactorum*, which is relatively rare in Australia today, would not provide an accurate description of its importance in controlling *Opuntia* in that country in previous decades. However, studies on ecosystem function may reveal clues as to the most sensitive components of food webs, and nutrient and energy flows. Research reveals that the most sensitive components of ecosystems are those in which the number of species performing a particular function is thought to be very small (Schindler, 1990).

Most ecologists recognize that some species play a more important functional role than others. But what does this imply in terms of the properties of an ecosystem? Land managers are principally concerned with ecosystem productivity and its variability. The next section provides a review of the relationship between biodiversity and the stability, resilience and productivity of ecosystems.

Importance of biodiversity in managed landscapes
The importance of biodiversity is associated with a controversial theoretical debate amongst ecologists that began in the early 1950s: does biodiversity affect the stability of an ecosystem? Holling (1973) refers to stability as a characteristic of the individual populations of an ecosystem. For example, stability is defined as the propensity of a population to return to some kind of equilibrium following a disturbance. The stability of ecosystems may be linked to their biodiversity and it has long been hypothesized that more diverse ecosystems are more stable. A clue as to why this may be the case is illustrated by a natural disturbance that deleted some species from the ecosystem. A diverse system might be little affected by the impact because other species with similar niches could perform similar functions to the missing species. Early advocates of this theory include MacArthur (1955) who postulated that a highly diverse ecosystem would change less upon the removal or addition of a species than would an ecosystem with fewer species. Elton (1958) also suggested that less diversity resulted in less ecological stability.

However these theories were not without their critics. May (1973) challenged this argument and showed that a highly connected system (higher biodiversity) may be less stable than simpler ones and more vulnerable to disturbance because all its components closely interact and are therefore subject to the effects of perturbations. A drought that eliminates key species in a complex ecosystem, for example, will have widespread repercussions on the animals that depend on them.

More recent work (Tilman, 1996) has shown that there exists an important distinction between the properties of a community and its individual species, so although diverse ecosystems are more stable than less diverse ecosystems the populations within them can have great variability. From this perspective, what matters is the stability of the community or ecosystem, not their individual populations. There is some experimental evidence to support these assertions. Tilman and Downing (1994) have shown that an ecosystem with many species is more likely to be stable even though the populations of individual species may experience considerable fluctuations.

Resilience is a further factor that refers to properties of the stability of a system. The traditional concept of resilience is a measure of the speed of return to an equilibrium state after an ecosystem has been disturbed (Pimm, 1984; O'Neill et al., 1986). Alternative definitions have been proposed by Holling (1973). He describes resilience as the propensity of an ecosystem to retain its functional and organizational structure following a disturbance. Expressed another way, resilience is the amount of disturbance that can be absorbed before the system changes its structure by changing the variables that control how the system behaves (Holling, 1973). A characteristic feature of 'Holling-resilience' then is that though the system parameters (net primary production, or system growth rates, species composition) may change after disturbance, a resilient community will return quickly to equilibrium after disturbance is removed. A resilient ecosystem does not necessarily imply that all of its component populations are stable. Environmental perturbation may result in the extinction of an individual species without affecting ecosystem function or resilience. Holling (1973) distinguishes between stability as a property associated with individual populations of an ecosystem, and resilience as a property of an ecosystem.

Early work by Holling (1973) has suggested that, in general, the resilience of an ecosystem is an increasing function of the diversity of that system. There is some empirical evidence to support this view. In a series of field experiments in drought-affected grasslands in Minnesota, Tilman (1996) has shown that species-poor plots were less productive in terms of biomass than species-rich plots (see Figure 4.1A). He also demonstrated that species-poor plots were more greatly harmed by drought (less resistant), took longer to return to pre-drought conditions (less resilient) and were less stable than species-rich plots. Tilman et al. (1997) also demonstrated that plots with lower functional diversity had lower productivity in biomass terms than plots with high functional diversity (see Figure 4.1B).

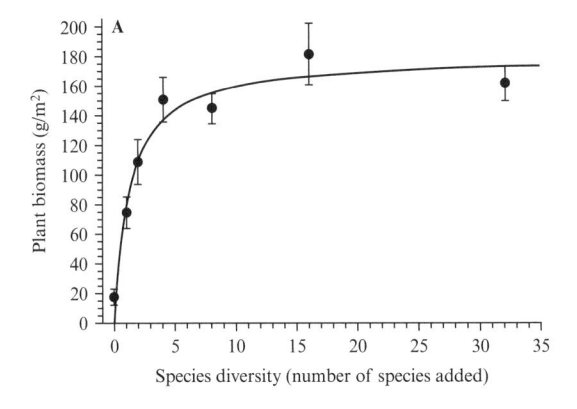

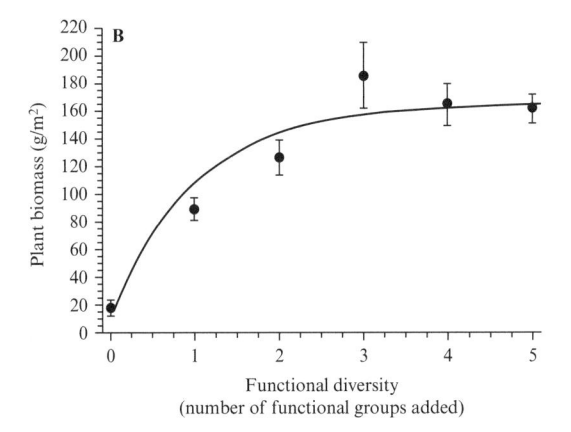

Note: Curves shown are simple asymptotic functions fitted to treatment means.

Source: Tilman et al. (1997).

Figure 4.1 (**A**) *Dependence of 1996 above-ground plant biomass (that is, productivity) (mean and standard error) on the number of plant species seeded into the 289 plots.* (**B**) *Dependence of 1996 above-ground plant biomass on the number of functional groups seeded into each plot*

Having considered some reasons why biodiversity might be important we now provide some evidence as to why managed landscapes might be so significant in supporting biodiversity conservation goals. Managed landscapes that are vital economically in supporting human populations can also make a very important contribution to biodiversity conservation. Biodiversity can coexist alongside human activities and economically valuable managed landscapes do not necessarily have to exclude biodiversity conservation goals. Indeed, there is a significant literature that suggests that many managed systems may actually augment biodiversity and that land abandonment or a decline in management activities can actually threaten biodiversity. We now turn to some examples of how land managers from around the world manage systems that conserve biodiversity. We also consider what can happen to biodiversity when land and management activities are abandoned.

Landscape management and biodiversity
In many managed landscapes good conservation practice succeeds because it is perceived to coincide with the interests of land managers whose support is vital for conservation initiatives. Such conservation practices may also have been developed to avoid over-utilization of the resource on which the human population depends. Consequently most biodiversity exists in human-dominated ecosystems (Pimmental et al., 1992).

First, some examples of where good conservation practice is coincidental with the interests of land managers are provided, and second, we look at abandonment. In areas where human populations have long been an integral part of the landscape and had much to do with its recent evolution, species may have adapted to 'managed' landscapes. For example, human impacts on biodiversity in the Mediterranean basin may play a positive role where current levels of biodiversity are in part maintained by continued human influence. Pignatti (1978) reports that domestic livestock, and an opening up of evergreen oak forests in the Mediterranean, provided new opportunities for speciation of herbaceous annual flora. For example, the *dehesas* of south-west Spain have evolved around a distinct and long history of anthropogenic influence. These open wood pasture systems are derived from ancient Mediterranean forests and are managed to support livestock production with some accompanying arable cultivation and silviculture but are widely recognized as being of high conservation value (Baldock et al., 1993; Telleria and Santos, 1995; Díaz et al., 1996).

Floristic diversity is high and *dehesa* grasslands are remarkable for maintaining some of the most species-rich grasslands outside the tropics, with as many as 60 plant species per square metre having been recorded (Marañon, 1986). A number of explanations have been advanced for the high floristic diversity associated with *dehesas*. The Mediterranean basin acts as a transitional biogeographical location. It has been suggested that its flora, which comprises several different genetic elements, has been enriched by historical climatic fluctuations during the Quaternary, by complexity of mountain relief and by altitudinal heterogeneity and historical human disturbance (Zohary, 1973; Whittaker, 1977; Marañon, 1986).

Defoliation by domestic herbivores and the occurrence of frequent fires in association with periodic droughts are also thought to have promoted plant diversification particularly of annual species and initiated adaptations to drought, fire and grazing (Pignatti, 1978; Naveh, 1994).

Many bird species from northern Europe overwinter in the *dehesas* and are reliant upon the *dehesas* as a food source. Telleria et al. (1992) provide evidence that suggests that *dehesas* may support more diverse communities of passerines than neighbouring stands of high forest. *Dehesa* habitat supports 64 per cent of the population of common cranes wintering in Spain (50 000–60 000 birds), which represents 70–85 per cent of the western European population (Alonso et al., 1990). The crane population is not widespread in other habitats as cranes rely on acorns, so holm oak *dehesas* are of great importance for this species, considered vulnerable because of their decreasing population trend (Tucker and Heath, 1994). It is also thought that the distribution of white stork is most strongly associated with open holm oak wood pasture and it has been suggested that conversion of this habitat could lead to a decline in its populations (Carrascal et al., 1993).

Managed woodland and grazing systems elsewhere in Europe are also important for maintaining biological diversity. The bird community of western oak woods in the United Kingdom, particularly the abundance of the wood warbler (*Phylloscopus sibilatrix*), pied flycatcher (*Ficedula hypoleuca*) and redstart has long been recognized as unique and some grazing may help to create the open conditions in the understorey and field layer favoured by these species. Some studies that focus on grazing even report that subspecies of grasses may develop according to the specific ecological conditions that occur in a grazed or mowed sward (Reinhammar, 1995). The development of species-rich raised coastal dune and bog habitats in north-western Europe, known as machairs, is also thought to be strongly associated with agriculture and human activity, particularly fire and grazing (Mate, 1992; Edwards, et al., 2005). Machairs, which are priority habitats under the European Habitats Directive, are unique ecosystems confined, in the northern hemisphere, primarily to west and north-west coasts of Ireland and Scotland. Machairs are priority habitats because of the high plant species richness which contain elements of calcareous grassland and sand dune plant communities. We now turn to the issue of abandonment.

Londo (1990) reports that, in the absence of management, semi-natural grassland communities revert by processes of natural succession to natural woodland and forest and the diversity of herbaceous species falls. Many traditional extensive farming practices have been shown to maintain plant and animal diversity (González Bernáldez, 1991; Naveh, 1994), and where these activities cease, susceptibility to disturbances, especially fire, can be increased. Fire in turn can have a negative effect on biodiversity (Faraco et al., 1993). Landscape homogenization can also result from the abandonment of agricultural/pastoral land (Alés et al., 1992). Without human management diverse plant communities in the Mediterranean basin, for example, become overgrown, and displaced by relatively few, shrubby unproductive species. Livestock may play a positive role in influencing the system.

Bokdam and Gleichman (2000) have suggested that abandonment is a major threat to traditional pastoral landscapes and their wildlife in Europe. They report that increased labour costs have undermined traditional herding systems, which are being replaced by free-ranging grazing systems leading to a decline in species-rich open heathland. The management of Mediterranean woodland has become an important issue in many areas because of the abandonment of large areas that were previously exploited by grazing. In many cases impenetrable thickets have developed with continuous accumulation of fuel leading to catastrophic wildfires. Valderrábano and Torrano (2000) evaluate goats as a potential management tool for controlling encroachment of *Genista scorpius* in black pine stands in the Spanish Pyrenees. They report that as a consequence of goat browsing and thinning, dense woodland was opened up and desirable tree growth and development was stimulated.

In traditional land husbandry, maintenance of biodiversity and economic outputs are closely intertwined. For example, the relationship between habitat characteristics, weather and spatial variation in animal behaviour was investigated by De Miguel et al. (1997). They suggest that shrub areas provide shelter and represent an important browse resource during winter and that this leads to the occurrence of a diversified landscape with different successional stages (from pastures to clear and dense woodlands) that occur in close proximity, which in turn leads to high levels of flora and fauna.

Clearly there is evidence to suggest that biodiversity can coexist in landscapes of economic importance but can land managers from around the world actually use biodiversity to support the productive process? This is the subject of the next section, which considers how land managers may conserve biodiversity because it supports the productivity, stability and resilience of the ecosystems they manage.

Landscape management and ecosystem properties
Maintenance of biodiversity may be coincident with management goals such as improved agricultural productivity under highly variable environmental and socioeconomic conditions. For example, pastoralists in Africa

deliberately maintain as many as a dozen breeds of camel in the Sudan because they are able to exploit the vegetation of extreme environments, including deserts and other uncultivated land. The loss of these hardy animal breeds therefore means a reduction in the area of human habitat (Köhler-Rollefson, 1993).

In Mediterranean *dehesa* systems good conservation practices that promote biodiversity have arisen because local farmers recognize that a diverse system helps reduce variation in productivity from year to year. Local farmers do not necessarily have biodiversity conservation goals in mind as a management aim. Nevertheless, biodiversity coincides with certain production goals such as improved stability of production under unpredictable environmental conditions. Large differences in climatic, geological and topographical gradients contribute to a considerable degree of variation in productivity across the regional landscape. Climatic factors are instrumental in dictating plant and animal dynamics and productivity. Consequently, a long history of anthropogenic influence has led to the development of a high level of management and functional complexity as a means of ensuring stability. Attempts to reach an understanding of the relationship between management practices and environmental variability have prompted research effort into the interactions between the individual components that comprise this complex ecosystem including tree, herbaceous and shrub, and livestock components.

For example, holm oak is preferred by *dehesa* farmers because it favours the growth of highly productive perennial herbaceous species through improved retention of soil moisture, modification of microclimate, improved nutrient availability and improved soil properties (Marañon, 1986; Joffre and Rambal, 1993). Marañon (1986) has observed a much higher phenological diversity in *dehesa* systems that include a tree component (Figure 4.2).

Groups of perennial herbaceous species may be significant for maintaining productivity because they are able to utilize nutrients and moisture more effectively. These include *Agrostis castellana, Dactylis glomerata, Lolium perenne* and *Phalaris aquatica*, which were all found more frequently beneath tree canopies than in the open field (Joffre et al., 1988). Joffre et al. (1988) hypothesized that differences in nitrogen utilization occurred between annual and perennial species and that the efficiency of nitrogen utilization by herbaceous species was affected by the tree canopy. They report higher nitrogen mineralization in grasslands with perennials compared with annuals and greater nitrogen mineralization below the tree canopy.

Farmers have evolved complex farming systems specifically to be able to exploit resources in species-rich environments. Diverse multispecies

86 *A handbook of environmental management*

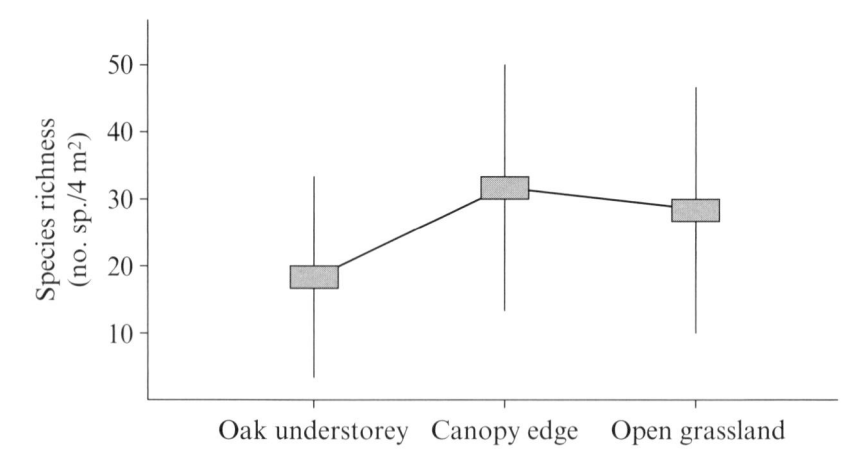

[Chart showing species richness across three habitats, y-axis: Species richness (no. sp./4 m²) from 0 to 50, x-axis: Oak understorey, Canopy edge, Open grassland]

Note: Mean standard error and range have been drawn.

Source: Marañón (1986).

Figure 4.2 Average species richness (number of species in 4m² plots) in three habitats: oak understorey, canopy edge and open grassland

herbivore systems, such as game ranches on the savannas of Africa, may include up to 20 different mammal herbivore species (Cumming, 1993). Managers exploit differences in grazing habits that can lead to a degree of complementarity in the use of forage resources where the total productivity of the range is seen to increase. Short grass (concentrate) grazers benefit from the modification of sward structure brought about by long grass (bulk) grazers, for example, sheep generally perform better when grazed in mixed systems than when grazed alone (Nolan and Connolly, 1977). This is usually only the case when large quantities of unpalatable poor-quality fodder are available. McNaughton (1984) reports that in the Serengeti, the larger bulk grazers consume long grass and these are then followed by smaller ungulates that create 'grazing lawns'. These lawns are sources of high-quality forage and so herbivores are seen to influence the quality and productivity of the grazing resource. Mixed species grazing systems may also be preferred to single species systems because they improve yields and do not over-exploit productive herbaceous species. It has been reported that sheep and cattle may affect the plant community in different ways. Bedell (1973) has shown that sheep can reduce the abundance of clover in a sward but they also increase the amount of *Poa trivialis* (Conway et al., 1972). In contrast, a high proportion of cattle has been shown to increase the amount of clover relative to grass. In this manner combined cattle and sheep grazing systems may be more productive than single species systems.

Examples of rangeland management systems that attempt to encourage diversity in herbivore populations to enhance resilience include replacing monocultures of domestic livestock with multispecies game systems and combined cattle/game ranches such as the CAMPFIRE (Communal Areas Management Programme for Indigenous Resources) programme in Zimbabwe (Cumming, 1993). Scholes and Walker (1993) have suggested that events such as fire and herbivory play an important role in maintaining the diversity and resilience of such systems. Here, the reduction of such perturbations is thought to reduce landscape diversity and the ability of the system to survive similar shocks in the future.

Moreover, in savannas (Walker et al., 1981) and *Agrostis-Festuca* grassland in Britain (Hulme et al., 1999) groups of grass species are important in maintaining the system's productivity. In other habitats, such as boreal and deciduous forests in North America and Europe, where insectivorous bird species are considered to be instrumental in controlling outbreaks of forest insect pests, overall species diversity is important for maintaining stability (Morris et al., 1958; Tinbergen, 1960; Campbell and Sloan, 1976; Holmes et al., 1979; Takekawa et al., 1982; Holling, 1988; Maquis and Whelan, 1994). There may also be indirect effects of diversity with some species influencing the survival of other species, such as key plant species determining the course of successional processes through the provision of so-called 'nurse effects'. Several studies have observed a greater number of seedlings beneath mature trees compared with more open areas (Griffin, 1971; Borchert et al., 1989; Espelta et al., 1995). Similarly, shrub species may influence seedling establishment, acorn consumption and the extent of browsing by herbivores (Morgan, 1991; Callaway, 1992; Herrera, 1995).

Despite the importance of biodiversity in contributing towards ecological services many ecosystems are undergoing profound change due to economic development. Heavily disturbed agricultural ecosystems in many parts of the world are threatened by human intervention, which has resulted in a loss of biodiversity, productivity and resilience. Land managers frequently need to know how biodiversity is affected by the level and intensity of management and what will be the result of that change. The significance of human-induced disturbance and environmental perturbation is the subject of the next section.

Ecosystem disturbance and biodiversity
Considerable insight into the understanding of conservation biology has been gained through knowledge of the effects of human-induced

disturbance on biodiversity (Wilson and Johns, 1982). There is a substantial literature that shows that human-induced disturbance and habitat degradation can result in a decline in biodiversity and species extinction.

Highly intensive agricultural practices that reduce spatial habitat complexity leading to homogenization of the landscape may lead to biodiversity loss. The decline in most of Europe's SPECS (Species of European Conservation Concern) has been linked to land use and management changes with agricultural intensification being cited as the most significant threat to bird populations (Tucker and Heath, 1994). Arable farming systems in parts of Europe are thought to have played a part in the decline of many species. For example, as a consequence of changing conditions in agricultural fields in Britain many bird species have undergone significant population declines. Fuller et al. (1991) report that many British farmland birds have declined dramatically over the last three decades as agricultural land use has altered, hedgerows have declined and farms have developed to form larger contiguous areas. A reduction in forest area due to agricultural expansion can also reduce species diversity. Studies of the avifauna of fragmented forests have shown that some species are absent or infrequent in very isolated sites and that smaller woodland size gives rise to less bird species diversity (Lynch and Whigham, 1984; Opdam et al., 1985; Ford, 1987; van Dorp and Opdam, 1987).

Data from censuses of domestic animals collected for tax purposes as well as from hunting statistics have been combined with palynological reconstructions of vegetation to demonstrate the long-term ecological effects of management practices. For example, hunting statistics for moose and roe deer in Sweden suggest dramatic recent population increases that have probably contributed to the decline of deciduous tree species (Ahlén, 1975). Peterken and Tubbs (1965) related fluctuating grazing regimes of horses, pigs and cattle in the New Forest, England, to waves of regeneration based on the age structure of existing trees. In Poland pollen data has enabled reconstructions of vegetation successions (Mitchell and Cole, 1998). This has been combined with data on herbivore densities for forests in eastern Poland over the last 200 years (Faliński, 1986) and shows that the proportion of conifers, principally *Picea abies*, increased dramatically at the expense of broadleaved species during the period of intensive grazing. Tree regeneration in the subsequent low-intensity grazing period was dominated by broadleaved taxa, initially *Betula, Populus* and *Caprinus* and, subsequently, *Tilia* and *Quercus*.

Jorritsma et al. (1999) used a dynamic simulation model FORGRA to evaluate the impact of grazing on Scots pine regeneration in the Netherlands. They showed that even low densities of ungulates could have a significant impact on forest development. Their model indicates that the presence of one cow per ha virtually eliminates recruitment entirely. Simulations of the model described above by Kienast et al. (1999) confirm these results since they also demonstrated that high browsing pressure does reduce recruitment and does alter the forest structure considerably, leading to high rates of tree mortality and more open forests. The spatial model developed by Weber et al. (1998) was used to determine the effects of grazing intensity and grazing heterogeneity applied to the southern Kalahari and shows that high levels of grazing lead to shrub invasion. Jeltsch et al. (1997) also reported that when grazing intensity reaches a critical level, shrub cover increases, drastically lowering the productivity of the range.

High levels of disturbance by wild or relatively unmanaged introduced animals are also thought to affect ecosystem productivity. The study of long-term grazing–vegetation interactions using palaeovegetation data in Ireland show that reasonably high populations of giant Irish deer imposed a high pressure on shrubby vegetation and had a profound effect on the change in vegetation communities from juniper scrub to grassland in Ireland during the Late-glacial Interstadial (11 000–12 000 BP) (Bradshaw and Mitchell, 1999). In Galicia, Spain, Hernandez and Silva-Pando (1996) report a decline in the abundance and diversity of shrub species after a period of three years' grazing by red (*Cervus elaphus*) and roe deer (*Capreolus capreolus*). Jane (1994) considered the long-term effects of browsing by red deer (*Cervus elaphus*) on mountain beech (*Nothofagus solandri*) in New Zealand and concluded that the impact of high deer densities on vegetation remains and can persist for many decades. She suggests that in critical high altitude areas, large reductions in deer densities were required to trigger the regeneration necessary for tree replacement.

Intensive levels of herbivory may reduce plant productivity, survival, reproduction and growth (Fay and Hartnett, 1991; Fox and Morrow, 1992; Relva and Veblen, 1998). In a long-term experiment to evaluate the impact of domestic livestock on tree species Hester et al. (1996) manipulated sheep stocking density and season in an upland broadleaved woodland in Cumbria, UK. They observed that growth and survival to the sapling stage was negatively correlated with grazing intensity, and suggest that, apart from plots grazed at the lowest animal densities, only a small proportion of saplings will attain canopy height. Other studies from around the world implicate browsing by domestic livestock as a cause of poor tree species recruitment (Kingery and Graham, 1991). Van Hees et al. (1996) employed an exclosure to determine the impact of roe (*Capreolus capreolus*) and red deer (*Cervus elaphus*) on beech (*Fagus silvatica*), pedunculate oak (*Quercus robur*) and silver birch (*Betula pendula*) in the Netherlands. They showed that browsing reduced sapling abundance, height and above-ground biomass of all three species. Some studies in the UK report a high number of seedlings within fenced enclosures compared with unfenced areas (Sykes and Horrill, 1979; Marrs and Welch, 1991; Staines, 1995). Historical records have also been used that suggest that deer may prevent natural regeneration of Scots pine (*Pinus silvestris*) in the United Kingdom. With respect to stocking densities, studies investigating the impact of ungulates foraging on upland heaths in Scotland suggest that red deer (*Cervus elaphus*) at stocking densities of >1 deer/20 ha can prevent tree regeneration (Staines et al., 1995).

Persistent high levels of disturbance are also thought to affect ecosystem function, particularly where these eliminate important functional groups that affect ecosystem processes. Groups of grass species may be significant in maintaining the productivity of savanna ecosystems (Walker et al., 1981). Walker et al. found that grasslands with persistent intensive grazing by settled peasant farmers had lower levels of productivity than moderate opportunistic grazing practices employed by nomadic pastoralists. In the former case, productive functional groups declined because herbivores showed a preference for the most palatable species, whilst in the latter case these preferred species were able to persist in the sward and adapt to change and instabilities caused by grazing and drought, thereby maintaining structural resilience.

Overgrazing may exacerbate the high inter-annual variation in productivity on many rangelands. Walker (1988) has observed a much higher phenological diversity in semi-arid systems not subject to heavy grazing compared with those that are intensively grazed. On lightly grazed areas he noted an even mix of early, mid- and late season grasses that were able to respond to rainfall wherever it occurred in the season. Heavy grazing leads to an absence of highly palatable early season species that are replaced by later growing species (Silva, 1987). The implication being that forage production was lower and more unstable on heavily grazed areas compared with lightly grazed land because the sward was not able to respond to early season rains. In the Serengeti, McNaughton (1985) has also shown that forage production was more stable where the number of species contributing to biomass was high compared with swards where relatively few species contributed to forage production.

Many complex ecosystems that aim to maximize heterogeneity (such as non-equilibrium systems) may be threatened by intensive grazing regimes that attempt to restrict livestock movements. This may have a negative impact on the stability and sustainability of the system. In areas where the fodder resource is widely dispersed seasonally and spatially, restrictions on stock movements by using paddocks can lead to land and vegetation degradation (Hoffman and Cowling, 1990). Increasing the connectivity of an under-connected system may also cause the system to change to a new stable state. An example of such a change has occurred in semi-arid shrublands, where the erection of fences has restricted animal movements, leading to localized feeding and degradation of vegetation and soil resources (Hoffman, 1988).

There is evidence that overgrazing may trigger the transition from one ecosystem to another, for example from forest to grassland or grassland to a shrubby semi-desert (Holling, 1973; Westoby et al., 1989). This evidence suggests that state and transition models (STM) are appropriate in many rangeland situations where vegetation is best described by a set of discrete 'states' and a set of discrete 'transitions' between states. During a transition the system jumps to another state if a threshold is exceeded. This can be triggered by fire, rainfall or grazing and the system never rests halfway through a transition. They use an STM in eastern Australia to show how once ecological thresholds are exceeded the system shifts from a woodland with a grass understorey to a less productive shrubby state. The concept of ecological thresholds suggests that there are limits to the ability of ecosystems to withstand environmental perturbation. If such limits are exceeded, ecosystems may shift to a less productive phase.

In some circumstances human management and spatial landscape change may undermine ecosystem processes. For example, some empirical studies have demonstrated that habitat fragmentation may reduce parasitism rates on herbivorous insects at different spatial scales (Kruess and Tscharntke, 1994; Roland and Taylor, 1997). Similarly, silvicultural practices may reduce parasitism rates of spruce bark beetle (*I. typographus*) in central Sweden (Weslien and Schroeder, 1999).

Clearly persistent intensive use of resources can degrade ecosystems and impact negatively on key functional groups. However, the level and nature of disturbance appears to be an important factor. Some examples of *managed* ecosystems indicate that an element of disturbance caused by human intervention may actually enhance biodiversity. In the next section we explore the impact of low and moderate levels of disturbance on biodiversity, productivity and resilience.

Moderate ecosystem disturbance and biodiversity
MacArthur and Wilson (1967) suggest that some disturbance can promote diversity because different species respond to disturbance in different ways. They first characterized species as either r- or K- strategists, which have evolved mechanisms to optimize resources in quite different environments. The former (r-strategist) refers to species that attempt to maximize growth in an unconstrained environment, reproduce quickly, disperse widely and are of smaller size and shorter lifespan. On the other hand K-strategists

include species that optimize growth in a climax successional phase or a crowded environment, are highly adapted to stable equilibrium conditions, are less flexible, more vulnerable to change, are generally longer lived and do not disperse as well. High levels of disturbance may lead to species-poor habitats since they favour the persistence of competitive, opportunistic *r* species better adapted to cope with disturbance (Miller, 1982). Conversely, undisturbed environments that do not undergo change may support less diversity because they favour the persistence of dominant *K*-strategists. Linder et al. (1997) examined the effects of fire history on stand structure and plant diversity in Swedish forest reserves. They concluded that the reintroduction of fire represents an important means of disturbance that was necessary to promote diversity of flora and fauna in the area. Continued fire suppression has changed successional patterns and altered stand structure. Late successional species such as spruce dominate due to lack of fire, and pioneer species such as pine, silver birch and aspen are decreasing in number because they require fire disturbance to regenerate. This appears to accord with MacArthur and Wilson's (1967) theory where undisturbed environments may therefore support less diversity because climax species are favoured. Linder et al. (1997) recommend prescribed burning to ensure a relatively wide range of successional stages to promote biodiversity over the longer term.

Higher habitat diversity due to moderate disturbance can also be explained by niche relations and the manner in which species divide up limited resources for their survival (Schmida and Wilson, 1985). They may divide up the available space (for example, by selecting different habitats) or energy resources (for example, by adopting different diets). Some studies serve to demonstrate that moderate levels of human activity may enhance biodiversity by opening up new niches, providing new food or protection from predators and by diversifying micro-habitats. For example, structural heterogeneity is thought to be important for bird species diversity and vegetation indexes have been developed to quantify structural diversity particularly in relation to bird species (MacArthur and MacArthur, 1961; Willson, 1974; Erdelen, 1984). Several studies indicate that a decline in structural diversity (James and Wamer, 1982; Terborgh, 1985; Ratcliffe, 1993; Telleria and Carrascal, 1994) and floristic diversity (Lynch and Whigham, 1984) leads to less bird species diversity. This is confirmed by Casey and Hein (1983) and Dambach (1944) who all reported that woodland heavily browsed by deer supported fewer bird species than woodland that was not grazed (although see DeGraaf et al., 1991).

Schemske and Brokaw (1981) provide empirical evidence to show that moderate disturbance in tropical forests caused by natural tree falls resulted in the greatest diversity of bird species. Clout and Gaze (1984) in

New Zealand found that the highest levels of bird diversity were recorded in disturbed productive forests while undisturbed mature forests contained less bird diversity, though they were populated predominantly by native bird species. Sternberg et al. (2000) conducted a four-year study on the response of a Mediterranean herbaceous community to grazing management in north-eastern Israel. Contrasting different grazing treatments they found that low and high grazing regimes reduced herbaceous diversity but that moderately grazed areas increased diversity.

The study of long-term grazing–vegetation interactions (10^2–10^6 years) using palaeovegetation data permits the reconstruction of vegetation and herbivore abundance and associations. Data from Jutland in Denmark from the Holocene about 5000–7000 years ago suggests that large forest herbivores did not have a significant influence on regional forest structure (Bradshaw and Mitchell, 1999). This is because either large predators held populations at modest levels, or the diversity of grazing species held populations at stable, low populations of individual species.

Moderate levels of insect herbivory may actually increase productivity. For example, Holling (1978) carried out an experiment on the defoliation of balsam fir (*Abies balsamea*) by spruce budworm. The larvae result in the death of mature trees aged 55–60 years though young trees are unaffected. Saplings grow rapidly after mature forest is damaged, and the forest is restored by its juvenile population. In the short term there is a shortfall in the production of timber, but over the longer term wood production remains unaffected. In fact, production rates of the juvenile forest remain above that of the mature forest because in a mature stand, most trees have passed their rapid-growth phase. Mattson and Addy (1975) reached similar conclusions in their study on the effects of forest tent caterpillars on aspen.

French et al. (1997) conducted a study on the development of Scots pine in the Cairngorm mountains in Scotland and found that recruitment is possible provided grazing/browsing pressure remains at a low level. Similarly Sun et al. (1997) evaluated the effects of cattle grazing and seedling size on the establishment of *Araucaria cunninghamii* in a silvo-pastoral system in north-east Australia. They report that grazing did not cause unacceptable mortality due to the fact that the tree has prickly needles, which prevented browsing by cattle. They recommend that grazing does not affect recruitment and can begin immediately after tree planting provided that a moderate stocking rate is used. Elsewhere, modelling has also shown that plant populations may be little affected by low levels of herbivory. Kienast et al. (1999) used a succession model – FORECE – to assess the long-term dynamics of alpine forests in Central Europe. They report that moderate levels of browsing posed no threat to the long-term survival of these forests and did not alter the successional sequence of forest development.

Recent developments on the functional complexity of ecosystems show that small disturbances may actually enhance ecosystem function and increase resilience. Holling (1986) and Holling et al. (1994) suggest that some natural disturbances initiated by fire, wind and herbivores are an inherent part of the internal dynamics of ecosystems and in many cases set the timing of successional cycles. These natural perturbations are part of ecosystem development and evolution, and seem to be crucial for maintaining ecosystem resilience and integrity (Costanza et al., 1993). In the absence of such shocks, the system will become highly connected and this will provoke even larger perturbations that are more destructive to the ecosystem because they reduce the ability of the system to survive similar shocks in the future (Scholes and Walker, 1993).

Some empirical studies reveal that herbivores may enhance a system's ability to resist environmental perturbation. For example, in their work on Florida mangroves, Simberloff et al. (1978) reported that the action of isopod and other invertebrate root borers resulted in new growth of roots at the point of attack. More extensive root systems in mangroves result in greater stability and resistance to storms and therefore benefit the plant. In Britain, Hulme et al. (1999) carried out a study to evaluate the effects of sheep grazing on the productivity of upland *Agrostis-Festuca* grassland. The experiment controlled sheep grazing at light, heavy and moderate levels. Both low and high levels of grazing resulted in the spread of less desirable species such as *Nardus stricta* and *Molinia caerulea.* Moderate levels of grazing maintained preferred species such as *Festuca rubra* and *Agrostis capillaris* and prevented the spread of *Nardus stricta* and *Molinia caerulea.*

Non-equilibrium rangeland systems as practised by nomadic pastoralists in parts of Africa have evolved opportunistic management regimes that employ moderate levels of grazing intensity that do not eliminate keystone elements but instead maintain the resilience of these components. Scholes and Walker (1993) have suggested that events such as fire and herbivory may play an important role in maintaining the diversity and resilience of such systems. Nutrient release following small fires supports a flush of new growth without destroying all of the old growth. Rangeland patches are affected but the forage resource remains intact. Small fires prevent the accumulation of forest biomass, which fuels very large fires that can decimate large areas of rangeland (Scholes and Walker, 1993) or whole forests (Holling et al., 1994). Such events may affect the parameters of the system and cause it to cross a threshold into an alternative state, which may alter the system's capacity to provide ecological services. For example, the Yellowstone National Park in the United States employed a 'natural burn' policy of management that culminated in catastrophic forest fires.

As described above, human-induced perturbation on managed ecosystems is a critical factor in maintaining biodiversity. However, the application of best scientific practice by land managers may not in itself be sufficient to achieve biodiversity conservation goals. This is because markets may fail to account for the value of biodiversity to society. It is essential therefore that land managers are aware of the limitations and opportunities of the market. The next section explores the economic linkages between markets and biodiversity.

Biodiversity in managed landscapes: economic issues
As reviewed above, biological resources in many of the world's low-intensity managed habitats represent a significant contribution to economic activity. However, many traditional low-intensity managed habitats are threatened by development – a change in land use management due to the prospect of increased private returns. According to the economic theory of general equilibrium, the search for opportunities for increased private returns can ensure that resources are allocated to the highest-value use available, so that economic efficiency is achieved. This result depends on a number of conditions. If these conditions are fully met, land use change motivated by private profit need not be a cause for concern. However, managed landscapes, in common with most environmental goods, have characteristics that ensure that the necessary conditions will never be fully met in practice. In general terms, this failure implies that any resulting allocation of resources is likely to be economically inefficient, meaning that it would be possible to reallocate resources in such a way as to make at least one member of society 'better off'.

Some mention of the distinction between economic efficiency and equity is worthwhile at this point. Economists place great emphasis on economic efficiency but this will not necessarily result in a fair outcome. For a society to be sustainable, its welfare should not be declining over time (WCED, 1987; Pezzey, 1989). In theory there are potentially a number of efficient time paths that are sustainable. However, efficiency does not necessarily guarantee sustainability between, say, current and future generations, in terms of the distribution of natural resources such as biodiversity (Common and Perrings, 1992; Perman et al., 2003).

In the case of managed landscapes, the danger is that land use change guided by market signals alone may lead not to beneficial development but to loss of valuable and possibly irreplaceable resources. The necessary conditions that must hold in order for market-led development to

96   *A handbook of environmental management*

be benign are typically discussed under two headings referring to their absence in practice. These are (1) market failure and (2) policy failure.

Causes of biodiversity loss: market and policy failure
Market failure occurs when private decisions based on a set of prices, or lack of them, do not generate an efficient allocation of resources (Hanley et al., 1997). With respect to biodiversity the concern is that market prices are not reliable indicators of social cost. Social cost refers to the opportunities forgone by society in committing resources in some way (Coase, 1960) and social cost in this study is taken to mean the true value that society as a whole places on natural resources. Private cost, on the other hand, refers simply to the financial cost faced by the private individual or firm undertaking the land use change, at current and expected market prices.

This divergence between private and social cost occurs because managed biodiverse landscapes generate benefits to society in addition to those that are transacted in the market system: external benefits. An absence of such external effects is one of the necessary conditions for market efficiency referred to in the previous section. Typically, the reason these benefits remain external to the market system is that they have the characteristics

Table 4.1   Characteristics of public and private goods

Pure Private Goods	Quasi-private Goods	Quasi-public Goods	Pure Public Goods
↓	↓	↓	↓
Rivalness in consumption; excludability; property rights, market prices	Regular payments in form of taxes or changes are made to finance supply	Up to some capability constraint (carrying capability), non-rivalness in consumption	Non-rivalness in consumption; non-excludability; or excessive costs of excludability
↓			↓
use values (market revealed)			non-use values (non-market revealed)

VALUATION DIFFICULTIES
Degree of 'familiarity' declining ─────────────────────────→
Problems of 'perception' increasing ─────────────────────────→
New information requirements increasing ─────────────────────→
Problems of reliability/validity increasing ──────────────────→
Reliance of interdisciplinary research effort and findings increasing ──→

Source: Turner (1993).

of public goods, in particular they are indivisible and perhaps also non-excludable, making their exchange in markets unlikely (Table 4.1).

The public good nature of biodiversity creates difficulties for its valuation. These will be discussed in some detail in the sections that follow. Because managed landscapes provide high levels of *unpriced* public benefits, in terms of wildlife and landscape quality, private agents will have no incentive to take account of these benefits in decisions over land use.

Conventional economic theory seeks to cast government in the role of an objective and well-informed 'third force' (in addition to individuals and firms), with some ability to intervene to correct for market failures. Government or policy failure occurs when policy decisions required to correct for market failure are not implemented and fail to fully recognize, or incorporate, the values associated with environmental resources. Policy failure may also arise where government decisions themselves induce economic inefficiencies. For example, agri-environment policies, through creating incentives for farmers to expand production, may result in a greater privately optimal level of degradation than would be the case in the absence of such policies. Poorly formulated policy instruments and incentives may distort the allocation of resources unintentionally. Simpson et al. (1998) suggest that high stocking rates are caused by incentives to graze moorland to achieve profit maximization, encouraged by support from the Common Agricultural Policy (CAP). They indicate that increases in the ewe flock across the Northern Isles (in Orkney from 37000 in 1983 to nearly 55000 in 1992, and in Shetland from 116000 in 1982 to 156000 in 1993) was in response to the EU's sheep meat regime introduced in the early 1980s. The scheme offered headage payments and a variable premium in fat lamb sales. They suggest that the policy has been sufficient to increase stocking levels, and hence heather utilization rates, across the Northern Isles. McNeely (1993) suggests that in Botswana, national and European Union subsidies have led to excessive uncontrolled grazing of rangelands and degradation of grazing savanna, which have affected the long-term productivity of the resource. Subsidies that aim to promote cash crops to secure export revenue may result in land degradation, soil nutrient losses and a reduction in the resilience of ecosystems (Grainger, 1990). Royalties in forestry can lead to excessive rates of deforestation (Repetto, 1989; Barbier et al., 1991).

The catch-all term 'market failure' is defined so as to refer to all situations where the market signals perceived by private individuals fail to coincide with social values (and fail to produce economic efficiency). However, some necessary conditions for market efficiency, which may be violated in practice, tend to be omitted from discussions of market failure, and are worth briefly mentioning here. The discussion above relates mainly to what might be called the 'complete set of markets' condition (Common, 1995).

98　*A handbook of environmental management*

Also important is the 'complete information' condition, requiring not only that prices be widely known, but also that they reflect the full implications of any reallocation of resources. It is clear from the discussion above that in the case of managed landscapes, such knowledge is available only in partial and uncertain form, and is not reflected in actual market prices. The effect of such uncertainty is discussed below.

The so-called 'rationality condition' may also be violated for environmental goods such as managed landscapes and associated biodiversity. The link between market efficiency and the (constrained) satisfaction of the wants of individuals and of the society of which they are members relies upon a number of assumptions about the nature of individuals' preferences. Rationality of preferences includes the ability and willingness always to make comparisons between goods. We will see below that stated preferences for environmental resources can include a refusal to do this, on the grounds that a biological resource should be preserved 'in its own right'. To the extent that individuals do not in fact have 'rational' preferences, market outcomes will tend to deviate from socially desired outcomes.

Valuing biodiversity
The main point that is frequently made by environmental economists working on valuation with regard to market and policy failure is that private resource users do not attribute sufficient weight to biodiversity. Valuation, it is argued, aims to redress this imbalance and sets out to determine what weight should be given to biodiversity in the interests of society as a whole.

The literature indicates that a variety of methods have been employed to estimate wildlife values in managed landscapes. Studies on wildlife value have focused on their use and non-use values. These values are based on an individual's willingness to pay (WTP) or willingness to accept (WTA) compensation. Gross willingness to pay might include the cost of travel, purchase of equipment to participate in the recreation activity, actual fees associated with the activity and consumer surplus. The concept of 'total economic value' (TEV) has been used to describe the components of value as shown in Figure 4.3. Use values associated with managed landscapes refer to the actual and/or planned use of a service by an individual and include recreational activities such as bird watching or hunting. Use values also include the following: option value, that is, the value of the option to guarantee use of the service by the individual in the future (Weisbrod, 1964); and quasi-option value, that is, the value of future information protected by preserving the resource now, given the expectation of future growth in knowledge relevant to the implications of development (Arrow and Fisher, 1974; Perman et al., 2003).

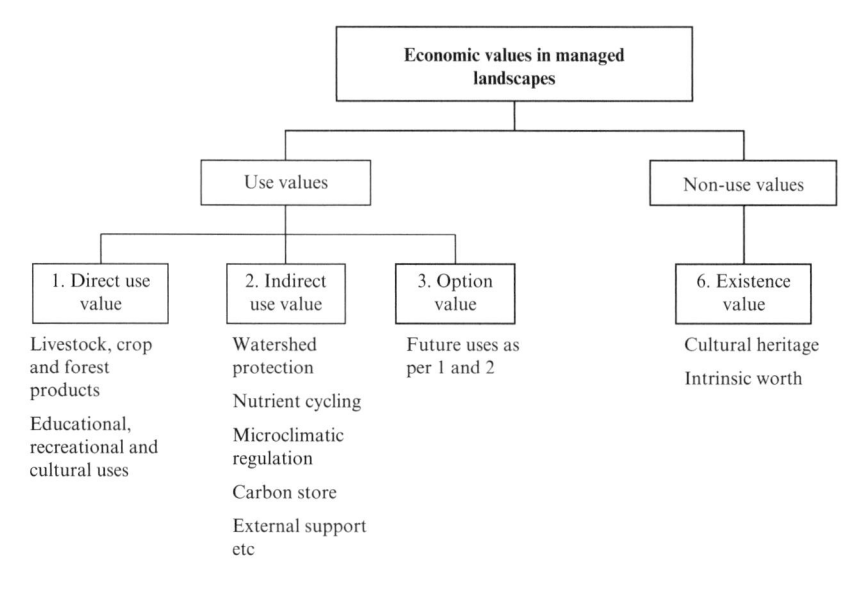

```
                    Economic values in managed
                              landscapes
                    ┌──────────────┴──────────────┐
                Use values                   Non-use values
      ┌─────────────┼─────────────┐                │
  1. Direct use  2. Indirect   3. Option      6. Existence
      value       use value      value           value

  Livestock, crop  Watershed    Future uses as  Cultural heritage
  and forest       protection   per 1 and 2
  products                                      Intrinsic worth
                   Nutrient cycling
  Educational,
  recreational and Microclimatic
  cultural uses    regulation

                   Carbon store

                   External support
                   etc
```

Source: Adapted from Barbier (1994).

Figure 4.3 Economic values in managed landscapes

More recently, empirical studies on wildlife values have placed emphasis on non-use values. These refer to situations where an individual knows a biological resource exists and will continue to exist, independently of any actual or prospective use by the individual and where that individual would feel a 'loss' if the resource were to disappear (Brown, 1990). Existence value arises when the utility function of a consumer is enhanced by the knowledge that a certain wildlife species exists. As indicated in Figure 4.3, non-use values thus refer to situations in which individuals would like to see a biological resource preserved 'in its own right'. Non-use values include the following: *bequest value*: the value of ensuring that the resource remains intact for one's future heirs (Krutilla, 1967); *existence value*: the value that arises from ensuring the survival of a resource (Pearce and Turner, 1990; Perman et al., 2003). Existence value is usually assumed to embody some form of altruism, either for other human beings or for a concern for non-human entities. For example, some of the literature distinguishes between philanthropic motives based on the provision of services to other people, and altruistic behaviour solely concerned with nature. The sum of all use values and non-use values is referred to as total economic value (TEV).

In the literature, two classes of use value are sometimes defined – direct use value and indirect use value. This is illustrated in Figure 4.3. Direct use value is the same as that outlined above and includes, for example, harvesting timber from a forest or the use of recreation services provided by a national park. Indirect use value on the other hand refers to the life support services provided by ecosystems. These include ecosystem functions such as flood control, catchment protection, nutrient cycling and carbon sequestration. The biological diversity of managed landscapes may serve an important role in maintaining ecosystem functions and thus serve to support the productive process.

Measuring biodiversity values
Early studies on wildlife values employed revealed preference methods, such as the so-called travel cost method (TCM). This technique was first proposed for use in recreation studies and was subsequently refined and applied in empirical studies by Clawson and Knetsch (1966). The method is based on the premise that it should be possible to infer values placed by visitors on environmental outdoor recreation services from the costs that they have incurred in order to experience these sites. Such costs include costs associated with travelling to a recreation site and the imputed value of people's time. A statistical relationship between observed visits and the cost of visiting a site is determined. This is then used to construct a demand curve from which consumer surplus can be measured. The current value of the resource and value of alternative policies affecting the resource can then be evaluated using consumer surplus calculated from the demand curves. The advantage of TCM is that the data collected involves actual consumer behaviour. Its chief disadvantage is that it does not accurately value trips for multiple purposes. TCM has been used extensively in the United Kingdom and United States for valuing the non-market benefits associated with national parks and managed landscapes including public forests (Bowes and Krutilla, 1989; Benson and Willis, 1991; Whiteman, 1991).

A second method of estimating wildlife value is a stated preference method, the contingent valuation method (CVM). This involves the construction of a hypothetical or simulated market for an environmental or wildlife resource. Contingent valuation techniques use surveys to elicit individuals' preferences for public goods by finding out what they would be willing to pay (WTP) for them, or what they would be willing to accept as compensation (so that they would not be worse off) for specified changes in them. This approach circumvents the absence of markets for public goods by presenting consumers with hypothetical markets in which they have the opportunity to purchase the good. Willingness to pay is determined either through a written questionnaire or using bidding games implemented by personal interviews. Demand curves are then constructed, and consumer surplus used as a measure of use and non-use value. The

CVM has the advantage of utilizing all the structural characteristics of demand analysis. Its chief disadvantage is that respondent bias may exist, pointing out the importance of the art of questionnaire design. Despite its widespread use CVM is extremely controversial and the values derived from the technique are treated with some scepticism by many economists (van Rensburg et al., 2002; Mill et al., 2007). Some go as far as to suggest that the technique should not be used as the basis for policy decisions (Hausman, 1993).

The contingent valuation method began to be used widely from the mid-1970s (Randall et al., 1974; Brookshire et al., 1976). Other detailed accounts of the method can be found in Mitchell and Carson (1989), Hanley and Spash (1993), Bateman and Willis (1995), van Rensburg et al. (2002), and Mill et al. (2007). Relatively few contingent valuation studies relate specifically to biodiversity (Diamond and Hausman, 1994; Hanemann, 1994; Portney, 1994).

The value of species and habitats
Many empirical studies applied to wilderness areas indicate that the value of recreational and other non-marketed direct values derived from areas of high nature conservation value can be significant and may compare favourably with competing commercial uses of the same resource. For example, Hanley and Craig (1991) contrasted the trade-offs implicit in permitting or prohibiting afforestation with respect to the flow country, in northern Scotland (the largest body of blanket peat bog in the northern hemisphere). The development would generate employment and produce timber but displace extensive populations of internationally rare breeding birds. They demonstrated that the total recreational value of the resource exceeded the benefits derived from afforestation at discount rates of 6, 4 and 3 per cent. Similarly, Willis (1991) established that the total recreational value of the Forestry Commission estate in the United Kingdom exceeded the value of timber sales.

Garrod and Willis (1997) carried out one of the few examples of contingent ranking techniques applied specifically to biodiversity. They employed a discrete choice contingent ranking approach to estimate the general public's WTP to increase the area of Forestry Commission forests managed under three forest management standards designed to offer increasing levels of biodiversity at the expense of commercial timber production. This method enables relative preferences for different forest management standards to be measured at the same time as WTP to promote biodiversity. They suggest that the benefits of changing forest management to meet these standards far outweigh the financial costs involved.

Some of the benefits associated with biodiversity can be deciphered from expenditure on the preservation of endangered species. Several empirical CVM studies have been used to determine values related to the conservation of individual and endangered species in protected areas (Stoll and Johnson, 1984; Brown and Henry, 1993). Research on endangered or threatened species includes the value of preserving the whooping crane (*Grus americana*) population at the Arkansas National Wildlife Refuge in Texas for viewers and non-viewers (Bowker and Stoll, 1988) at about US$6 per person per year. Similarly, Boyle and Bishop (1987) estimated the value of preserving the bald eagle at US$17.46 per person per year.

A study conducted by Brown et al. (1994) values the northern spotted owl and its ancient old growth forest habitat using the contingent ranking approach. In this study, respondents were offered five different policies. Associated with each policy were the cost of the policy, the area preserved, the estimated number of owl pairs preserved and their probability of survival. They estimated existence values for conserving the northern spotted owl at about US$20 per person per year. Probabilistic theoretical models have been used to determine the benefits of important wildlife species such as the northern spotted owl in old naturally regenerated redwood forests and have demonstrated the high marginal cost of preservation (Montgomery et al., 1994). Estimates based on the probability of survival and a reduction in timber stumpage supply give an estimated welfare cost of US$21 billion to ensure an 82 per cent chance of the species surviving. Increasing the chance of survival from say 90 per cent to 95 per cent was estimated to cost an additional US$13 billion.

Indirect use values and ecosystem function
Much of the discussion in the second section of this chapter dealt with the properties of ecosystems including their productivity, resilience and stability. There is a significant literature on the value of ecosystem services including indirect values (Ellis and Fisher, 1987; Barbier, 1994; Bell, 1997; Daily, 1997; Barbier, 2000; Daily et al., 2000). Indirect values associated with biodiversity can be measured using surrogate market approaches using the production function approach. Information about a marketed good (timber, crops or livestock sales) is used to infer the value of a related non-marketed good (for example, forest, agricultural or wetland habitat). The basic assumption underlying this approach is that, if, for instance, biodiversity supports agricultural or forest production, then this ecological service provides an additional environmental input into the agricultural or forest enterprise.

For example, the stability of a managed ecosystem constitutes an indirect use value and represents an important function to land managers. As seen above, biodiversity may mitigate large inter-annual variation in productivity (McNaughton, 1985; Walker, 1988). For instance, the economic value of a change in diversity can be evaluated from the change in livestock liveweight gain associated with a decline in forage biomass as a result of a decline in grassland diversity.

A number of studies in the applied economics literature have used the stochastic production function approach suggested by Just and Pope (1978) to capture the value of crop diversity. These studies indicate that genetic variability within and between crop species confers the potential to resist stress, provide shelter from adverse conditions and increase the resilience and sustainability of agro-ecosystems. Plot studies indicate that intercropping can reduce the probability of absolute crop failure and that crop diversification increases crop income stability (Walker et al., 1983). Therefore, the greater the diversity between and/or within species and functional groups, the greater is the tolerance to pests. This is because pests easily spread through crops with the same genetic base (Sumner et al., 1981; Altieri and Liebman, 1986).

Crop diversity may enhance farm productivity, stabilize farm income and reduce the risk of outright crop failure (Long et al., 2000). The existence of a limited number of crops grown in an area makes these crops more vulnerable to diseases and pests. By maintaining proper crop rotations diversity can improve soil productivity and reduce the need for agrochemical applications. Land managers also recognize that soil and climatic conditions can vary considerably. In such circumstances, growing different crops and crop varieties can lead to more efficient use of resources. Some crops can be grown on fertile land while others can utilize marginal areas. Therefore, the greater the variability of soil and climatic conditions, the greater the impact biodiversity will have on improved agricultural production.

For example, Smale et al. (1998), report that crop diversity is positively related to the mean of yields and negatively correlated with the variance of yields in rain-fed districts of the Punjab in Pakistan. Di Falco and Perrings (2003, 2005) found cereal diversity to be positively correlated with yields and negatively correlated with revenue variability in two studies in southern Italy. Di Falco and Chavas (2006) point out increased crop diversity may also reduce the likelihood of complete crop failure. Diversity is important also for commercial farmers, since they are dependent on diversity in the breeding pool, regardless of whether it is provided on or off farm.

Other examples of indirect values associated with diversity include mycorrhiza, which are important for the functioning of ecosystems and can be considered as a complementary input to timber production. They represent an indirect use value. Silvicultural practices that eliminate mycorrhiza from the system will involve the loss of timber revenue. Although mycorrhiza are not consumed themselves, they are essential for the growth of many timber species that are harvested and they are necessary to support the production process that produces goods and services that are consumed directly.

Similarly, the importance of bird species used as a biological control agent can be captured from increased timber sales associated with insect pest reduction. Takekawa and Garton (1984) used the substitution method to determine the value of a bird species, the evening grosbeak (*Hesperiphona vespertina*) in controlling spruce budworm populations affecting stands of Douglas fir (*Pseudotsuga menziesii*) in Washington. They substituted the costs of insecticide to produce the same mortality that birds cause and established that it would cost at least $1820 per square km per year over a 100-year rotation.

Biodiversity in managed landscapes: policy issues
Earlier sections indicated why markets may fail to protect biodiversity and give some examples of how economic tools can provide a useful means by which to measure the non-benefits and costs associated with biodiversity and thereby go some way towards dealing with market failure. A further solution is to develop economic incentives and instruments that correct for market failure. In the section that follows we consider the importance of policy with respect to biodiversity conservation.

The aims of a society may be formulated within the framework of national environmental policy. Policy can be regarded as 'the compendium of statements, laws and other actions concerning government's intentions for a particular human activity under its jurisdiction' (Miller, 1999).

Objectives concerning natural resources are not necessarily static. History indicates that environmental policies have changed progressively with time in response to changes in society. This has led to changes in the public demands placed on environmental resources. Human populations are concerned with using environmental resources as a means of survival but also increasingly to meet recreation and conservation goals.

In managed landscapes there is also public concern about the importance of ecological functions – water quality, biodiversity, aesthetic values and international and national organizations are under increasing public pressure to take action to develop economic incentives to protect public values on privately managed land (WCED, 1987).

Economic instruments
In what follows we outline two types of policy instruments of relevance to land managers – economic incentives and command and control regulations. We discuss economic incentives first.

McNeely (1988, p. 39) has defined incentives as 'an inducement, which is specifically intended to incite or motivate governments, local people and international organisations to conserve biological diversity'. The idea behind economic incentives is to increase the cost of non-compliance with environmental standards yet allow the producer the flexibility to employ the least-cost method of meeting these standards. By increasing the cost of non-compliance the producer has a private incentive to meet the standards set by the policy instrument. One of the advantages of incentive systems is that they are seen by economists as a cost-effective alternative to inflexible command and control environmental regulations (Hanley et al., 1997). However, in practice subsidies are much more widely used because of the resistance to other instruments by the agricultural sector (Hanley and Spash, 1993).

Many incentives are based on the level of opportunity costs or *financial costs* forgone by the producer. For example, the financial costs of conservation as estimated by Willis and Benson (1988) in the United Kingdom are offered to farmers as compensation for not developing their land. This is based on profits forgone under a management agreement. The current financial cost is the difference between the value of the output (less inputs) of the land under intensive management minus the value of output (less inputs) under a conservation regime. A complete financial evaluation of conservation also needs to include administrative costs, legal fees, labour and material costs for the maintenance of habitats (Willis and Benson, 1988).

Once the specific costs to the producer are known, policy instruments can be formulated that are targeted at the producer and that persuade producers to achieve the desired environmental objectives. Typically, agri-environmental policies employ market-based instruments such as subsidies that create economic incentives that allow individual producers to choose freely to adjust their activities thereby producing an environmental improvement (Barbier et al., 1994). Taxes as opposed to subsidies are generally preferred by economists because the latter inject income and lead to expansion of the sector under consideration. Subsidies can attract new entrants that may lead to greater aggregate levels of environmental damage and to other market distortions (Hanley et al., 1997).

An example of such a broad appraisal is agri-environment policy used to maintain ecologically important habitats such as the Environmentally Sensitive Area (ESA) scheme in the United Kingdom. Specific areas of land providing habitats for valuable species are identified as conservation areas under which agricultural management practices are regulated. Typically the policy is aimed at the farmer or forester to meet the desired environmental objectives where, for example, farmers might be expected to employ 'traditional' agronomic practices. The producer is then expected to change his or her management methods in accordance with certain regulations specified under a 'management agreement'. Such management agreements usually involve an identification of, for example, the farming practices necessary to achieve environmental objectives and then stipulate how they should be put into practice. In order to specify guidelines for 'good environmental practice', policy-makers need to understand the relationship between management practices and the species, population or community concerned; for example, the specific relationships between farm management methods and the species composition of grasslands.

An example of this includes the use of stocking restrictions to encourage heather moorland in the United Kingdom. The model developed by Simpson et al. (1998) crucially relates heather productivity and survival to varying intensities of the management variable (in this case, the stocking rate). Reductions in stocking rate can then be used to target farmers who are able to manage heather sustainably under, say, a management agreement.

Typically, agri-environment policy under a management agreement involves reductions in farm intensity in exchange for compensatory payments. In order to do this a precise estimate of the changes in management intensity to meet environmental objectives is required. This then enables the specific costs to the producer to be calculated based on opportunity cost pricing procedures. In the example described above, Simpson et al. (1998) suggest that in order to meet conservation guidelines for heather conservation ewe stocking rates will have to be reduced to between 13 and 91 per cent on Orkney and between 5 and 89 per cent on Shetland. They report that such a reduction would in most cases result in major financial losses to farmers who would need to be compensated if they were to comply with their recommendations.

This process of European agricultural reform has influenced the objectives of the CAP, which have undergone significant changes in recent years. The aims of the CAP are now strongly oriented towards environmental conservation rather than agricultural productivity. The development of these initiatives has provoked many EU countries to adopt environmental policies specifically aimed at encouraging producers to adopt less intensive agronomic and silvicultural practices (Hanley, 1995). The status of environmental objectives therefore is increasingly recognized to be as important as other goals such as rural income stability, employment and support for agricultural commodities. As a consequence, the monitoring and evaluation of environmental policy includes an increasing environmental component. The appraisal of agri-environment policy needs to include an assessment of physical economic targets but also needs to meet environmental objectives.

The second type of instrument includes command and control regulations. Situations may occur where economic activities need to be restrained in areas that are especially rich in biodiversity to protect the resource for present and future generations and yet it may not be possible to control market behaviour using incentives. Command and control environmental regulations may be used in such circumstances. Regulatory control involves the direct limitation or reduction of activities that degrade an environmental resource in accordance with some legislated or agreed standard (Barbier et al., 1994).

This is especially important where development initiatives that threaten biological diversity involve uncertainty. In the case of risk, as opposed to uncertainty, it is possible to completely list the range of possible outcomes, and to assign an estimated probability to each outcome. Given this information, and preferences over risk and return, rational decision-making is possible. In circumstances of uncertainty, however, where the range of possible outcomes is unknown, it is not possible to determine the expected profitability of a project. Although in the case of species extinctions a probability can be attached to the loss of species, the total consequence of this in terms of the loss of environmental services and ecosystem support and duration of these effects cannot be known with certainty. Decision-making in the presence of uncertainty relies not on rational comparison of all options, but on adoption of some decision rule that has appealing properties (Common, 1995). It has been argued that a precautionary approach to the conservation of biological resources should be adopted.

The policy of taking action before uncertainty about possible environmental damage is resolved has been referred to as the 'precautionary principle'. One justification for this is that the costs of damage to biological resources may exceed the costs of preventative action (Jackson and Taylor, 1992). Also, irreversible damage may occur such as species extinctions. The emphasis is thus on avoiding potentially damaging situations in the face of uncertainty over future outcomes. It has been proposed for decisions taken over the Convention on Biodiversity and has been used in conjunction with the Montreal Protocol (Myers, 1992; Haigh, 1993).

Ciriacy-Wantrup (1968) and Bishop (1978) have proposed 'the safe minimum standards' approach, which involves setting quantitative, and qualitative limits for, say, the preservation of species and their habitats. A programme is developed to maintain such limits unless the costs of doing so are 'unacceptably high'. Hanley et al. (1991) indicate that sites of special scientific interest (SSSI) in the United Kingdom provide an example of this approach in practice. These sites may be lost if the costs of conservation are prohibitive in terms of the government's conservation budget, but they are still protected regardless of any cost–benefit analysis having been undertaken.

Conclusions
Rapidly increasing human populations and associated economic development around the world have imposed real pressures on natural habitat and its biodiversity. This is a subject of major concern to policy-makers and the public at large because it is recognized that biodiversity loss could seriously diminish the options open to future generations. All too often market and policy imperfections obscure the social costs of managed lands, giving rise to inefficient land use and biodiversity loss.

Protected areas represent a high cost solution to biodiversity conservation in many areas. They impose considerable costs on producers, limit future development options, reduce the supply of market produce and they fail to engage land managers in conservation initiatives.

Joint production of commercial goods and biodiversity in managed landscapes represents an important alternative to reserves. Indeed there is evidence to suggest that biodiversity can coexist in landscapes of economic importance and that it is important in supporting productive processes in managed areas. However, highly intensive managed systems may pose a threat to biodiversity in some areas and it is vital that managers and policy-makers work together to develop strategies to avoid such losses.

Policy-makers should contribute to this process by developing instruments that internalize biodiversity values into market behaviour. This will help to avoid intervention failure and perverse incentives that lead to biodiversity loss, ensuring that biodiversity values are protected and provided efficiently.

Uncertainty over the benefits and costs of biodiversity and its role in the functioning of ecosystems point towards the need for a diversified strategy that includes protected areas as well as privately managed land used for production. In the absence of a concerted effort by policy-makers and land managers, the opportunity to develop initiatives that include private lands in such a strategy to achieve biodiversity conservation goals will be missed.

Note
1. Biodiversity thus represents the diversity of all life as being a characteristic property of nature, rather than a resource. The term also has a broader meaning for the set of organisms themselves. For example, a biodiverse tropical rainforest, therefore, refers to the quality or range of diversity within it.

Bibliography

Ahlén, I. (1975), 'Winter habitats of moose and deer in relation to land use in Scandinavia', *Viltrevy*, **9**(3), 1–88.
Alés, Fernandez R., A. Martin, F. Ortega and E.E. Alés (1992), 'Recent changes in landscape structure and function in a Mediterranean region of SW Spain (1950–1984)', *Landscape Ecology*, **7**(1), 3–18.
Alonso, J.A., J.C. Alonso and R. Muñoz-Pulido (1990), 'Areas de invernada de la grulla común *Grus grus* en España', in J.A. Alonso and J.C. Alonso (eds), *Distribución y demografia de la grulla común (Grus grus) en España*, Madrid: ICONA.
Altieri, M. and M. Liebman (eds) (1986), 'Insect, weed, and plant disease management in multiple cropping systems', in C. Francis, *Multiple Cropping Systems*, New York: Macmillan.
Arrow, K.J. and K.C. Fisher (1974), 'Environmental preservation, uncertainty and irreversibility', *Quarterly Journal of Economics*, **88**, 312–19.
Baldock, D., G. Beaufoy, G. Bennet and J. Clark (eds) (1993), *Nature Conservation and New Directions in the EC Common Agricultural Policy*, Arnhem: Institute for European Environmental Policy, HPC.
Barbier, E.B. (1994), 'Valuing environmental functions: tropical wetlands', *Land Economics*, **70**, 155–73.
Barbier, E.B. (2000), 'Valuing the environment as input: review of applications to mangrove–fishery linkages', *Ecological Economics*, **35**(1), 47–61.
Barbier, E.B., J.C. Burgess and A. Markyanda (1991), 'The economics of tropical deforestation', *Ambio*, **20**(21), 55–8.
Barbier, E.B., J.C. Burgess and C. Folke (1994), *Paradise Lost? The Ecological Economics of Biodiversity*, London: Earthscan.
Bateman, I. and K. Willis (1995), *Valuing Environmental Preferences: Theory and Practice of the Contingent Valuation Method*, Oxford, UK: Oxford University Press.
Bedell, T.E. (1973), 'Botanical composition of a subclover-grass pasture as affected by single and dual grazing by cattle and sheep', *Agronomy Journal*, **65**, 502–4.
Bell, F.W. (1997), 'The economic value of saltwater marsh supporting marine recreational fishing in the Southeastern United States', *Ecological Economics*, **21**(3), 243–54.
Benson, J. and K. Willis (1991), 'The demand for forests for recreation', in *Forestry Expansion: A Study of Technical, Economic and Ecological Factors*, Edinburgh: Forestry Commission.
Bishop, R. (1978), 'Endangered species and uncertainty: the economics of a safe minimum standard', *American Journal of Agricultural Economics*, **60**(1), 10–18.
Bokdam, J. and M. Gleichman (2000), 'Effects of grazing by free ranging cattle on vegetation dynamics in a continental north-west European heathland', *Journal of Applied Ecology*, **37**(5), 415–31.
Borchert, M.I., F.W. Davis, J. Michaelsen and L.D. Oyler (1989), 'Interactions of factors affecting seedling recruitment of blue oak (*Quercus douglasii*) in California', *Ecology*, **70**(2), 389–404.
Bowes, M. and J. Krutilla (1989), *Multiple Use Management: The Economics of Public Forestlands*, Washington, DC: Resources for the Future.
Bowker, J. and J.R. Stoll (1988), 'Use of dichotomous choice nonmarket methods to value the whooping crane resource', *American Journal of Agricultural Economics*, **70**(2), 372–81.
Boyle, K.J. and R.C. Bishop (1987), 'Valuing wildlife in benefit–cost analyses: a case study involving endangered species', *Water Resources Research*, **23**(5), 943–50.
Bradshaw, R. and F.J.G. Mitchell (1999), 'The palaeoecological approach to reconstructing former grazing–vegetation interactions', *Forest Ecology and Management*, **120**(1), 3–12.
Brookshire, D., B. Ives and W. Schulze (1976), 'The valuation of aesthetic preferences', *Journal of Environmental Economics and Management*, **3**(4), 325–46.
Brown, G.M. (1990), 'Valuation of genetic resources', in G.H. Orians, G.M. Brown, W.E. Kunin and J.E. Swierbinski (eds), *The Preservation and Valuation of Biological Resources*, Seattle: University of Washington Press.

Brown, G.M. and W. Henry (1993), 'The economic value of elephants', in E.B. Barbier (ed.), *Economics and Ecology: New Frontiers and Sustainable Development*, London: Chapman and Hall.

Brown, G.M., D. Layton and J. Lazo (1994), 'Valuing habitat and endangered species', Discussion Paper No. 94-1 (January), Institute for Economic Research, University of Washington.

Callaway, R.M. (1992), 'Effect of shrubs on recruitment of *Quercus douglasii* and *Quercus lobata* in California', *Ecology*, **73**(6), 2118–28.

Campbell, R.W. and R.J. Sloan (1976), 'Influence of behavioural evolution on gypsy moth pupal survival in sparse populations', *Environmental Entomology*, **5**(6), 1211–17.

Carrascal, L.M., L.M. Bautista and E. Lázaro (1993), 'Geographical variation in the density of the white stork *Ciconia ciconia* in Spain: influence of habitat structure and climate', *Biological Conservation*, **65**(11), 83–7.

Casey, D. and D. Hein (1983), 'Effects of heavy browsing on a bird community in deciduous forest', *Journal of Wildlife Management*, **47**(3), 351–96.

Ciriacy-Wantrup, S. (1968), *Resource Conservation: Economics and Policies*, Berkeley: University of California.

Clawson, M. and J. Knetsch (1966), *Economics of Outdoor Recreation*, Baltimore: Johns Hopkins University Press.

Clout, M.N. and P.D. Gaze (1984), 'Effects of plantation forestry on birds in New Zealand', *Journal of Applied Ecology*, **21**(3), 795–815.

Coase, R. (1960), 'The problem of social cost', *Journal of Law and Economics*, **3**(1), 1–44.

Common, M. (1995), *Sustainability and Policy: Limits to Economics*, Cambridge, UK: Cambridge University Press.

Common, M. and C. Perrings (1992), 'Towards an ecological economics of sustainability', *Ecological Economics*, **6**(1), 7–34.

Conway, A., A. McLoughlin and W.E. Murphy (1972), 'Development of a cattle and sheep farm', *Animal Management Series No. 2*, An Foras Taluntais, Dublin.

Costanza, R., W.M. Kemp and W.R. Boynton (1993), 'Predictability, scale and biodiversity in coastal and estuarine systems', in C. Perrings, K.-G. Mäler, C. Folke, B.O. Jansson and C.S. Holling (eds), *Biodiversity Loss: Ecological and Economic Issues*, Cambridge, UK: Cambridge University Press.

Cousins, S.H. (1991), 'Species diversity measurement: choosing the right index', *Tree*, **6**(6), 190–2.

Crozier, R.H. (1992), 'Genetic diversity and the agony of choice', *Biological Conservation*, **61**(1), 11–15.

Cumming, D.H.M. (1993), 'Multispecies systems: progress, prospects and challenges in sustaining range animal production and biodiversity in east and southern Africa', *Proceedings of the 7th World Conference on Animal Production*, Edmonton, Alberta.

Daily, G.C. (ed.) (1997), *Nature's Services: Societal Dependence on Natural Ecosystems*, Washington, DC: Island Press.

Daily, G.C., P.R. Ehrlich and N.M. Haddad (1993), 'Double keystone bird in a keystone species complex', *Proceedings of the Natural Academy of Sciences*, **90**(2), 592–4.

Daily, G.C., T. Soderqvist, S. Aniyar, K. Arrow, P. Dasgupta, P.R. Ehrlich, C. Folke, A. Jansson, B.O. Jansson, N. Kautsky, S. Levin, J. Lubchenco, K.G. Maler, D. Simpson, D. Starrett, D. Tilman and B. Walker (2000), 'The value of nature and the nature of value', *Science*, **289**(5478), 395–6.

Daily, G.C., P.R. Ehrlich and G.A. Sanchez-Azofeifa (2001), 'Countryside biogeography: use of human dominated habitats by the avifauna of southern Costa Rica', *Ecological Applications*, **11**(1), 1–13.

Dambach, C.A. (1944), 'A ten-year ecological study of adjoining grazed and ungrazed woodlands in northeastern Ohio', *Ecological Monographs*, **14**(3), 69–105.

De Miguel, J.M., M.A. Rodriguez and A. Gómez-Sal (1997), 'Determination of animal behavior–environment relationships by correspondence analysis', *Journal of Range Management*, **50**(1), 85–93.

Debussche, M. and P. Isenmann (1994), 'Bird-dispersed seed rain and seedling establishment in patchy mediterranean vegetation', *Oikos*, **69**(33), 414–26.
DeGraaf, R.M., W.M. Healy and R.T. Brooks (1991), 'Effects of thinning and deer browsing on breeding birds in New England oak woodlands', *Forest Ecology and Management*, **41**(3–4), 179–191.
Di Falco, S. and J.-P. Chavas (2006), 'Crop biodiversity, farm productivity and the management of environmental risk in rainfed agriculture', *European Review of Agricultural Economics*, **33**(3), 289–314.
Di Falco, S. and C. Perrings (2003), 'Crop genetic diversity, productivity and stability of agroecosystems: a theoretical and empirical investigation', *Scottish Journal of Political Economy*, **50**(2), 207–16.
Di Falco, S. and C. Perrings (2005), 'Crop biodiversity, risk management and the implications of agricultural assistance', *Ecological Economics*, **55**(4), 459–66.
Diamond, P.A. and J.A. Hausman (1994), 'Contingent valuation: is some number better than no number?', *Journal of Economic Perspectives*, **8**(4), 19–44.
Díaz, M., E. González, R. Munoz-Pulido and M.A. Naveso (1996), 'Habitat selection patterns of common cranes *Grus grus* wintering in holm oak *Quercus ilex dehesas* of central Spain: effects of human management', *Biological Conservation*, **75**(22), 119–23.
Dobzhansky, T. (1970), *Genetics of the Evolutionary Process*, New York: Columbia University Press.
Edwards, K.J., G. Whittington and W. Ritchie (2005), 'The possible role of humans in the early stages of machair evolution: palaeoenvironmental investigations in the Outer Hebrides, Scotland', *Journal of Archaeological Science*, **32**(3), 435–49.
Ellis, G.M. and A.C. Fisher (1987), 'Valuing the environment as an input', *Journal of Environmental Management*, **85**(2), 149–56.
Elton, C. (1958), *The Ecology of Invasions of Animals and Plants*, London: Chapman and Hall.
Erdelen, M. (1984), 'Bird communities and vegetation structure: I. Correlations and comparisons of simple and diversity indices', *Oecologia*, **61**(2), 277–84.
Espelta, J.S., M. Riba and J. Retana (1995), 'Patterns of seedling recruitment in West-Mediterranean *Quercus ilex* forests influenced by canopy development', *Journal of Vegetation Science*, **6**(4), 465–72.
Faliñski, J.B. (1986), *Vegetation Dynamics in Temperate Lowland Forests*, Dordrecht: Junk Publishers.
FAO (2004), http:www.faostat.fao.org/corp/statistics/en, accessed 13 August 2009.
Faraco, A.M., F. Fernandez and J.M. Moreno (1993), 'Post-fire dynamics of pine woodlands and shrublands in the *Sierra de Gredos*', in L. Traubaud and R. Prodon (eds), *Fire in Mediterranean Ecosystems*, Ecosystems Research Report No. 5, pp. 101–13.
Fay, P.A. and D.C. Hartnett (1991), 'Constraints on growth and allocation patterns of *Silphium integrifolium* (*Asteraceae*) caused by a cynipid gall wasp', *Oecologia*, **88**(2), 243–50.
Ford, H.A. (1987), 'Bird communities on habitat islands in England', *Bird Study*, **34**(3), 205–18.
Fox, L.R. and P.A. Morrow (1992), 'Eucalypt responses to fertilization and reduced herbivory', *Oecologia*, **89**(2), 214–22.
French, D.D., G.R. Miller and R.P. Cummins (1997), 'Recent development of high-altitude *Pinus sylvestris* scrub in the Northern Cairngorm mountains, Scotland', *Biological Conservation*, **79**(2–3), 133–44.
Fuller, R.J., D. Hill and G.M. Tucker (1991), 'Feeding the birds down on the farm: perspectives from Britain', *Ambio*, **20**(6), 232–7.
Garrod, G.D. and K.G. Willis (1997), 'The non-use benefits of enhancing forest biodiversity: a contingent ranking study', *Ecological Economics*, **21**(1), 45–61.
González Bernáldez, F. (1991), 'Ecological consequences of the abandonment of traditional land use systems in central Spain', in J. Baudry and R.G.H. Bunce (eds), *Land Abandonment and its Role in Conservation*, Options Mediterranéennes Ser. A15, pp. 23–9.

Gourlay, D. (1996), 'Loch Lomond and Stewartry ESAs: a study of public perceptions of policy', unpublished PhD dissertation, School of Agriculture, University of Aberdeen.

Grainger, A. (1990), *The Threatening Desert – Controlling Desertification*, London: Earthscan.

Griffin, J.R. (1971), 'Oak regeneration in the upper Carmel valley, California', *Ecology*, **52**(5), 862–8.

Haigh, N. (1993), *The Precautionary Principle and British Environmental Policy*, London: Institute for European Environmental Policy.

Hanemann, W.M. (1994), 'Valuing the environment through contingent valuation', *Journal of Economic Perspectives*, **8**(4), 19–44.

Hanley, N. (1995), *Rural Amenities and Rural Development: Empirical Evidence*, Synthesis Report to the Rural Development Programme, OECD, Paris.

Hanley, N. and S. Craig (1991), 'Wilderness development decisions and the Krutilla–Fisher model: the case of Scotland's flow country', *Ecological Economics*, **4**(2), 145–64.

Hanley, N. and C. Spash (1993), *Cost–Benefit Analysis and the Environment*, Aldershot, UK and Brookfield, VT, USA: Edward Elgar.

Hanley, N., A. Munro, D. Jamieson and D. Ghosh (1991), 'Environmental economics and sustainable development in nature conservation', unpublished NCC Report.

Hanley, N., J.F. Shogren and B. White (1997), *Environmental Economics: In Theory and Practice*, London: Macmillan.

Hausman, J.A. (ed.) (1993), *Contingent Valuation: A Critical Assessment*, Contributions to Economic Analysis, Amsterdam: North Holland Publishers.

Hernandez, M.P.G. and F.J. Silva-Pando (1996), 'Grazing effects of ungulates in a Galician oak forest (Northwest Spain)', *Forest Ecology and Management*, **88**(1–2), 65–70.

Herrera, J. (1995), 'Acorn predation and seedling production in a low-density population of cork oak (*Quercus-suber L.*)', *Forest Ecology and Management*, **76**(1–3), 197–201.

Hester, A.J., F.J.G. Mitchell and K.J. Kirby (1996), 'Effects of season and intensity of sheep grazing on tree regeneration in a British upland woodland', *Forest Ecology and Management*, **88**(1–2), 99–106.

Hoffman, M.T. (1988), 'Rationale for Karoo grazing systems: criticisms and research implications', *South African Journal of Science*, **84**, 556–9.

Hoffman, M.T. and R.M. Cowling (1990), 'Vegetation change in the semi-arid eastern Karoo over the last 200 years: an expanding Karoo – fact or fiction?', *South African Journal of Science*, **86**, 286–94.

Holling, C.S. (1973), 'Resilience and stability of ecological systems', *Annual Review of Ecological Systems*, **4**(1), 1–24.

Holling, C.S. (ed.) (1978), *Adaptive Environmental Assessment and Management*, Chichester, UK: John Wiley and Sons.

Holling, C.S. (1986), 'Resilience of ecosystem: local surprise and global change', in E.C. Clark and R.E. Munn (eds), *Sustainable Development of the Biosphere*, Cambridge, UK: Cambridge University Press.

Holling, C.S. (1988), 'Temperate forest insect outbreaks, tropical deforestation and migratory birds', *Memoirs of the Entomological Society of Canada*, No. 146, 21–32.

Holling, C.S. (1992), 'Cross-scale morphology, geometry and dynamics of ecosystems', *Ecological Monographs*, **62**(4), 47–52.

Holling, C.S., D.W. Schindler, B.W. Walker and J. Roughgarden (1994), 'Biodiversity in the functioning of ecosystems: an ecological primer and synthesis', in C. Perrings, K.-G. Mäler, C. Folke, B.O. Jansson and C.S. Holling (eds), *Biodiversity Loss: Ecological and Economic Issues*, Cambridge, UK: Cambridge University Press.

Holmes, R.T., J.C. Schultz and P. Nothnagle (1979), 'Bird predation on forest insects: an exclosure experiment', *Science*, **206**(4417), 462–3.

Hulme, P.D., R.J. Pakeman, J.M. Torvell, J.M. Fisher and I.J. Gordon (1999), 'The effects of controlled sheep grazing on the dynamics of upland *Agrostis-Festuca* grassland', *Journal of Applied Ecology*, **36**(66), 886–900.

Hutchinson, G.E. (1961), 'The paradox of the plankton', *American Naturalist*, **95**(882), 137–47.

IUCN (1988), 'General Assembly of the International Union for Conservation of Nature and Natural Resources', Costa Rica.
Jackson, T. and P.J. Taylor (1992), 'The precautionary principle and the prevention of marine pollution', *Chemistry and Ecology*, **71**(1–4), 123–34.
James, F.C. and N.O. Wamer (1982), 'Relationships between temperate forest bird communities and vegetation structure', *Ecology*, **63**(1), 159–71.
Jane, G.T. (1994), 'The impact of browsing animals on the stand dynamics of monotypic mountain beech (*Nothofagus solandri*) forests in Canterbury, New Zealand', *Australia Journal of Botany*, **42**(22), 113–24.
Jeltsch, F., S.J. Milton, W.R.J. Dean and N. van Rooyen (1997), 'Analysing shrub encroachment in the southern Kalahari: a grid-based modelling approach', *Journal of Applied Ecology*, **34**(6), 1497–1509.
Joffre, R. and S. Rambal (1993), 'How tree cover influences the water balance of Mediterranean rangelands', *Ecology*, **74**(2), 570–82.
Joffre, R., J. Vacher, C. de los Llanos and G. Long (1988), 'The *dehesa*: an agrosilvopastoral system of the Mediterranean region with special reference to the Sierra Morena area of Spain', *Agroforestry Systems*, **6**(1), 71–96.
Jorritsma, I.T.M., A.F.M. van Hees and G.M.J. Mohren (1999), 'Forest development in relation to ungulate grazing: a modelling approach', *Forest Ecology and Management*, **120**(1–3), 23–4.
Just, R.E. and R.D. Pope (1978), 'Stochastic representation of production functions and econometric implications', *Journal of Econometrics*, **7**(1), 67–86.
Kienast, F., J. Fritschi, M. Bissegger and W. Abderhalden (1999), 'Modelling successional patterns of high-elevation forests under changing herbivore pressure – responses at the landscape level', *Forest Ecology and Management*, **120**(1–3), 35–46.
Kingery, J.L. and R.T. Graham (1991), 'The effect of cattle grazing on ponderosa pine regeneration', *Forestry Chronicle*, **67**(3), 245–8.
Köhler-Rollefson, I. (1993), 'Traditional pastoralists as guardians of biological diversity', *Indigenous Knowledge and Development Monitor*, **1**(3), 14–16.
Kruess, A. and T. Tscharntke (1994), 'Habitat fragmentation, species loss and biological control', *Science*, **264**(5165), 1581–84.
Krutilla, J.V. (1967), 'Conservation reconsidered', *American Economic Review*, **57**(4), 778–86.
Lindberg, K. (1991), *Policies for Maximizing Nature Tourism's Ecological and Economic Benefits*, Washington, DC: World Resources Institute.
Linder, P., B. Elfving and O. Zackrisson (1997), 'Stand structure and successional trends in virgin boreal forest reserves in Sweden', *Forest Ecology and Management*, **98**(1), 17–33.
Londo, G. (1990), 'Conservation and management of semi-natural grasslands in Northwestern Europe', in U. Bohn and R. Neuhäusl (eds), *Vegetation and Flora of Temperate Zones*, The Hague: Academic Publishing.
Long, J., E. Cromwell and K. Gold (2000), *On-farm Management of Crop Diversity: An Introductory Bibliography*, London: Overseas Development Institute for ITDG.
Lovejoy, T.E. (1988), 'Diverse considerations', in E.O. Wilson and F.M. Peter (eds), *Issues in Risk Assessment*, Washington, DC: National Academy Press.
Lynch, J.F. and D.F. Whigham (1984), 'Effects of forest fragmentation on breeding bird communiites in Maryland, USA', *Biological Conservation*, **28**(44), 287–324.
MacArthur, R.H. (1955), 'Fluctuations of animal populations and a measure of community stability', *Ecology*, **36**(3), 533–6.
MacArthur, R.H. and J.W. MacArthur (1961), 'On bird species diversity', *Ecology*, **42**(3), 594–8.
MacArthur, R.H. and E.O. Wilson (1967), *The Theory of Island Biogeography*, Princeton: Princeton University Press.
Main, A.F. (1982), 'Rare species: precious or dross?', in R.H. Groves and W.D.L. Ride (eds), *Species at Risk: Research in Australia*, Canberra: Australian Academy of Science.
Maquis, R.J. and C.J. Whelan (1994), 'Insectivorous birds increase growth of white oak through consumption of leaf-chewing insects', *Ecology*, **75**(7), 2007–14.

Marañon, M. (1986), 'Plant species richness and canopy effect in the savannah-like *dehesa* of south-west Spain', *Ecologia Mediterranea*, **12**(1), 131–41.

Marrs, R.H. and D. Welch (1991), *Moorland Wilderness: The Potential Effects of Removing Domestic Livestock, Particularly Sheep*, ITE Report to the Department of the Environment.

Mate, I.D. (1992), 'The theoretical development of machair in the Hebrides', *Scottish Geographical Magazine*, **108**(1), 35–8.

Mattson, W.J. and N.D. Addy (1975), 'Phytophagous insects as regulators of forest primary production', *Science*, **190**(4214), 515–22.

May, R.M. (1973), *Stability and Complexity in Model Ecosystems*, Princeton: Princeton University Press.

McNaughton, S.J. (1984), 'Grazing lawns: animals in herds, plant form and co-evolution', *American Naturalist*, **124**(6), 863–86.

McNaughton, S.J. (1985), 'Ecology of a grazing ecosystem: the Serengeti', *Ecological Monographs*, **55**(3), 259–94.

McNeeley, J.A. (1988), *Economics and Biological Diversity*, Gland, Switzerland: IUCN.

McNeeley, J.A. (1993), 'Economic incentives for conserving biodiversity: lessons for Africa', *Ambio*, **22**(2–3), 144–50.

Metrick, A. and M.L. Weitzman (1994), 'Patterns of behavior in biodiversity preservation', Policy Research Working Paper No. 1358, Environment, Infrastructure, and Agriculture Division, The World Bank, Washington, DC.

Mill, G.A., T.M. van Rensburg, S. Hynes and C. Dooley (2007), 'Preferences and multiple use forest management in Ireland: a contingent valuation approach', *Ecological Economics*, **60**(3), 642–53.

Miller, H.G. (1999), *Forest Policy: The International and British Dimensions*, Department of Forestry, University of Aberdeen.

Miller, K.R. (1996), 'Conserving biodiversity in managed landscapes', in R.C. Szaro and D.W. Johnston (eds), *Biodiversity in Managed Landscapes: Theory and Practice*, Oxford, UK: Oxford University Press.

Miller, T.E. (1982), 'Community diversity and interactions between the size and frequency of disturbance', *American Naturalist*, **120**(4), 533–42.

Mitchell, F.J.G. and E. Cole (1998), 'Reconstruction of long-term successional dynamics of temperate woodland in Bialowieza forest, Poland', *Journal of Ecology*, **86**(6), 1042–59.

Mitchell, R.C. and R.T. Carson (1989), *Using Surveys to Value Public Goods: The Contingent Valuation Method*, Washington, DC: Resources for the Future.

Montgomery, C.A., G.M. Brown and D.M. Adams (1994), 'The marginal cost of species preservation: the northern spotted owl', *Journal of Environmental Economics and Management*, **26**(2), 111–28.

Moore, A.D. (1990), 'The semi-Markov process: a useful tool in the analysis of vegetation dynamics for management', *Journal of Environmental Management*, **30**(2), 111–30.

Morgan, R.K. (1991), 'The role of protective understorey in the regeneration system of a heavily browsed woodland', *Vegetatio*, **92**(2), 119–32.

Morris, R.F., W.F. Cheshire, C.A. Miller and D.G. Mott (1958), 'The numerical response of avian and mammalian predators during a gradation of the spruce budworm', *Ecology*, **39**(3), 487–94.

Myers, N. (1992), 'Population/environment linkages: discontinuities ahead', *Ambio*, **21**(1), 116–18.

Naveh, Z. (1994), 'The role of fire and its management in the conservation of Mediterranean ecosystems and landscapes', in J.M. Moreno and W.C. Oechel (eds), *The Role of Fire in Mediterranean-type Ecosystems*, New York: Springer Verlag.

Naveh, Z. and R.H. Whittaker (1979), 'Structural and floristic diversity of shrublands and woodlands in northern Israel and other Mediterranean areas', *Vegetatio*, **41**(3), 171–90.

Nolan, T. and J. Connolly (1977), 'Mixed stocking by sheep and steers – a review', *Herbage Abstracts*, **47**(11), 367–79.

O'Neill, R.V., D.L. DeAngelis, J.B. Waide and T.F.H. Allen (1986), *A Hierarchical Concept of Ecosystems*, Princeton: Princeton University Press.

Opdam, P., G. Rijsdijk and F. Hustings (1985), 'Bird communities in small woods in an agricultural landscape: effects of area and isolation', *Biological Conservation*, **34**(4), 333–51.

Pearce, D.W. and R.K. Turner (1990), *Economics of Natural Resources and the Environment*, Hemel Hempstead, UK: Harvester Wheatsheaf.

Perman, R., Y. Ma, J. McGilvray and M. Common (2003), *Natural Resource and Environmental Economics*, New York: Pearson Education, Longman.

Peterken, G.F. and C.R. Tubbs (1965), 'Woodland regeneration in the New Forest, Hampshire, since 1650', *Journal of Applied Ecology*, **2**(1), 159–70.

Pezzey, J. (1989), 'Economic analysis of sustainable growth and sustainable development', Environment Department Working Paper No. 15, The World Bank, Washington, DC.

Pignatti, S. (1978), 'Evolutionary trends in Mediterranean flora and vegetation', *Vegetatio*, **37**(3), 175–85.

Pimm, S.L. (1984), 'The complexity and stability of ecosystems', *Nature*, **307**(26), 321–6.

Pimmental, D., U. Stachow, D.A. Takacs, H.W. Brubaker, A.R. Dumas, J.J. Meaney, J.A. O'Neill, D.E. Onsi and D.B. Corzilius (1992), 'Conserving biological diversity in agricultural/forestry systems', *Bioscience*, **42**(5), 354–62.

Polasky, S., E. Nelson, E. Lonsdorf, P. Fackler and A. Starfield (2005), 'Conserving species in a working landscape: land use with biological and economic objectives', *Ecological Applications*, **15**(4), 1387–1401.

Portney, P.R. (1994), 'The contingent valuation debate: why economists should care', *Journal of Economic Perspectives*, **8**(4), 3–18.

Power, M.E., D. Tilman, J.A. Estes, B.A. Menge, W.J. Bond, L.S. Mills, G. Daily, J.C. Castilla, J. Lubchenko and R.T. Paine (1996), 'Challenges in the quest for keystones', *BioScience*, **46**(8), 609–20.

Randall, A., B. Ives and C. Eastman (1974), 'Bidding games for the valuation of aesthetic environmental improvements', *Journal of Environmental Economics and Management*, **1**(2), 132–49.

Ratcliffe, P.R. (1993), *Biodiversity in Britain's Forests*, Edinburgh: The Forest Authority.

Raven, R.H., K. Beese, T. Eisner, N. Morin, T. Duncan, S.L.A. Hobbs, J. Hodges, W.F. Rall, R.R. Colwell and J. Swings (1992), 'Biotechnology and genetic resources: United States–Commission of the European Communities Workshop', 21–22 October, Airlie, Virginia, USA; Brussels, Belgium: Commission of the European Communities.

Reid, W.V. (1996), 'Beyond protected areas: changing perceptions of ecological management objectives', in R.C. Szaro and D.W. Johnston (eds), *Biodiversity in Managed Landscapes: Theory and Practice*, Oxford, UK: Oxford University Press.

Reinhammar, L.-G. (1995), 'Evidence for two distinct species of *Pseudorchis* (Orchidaceae) in Scandinavia', *Nordic Journal of Botany*, **15**(5), 469–81.

Relva, M.A. and T.T. Veblen (1998), 'Impacts of introduced large herbivores on *Austrocedrus chilensis* forests in northern Patagonia, Argentina', *Forest Ecology and Management*, **108**(1–2), 27–40.

Repetto, R. (1989), 'Economic incentives for sustainable production', in G. Schramme and J.J. Warford (eds), *Environmental Management and Economic Development*, Baltimore: Johns Hopkins for the World Bank.

Roland, M.A. and P.D. Taylor (1997), 'Insect parasitoid species respond to forest structure at different spatial scales', *Nature*, **386**(6636), 710–13.

Rosenzweig, M.L. (2003), *Win-Win Ecology: How the Earth's Species Can Survive in the Midst of Human Enterprise*, Oxford, UK: Oxford University Press.

Schemske, D.W. and N. Brokaw (1981), 'Treefalls and the distribution of understorey birds in a tropical forest', *Ecology*, **62**(4), 938–45.

Schindler, D.W. (1990), 'Experimental perturbations of whole lakes as tests of hypotheses concerning ecosystem structure and function', *Proceedings of 1987 Crafoord Symposium*, *Oikos*, **57**(1), 25–41.

Schindler, D.W., S.E.M. Kasian and R.H. Hesslein (1989), 'Biological impoverishment in lakes of the northeastern United States from acid rain', *Environmental Science and Technology*, **23**(5), 573–80.

Schmida, A. and M.V. Wilson (1985), 'Biological determinants of species diversity', *Journal of Biogeography*, **12**(4), 1–20.

Scholes, R.J. and B.H. Walker (1993), *An African Savanna: Synthesis of the Nylsvlei Study*, Cambridge, UK: Cambridge University Press.

Silva, J. (1987), 'Responses of savanna to stress and disturbance: species dynamics', in B.H. Walker (ed.), *Determinants of Tropical Savannas*, Oxford, UK: IRL Press.

Simberloff, D., B.J. Brown and S. Lowrie (1978), 'Isopod and insect root borers may benefit Florida mangroves', *Science*, **201**(4356), 630–32.

Simpson, I.A., A.H. Kirkpatrick, L. Scott, J.P. Gill, N. Hanley and A.J. MacDonald (1998), 'Application of a grazing model to predict heather moorland utilization and implications for nature conservation', *Journal of Environmental Management*, **54**(3), 215–31.

Smale, M., J. Hartell, P.W. Heisey and B. Senauer (1998), 'The contribution of genetic resources and diversity to wheat production in the Punjab of Pakistan', *American Journal of Agricultural Economics*, **80**(3), 482–93.

Solow, A., S. Polasky and J. Broadus (1993), 'On the measurement of biological diversity', *Journal of Environmental Economics and Management*, **24**(1), 60–68.

Soulé, M.E. and B.A. Wilcox (eds) (1980), *Conservation Biology, an Evolutionary Ecological Perspective*, Sunderland, MA: Sinauer Associates.

Staines, B.W. (1995), 'The impact of red deer on the regeneration of native pinewoods', in *Forestry Commission, Our Pinewood Heritage*, Conference proceedings, Inverness, 1994, Farnham, UK: Forestry Commission, RSPB, SNH.

Staines, B.W., R. Balharry and D. Welch (1995), 'The impact of red deer and their management on the natural heritage in the uplands', in D.B.A. Thompson, A.J. Hester and M.B. Usher (eds), *Heaths and Moorlands: Cultural Landscapes*, Edinburgh: Scottish Natural Heritage.

Sternberg, M., M. Gutman, A. Perevolotsky, E.D. Ungar and J. Kigel (2000), 'Vegetation response to grazing management in a Mediterranean herbaceous community: a functional group approach', *Journal of Applied Ecology*, **37**, 224–37.

Stoll, J.R. and L.A. Johnson (1984), 'Concepts of value, non-market valuation and the case of the whooping crane', *Transactions of the Forty Ninth American Wildlife and Natural Resources Conference*, **49**, 382–93.

Stowe, T.J. (1987), 'The management of sessile oakwoods for pied flycatchers', *RSPB Conservation Review*, **1**, 78–83.

Sumner, D.R., B. Doupnik and M.G. Boosalis (1981), 'Effects of reduced tillage and multiple cropping on plant diseases', *Annual Review of Phytopathology*, **19**(3755), 167–87.

Sun, D., G.R. Dickinson and A.L. Bragg (1997), 'Effect of cattle grazing and seedling size on the establishment of *Araucaria cunninghamii* in a silvo-pastoral system in Northeast Australia', *Journal of Environmental Management*, **49**(4), 435–44.

Sykes, J.M. and A.D. Horrill (1979), 'Regeneration of native pinewoods', *Institute of Terrestrial Ecology Annual Report 1978*, pp. 103–6.

Takekawa, J.Y. and E.O. Garton (1984), 'How much is an evening grosbeak worth?', *Journal of Forestry*, **82**(7), 426–8.

Takekawa, J.Y., E.O. Garton and L.A. Langelier (1982), 'Biological control of forest insect outbreaks: the use of avian predators', *Transactions of the 47th North American Wildlife Conference*, pp. 393–409.

Telleria, J.L. and L.M. Carrascal (1994), 'Weight–density relationships between and within bird communities: implications of niche space and vegetation structure', *The American Naturalist*, **143**, 1083–92.

Telleria, J.L. and T. Santos (1995), 'Effects of forest fragmentation on a guild of wintering passerines: the role of habitat selection', *Biological conservation*, **71**(1), 61–7.

Telleria, J.L., T. Santos, A. Sanchez and A. Galarza (1992), 'Habitat structure predicts bird diversity distribution in Iberian forests better than climate', *Bird Study*, **39**(1), 63–8.

Terborgh, J. (1985), 'Habitat selection in Amazonian birds', in M.L. Cody (ed.), *Habitat Selection in Birds*, Orlando: Academic Press.
Tilman, D. (1996), 'Biodiversity: population versus ecosystem stability', *Ecology*, **77**(3), 350–63.
Tilman, D. and J.A. Downing (1994), 'Biodiversity and stability in grasslands', *Nature*, **367**, 363–6.
Tilman, D., J. Knops, D. Wedin, P. Reich, M. Ritchie and E. Siemann (1997), 'The influence of functional diversity and composition on ecosystem processes', *Science*, **277**(5330), 1300–02.
Tinbergen, L. (1960), 'The natural control of insects in pinewoods. I. Factors influencing the intensity of predation by songbirds', *Archives Neerlandaises de Zoologie*, **13**(3), 265–43.
Tucker, G.M. and M.F. Heath (1994), *Birds in Europe: Their Conservation Status*, Cambridge, UK: BirdLife International.
Turner, R.K. (ed.) (1993), *Sustainable Environmental Economics and Management*, London: Belhaven Press.
UNEP (1995), *Global Biodiversity Assessment*, Cambridge, UK: Cambridge University Press.
Valderrábano, J. and L. Torrano (2000), 'The potential for using goats to control *Genista scorpius* shrubs in European black pine stands', *Forest Ecology and Management*, **126**(33), 377–83.
Van Dorp, D. and P.F.M. Opdam (1987), 'Effects of patch size, isolation and regional abundance on forest bird communities', *Landscape Ecology*, **1**(1), 59–73.
Vane-Wright, R.I., C.J. Humphries and P.H. Williams (1991), 'What to protect? Systematics and the agony of choice', *Biological Conservation*, **55**(3), 235–54.
Van Hees, A.F.M., A.T. Kuiters and P.A. Slim (1996), 'Growth and development of silver birch, pedunculate oak and beech as affected by deer browsing', *Forest Ecology and Management*, **88**(1–2), 55–63.
Van Rensburg, T.M., G.A. Mill, M. Common and J.C. Lovett (2002), 'Preferences and multiple use forest management', *Ecological Economics*, **43**(2–3), 231–44.
Walker, B.H. (1988), 'Autoecology, synecology, climate and livestock as agents of rangelands dynamics', *Australian Range Journal*, **10**(2), 69–75.
Walker, B.H. (1992), 'Biodiversity and ecological redundancy', *Conservation Biology*, **6**, 18–23.
Walker, B.H., D. Ludwig, C.S. Holling and R.M. Peterman (1981), 'Stability of semi-arid savanna grazing systems', *Ecology*, **69**(2), 473–498.
Walker T.S., R.P. Singh and N.S. Jodha (1983), *Dimensions of Farm-level Diversification in the Semi-arid Tropics of Rural South India*, Economic Program Progress Report, 51 ICRISAT, Patancheru, India.
WCED (1987), *Our Common Future*, The World Commission on Environment and Development, Oxford: Oxford University Press.
Weber, G.E., F. Jeltsch, N. van Rooyen and S.J. Milton (1998), 'Simulated long-term vegetation response to grazing heterogeneity in semi-arid rangelands', *Journal of Applied Ecology*, **35**(5), 687–99.
Weisbrod, B. (1964), 'Collective consumption services of individual consumption goods', *Quarterly Journal of Economics*, **78**(3), 471–7.
Weitzman, M.L. (1998), 'The Noah's ark problem', *Econometrica*, **66**(6), 1279–98.
Weslien, J. and M. Schroeder (1999), 'Population levels of bark beetles and associated insects in managed and unmanaged spruce stands', *Forest Ecology and Management*, **115**(2–3), 267–75.
Westoby, M., B.H. Walker and I. Noy-Meir (1989), 'Opportunistic management for rangelands not at equilibrium', *Journal of Range Management*, **42**(4), 266–74.
Whiteman, A. (1991), *A Cost–Benefit Analysis of Forest Replanting in East England*, Edinburgh: Forestry Commission Development Division.
Whittaker, R.H. (1977), 'Evolution of species diversity in land communities', *Evolutionary Biology*, **10**(1), 1–67.

Willis, K.G. (1991), 'The recreational value of the Forestry Commission estate in Great Britain: a Clawson–Knetsch travel cost analysis', *Scottish Journal of Political Economy*, **38**(1), 58–75.
Willis, K.G. and J.F. Benson (1988), 'Financial and social costs of management agreements for wildlife conservation in Britain', *Journal of Environmental Management*, **26**(1), 43–63.
Willson, M.F. (1974), 'Avian community organization and habitat structure', *Ecology*, **55**(5), 1017–29.
Wilson, E.O. (1985), 'The biological diversity crisis', *BioScience*, **35**(11), 700–706.
Wilson, E.O. (ed.) (1988), *Biodiversity*, Washington, DC: National Academy Press.
Wilson, W.L. and A.D. Johns (1982), 'Diversity and abundance of selected animal species in undisturbed forest, selectively logged forest and plantations in east Kalimantan, Indonesia', *Biological Conservation*, **24**(3), 205–18.
World Resources Institute (WRI) (2005), 'Earth trends: the environmental information portal', http://earthtrends.wri.org/, accessed 13 August 2009.
Zohary, M. (1973), *Geobotanical Foundations of the Middle East. Vol 1*, Stuttgart: Gustav Fisher Verlag.

5. How do institutions affect the management of environmental resources?
Bhim Adhikari

Introduction

In recent years, institutions and institutional arrangements have become central in the study of the success or failure of environmental resource management. The enforcement of institutions such as contracts[1] and property rights plays a crucial role in managing natural resources, affecting the equity and efficiency of resource management regimes. The centrality of contracts and property rights in understanding the diversity of institutional arrangements began in the research programme initiated by Coase (1937, 1960) and implemented by new institutional economics (NIE), and is now widespread throughout the economics literature (Menard, 2000). NIE focuses on explaining the determinants of institutions and their evolution over time and evaluates their impact on economic performance, efficiency and distribution (Nabil and Nugent, 1989). The theoretical framework of NIE has been used in many disciplines, ranging from sociology, anthropology and legal studies to applied fields such as policy analysis, planning and organizational development. It is recognized that 'institutions matter' and that the associated incentive structure in a particular form of institution substantially influences economic performance (Bardhan, 1999).

NIE provides a coherent theory of how contracts and collective action can be seen as the logical outcome of rational individuals' utility maximization and how institutional changes alter the pattern of individual choice and incentive directions. Lin and Nugent (1995) divide NIE into two broad categories, one studying the demand for institutions and one studying the supply of institutions. So the institutional analysis has adopted two inter-related approaches: (1) the transaction costs and information costs approach and (2) the collective action approach (Nabil and Nugent, 1989). The former is concerned with the role of transaction costs in economic organizations. The general hypothesis is that institutions are transaction-cost-minimizing arrangements, which may change and evolve with changes in the nature and sources of transaction costs and the means for minimizing them. The transaction costs approach is thus suitable for analysing the functional role of common property institutions (that is, demand for institutions). A second theme is the collective action approach, which focuses on property rights. The existence of property rights facilitates cooperation, which significantly reduces the costs of transaction. In this way, along with technology and other traditional constraints, institutional constraints enter into the decision process of individuals (Aredo, 1999). The collective action theory is the appropriate framework for understanding institutional change (that is, the supply of institutions). The law and economics branch of NIE studies the behaviour of rational individuals in a setting where the rule of law imposes prices on various non-market decisions (Posner, 1987). A closely related area is the economics of property rights and contracts (Alchian, 1965).

Institutions are an integral part of an economic system. They help to guide human behaviour and act as a key to economic performance. Institutions are the rules of the game, both in game theory settings and in the arena where individuals exchange goods and services. Institutions serve a number of important economic functions like facilitating market and non-market transactions, coordinating the formation of expectations, encouraging cooperation and reducing transaction costs. Apart from being behavioural constraints, institutions also serve as a kind of knowledge in a world of imperfect information and imperfectly rational individuals (Olsson, 1999). In the context of common pool resource (CPR) management,[2] institutions can be more specifically defined as a set of accepted social norms and rules for making decisions about resource use: these define who controls the resource, how conflicts are resolved and how the resource is managed and exploited (Richards, 1997). Institutions are often subdivided into formal and informal institutions. Formal institutions include laws, contracts, political systems, organization and markets; informal institutions are informal rules of conduct like norms, traditions, ethics, value systems, religion and ideologies. The former include informal cooperation and exchange, and moral or spiritual controls, often based on traditional heads, organized user groups, village committees and district councils. CPRs usually depend on a mix of both types of institutions. An institutional arrangement[3] is basically an arrangement between two different economic units that govern and shape the way in which each economic agent can negotiate and cooperate (Kherallah and Kirsten, 2001).

The role of property rights in managing local public goods[4] is examined in the NIE literature. Property rights are social institutions that define or delimit the range of privileges granted to individuals or groups to specific assets, such as parcels of land, water or forest (Libecap, 1989). According to Coase (1960), externalities that arise from the use of public goods can be internalized through bargaining and negotiation if property rights are well established and transaction costs are zero. That is, voluntary negotiation will lead to a fully efficient outcome providing that (1) rights are well defined, (2) transactions are costless and (3) there is no income effect (Baland and Platteau, 1996). The Coase theorem was later criticized for the following reasons. First, in a game involving more than two individuals (parties), the solution that leads to an efficient outcome through decentralized bargaining depends on the initial assignment of rights. Second, the negative effects of many resource and environmental problems occur in the distant future. The Coase theorem is not clear about how the concern of future generations is taken into account in the bargaining process. Despite some crucial difficulties, there is no doubt that the Coase theorem demonstrates the importance of property rights and transaction costs in order to internalize the externality associated with public good management. The basic motivations for contracting property rights are as follows. First, individuals do not have to consider the full social costs of their activities with respect to resource use in the absence of well-defined property rights. Second, resources will be undervalued because reallocation of the resources to higher-value uses becomes more costly and not feasible if property rights are absent. I shall clarify this argument in the subsequent discussion.

This chapter proceeds as follows. The second section focuses on the current debate concerning property rights transformation, placing the problem in a broader theoretical context. I then analyse the notion of common property resources and show how open access resources differ fundamentally from resources held under community ownership. The role of transaction costs is then examined before considering the impact of group heterogeneity.

Property rights transformation and resource management
Property rights are social institutions, including formal legal codes and informal social norms, which define and enforce the range of privileges granted to an individual or group of individuals with respect to specific economic resources (see Barzel, 1997). The assignment and enforcement of property rights is thus a legal mechanism that institutionalizes ownership of resources to a particular agent (however, property rights may also be informal institutions). According to the property rights school of thinking, the problem of over-exploitation and degradation of CPRs can be resolved only by creating and enforcing private property rights (Demsetz, 1967; Cheung, 1970; Johnson, 1972; Smith, 1981). Private property is considered to be the most efficient way to internalize the externalities generated from the over-exploitation of the commons. On this basis, restructuring property rights remains one of the top priorities in land reform and natural resource policy in many developing countries. A common practice is to clarify poorly defined property rights over these resources, especially

exerting state ownership over communally managed resources to maximize long-term economic rent.[5] However, experience from different parts of the world indicates that the efficiency of changing property rights cannot be guaranteed simply by enacting new legal rules, since the very notion of property rights is largely embedded in prevailing social customs (informal institutions) that guide individual behaviour in respect of environmental resource use. The nature of this transformation in property rights determines the parameters for the use of scarce resources and assigns incentive structures and costs to economic actors. Since property rights institutions range from formal arrangements, including constitutional provisions, statutes and judicial rulings, to informal conventions and customs regarding the allocation and use of property (Libecap, 1989), transforming property rights requires complementary changes in social norms. Together with new formal rules and other constraints, these changes in social norms redefine economic opportunity and redraw the rules of the game that govern the actions of economic actors, including business organizations and individual entrepreneurs, in their pursuit of economic gain (Wang, 2001). Correspondingly, the new structure of property rights does not necessarily ensure efficient utilization of environmental resources until economic actors adjust their expectation and behaviour in response to underlying changes in property rights.

Contemporary theoretical debate on property rights changes is broadly dominated by two schools of thoughts, the 'economic school' and the 'distribution school' (Coase, 1960; Demsetz, 1967; North 1981; Eggertsson, 1990). The basic theme of economic reasoning in the domain of institutional changes is that the propensity to achieve economic efficiency in the allocation of resources is the main force propelling such changes. Individuals will work to establish new rights or reallocate existing rights only if the benefits from doing so exceed the costs (Libecap, 1989). Libecap further argues that 'capturing a portion of the aggregate gains from mitigating common pool losses is a primary motivating force for an individual to bargain, to install or to modify property rights arrangements' (Libecap, 1989 p. 19). An institution established to achieve such an objective remains in place as long as it serves the purpose. Whenever the underlying economic relations change, the existing institution cannot serve the purpose efficiently and so provides the motivation to change the existing worn out institution. The basic theoretical thrust of the economic school is to view property rights evolution as a response to changes of relative prices, either direct or indirect, via the opening of new markets, population change, technological innovation and so forth (Wang, 2001). Changing economic conditions create new opportunities that could not be captured by the existing property rights structure, which is thus under pressure to change (Demsetz, 1967). Demsetz provided a classic example of economic reasoning driving property rights changes in relation to land following the opening of the fur trade. Since the existing property rights arrangements could not guarantee the maintenance of minimum stock required to conserve fur-bearing animals, private property rights over land resource emerged in response to this economic change to generate long-term commercial gain (Wang, 2001).

The distribution school, on the other hand, emphasizes the role of economic actors in changing the property rights structure of any particular resource system. This school argues that distributional issues can complicate the process of evolution of new systems of property rights since efficiency of improvement is blocked by distributional inequality. Built upon a perceived weakness of the economic school, the distribution school recognizes that distributional issues can complicate the process of property rights evolution (Wang, 2001). Shifts in the political influence of potential claimants can lead to the transformation of property rights. The agents who benefit from such a change have a larger stake in new systems of property rights and therefore have a greater incentive to act in favour of such institutional change. However, the same change has a corresponding disincentive effect on the other agents whose endowments have been reduced. Economic inequality among competing resource users may give rise to conflicts and can severely block any institutional response to CPR problems. Libecap (1989) argues that distributional conflicts inherent in any property rights arrangements, even those with important efficiency implications, hinder the adoption of new institutions. In criticizing the economic school as being naive and single-minded in emphasizing the 'demand side of institutional change (that is, gain from the change), the distribution school correctly points out factors such as distributional conflicts on the "supply side" can block property rights change' (Wang, 2001, p. 419).

Understanding institutions and institutional change is therefore a prerequisite for optimal policy prescription. Since an institution creates a social equilibrium, institutional change is thus a move from one equilibrium to another. The success of institutional change in this regard can be judged by whether this equilibrium is able to establish stable human expectation with respect to the use of scarce resources. What is critical in this regard is whether the very notion of different forms of institutions for CPRs is understood while creating and enforcing new institutions for sustainable resource management. From the perspective of institutional approaches to resource management, Libecap (1989) suggested that common pool losses are the primary motivation for contracting property rights (institutional change). Libecap further hypothesized that, all things being equal, the greater the losses of the common pool then the greater the size of the anticipated aggregate benefits of institutional change and the more likely new property rights will be sought and adopted. Nationalization of environmental resources in many developing countries seems to be influenced by this hypothesis as well as 'the tragedy of the commons' metaphor. In this metaphor, resources managed under common property rights and open accesses were frequently viewed as synonymous. It was thought that a resource held under a CPR regime is inherently inefficient since individuals do not get proper incentives to act in a socially efficient way. As a consequence scholars have long questioned the incentive for efficient use of CPRs under common property regimes (Gordon, 1954; Scott, 1955; Hardin, 1968) and solutions have been proposed, such as state control and management (Hardin, 1968) or privatization of the commons (Demsetz, 1964). In explaining the tragedy of the commons, the economic theory focuses less upon the weakness of the individuals and more upon the imperfections in the social systems to which they respond. In fact, economists are less convinced of the importance of human failing in determining social outcomes, simply because economists have believed in a special form of social synergism since the time of Adam Smith. However, more careful analysis of the foundation of common property regimes in developing countries has shown that local institutional arrangements, including customs and social conventions designed to induce cooperative solutions, can overcome the collective action problems and help achieve efficiency in the use of such resources (Gibbs and Bromley, 1989; Ostrom, 1990). In the following section, I present a critical analysis of the common property and open access resource systems and explain how common property institutions were misunderstood.

Property regimes: open access versus common property
The terms 'open access resource' and 'common property resource' are often confused and sometimes mistakenly used interchangeably. While the first term refers to resources over which no defined property rights exist, the second specifies the resources managed under community ownership. Hardin (1968) saw over-exploitation as an inevitable outcome of the use of common resources, even when the individuals sharing the benefits of such resources act in an economically rational way. Hardin was neither alone nor novel in making the argument about open access resources. An article on the open access problem can be traced back to the early twentieth century. A little-known article by Jens Warming written in 1911 (cited in Stevenson, 1991) is perhaps the earliest more or less accurate description of the open access problem. Two modern resource economists, Gordon (1954) and Scott (1955) provided the general conventional theory

of common resources in their articles on fishery economics. Building on Gordon and Scott, Smith (1968) provided a general theory of production and consumption of common property resources using an algebraic model. Smith applied his general model to fisheries, mining and timber resources and showed how differences in production externalities,[6] either from production scarcity or from crowding by procedures, lead to different outcomes. Most of these articles, however, advocated that a resource held under a common property regime is not efficient since individual resource users do not get proper incentives to act in a socially efficient way.

Hardin's arguments have been formalized later on in the form of the 'Prisoner's Dilemma Game' (Runge, 1981). The prisoner's dilemma game is conceptualized in a non-cooperative game theory in which all players are assumed to have complete information about the game to be played. Each player has a dominant strategy in the sense that the player is always better off choosing a dominant strategy – to defect – no matter what the other player does. The disturbing conclusion of prisoner's dilemma is that rational people cannot achieve collective outcomes. However, where the situation is a recurrent one, for example, as in the case of a repeated game, the logic changes (Axelrod, 1981). Free-riding in this circumstance remains a possibility but not an imperative as described in the 'Simple Prisoner's Dilemma Game' (Runge, 1984; Sugden, 1984). Olson (1965) also discusses the difficulty of getting individuals to act in such a way that it increases their joint welfare. Olson challenges proponents of 'group theory' who believe that individuals with common interests would voluntarily act to maximize the collective benefits. Olson argues that unless the number of individuals is quite small, or unless there is coercion or some other special device to make individuals act towards the overall common interest, rational, self-interested individuals will not act to achieve their own common interest (Olson, 1965). The tragedy of the commons, the prisoner's dilemma and the logic of collective action are closely related concepts in the models that have defined the accepted way of viewing many problems that individuals face when attempting to achieve collective benefits (Ostrom, 1998).

The tragedy of the commons metaphor confused common property regimes with open access regimes. It did not understand the very essence of community wisdom to act together and institute checks and balances, rules and sanctions, for sustainable management and utilization of environmental resources. In other words, followers of this concept after Hardin did not understand the fact that many resources used by rural communities are not open access but are managed under community ownership. Scholars of the commons argued that Hardin confused common property with open access, failing to distinguish between collective property rights

Table 5.1 Types of property rights regime relevant to common pool resource

Regime	
Open access (*res nullius*)	Free for all; no defined group of users or owners and benefit stream is available to anyone interested in entering into harvesting the resource, resource use rights are neither exclusive nor transferable
State property (*res publica*)	Ownership and management control held by the nation state, state has right to determine use/access rules (but in real world, use rights and access rights are often not enforceable without high cost)
Common property (*res communes*)	Use rights for the resource are controlled by an identifiable group and are not privately owned or managed by government; rules exist concerning who may use the resource, who is excluded from using the resource, and how the resource should be used; the co-owner has both rights and duties with respect to use rate and maintenance of the property owned
Private property	Owned by individual; individuals have right to undertake socially acceptable uses; claim rests with the individual or the corporation

Sources: Berkes and Taghi Farvar (1989); Bromley (1991).

and no property rights (Ciriacy-Wantrup and Bishop, 1975). One possible reason behind this misunderstanding is a lack of clarity between property and resource regime. While 'property' refers to rights and duties or social relationships between individuals in respect to a resource, 'regime' refers to the type of ownership under which these resources are managed. The economics literature describes four different types of property rights regimes relevant to CPRs. Table 5.1 presents the four basic types of property regimes, which will facilitate the subsequent discussion on open access and common property resource regimes.

Before turning to the essential difference between common property and open access, a brief statement about problems associated with common goods deserves mentioning. According to an economic definition, a common good is located between a 'pure private good' and 'pure public good'. The differences between these two goods reside in the concepts of jointness and exclusion. A pure private good is a good with the property of exclusivity, which means that the consumption of the good in question by one individual prevents its consumption by another. The owner of such a good can dispose of it as desired and can exclude other people from its use. On the other hand, a public good can be jointly consumed with others and is therefore non-exclusive. The rate of consumption of such goods is independent of the number of consumers and how the good is utilized (Oakerson, 1986). A common good has characteristics of both private (subtractable) and public goods (difficult to exclude) since it contains a certain degree of subtractability and excludability. This implies that the consumption of a common good by one individual will reduce another individual's ability to consume the same good. The rate of consumption of a common good, however, varies according to the number of users and the type of use. It appears that it is difficult to exclude anyone from utilizing the benefits generated by common goods, which characterizes the problem of common property as the provision of these benefits becomes problematic.

However, it is not impossible for co-owners to jointly benefit from common resources, as long as there are mechanisms to exclude non-owners from their use. Common property resources are partly joint and partly exclusive, which means that a type of property regime has to apply to exclude certain sections of society from entering into resource appropriation and exploitation. Communal or common property regimes try to achieve this exclusion by making the resource accessible only to an identifiable group or members, who devise certain mechanisms to regulate the pattern of resource use. Many policy prescriptions towards centralized management of CPRs in the developing world stem from a fundamental misunderstanding of possible resource regimes. Due to this confusion, common property carries the false and misplaced burden of 'inevitable' resource degradation that properly lies with situations of open access (Bromley, 1991). Policy-makers without complete knowledge of tenurial differences and systems of customary rights quite often advocate the argument that the efficient utilization of CPRs is only possible under state property regimes. Nationalization of Nepal's forests is an example of transformation of private/communal systems of property rights into state property (de facto open access), which actually upset centuries of social arrangements adopted by villagers to overcome resource degradation and make common property regimes viable. In the following section I will take up further discussions that underscore theoretical aspects of open access and common property regime systems.

Open access
Open access conditions occur when there is no exclusively defined access and use right to a particular resource system. Resources that fall into this category are subject to use by any person who has the capability and desire to harvest or extract the resource. This situation often results from the

absence, or the breakdown, of management and authority systems whose very purpose was to introduce and enforce a set of norms of behaviour among resource users with respect to that particular resource. Bromley (1991) considers the open access situation as a resource regime in which there are no property rights (*res nullius*). There is no defined group of users or owners and the benefit from the resource is available to anyone. All individuals have both privileges and no rights; no user has the right to preclude use by any other party (Bromley, 1991). Resources held under an open access situation are doomed to over-exploitation since each resource user places immediate self-interest above social interest. In the absence of informal/formal management institutions, there is a consensus that CPRs are typically subject to Hardin's tragedy of the commons. Since all members of a resource-using group are assumed to behave in a socially inefficient way, the carrying capacity of a resource system will eventually exceed its rate of regeneration. If property and management arrangements are not determined, and if investment is in the form of capital assets such as an improved tree species or range revegetation, the institutional vacuum of open access ensures that use rates will eventually deplete the asset (Bromley, 1991).

Under open access, a right of inclusion is granted to anyone who wants to use the resource and such property systems are likely to generate negative externalities (Baland and Platteau, 1996). Some CPRs are fugitive (that is, move from one property to another, such as water) and can be depleted, so are characterized by rivalry in exploitation (Stevenson, 1991). The rivalry in consumption of a CPR indicates that extraction by one user of the resources precludes another's possession. For example, if one user cuts a tree, another cannot use the same tree. However, for some ubiquitous CPRs, such as the air, the relevance of rivalry might not be applicable until they are consumed (or polluted) at a very high rate. Rivalry in extraction indicates that a CPR is not a pure public good at all potential use rates. As a community size grows, and therefore the number of rights holders increases, the higher use rate will ultimately exceed the resource's regenerative capacity. Depletability of a CPR indicates that, along with rivalry in consumption, resource supply might reduce to zero at some use rates. This is true both of strictly exhaustible resources, such as oil and minerals, and of renewable resources, such as fish and trees (Stevenson, 1991). Simple physical or economic exhaustion can reduce the former's supply to zero, and a sufficiently high use rate can extinguish the latter's capability to reproduce (Dasgupta and Heal, 1979).

The fugitive nature of some CPRs under open access means that they must be 'reduced to ownership by capture' (Ciriacy-Wantrup, 1952: noted by Stevenson, 1991). There are no formal property rights over the resource in an in situ condition. This means that a physical unit of the resource in its in situ or fugitive state cannot be associated with a particular owner unlike a private property regime where an in situ resource can be said to belong to a particular real or legal person. So in an open access condition anyone who possesses the social and physical capital to exploit the resource and the desire to enter into resource harvest can do so.

Another fundamental problem associated with an open access regime is that the over-exploitation of CPRs under open access will then result in symmetric or asymmetric negative externalities. A symmetric externality is present in an open access regime where each entrant to resource use imparts a negative externality to all other producers. The new entrant, in turn, simultaneously has a negative externality imposed on them by the others. The externality is reciprocal or symmetric. Common examples include fisheries, wildlife, open grazing land, ground water, unregulated woodland and forests, and common oil and gas pools. On the other hand, an asymmetric externality occurs when the production or consumption decisions of actors enter the production or utility functions of others while the recipients of the externality do not cause any reciprocal effects (Stevenson, 1991). A typical example includes the classic case of a smoking factory dirtying a nearby laundry's clothes. Most of the literature on resources held under open access has concentrated on the symmetric externality, however, the concept can easily be extended into both types of externality.

There is a vast literature developing on the issue of resource exploitation as characterized by an open access regime. In his analysis of deforestation in Nepal, Wallace (1981) reached several important conclusions in respect to exploitation of forest resources under an open access condition. First, resource users over-consume the resource in two different ways: they over-use the resource relative to other goods, and they over-use the resource this year relative to next year. Both kinds of over-exploitation occur because the cost of resource use to each individual is less than the cost to society. For each user, the cost of a particular product from the commons next year depends mostly on this year's consumption by other users. The benefit of the resource next year is assumed to be independent of this year's consumption. Unable to influence this year's consumption by other harvesters, each user will consume the resources until this year's marginal benefit equals this year's marginal cost. Second, with two substitutable resources, resource users may consume too much of one and too little of another, even if total use is efficient. This unbalanced consumption mix also results from the divergence of private and social costs of resource consumption. Resource users in open access regimes tend to react to average rather than marginal costs and the unbalanced consumption mix is the result of different average and marginal costs (Wallace, 1981).

Third, resource users may use inefficient methods to harvest a resource as in the case of unregulated common property resources. Competing resource users over-use capital-intensive harvesting methods in their attempts to out-harvest each other. In general, they do not consider the negative impact of their activity on the welfare of another. Fourth, resource users under open access regimes are not likely to invest in resource conservation, even if they know that investment improves the productivity of the commons. No one has any incentive to invest unless there is an assurance that other users will also invest in order to enhance the productivity of the commons. This situation is similar to under-investment in public goods such as clean air. Open access resource users invest in replenishing the forest only until the marginal costs equal a fraction of the marginal benefit. This under-investment results from a divergence between those who invest in the improvements and those who reap the benefits. The divergence results from a mismatch of the scale of some investments and the amount of potential individual benefit, and from a lack of incentive to invest in the resource for future benefits because of a competitive rush for the resource exploitation in the present (Stevenson, 1991).

People who may have strong incentives to invest in protecting trees for fodder or timber will have much less incentive to do so for public goods like clean air and soil conservation because they fear others will 'free-ride' on their efforts or because they can free-ride themselves. Those who do not invest because they see little direct benefit are still able to gain from the investments by others (Varughese, 1999). This inefficiently low investment by resource users imposes a welfare loss on the group of community members. Finally, users of CPRs in open access regimes under-invest in information about the resource since they have no incentive to acquire knowledge about planting methods, growth rates, or optimal cutting techniques and so on. A person who has perfect information about a CPR under an open access regime would not change his or her behaviour regarding the resource use, because other users would capture most of the benefits of any potential change. Thus, no one has any incentive to gather the information necessary to increase the productivity of commons held under open access regimes (Wallace, 1981), which seriously threaten the long-term sustainability of natural resources.

Common property
Common property refers to resources for which there are communal arrangements for the exclusion of non-owners and allocation among co-owners. Common property exists when a defined group of resource users holds property rights to natural resources and there is a restriction on the number of people who can reap a benefit from the commons. As I noted

earlier, this is a resource regime, where exclusion is difficult and joint use involves subtractability. On this front, they share the first attribute with pure public goods and the second attribute with pure private goods. Ciriacy-Wantrup and Bishop (1975) provide the first accurate description of the concept of common property by specifying two fundamental characteristics associated with common property regimes. First, the distribution of property rights among co-owners of common property is equal in terms of their rights to use the resource. Second, potential resource users who are not legitimate members of a resource-using group of co-equal owners are excluded. This implies that, within the community, rights to the resources are unlikely to be either exclusive or transferable; they are often rights of equal access and use (Feeny et al., 1998). Common property does not imply communal ownership, which has been described as a right that can be exercised by all members of the community (Demsetz, 1967), nor does it imply free access by all to the resource (North and Thomas, 1977). Bromley (1991) argues that a common property regime (*res communes*) represents private property for the group of co-owners (since all others are excluded from use and decision-making) and individuals have rights (and duties) with respect to the resource in question.

Common property is said to be similar to private property in a sense that there is exclusion of non-owners. The property-owning group varies in nature, size and internal structure across a broad spectrum, but it is a social unit with definite membership and boundaries, with certain common interests, with at least some interaction among members, with some common cultural norms and often their own endogenous authority system (Bromley, 1991). The management group (the 'owners') has the right to exclude non-members, and non-members have a duty to abide by exclusion. Individual members of the management group (the 'co-owners') have both rights and duties with respect to use rates and maintenance of the property owned (Bromley, 1991). The rights of the group may be legally recognized or in some cases they may be de facto rights. The fundamental difference between open access and common property is that in open access situations, every potential user has a privilege with respect to use of the resource since no one else has the legal ability to keep the person out. Therefore, an open access situation is one of mutual privilege and no rights (Bromley, 1991). In contrast, a common property regime indicates a resource system in which there are rules and regulations defining who is the legitimate user and who is not. Many misunderstandings found in the literature may be traced to the assumption that common property regime is the same as open access. Hardin's prediction of the inevitability of overexploitation follows this assumption (Feeny et al., 1998).

For almost two decades after Hardin's article, CPRs managed under communal property and open access were frequently viewed as synonymous. It was thought that common property was inherently unstable and pressures from free-riders were inevitable, leading natural resources to be degraded in the 'tragedy of the commons'. However, in many cases this is not true. Evidence suggests that successful exclusion under communal property regimes is the rule rather than the exception. More careful analysis of the foundation of common property regimes, combined with closer investigation of the management of collective goods in the developing world, suggests that common property regimes are not only viable, but in some circumstances are essential (Gibbs and Bromley, 1989). Even the common grazing lands in Hardin's classic 'tragedy of the commons' were well looked after for many centuries, before they declined for reasons unrelated to any inherent flaw in the commons system (Cox, 1985). The tragedy tends to be related to the breakdown of existing commons systems due to disruptions that have originated externally to the community (Berkes, 1989). Hardin's tragedy of the commons often results, not from any inherent failure of common property, but from institutional failure to control access to resources, and to make and enforce internal decisions for collective use. Institutional failure could be due to internal reasons, such as the inability of the users to manage themselves, or it could be due to external reasons, for example an incursion of outsiders (Dove, 1993; Berkes and Folke, 1998). Pressure on the resource because of human population growth, technological change, or economic change, including new market opportunities, may contribute to the breakdown of common property mechanisms for exclusion (Feeny et al., 1998). The social and political characteristics of the users of the resource and how they relate to the larger political system affects the ability of local groups to organize and manage communal property (Ostrom, 1987).

Stevenson (1991) noted seven different characteristics of common property resources, which he regards as necessary to manage common property successfully. The conditions are individually necessary because a resource managed under common property must meet all seven of them and the conditions are jointly sufficient for common property because all other resource use regimes (in particular, various forms of open access and private property) fail to meet at least one of the conditions (Stevenson, 1991). Based on the analysis of Ciriacy-Wantrup (1971) and Ciriacy-Wantrup and Bishop (1975) on the distinction between open access and common property resources, Stevenson (1991) described the following characteristics as a form of resource ownership under common property regimes:

1. The resource unit has bounds that are well defined by physical, biological and social parameters.

2. There is a well-delineated group of users, who are distinct from persons excluded from resource use.
3. Multiple included users participate in resource extraction.
4. Explicit or implicit well-understood rules exist among users regarding their rights and their duties to one another about resource extraction.
5. Users share joint, non-exclusive entitlement to the in situ or fugitive resource prior to its capture or use.
6. Users compete for the resource, and thereby impose negative externalities on one another.
7. A well-delineated group of rights holders exists, which may or may not coincide with the group of users.

The first point indicates that a resource held under common property must be defined either by biological, physical or social conventions, or a combination of these. At this point I again emphasize the clear distinction between the resource and common property regime. As was shown above, common property refers to management institutions that underline the relation between individuals, which differ from physical objects. In contrast, the resource is the physical or intangible asset that a group can own and manage as common property. Since the institution cannot exist without the resource that it controls, demarcation of resources, nonetheless, must be included in the definition of the social institution of common property (Stevenson, 1991).

The second point specifies that there are two groups in respect to the resource: included users and excluded persons. The first group consists of an identifiable and countable number of users. The second set of persons do not have the right to use the resource (Stevenson, 1991). This is in contrast to open access where everyone is a potential user. Third, more than two users utilize a common property resource. This clearly distinguishes common property from private property, where a single person is considered to be the sole owner. Fourth, the existence of rules (formal/informal) regarding resource appropriation and exploitation in order to guide the groups of resource users is the main characteristic that helps distinguish common property from an open access condition. This includes how rights are transferred, what financial obligation a user has to the group, what contribution he or she makes and how the rules themselves are changed. The rules may be formal and explicit or they may be informal and implicitly accepted (Stevenson, 1991).

The fifth point provides an essential difference between common and private property. It also highlights the relationship between common property and a public good. Unlike common property, in private property the in situ resource belongs to a particular owner. However, under a common property regime, the user may have a secure expectation of getting certain amounts of product from the commons, but not particular physical units. The joint, non-exclusive entitlement condition means that participants in common property arrangements have simultaneous, ex ante claims on any particular unit of the resource and it can be argued that an essential step in the use of common property resources (except the resources that have a pure public good characteristic) is that they be 'reduced' to sole ownership by capture (Stevenson, 1991). As indicated earlier, point five also provides some basis to distinguish between common property and public goods. First, some common property resources like national parks, reserves and so on have public good characteristics that do not exhibit rivalry at low or moderate levels of use. Reducing the resource to sole ownership through capture does not apply in the way that it does to resources that exhibit rivalry in extraction. Second, these resources exhibit joint, non-exclusive entitlement, because all participants who use the resource have an ex ante claim to benefits from the resource. For these reasons, reduction to sole ownership through capture is not a necessary condition for common property, but joint, non-exclusive entitlement is (Stevenson, 1991).

Point six indicates that, though multiple users compete for the resource appropriation and exploitation in common property, they undertake mutual capital investments in resource conservation. To better understand this idea, reconsider the problem of common goods. As in an open access condition, extraction by one user of the resource in a common property regime may generate negative externalities for other users. However, the difference lies in the extent to which externalities are generated. Point seven recognizes that the resource users and resource owners do not always coincide in a common property regime. Common property rights holders may rent their resource use rights to the actual users subject to the condition that the rights holders are a group of people who fulfil the institutional criteria of common property regimes (Stevenson, 1991). This is not meant to preclude the situation in which a government entity coordinates or imposes rules regarding resource extraction on users and rights holders. A common property resource, therefore, is a resource held by an identifiable community of interdependent users in which these users exclude outsiders while regulating use by members of the local community (Feeny et al., 1998).

Transaction costs and natural resource management

Insights from the economics literature
Transaction costs have been a subject of discussion in the literature on externalities over the past few decades. In his seminal article 'The

Nature of the Firm', Coase (1937) discussed why firms exist and sowed the seed of the concept of transaction costs in the economics literature. Coase underlines the important role of transaction costs in the organization of firms and other contracts. Coase's theory was later developed by many scholars (Coase, 1960; Cheung, 1969, 1983; Alchian and Demsetz, 1972; Williamson, 1985). Coase (1960) observed that identifying relevant parties, collecting information, undertaking negotiation, enforcing agreements and so on could be sufficiently costly to prevent many transactions from being achieved. When two or more parties agree to exchange or transfer goods or services, the transaction can be seen as a form of contract (Drennan, 2000). Arrow (1969) defines transaction costs as the costs of running the economic system. Transaction costs are the costs of arranging, monitoring, or enforcing agreements; the cost associated with all the exchanges that take place within an economy (Eggertsson, 1990; North, 1990). North (1997, p. 150) points out, 'the study of transaction costs, in addition to giving us insights into static economic analysis, also holds the key to unlocking the doors to an improved understanding of economic and societal performance through time'.

The importance of transaction costs can be studied through 'transaction cost economics (TCE)', which is a new type of economics first mentioned in Williamson (1991). According to him, TCE is focused on reducing maladaptation costs through ex ante selection of governance structure for the antecedent conditions. Its working hypothesis is that economic organization is really an effort to 'align transactions, which differ in their attributes, with governance structures, which differ in their costs and competencies, in a discriminating (mainly, transaction cost economizing) way' (Williamson, 1991). TCE tries to explain how trading partners choose, from the set of feasible institutional alternatives, the arrangements that offer protection for their relationship-specific investments at the lowest total cost (Shelanski and Klein, 1995). TCE maintains that in a complex world, contracts are typically incomplete because agents are boundedly rational, or because certain quantities or outcomes are unobservable. Due to this incompleteness, parties who invest in relationship-specific assets expose themselves to a risk: if circumstances change, their trading partners may try to expropriate the rents accruing to the specific assets. A variety of governance structures could be adopted in order to avoid this risk. Nevertheless, the appropriate one depends on the particular characteristics of the relationship. In this respect, TCE can be seen as the study of alternative institutions of governance.

Transaction costs are incurred in different stages of production and exchange. Dahlman (1979) separates transaction costs into three broad categories: (1) search and information costs, (2) bargaining and decision costs and (3) policing and enforcement costs. He documented that all of these costs represent resource losses due to lack of information. Transaction costs are a real and unavoidable aspect of any economic system. It is not even possible to eliminate transaction costs by prohibiting all transactions because such a decree would have to be deliberated and enforced and other institutions would emerge to replace banned markets. Libecap (1989) points out that having lower transaction costs is a necessary rather than a sufficient condition for adoption of an institutional arrangement. The inevitability of transaction costs means that any notion of Pareto optimality[7] is incomplete until transaction costs are incorporated (Griffin, 1991). It is therefore appropriate to examine transaction costs when evaluating the potential of new institutions as alternatives to the existing one.

Transactions differ in a variety of ways. Williamson (1985) identifies asset specificity, uncertainty and frequency as three major attributes of transactions that have direct cost implications. These three attributes of the transactions are also relevant to many natural resource systems. Asset specificity can be defined as investment in assets specifically relevant for a particular transaction. This is generally regarded as the most important of the three attributes because it creates market imperfections and allows asset owners to earn rents (Drennan, 2000). It can take the form of a physical asset related to location, an asset dedicated to a particular consumer, or a tangible asset that can be easily duplicated. It turns out to be the case that asset specificity creates information asymmetry, which permits owners to earn economic rents. Having private information makes it possible for people to behave opportunistically towards others who contract for their services. However, asset specificity also reduces the mobility of assets into alternative uses, with consequent market imperfections (Drennan, 2000). Most of the physical assets in the form of natural resources (forests, wildlife, fish and so on) are very site specific. Based upon this site specificity, species are considered endemic or endangered in their natural habitat. Transactions in natural resource systems also differ in the required specificity of human resources. It is useful to distinguish between idiosyncratic site-specific knowledge, especially the indigenous knowledge of a local community, and scientific knowledge required for planning and implementation of resource management activities (Birner and Wittmer, 2000).

The second attribute of transactions with costly implications for contracting is the extent of uncertainty surrounding the contract. Transactions surrounded by little or no uncertainty require minimal governance because ex ante and ex post information about the transaction is available to all concerned parties while uncertainty in defining and observing performance makes it difficult to place contracts on that performance (Drennan,

2000). However, natural resources are subjected to various types of threats according to their causes (human-made vs. natural) and effect (irreversible vs. reversible). The threat of irreversible loss of natural resources is considered to be a particular problem since it is associated with extinction of endemic and endangered species. Whether the threats to natural resources are caused by community members or outsiders also influences the appropriate governance mechanisms: using the instruments of social control by communities or user groups is usually more effective in dealing with threats caused by insiders than with those caused by outsiders (Meinzen-Dick and Knox, 1999).

The last attribute of transaction, frequency, also has implications for the extensiveness of governance investment. Frequent transactions justify an investment in governance because the cost can be amortized across many transactions and is therefore more likely to be recovered while infrequent transactions are more likely to be controlled by ad hoc arrangements put in place if and when the need arises (Drennan, 2000). In the case of natural resource management, one can distinguish frequent day-to-day management decisions and less frequent strategic decisions, for example, on the establishment of protected areas. Most activities carried out to implement management decisions are frequent, ranging from daily to seasonal (Birner and Wittmer, 2000), which implies high transaction costs of management.

Birner and Wittmer (2000) discussed 'public relevance' as an additional attribute of transaction, which influences the necessity of public sector involvement in commons management. Public sector engagement or intervention is required to deal with various aspects of externalities that arise while using common goods. From the outset, we have seen that negative externalities occur due to the public nature of environmental goods that affect the interest of society at large, especially concerning environmental protection. So, public relevance is an equally important attribute of transactions in the issue of intergenerational equity. For example, formulation of biodiversity policy needs to pay adequate attention to how the interests of future generations are taken into account while making current conservation/resource utilization decisions.

Fenoaltea (1984) considered 'care- or effort-intensity' as another key attribute of transaction. Effort-intensive transactions are easier to monitor and typically relate to production activities (for example, seedling production, plantation, felling trees and so on) in natural resource management, while care-intensive transactions are difficult to measure because they leave ample room for shirking and even sabotage (Fenoaltea, 1984), which are characteristic of protection activities. The monitoring problem of care-intensive activities is aggravated if the outcome of a transaction is difficult to measure. From the theoretical framework, it revealed that protecting the relationship of the contracting parties from opportunistic behaviour (as well as uncertainty, opportunism, asset specificity and frequency) justifies the existence of an appropriate governance structure. Therefore, governance structures offer various remedies for protection against the uncertainty and opportunism of economic relations. As pointed out by Pelletier-Fleury et al. (1997, p. 5), 'exchange requiring investment in specific assets, in a context of uncertainty and strong information asymmetries, justifies the recourse to vertical integration as opposed to market coordination, as this allows transaction costs to be contained'. Moreover, particular types of transaction are thus handled under particular governance structures depending on the relative costs of production and transaction (Pelletier-Fleury et al., 1997).

Transaction costs and collective action
At this point, it became clear that resource users enter into various kinds of explicit and implicit agreements (contracts) in order to initiate collective action or agree to exchange or transfer goods or services. The process of contracting involves two parties agreeing to ex ante and ex post situations and these contributions will be in the form of negotiation, monitoring and enforcement, where substantial amounts of costs are incurred. Ex ante costs involve the search costs of finding partners, setting up the agreement and costs incurred negotiating with the potential partners. Ex post costs are needed to ensure that exchange is carried out, or monitoring and if necessary enforcing its performance. Hanna (1995) describes four different resource management stages in which variable transaction costs are inevitable. These four different stages are: (1) the description of the resource context, (2) regulatory design, (3) implementation and (4) enforcement of agreed-upon rules. Description of the resource context includes a description of resource users, processors, markets and the analysis of associated socioeconomic characteristics. Designing the regulation requires information describing the resource context, which in turn depends on the quality of contextual information provided. Implementation of a regulation is a critical test of a regulation's fit to its contexts. Monitoring and enforcement of a regulation is the final area of transaction costs. Monitoring compliance with regulations will be excessively costly if monitoring systems are not designed to be consistent with resource dynamics or a user's operation (Hanna, 1995).

Despite the inevitability of transaction costs of resource management, at this stage I would like to point out that empirical discussion regarding transaction costs associated with different forms of governance structure is very sparse. Some scholars argued that the costs of privatization of communal resources, for example costs incurred for fencing, measurement, title insurance, record-keeping and so on are greater than those of collective management. However, when various hidden costs of resource management are incorporated into the analysis, a somewhat modified picture appears. A common property regime would not have the need for extensive records on boundaries and sales, but instead require meetings and discussions where the co-owners decided their strategies for the coming period (Bromley, 1991). This may constitute a significant portion of costs of resource management. In many field settings, efficient management of common property resources is often challenged by various sources of uncertainty that result in high levels of transactions costs. For individual resource users, the transaction costs of resource management are related to participation, opportunity cost of time involved in meetings, time required to acquire information and to communicate and direct monetary expenses for travel, communication and information. These costs are directly related to the effectiveness and efficiency of a collective action; and at the community level these costs may well be borne by poor community members (Adhikari et al., 2004; Adhikari and Lovett, 2006a; Meshack et al., 2006). It may therefore be that benefits from collective action are exceeded by transaction costs (Hanna, 1995). Thus, ignoring transaction costs in policy design and evaluation leads to the risk of producing suboptimal policy recommendations. A starting point for analysing this lies in an examination of what transactions occur, and what interactions are needed as the bare minimum for effective policy operation (Falconer, 2000).

High transaction costs, whether perceived or actual, related to entry into collective action may pose significant constraints on participation. The private transaction costs incurred by individuals may be so high that they might limit participation of poorer sections of society in some form of collective action. The existence of transaction costs may also have important distributional aspects. For example, sizeable fixed transaction costs related to participation (that is, mandatory start-up costs to be contributed by each participating community member at the initial stage of collective action) may discourage poor community members entering into community-based management regimes, as their income from the management of CPRs is relatively small. Room (1980) argues that economic studies of participatory forest management have been biased towards measuring benefits as opposed to costs, especially the likely major transaction costs of management for local forest users. In most of the community-based resource management systems with an initially degraded resource base, the costs associated with management are reported to be higher than the expected benefits. Nonetheless, in many economic models, physical input and property rights are taken as variables and transaction costs of resource management are seldom incorporated in the 'price' of resource consumption, though they can be a significant component of resource use. These costs vary with the attributes of the resource, the nature of use rights and the socioeconomic circumstances of the local communities. However, there has been very little attention paid to the socioeconomic significance of such costs. Transaction costs are largely invisible and there has been little attempt to quantify them (Falconer, 2000). While there has been a recent growth in theoretical research in this area, systematic empirical research is still lagging behind (Aggarwal, 2000).

Despite the importance of transaction costs in resource management, there are very few empirical estimates of transaction costs. Moreover, empirical estimates of transaction costs suffer from various uncertainties that hinder their quantification. Benham and Benham (2000) proposed four different arguments, which make empirical measurement of transaction costs difficult. First, there is lack of clarity on definitions of transaction costs in existing literature. These definitions offer powerful conceptual insights, but they have not been translated into widely accepted operational standards. Second, there is a problem in separating transaction costs from production costs since quite often they are jointly determined. This leads to formidable difficulties in estimating transaction costs separately. Third, if the cost of transacting is very high, many forms of transaction may not take place in any economic system of interest. Even if some specific form of transaction does occur, it may not take place in the form of monetary exchange. Therefore, of all potential transactions, very few may appear in the market place. To understand why a particular transaction is likely to be adopted by an individual requires knowledge of the opportunity costs faced by this individual. To understand the choices made, we may need to estimate the cost of transactions that did not actually occur (Benham and Benham, 2000). Fourth, since individuals and groups in any given society face different opportunity and thus transaction costs, many estimates may be needed. Other things being equal, an individual's political connections, ethnicity, endowments and other characteristics will affect the opportunity costs of a particular exchange (Benham and Benham, 2000). Though a few measurements of transaction costs exist in natural resource management, hardly any of these estimate their magnitude and variation. Transaction costs in the public sector can be estimated on the basis of information to be acquired from public agencies, which are typically part of the organization's budget. However, measuring the transaction costs of resource users is often difficult (Birner and Wittmer, 2000) since most of these costs are incurred indirectly, for example, time spent on meetings, carrying out protection work and other daily activities. In this connection, an adequate theory of natural resource use should incorporate the role of institutional structures associated with different management regimes and their associated transaction costs (Kant, 2000).

Transaction costs, governance structures and natural resource management
In the previous sections, I have discussed the underlying theory of transaction costs and its relevance to the natural resource sector. I now turn to appropriate governance structures for natural resource management from the perspective of transaction cost economics. Menard (2000, p. 240) points out that 'variability in contractual arrangements results from the necessity of setting up and monitoring transactions that have distinctive characteristics, particularly with regard to the degree of uncertainty of the environment in which transactions take place and to the degree of specificity of assets that they mobilize'. Indeed, the different conditions demand differing governance structures (enforcement procedures) in managing the CPRs if the so-called tragedy of the commons is to be avoided. Drawing heavily on Birner and Wittmer (2000), to which I return below, I shall offer some explanation of how to go about choosing efficient governance structures for CPR management. I have made clear that asset specificity and uncertainty provide the basis for selecting appropriate governance structures. In the view of Menard (2000, pp. 240–41), 'while specificity of assets plays a predominant role in the search for an efficient governance structure, uncertainty is the key factor in choosing the enforcement procedures of contractual arrangements'. Though I will not go further on uncertainty and asset specificity issues (see Williamson, 1985 for more detail), I will review the literature that uses the analytical apparatus of transaction cost economics, which helps to understand the type of appropriate governance structure for natural resource management.

Birner and Wittmer (2000) divided transaction costs of natural resource management into two different parts: transaction costs of decision-making (TC_D) and transaction costs to implement those decisions (TC_I). The decision costs (TC_D) are incurred during the process of acquiring various information prerequisite to making appropriate decisions and include costs of coordinating activities, such as resources spent on meetings, settling conflicts and costs arising due to delayed decisions. Transaction costs of implementation (TC_I) are influenced by the incentives of those carrying out implementation activities to comply with the management decision made, the presence of asymmetrical information and the measurability of the outcome, the possibilities to use social control for monitoring and the damage caused in the case of non-compliance (Birner and Wittmer, 2000). The economics literature suggests that the incentive for compliance depends on direct benefits from compliance as compared with defection. Moreover, the incentive for compliance is also influenced by the value that resource users put on management decisions and the degree of members' compliance with these obligations. The extent of group obligation depends positively upon (1) the cost of producing the joint goods and (2) the degree of dependency among members. The degree of members' compliance with these obligations depends positively upon the monitoring and sanctioning capabilities of the group. Since costs of monitoring and sanctioning can be high, the degree of cooperative success will depend on the mechanisms the group adopts to economize on such costs (Hechter, 1990 noted by Molinas, 1998). Encouraging user participation in decision-making processes, which possibly creates the legitimacy required for compliance, can minimize monitoring costs. User participation in common property regimes is the cornerstone to the sustainable management of the commons in most developing countries. Because resource dependency is very high, and the number of users is comparatively large, spatial extension and poor infrastructure make monitoring costly. Conservation measures are also care-intensive and resources are prone to irreversible damage due to a high degree of dependency (Birner and Wittmer, 2000).

I now turn to the graphical representation of the nature and extent of transaction costs as described in Birner and Wittmer (2000), who presented a very subtle analysis of transaction costs of different governance structures. Transaction costs of public sector governance (TC^p) and co-management (TC^{cc}) are presented in Figure 5.1a and b respectively. Transaction costs of decisions and implementation change accordingly

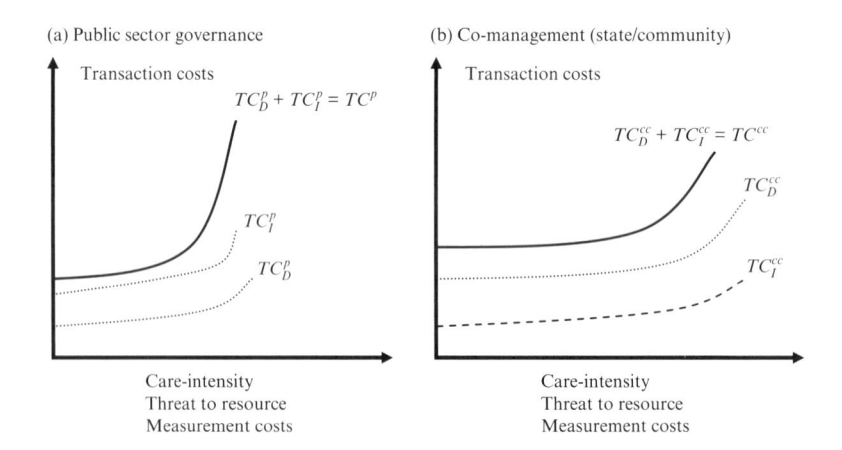

Source: Birner and Wittmer (2000).

Figure 5.1 Decision and implementation costs of public sector governance and co-management

in relation to the variable cost of resource management. Variable costs described here represent the care-intensity of the implementation activities, the costs of measuring the outcomes and/or the intensity of the threats to the natural resource in question. It is clear that the decision costs of state governance (TC_D^p) are lower than those of co-management (TC_D^{cc}) with lower variable costs (that is, little care-intensity, little threat to resources and measurement of outcome is possible). This implies that transaction costs are higher for a co-management strategy since a higher degree of effort is required for joint decision-making and coordination. The decision-making procedure becomes more complicated and costly with increasing group size since time and effort required appear to be rapidly increasing functions of the size of the group. This is, indeed, in line with Olson's hypothesis, which maintains that the smaller the group size, the greater the likelihood of collective action. When the consent of every party participating in collective action is required for agreement, these costs may be very high indeed (Baden, 1998). Nonetheless, with increasing variable cost, the decision costs under public sector governance are assumed to increase more rapidly than those under co-management, because the probability of making wrong decisions is assumed to be higher due to a lack of idiosyncratic knowledge that further increases the decision costs (Birner and Wittmer, 2000).

In contrast, transaction costs of implementation seem to be higher for public sector governance than those of co-management. This is due to the fact that community-based management systems have the potential for solving the commons dilemma by internalizing transaction costs of resource management within the community. The community has a kind of built-in incentive of social capital[8] and idiosyncratic knowledge that can be used to make resource-specific decisions or overcome factors such as social heterogeneity in the group (Adhikari and Lovett, 2006b). The community can overcome the problem caused by asymmetrical information through informal institutions and lower the opportunity costs of their time which considerably reduces the extent of transaction costs. The community also has at its disposal the requisite social coercive mechanisms to force compliance with expected harvest (Grima and Berkes, 1989). Though Figure 5.1 does not specify how the transaction costs are distributed between state government and community, co-management seems to be crucial to shift transaction costs from state agencies to local users (Birner and Wittmer, 2000).

Transaction costs of public sector governance and co-management are compared with those of hybrid private sector governance (TC^{hp}) in Figure 5.2. The state government's involvement in common property resource management seems to be essential for various types of conservation

144   *A handbook of environmental management*

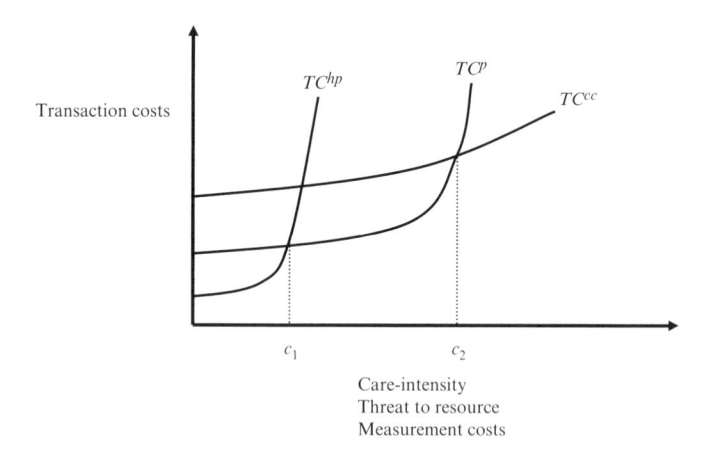

Source: Birner and Wittmer (2000).

Figure 5.2   Comparative efficiency of different governance structures

initiatives like biodiversity conservation. The graph tries to deal with 'public relevance' attributes of transaction costs as described in the previous section. This specifically represents the resulting efficient governance structure in relation to the variable costs. Hybrid private sector governance is assumed to have the lowest transaction costs for low values of variable costs because (1) decision costs are low, (2) there is no need to overcome collective action and coordination problems and (3) a private enterprise can typically use stronger incentives than the state (Birner and Wittmer, 2000). The figure demonstrates that contracted or regulated private sector governance is comparatively efficient if c is smaller than c_1, pure state governance is comparatively efficient for $c_1 < c < c_2$ and co-management (public and collective action sector) is comparatively efficient if $c > c_2$ (Birner and Wittmer, 2000).

At this particular juncture, it is relevant to show how efficiency of governance structure is influenced by the level of state capability and social capital. This relationship is presented in Figure 5.3. As I noted earlier, social capital is very productive since it makes possible the achievement of certain outcomes that would not be attainable otherwise. Peer monitoring can considerably reduce the cost of monitoring and this is one reason why local informal institutions of resource management are able to perform better than centralized government-mandated institutions (Aggarwal, 2000). The transaction costs of hybrid private governance and of state governance increase more rapidly with increasing values of variable cost if state capability is low. Experience from different parts of the world

Figure 5.3 Impact of state capability and social capital on governance structure

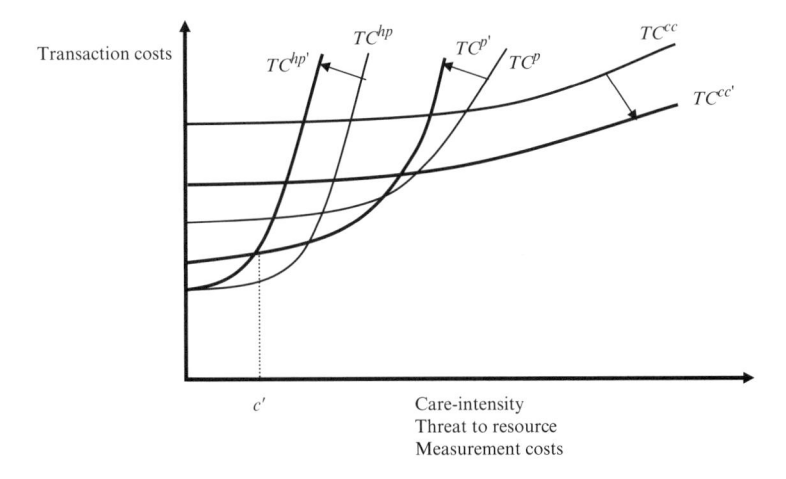

Source: Birner and Wittmer (2000).

suggests that sole state ownership over natural resources is less likely to be efficient in protecting these resources. For example, biodiversity losses may increase more rapidly due to difficulties faced by state governance in preventing over-exploitation of biological resources due to high enforcement and monitoring costs. As discussed earlier, increased social capital reduces the transaction costs of collective action through coordinating the resource users and implementation of instruments of social control. Figure 5.3 shows that hybrid private governance is comparatively efficient for $c > c'$ and co-management is the optimal choice for $c > c'$ (Birner and Wittmer, 2000).

Table 5.2 provides a summary of the institutional choice and governance structure for natural resource management under different state capability and social capital.

If both state capability and social capital are low, private sector management with state regulation is superior to community-based management. This significantly reduces the transaction costs of resource management. Participatory management is especially suited to cases where there is a high probability of strong community participation. User involvement in decision-making processes enhances compliance with resource use regulation. Moreover, community-based management is best suited where equity issues need to be taken into account. Co-management may be the optimal choice where governance structure places less demand on social

Table 5.2 Role of social capital and state capability in selecting appropriate institutions for commons management

State Capability	Social Capital	
	Low	High
Low	(Hybrid: private sector management under contract or regulation)	Community-based management
High	Public sector management	Hybrid: co-management involving government agencies and local communities

Source: Birner and Wittmer (2000).

capital (state assistance in maintaining this governance structure) and state capability (transparency may help to reduce corruption in the public sector) (Birner and Wittmer, 2000). Birner and Wittmer (2000) summarize the attributes of the most important governance structures for common property resource management in Table 5.3.

Because transaction costs are a function of the chosen governance structure and physical characteristics of the resource, ex ante evaluation of possible resource regime and associated transaction costs is thus an apparent challenge. Since measuring transaction costs is itself a very difficult task, estimating the expected gain from selecting a particular governance structure is also troublesome. Vätn (2001, p. 671) points out, 'when setting up a regime one must also evaluate or make qualified guesses about the future development of external costs, for example, costs due to technological change, new products, etc, implying changes in matter cycles and habitat interactions'. Theoretical insights presented in this section help to evaluate and design appropriate forms of governance structure for natural resource management in these circumstances.

Empirical studies
Having described transaction cost theory and the criteria in selecting appropriate governance structures I now provide a brief overview of empirical attempts that try to measure the transaction costs of collective action. The concern in this section is with three types of transaction costs: (1) information costs, (2) bargaining costs and (3) enforcement costs, as described in the previous section. Very few empirical studies have attempted to measure the transaction costs of community-based resource management. Davies and Richards (1999) extensively reviewed the literature on economic analysis of participatory forest management to

Table 5.3 Important attributes of different property rights' structure in natural resource management

Governance Structure Attributes	Centralized/ Concentrated Public Agency	Community Management	Private Sector Management	Regulated Community Management	Regulated Private Sector Management	Co-management (State/ Communities)
Instruments						
Incentive intensity	−	+	++	+	+	+
Administrative control	++	−	−	+	+	−/+*
Participation, social control	−	++	−	+	−	+
Demand on frame conditions						
State capability	++	−	−	+	+	+
Social capital	−	++	−	+	−	+
Accommodation of public interests						
Local	−	++	−	+	−	++
Higher level/Sovereign	++	−	−	++	++	+

Note: ++ = strong, + = semi-strong, − = weak, * depending on the type of co-management arrangements.

Source: After Birner and Wittmer (2000); based on Williamson (1999).

understand the incentives for different stakeholders. They found that these studies tend to be biased towards benefits in general as opposed to costs like transaction costs and ex ante studies for project preparation as opposed to ex post monitoring and impact analysis. Cheung (1987) suggests that the challenge to economists is to specify and identify what these transaction costs are and how they will vary under differing circumstances.

Despite this, few empirical studies on transaction costs exist. For example, Crocker (1971) conducted an empirical analysis of the role of transaction costs in natural resource transfer using the case of the impact of air pollution on agricultural land use. He concluded that transaction costs for affected farmland owners to bargain with polluters were very high. Leffler and Rucker (1990) applied transaction costs analysis to the structure of timber harvesting contracts and established empirical evidence for the influence of specific types of transaction costs on contractual provisions. Kumm and Drake (1998) estimated the private transaction costs incurred in relation to participation in the Swedish agri-environmental programme, using data from a survey of 90 randomly selected farmers. Transaction costs were defined to include expenditure for assistance from agriculture or conservation consultants, mapping, communication costs related to participation and time inputs (individuals' working hours). On average, consultants' costs accounted for approximately one-third of the total costs and the individuals' labour accounted for approximately two-thirds. Transaction costs, as a share of actual compensation received, are typically around 12 per cent and private transaction costs have risen over recent years.

Drake et al. (1999) carried out a pan-European survey to determine the cause of participation and non-participation of farmers in agri-environmental programmes in eight EU member states. They outlined a theoretical econometric participation function related to parameters like the direct resource costs of conservation (in terms of reduced production levels), the direct utility of the farmer derived from conservation activities, and the transaction costs borne by farmers in relation to participation. They found that transaction costs borne by farmers might pose constraints on participation. Information-gathering, for example, on the economics of organic conversion and how to change management practices, is often a key component of the transaction costs incurred by farmers wishing to participate in conservation schemes (Drake et al., 1999).

Aggarwal (2000) undertook a case study of group-owned wells in southern India in an attempt to understand the possibilities and limitations to cooperation in small groups by looking at the transaction costs associated with these activities. He observed that start-up costs of well construction in these villages included the costs associated with digging, pump installation, construction of waterways and electricity connection. The major operating expenses were the costs of electricity used in pumping, the cost of pump repair and maintenance and costs associated with periodic removal of silt from the well. The community also invests substantial amounts of money periodically for the expansion of the wells. He observed that costs of negotiating are likely to be higher in the case of expansion activities, particularly in groups where heterogeneity among members in terms of their endowments and needs is high. Because of the higher stakes involved in the case of expansion activities, a higher peer pressure is required to enforce collective arrangements (Aggarwal, 2000). Moreover, he also observed that instead of community-owned wells, most of the villagers prefer to have their own private well. Their reluctance for joint investment can be understood in terms of the high transaction costs of negotiating and enforcing a complete contract that outlines the obligations of each member under the different possible contingencies that can arise (Aggarwal, 2000).

Richards et al. (1999) undertook a participatory economic analysis of community forestry in Nepal in an attempt to improve donor and project understanding of the economic incentives faced by different stakeholders, and in particular local forest users. The case study particularly seeks to contribute to efforts to improve equity in the forest user groups and to understand the role of recurrent annual transaction costs faced by community members. Transaction costs were simply measured in terms of the opportunity costs of time spent in obligatory forest activities (planting, protection, weeding and so on) and in various community meetings. They found that transaction costs of resource management as a percentage of total costs were significantly higher for the less forest-dependent communities than for more dependent forest user groups. The study concluded that it is very important to include transaction costs in economic studies of community-based resource management (Richards et al., 1999).

Empirical estimation of transaction costs is also documented in Kuperan et al. (1998), who attempted to analyse a fisheries co-management system in San Salvador Island in the Philippines. The transaction costs of fisheries management were categorized into three major cost items: (1) information costs, (2) collective fisheries decision-making costs and (3) collective operational costs. They found that the difference in the total costs of fisheries management between centralized government management and co-management is significant; there is significant difference in the costs at the different stages of management. For the first two stages, which are the stages of initiating a new management regime and community education, the costs are higher for the co-management approach compared with the centralized approach. Nonetheless, transaction costs are lower in the latter stage for a co-management approach when monitoring and enforcement and conflict resolution become more important. This is because the costs of monitoring and enforcement are likely to be lower as resource users are more likely to comply with community-devised rules and regulations as opposed to regulation imposed by a centralized government authority. Since monitoring costs are the major transaction costs, and monitoring is undertaken by the community, there is an opportunity for these costs to decline over time as community acceptance of the rules and regulations for managing the common property increases with a greater moral obligation to obey those rules and regulations (Kuperan et al., 1998), that is, the costs are internalized. They further argue that monitoring activities emerge as the activity accounts for more than 50 per cent of the total costs of all the activities involved in co-management. It takes up the bulk of the time as it is a continuous day-to-day activity and it is a crucial activity for the maintenance of institutions (Kuperan et al., 1998).

The impact of group heterogeneity
I now turn to a slightly different issue, heterogeneity, which has direct policy implications for the emergence of local management institutions. One of the issues related to successful collective action is the contested role of group heterogeneity, which is assumed to have something to do with the way institutions evolve. Particularly important among these issues is the question of socioeconomic, ethnic and political heterogeneity and their effect on local-level collective action and resource management (Keohane and Ostrom, 1995; Baland and Platteau, 1996, 1998; Schlager and Blomquist, 1998; Uphoff, 1998; Bardhan and Dayton-Johnson, 2000; Velded, 2000; Varughese and Ostrom, 2001). The assumption is that socioeconomic differentiation and group heterogeneity make cooperative arrangements more difficult. On the one hand, there is widespread realization that productivity-enhancing CPR governance is difficult when appropriators are heterogeneous in regard to their socioeconomic endowments. A large and influential component of the literature claims that heterogeneity inhibits innovation of local management institutions since it creates distrust and suppresses the level of mutual understanding among community members. In such communities, the process of crafting rules and regulations with respect to how a resource should be managed can involve high levels of local dispute. Some economic and social science literature emphasizes that homogeneity or heterogeneity among agents in any society reflects the levels of trust, which influences the emergence of local management institutions through its impact on costs of transactions (Zak and Knack, 2001). Some other scholars, on the other hand, posit that heterogeneity is not necessarily bad for collective action. Baland and

Platteau (1996, p. 301), point out that, too often, 'heterogeneity is blamed as a matter of principle without enough effort being devoted to spelling out the precise conditions under which it undermines collective action'. Indeed, the role of group heterogeneity in the evolution of local management institutions is ambiguous. What deserves credit in this respect is to clarify the various forms of heterogeneity and the way they bear upon collective action dilemmas. In the subsequent discussion I will critically examine group heterogeneity and the success or failure of collective action drawing upon evidence from theoretical and empirical work from both the economic and social science literatures.

The source of heterogeneities are diverse, and include differences in asset holding, appropriation skills and access to technology, caste, gender, ethnicity, opportunity cost, political influence and other local differences, which might influence the equity of resource distribution and thus performance of collective action. Kant (2000, p. 288), points out, 'resource users will often have somewhat different preferences regarding resource management, or assign different priorities to the various objectives of resource management, either because of differing personal interest in the resource or differing degree of involvement in the social group'. The heterogeneity of individual interest with respect to how a resource is managed affects individual incentives and thus economic use of the local commons. As rightly pointed out by Seabright (1993), the degree of trust economic agents, participating in some form of collective action, have in one another serves a crucial role in common property regimes. He developed a model of 'habit forming' cooperation. The model revealed the fact that players' belief about each other's trustworthiness is confirmed and contributes to cooperative behaviour. Such a 'habit-forming' process is, however, unlikely to work in a community which starts with a high level of heterogeneity with respect to resource management preference (Kant, 1998). Moreover, economic inequality or socioeconomic heterogeneity among the members of a resource-using group might be associated with different degrees of access to, and control over, the local commons. Intra-group heterogeneity in terms of private payoffs, therefore, can often lead to conflicts of interest and thus hinder the emergence of egalitarian institutions and the performance of any cooperative arrangements. Ostrom (1990) argues that none of the successful CPR situations involves participants who vary greatly in regard to ownership of assets, skills, ethnicity, race, or other variables that could strongly divide a group of individuals.

Socioeconomic heterogeneity, relative income effect and the political framework in which policy decisions are taken determine the benefit-sharing arrangements in commons management. Income inequality produces the gap between poor individuals' ability to pay and willingness to pay for collective action. The level of the poorer users is so low that they might not be able to participate in community-based management systems, because they cannot meet the costs involved. For example, groups that depend heavily on daily wage labour sometimes find it difficult to contribute their share of resource management costs. As a result they are deemed ineligible for the benefits. McKean (2000) observes that common property regimes do not always serve to equalize income within the user group. Communities may vary enormously in how equally or unequally they distribute the products of the commons to eligible users. Decision-making rights tend to be egalitarian in the formal sense; however, richer households may actually have additional social influence on decisions due to the relative strength of their socioeconomic position. Greater homogeneity, on the other hand, promotes both equity and efficiency by facilitating the adoption of more coordination and cooperative arrangements at the local level. In a similar vein, Guggenheim and Spears (1991) argue that in light of likely scenarios of sociopolitical heterogeneity within spatially defined community groupings, participation in community decision-making can be skewed in favour of more powerful subgroups. Asymmetries among participants facing CPR provision and appropriation problems, therefore, can present substantial barriers to overcoming the disincentives associated with collective action (Ostrom and Gardner, 1993).

The sources of heterogeneities are diverse and there are several possible definitions of heterogeneity. A number of researchers seem to be much focused on economic inequality as a major source of heterogeneity, that is, inequality in wealth or income among the community members involved in collective action. However, Bardhan and Dayton-Johnson (2000) also noted other kinds of inequality within a resource-dependent community such as ethnic and social heterogeneity, and environmental or state variables like low levels of trust or social cohesion. They noted a U-shaped relationship between inequality and commons management: very high and very low levels of inequality are associated with better commons performance, while mid-range levels of inequality are associated with poor outcomes. Social heterogeneity increases the cost of negotiation and bargaining inherent in the process of crafting institutions. Economic inequality, combined with other constraints, severely limits the possible bargaining outcomes available to resource users. In this regard, interrelated 'commons outcomes' that might be affected by inequality include resource conservation, maintenance of infrastructure, the supply of local institutions, monitoring and enforcement of regulations and conflict resolution (Bardhan and Dayton-Johnson, 2000).

Economic opportunities available to resource users outside the local commons are considered to be a complicated form of heterogeneity that

Bardhan (1999) considers 'exit options'. He argues that the effect of exit options on conservation is complicated. Nonetheless, this depends in part on the nature of the relationship between wealth and the exit options. If a resource harvester's exit option is a concave function of wealth – meaning that the value of the outside option rises very quickly with wealth at low levels of wealth, but increases more slowly at higher wealth levels – then increases in inequality, starting from relatively equal wealth distributions, will reduce conservation and thus performance of cooperative arrangements. In that case, the relatively poorer harvesters will not be interested in optimal conservation; as her/his wealth declines, their gain from conservation falls off more rapidly than the gain from exercising her/his exit option. On the other hand, if the exit option is a convex function of the wealth level then increased inequality might either enhance or damage the prospects for conservation: the effect is intermediate. If resource users have relatively lucrative earnings outside the commons, this can affect their individual incentives. Wealthier households might have reduced incentives to participate in collective action, as their wealth endowments afford them attractive outside earning opportunities. For those enjoying such opportunities, the discounted value of their future income flows from the CPR may fall below that of alternative incomes available. On the other hand, users lacking such outside opportunities attach a higher value to the future state of the resources, which may enhance the likelihood of collective action.

In his simple numerical model, Kanbur (1992) investigated the role of group heterogeneity in the success or failure of common property resource management. He argues that cooperative arrangements are less likely when agents are highly heterogeneous. Moreover, existing arrangements are also likely to break down as a group becomes more heterogeneous. Drawing upon the conclusions from various case studies, he vividly presents the fact that greater equity (at least greater homogeneity) promotes greater efficiency by facilitating adoption of cooperative arrangements, in situations where externalities make non-cooperative outcomes inefficient. Tang (1992) presents a synthesis based on careful analysis of 47 case studies of community irrigation systems and found that a low variance of average family income among irrigators tends to be associated with a high degree of rule conformance and good maintenance: 72 per cent more of the cases with low-income variance are characterized by both a high degree of rule conformance and good maintenance than the cases with high income variance. This finding may throw new light on the fact that successful schemes are to be found in relatively homogeneous communities, and it is precisely in these communities that one is unlikely to find arrangements that favour one subgroup over another (Kanbur, 1992).

In their theoretical modelling, Baland and Platteau (1999) investigated the impact of inequality on the ability of resource users to undertake successful collective action with special reference to over-exploitation of common property resources. They argued that in the voluntary provision problem, inequality has an ambiguous impact on efficient outcomes, while in regulated settings it tends to reduce the acceptability of available regulatory schemes, which makes collective action more difficult. Changes that disequalize distribution of access rights through the definition or redefinition of property rights will have two different effects. First, the agents who benefit from such a change have a larger stake in the common property resources and therefore have a greater incentive to take part in conservation efforts. The same change has a corresponding disincentive effect on the other agents whose endowments have been reduced. Since the increasing inequality redistributes incentives in different directions, it may have ambiguous effects on the ability of users to initiate collective action. Under such circumstances, a homogeneous group may better succeed in designing and enforcing equitable conservation measures than a heterogeneous one.

In contrast to what I have discussed earlier, heterogeneity and income inequality in community-based property rights structures are also said to be conducive to the successful outcome of collective action. Olson (1965) hypothesized the possibility that groups where considerable heterogeneities exist may be privileged if those with the most economic interests and power were to initiate collective action to protect their own interests. Wealthy users may be more concerned with resource conservation since they can greatly internalize the benefits generated from the commons and therefore have more incentive to contribute to the local commons. Olson further argued that the greater the share in the benefits of a collective action for any single member, the greater the propensity of this 'large' member to bear the costs involved in commons management. In a related vein, Baland and Platteau (1999) revealed that increasing inequality could stimulate the incentive of the big users to voluntarily contribute and simultaneously encourage the small users to free-ride on the former's contribution. As a result, the net impact of inequality on collective action will hinge upon the respective strengths of these two opposite effects. However, the fact that rich users are more inclined to contribute does not imply that increased inequality favours collective action. Again, inequality has an ambiguous impact on collective action.

The Olson hypothesis is likely to make sense when management of a CPR involves important 'non-convexities' in its production function. Non-convexities indicate the large start-up costs, which are likely to be incurred in setting up a commons management regime. These costs might be labour costs of constructing irrigation infrastructure, fencing around the pastures and maintenance costs like fire prevention measures in community forestry and cleaning of canals in community-managed irrigation systems. Furthermore, start-up costs also involve organizational effort to collectively mobilize community members towards collective action. Bardhan and Dayton-Johnson (2000) considered that benefits from local-level collective action are a non-convex function of the effort provided to produce those benefits if there is a threshold level of aggregate effort that must be supplied before any benefits are realized. As effort increases beyond the threshold, however, benefits to the group begin to increase. It is impossible for poorer members to contribute and bear a significant part of these start-up costs in order to initiate community-based resource management. In this respect, wealthier and better-endowed users may be able to mobilize the capital necessary to support and materialize the collective action. Baland and Platteau (1997) also confirm the theoretical possibility of this Olson effect in the presence of non-convexities.

Theoretical literature on heterogeneity is also supplemented by empirical studies undertaken in different parts of the world, especially in developing countries. Drawing upon recent theoretical advances in the analysis of cooperation, Molinas (1998) undertook an econometric analysis of the determinants of successful collective action based on a survey of 104 peasant cooperative institutions in Paraguay. This study shows that controlling as much as possible for the specific characteristics of the community and peasant committee, the relationship between community inequality and cooperative performance is an inverted U. It was evident that community members with bigger land holdings are expected to benefit proportionally more from the committee's activities than members with smaller holdings. Since the provision of community infrastructure will increase the average price of land in the community, users who own more land naturally benefit more. In addition, the benefits accruing from joint commercialization of the outputs and collective buying of inputs are proportional to the scale of production, which in turn depends upon the size of the land holding. The study concludes that local-level cooperation is not monotonically related to either the degree of inequality of endowments within the community or the local intervention; rather, it is of an inverted U-shape form. Dayton-Johnson (2000) develops a simple model of individual households' incentives to provide collective maintenance effort in a communally owned irrigation system in Mexico. He found that economic inequality and social heterogeneity are consistently and significantly associated with lower levels of infrastructure maintenance and reduce the performance of collective action, while inequality in landholdings has a negative, though complicated, effect on maintenance. Similar to

Molinas' findings, his model shows that the effect of economic inequality is complicated and not monotonic (Dayton-Johnson, 2000).

Varughese and Ostrom (2001) investigate the role of heterogeneity in affecting the likelihood of collective action in a study of 18 forest user groups in Nepal. Their study focuses on some important community attributes that affect users' incentives to cooperate in resource management like the size of the community of resource users, locational differences among users with respect to forest areas, differences in forest users' income and presence or absence of economic/social/ethnic disparities and the availability of alternative forest resources. They, however, conclude that social and cultural heterogeneity does not have a determinant impact on the likelihood or success of collective action. Successful groups overcome stressful heterogeneities by crafting innovative institutional arrangements well matched to their local circumstances within which user groups decide how to organize themselves and which rules to adapt to allocate rights and duties as well as costs and benefits. They further argue that communities would collectively manage local-level natural resources where there are very substantive benefits to be obtained through collective action. Heterogeneity is not a variable with a uniform effect on the likelihood of organizing self-governing enterprises, as communities put effort into devising rules to cope with heterogeneities and they may be able to invest more heavily in finding effective rules that are considered fair, effective and efficient to most users (Varughese and Ostrom, 2001).

A study of CPR management in Fulanui village in Mali (Velded, 2000) demonstrated little direct relationship between the degree of heterogeneity and the success of collective action. It was found that homogeneity among elite groups enhanced capacity for collective action. However, when heterogeneity in economic interests between elite groups intensified and coincided with other dimensions of heterogeneity, such as heterogeneity in economic wealth, access to land and CPRs and agreement over authority of the leadership, the collective action became difficult to achieve. He further observed that heterogeneity of wealth and ethnicity do not prevent 'common interest' among elite and subordinate groups in collective action. Rather, Velded argues that the political elite of the community can assume leadership in local organization of CPR management and provide an authority structure for monitoring and rule enforcement. On the other hand, conflict was observed in a similar setting when economic interests differed with regard to use of CPRs. Moreover, in terms of resource use, access to local CPR grazing is more exclusively for the noble elite and influential section of the community. This study suggests that as long as there is reasonable homogeneity among elite groups, cooperative outcomes can be relatively persistent even among heterogeneous social groups (Velded, 2000). However, in line with Baland and Platteau (1995, 1996), Velded emphasized that there is a need to distinguish between various forms of heterogeneity in analysis and explanation of collective action, especially in relation to heterogeneity in endowments, entitlements (wealth) and economic interests.

Kant (2000) developed an optimal control model for a dynamic approach to forest regimes in developing economies integrating the natural system and socioeconomic factors of the resource users. The model demonstrates that the dynamics of the optimal forest regimes depend on the change in natural factors, socioeconomic factors (user group heterogeneity) and on the interaction between natural and socioeconomic factors. The model defines cultural, economic, ethical and other social differences as a basic level of heterogeneity. Due to this basic heterogeneity, members of the user groups may have diverse preferences for timber and non-timber products and hence prefer a different mix of products, which is considered as second-level heterogeneity. Preference for diversified forest products often leads to different preferences for resource management regimes, which can be further described as a third level of heterogeneity. Therefore, heterogeneity with respect to resource regimes is treated as a function of the product preference differences, which in turn is treated as a function of cultural, economic and social heterogeneity. Though the model is crucial to understanding optimal resource regimes, their linkages with socioeconomic factors (heterogeneity), and the dynamics of these optimal regimes and socioeconomic factors, it is explicit about how and to what extent intra-community heterogeneity or wealth endowment affects equitable benefit-sharing mechanisms among resource users.

While there has been a great deal of work on the management of local environmental resources in recent years, there has been surprising little work, either theoretical or empirical, on how inequality helps or hinders cooperative management of the local commons (Bardhan, 1999). Not much is understood on how socioeconomic status of resource users promotes or discourages participation in CPR management and consequently the equity of resource distribution. Nonetheless, evidence from South Asia suggests that the socioeconomic status of resource users may place stringent limits on the extent to which certain groups are able to participate and benefit from the management of CPRs. The landless, agropastoralists, and other politically and economically marginalized user groups may not be able to take advantage of incentives for tree growing (Guggenheim and Spears, 1991). Moreover, participatory forest management in South Asia illustrates the sharp equity problem elsewhere since social structure itself is for the most part inherently unequal. These resource management initiatives were supposed to help the poorest of the village population, lighten women's

158 *A handbook of environmental management*

workload and relieve family budgets. The income of many such projects, however, has been invested in other community development activities like village schools, rural roads and small-scale irrigation and very often productivity gained from such investments is captured by relatively well-off farmers. Entitlement to products of the commons varies to a surprising extent (McKean, 1992a). Hill and Shields (1998) observed that the community incentives in joint forest management (JFM) in India are not so clear-cut, however, the main losses in JFM are fuel wood head loaders who are often from the poorest subgroup within the village. Many studies have shown that the poor and the disadvantaged do not necessarily benefit from community-based forest management (Bhatia, 1999). While the aggregate gains from reducing common pool problems or promoting economic growth through the definition or redefinition of property rights are unlikely to be controversial, the distribution of wealth and political power inherent in the proposed rights structure will be a source of dispute (Libecap, 1989).

The recent literature on common property resource management, however, indicates that sustaining environmental resources is not dependent on a particular structure of property rights regime, but rather on a well-specified property rights regime and a congruence of that regime with its ecological and social context (Hanna and Munasinghe, 1995). Success of the property rights regime depends upon the congruence of ecosystem and governance boundaries, the specification and representation of interests, the matching of governance structure to ecosystem characteristics, the containment of transaction costs and the establishment of monitoring, enforcement and adoption processes at the appropriate scale (Eggertsson, 1990; Ostrom, 1990; Bromley, 1991 and Hanna et al., 1995). More importantly, the equity and distributional aspects of the CPR regime are considered to be one of the major determinants of long-term sustainability of CPR institutions. Since the nature of property rights regimes and the distribution of access to natural resources affect both the level of poverty and the quantity and quality of the environmental resource base in the long run (Dasgupta and Maler, 1991), property rights institutions will have to be more egalitarian in order to avoid unilateral appropriations of the commons. Empirical regularities that link inequality to better or worse outcomes of community-based resource management would be a basis for asset redistribution programmes including land reform and poverty alleviation that target communities based on the level of inequality (Bardhan and Dayton-Johnson, 2000).

Conclusions
This chapter explores the role of institutions in managing local-level environmental resources. The basic purpose was to explain the emergence of

local management institutions and their impact on successful management of common property resources at the local level. The analytical approach of this chapter builds on insights from 'new institutionalism' and theoretical and empirical literature from new institutional economics that underscore the role of formal and informal institutions for a solution to the CPR problem. Institutions are humanly devised constraints that shape human interaction and that ultimately affect the performance of the economy by their effects on the costs of exchange (North, 1990). Institutions serve a number of important economic functions like facilitating market and non-market transactions, coordinating the formation of expectation, encouraging cooperation and reducing transaction costs. From the economic perspective, it became clear that the problem of environmental degradation is to get the institution right. The underlying causes of resource degradation are found in those problems that systematically result in institutional failures. Since the basic theme of economic reasoning in the domain of institutional changes is to achieve equity and economic efficiency in the allocation of resources, an institution established to achieve such objectives remains in place as long as it serves the purpose. Whenever the underlying economic relations change, new institutions will emerge in response to underlying economic circumstances of the community. However, North (1990) pointed out that not all institutions are efficient and even inefficient institutions can persist for a long time. Institutions do not always decrease transaction costs but might actually, when they are inefficient, increase transaction costs (Olsson, 1999). Because transaction costs are a function of the chosen governance structure and physical characteristics of the resource, designing the governance structure depends on the attributes of transaction, that is, asset specificity, uncertainty and frequency of transaction associated with making various management decisions. Because of the public nature of environmental goods, public relevance also needs to be taken into account while selecting particular types of management regime.

The chapter has analysed different resource management regimes and addressed appropriate governance structures for environmental resource management from institutional perspectives. I revisited and analysed open access and common property resource systems and associated transaction costs. It is shown that open access exploitation results from the absence of well-defined property rights. Access to the resource system is unregulated and is free and open to everyone interested in resource appropriation and exploitation. Rent is completely dissipated at open access equilibrium. In another words, zero average net revenue characterizes the equilibrium of an open access regime. There is over-use resulting from resource users ignoring the effects of their consumption on the costs faced by other users.

Similarly, there is over-use resulting from users ignoring the effect of their consumption this year on the costs they will face next year. On the supply side, CPRs held under open access regimes are like public goods. Individuals cannot capture the benefits of their investments in these resources, and as a result investment is inefficiently low, resources are misallocated, and there is under-investment in information (Wallace, 1981). Governance of natural resources can thus be conceptualized as a collective endeavour of individuals organizing for the provision of, and appropriation from, resources that have public good characteristics. Since individual interests are unlikely to lead to sustainable management of CPRs in an open access regime, the design of governance for resource management has to include some elements of support from government to modify the incentives for individual resource users (Varughese, 1999).

Analysis of CPR management under common property regimes indicates that resource management under community ownership is not operating in a vacuum. Instead, common property is a form of resource management regime in which a well-delineated group of competing users participates in extraction or use of a jointly held, fugitive resource according to explicitly or implicitly understood rules about who may take how much of the resource (Stevenson, 1991). The confusion in the conventional literature over the tragedy of the commons arises from a failure to understand the concept of property, and therefore to fail to understand common property regimes (Bromley, 1991). The economics literature also discusses the problem associated with common property, which results from some type of adverse interaction among resource users and unrestricted access to the resource system by all who care to use it. As discussed earlier, common property resources share two important characteristics. First, exclusion of resource users from these resources is difficult. Second, the use of resources by one person subtracts from the welfare of other users. Natural products like trees, water and wildlife are subtractable, and in most cases, exclusion will be problematic and costly. If one individual uses more, then less remains for another. These resources are therefore potentially subject to depletion or degradation, that is, use that is pushed beyond the limits of sustainable yields. So the problem raised by common property is usually represented in the formal framework of the 'prisoner's dilemma'.

This chapter also attends to the transaction costs associated with community-based resource management. An institution's primary purpose is to reduce transaction costs and thereby enhance economic performance. Neo-classical economic analysis tends to be preoccupied with production costs, largely ignoring the transaction costs associated with production and economic exchange. Despite the importance of transaction costs in functioning of an economic system, there are very few empirical estimates of transaction costs and very few comparative estimates. The analysis emphasized that transaction costs of community-based resource management can be a significant part of resource management costs, which is generally ignored in economic analysis of participatory forest management.

The chapter also discussed the role of heterogeneity in the performance of collective action in general and distributional implications of such regimes in particular. On the one hand, it turned out to be the case that wealth might affect private benefits from the commons indirectly, through social relations between one group and another, when users attempt to deal with collective action dilemmas associated with joint use of common resources. However, there is still disagreement among social scientists and economists regarding the effect of heterogeneity on the capabilities of social groups to undertake successful collective action. In fact, a general finding from studies on the management of common property systems is that entitlements to products of the commons are almost always based on private holdings (Dasgupta, 1999). McKean (1992b) reaches a similar conclusion that the distribution of benefits in collective action reflects inequalities in private wealth. Drawing upon several theoretical and empirical studies in Asia and Africa, Baland and Platteau (1996) provide several examples of homogeneity/heterogeneity and claim that equitable access to resources is possible even in heterogeneous village society in Japan (McKean, 1992a, 1992b) and an ethnically diverse village in India (Bardhan, 1993a, 1993b), as well as most pastoral societies of Africa. In most of these villages, scholars observed a kind of equitable access to locally managed CPRs with local rules, norms and customary law. It seems that the impact of heterogeneity is ambiguous. Analysis of heterogeneity and distributive issues of collective action have important policy implications for community-based resource management initiatives, especially those aimed at shaping resource use decisions by households and poverty reduction through better management of local commons.

Notes

1. I borrow the idea of Libecap (1989) to conceptualize the process of defining or changing property rights in terms of contracting. Contracting includes both private bargaining to assign or adjust informal ownership arrangements and lobbying efforts among private claimants, politicians and bureaucrats to define, administer and modify more formal property institutions (Libecap, 1989).
2. Common pool resources refer to the natural resource that is available in the commons, whereas common property regime refers to the institutional property rights arrangement by which the resource is managed. The two are often confused as the same acronym, CPR, is used for both.
3. Williamson (2000) notes that NIE operates at two different levels: macro and micro. The macro level deals with the institutional environment, or rules of the game, which affect the behaviour and performance of economic actors in which organizational forms and transactions are embedded. Williamson discusses it as the set of fundamental political,

social and legal ground rules that establish the basis for production, exchange and distribution. The micro-level analysis, on the other hand, also known as the institutional arrangements, deals with the institution of governance. These refer to the modes of managing transactions including market, quasi-market and hierarchical modes of contracting (Williamson, 2000).
4. Public goods are resources that can be accessed by anyone and the consumption of the resource by one person does not affect the consumption by anyone else. Enjoyment of a sunrise is a classic example. However, many natural resource public goods can be readily over-exploited, which is why property rights need to be defined and consumption regulated.
5. Economic rent is the difference between the actual cost of bringing something into production and the amount that is paid for the product. Economic rents are low if there is a lot of competition and tend to be higher when there are monopolies.
6. Production externalities are costs of production that are incurred by someone other than the producer. For example, a logging company exploiting old growth forest will pay stumpage costs on each tree extracted to the owner of the land, but does not compensate society for the loss of a valuable habitat.
7. Pareto optimality is when goods are exchanged until no one can be made better off without someone being made worse off. Under these conditions the market is efficient in that it is responsive to supply and demand.
8. Social capital comprises features of social organization such as networks, norms and trust that can improve the efficiency of a community by facilitating cooperation and coordination of relations between actors and among actors (see Coleman, 1990; Putnam, 1993). Social capital is also productive since it makes possible the achievement of certain outcomes that would not be attainable otherwise (Coleman, 1990; Molians, 1998) and it also reduces the transaction costs of collective action.

References

Adhikari, B. and J.C. Lovett (2006a), 'Transaction Costs and Community-based Natural Resource Management in Nepal', *Journal of Environmental Management*, **78**(1), 5–15.

Adhikari, B. and J.C. Lovett (2006b), 'Institutions and Collective Action: Does Heterogeneity Matter in Community-based Resource Management?', *Journal of Development Studies*, **42**(3), 426–45.

Adhikari, B., S. Di Falco and J.C. Lovett (2004), 'Household Characteristics and Forest Dependency: Evidence from Common Property Forest Management in Nepal', *Ecological Economics*, **48**(2), 245–57.

Aggarwal, R.M. (2000), 'Possibilities and Limitations to Cooperation in Small Groups: The Case of Group-owned Wells in Southern India', *World Development*, **28**(8), 1481–97.

Alchian, A.A. (1965), 'Some Economics of Property', *Il Politico*, **30**(4), 816–29.

Alchian, A.A. and H. Demsetz (1972), 'Production, Information Costs, and Economic Organization', *American Economic Review*, **62**(5), 777–95.

Aredo, D. (1999), 'Institution and Development: A Note on Conceptual Framework', *Economic Focus*, **2**(2).

Arrow, K.J. (1969), 'The Organization of Economic Activity: Issues Pertinent to the Choice of Market versus Non-market Allocation', in *The Analysis and Evaluation of Public Expenditure: The PBB-System*, Joint Economic Committee, 91st Congress, 1st Session, Vol 1, Washington, DC.

Axelrod, R. (1981), 'The Emergence of Co-operation among Egoists', *American Political Science Review*, **75**(2), 306–18.

Baden, J.A. (1998), 'A New Primer for the Management of Common-pool Resources and Public Goods', in J.A. Baden and D. Noonan (eds), *Managing the Commons*, Bloomington, IN: Indiana University Press.

Baland, J. and J. Platteau (1995), 'Does Heterogeneity Hinder Collective Action?', Working

Paper, Centre for Research in Economic Development (CRED), University of Namur, Belgium.
Baland, J. and J. Platteau (1996), *Halting Degradation of Natural Resources: Is There a Role for Rural Communities?*, Oxford, UK: FAO and Clarendon Press.
Baland, J. and J.P. Platteau (1997), 'Wealth Inequality and Efficiency in the Commons, Part I: The Unregulated Case', *Oxford Economics Papers*, 49(4), 451–82.
Baland, J. and J.P. Platteau (1998), 'Division of the Commons: A Partial Assessment of the New Institutional Economics of Land Rights', *American Journal of Agricultural Economics*, **80**(3), 644–50.
Baland, J. and J.P. Platteau (1999), 'The Ambiguous Impact of Inequality on Local Resource Management', *World Development*, **27**(5), 773–88.
Bardhan, P. (1993a), 'Symposium on Management of Local Commons', *Journal of Economic Perspectives*, **7**(4), 87–92.
Bardhan, P. (1993b), 'Analytics of the Institutions of Informal Cooperation in Rural Development', *World Development*, **21**(4), 633–9.
Bardhan, P. (1999), 'Distributive Conflicts, Collective Action, and Institutional Economics', Department of Economics, University of California at Berkeley, mimeo.
Bardhan, P. and J. Dayton-Johnson (2000), 'Heterogeneity and Commons Management', Department of Economics, University of California, Berkeley, mimeo.
Barzel, Y. (1997), *Economic Analysis of Property Rights*, New York: Cambridge University Press.
Benham, A. and L. Benham (2000), 'Measuring the Transaction Costs', in C. Menard (ed.), *Institutions, Contracts and Organizations: Perspectives from New Institutional Economics*, Cheltenham, UK and Northampton, MA, USA: Edward Elgar, pp. 367–75.
Berkes, F. (ed.) (1989), *Common Property Resources – Ecology and Community Based Sustainable Development*, London: Belhaven Press.
Berkes, F. and C. Folke (1998), 'Linking Social and Ecological Systems for Resilience and Sustainability', in F. Berkes and C. Folke (eds), *Linking Social and Ecological Systems: Management Practices and Social Mechanisms for Building Resilience*, Cambridge, UK: Cambridge University Press.
Berkes, F. and M. Taghi Farvar (1989), 'Introduction and Overview', *Common Property Resources: Ecology and Community-based Sustainable Development*, London: Belhaven Press.
Bhatia, A. (1999), 'Governance and Management of Community Forestry', *Management of Mountain Commons in the Hindu Kush Himalayas*, Newsletter No. 35, Kathmandu, Nepal: International Centre for Integrated Mountain Development.
Birner, R. and H. Wittmer (2000), 'Co-management of Natural Resources: A Transaction Cost Economic Approach to Determine the Efficient Boundary of the State', in *Proceedings of International Symposium of New Institutional Economics*, 22–24 September 2000, Tubingen, Germany.
Bromley, D.W. (1991), *Environment and Economy: Property Rights and Public Policy*, Oxford, UK: Oxford University Press.
Cheung, S.N. (1969), 'Transaction Costs, Risk Aversion and the Choice of Contractual Arrangements', *Journal of Law and Economics*, **12**(1), 23–45.
Cheung, S.N. (1970), 'The Structure of a Contract and the Theory of a Non-exclusive Resource', *Journal of Law and Economics*, **13**(1), 49–70.
Cheung, S.N. (1983), 'The Contractual Nature of the Firm', *Journal of Law and Economics*, **26**, 1–22.
Cheung, S.N. (1987), 'Common Property Rights', in J. Eatwell, M. Milgate and P. Newman (eds), *The New Palgrave: A Dictionary of Economics*, volume 1, London: Macmillan.
Ciracy-Wantrup, S.V. (1971), 'The Economics of Environmental Policy', *Land Economics*, **47**(1), 36–45.
Ciracy-Wantrup, S.V. and R.C. Bishop (1975), 'Common Property as a Concept in Natural Resource Policy', *Natural Resource Journal*, **15**(4), 713–27.
Coase, R. (1937), 'The Nature of the Firm', *Economica*, **4**(16), 386–406.
Coase, R. (1960), 'The Problem of Social Cost', *Journal of Law and Economics*, **3**(1), 1–44.

Coleman, J.S. (1990), *Foundations of Social Theory*, Cambridge, MA: Belknap Press.
Cox, S.J.B. (1985), ' No Tragedy on the Commons', *Environmental Ethics*, **7**(1), 49–61.
Crocker, T.D. (1971), 'Externalities, Property Rights and Transactions Costs: An Empirical Study', *Journal of Law and Economics*, **14**(2), 451–64.
Dahlman, C.J. (1979), 'The Problem of Externality', *Journal of Law and Economics*, **22**(1), 141–62.
Dasgupta, P. (1999), 'Economic Progress and the Idea of Social Capital', in P. Dasgupta and I. Serageldin (eds), *Social Capital: A Multifaceted Perspective*, Washington, DC: The World Bank.
Dasgupta, P.S. and G.M. Heal (1979), *Economic Theory and Exhaustible Resources*, Cambridge, UK: Cambridge University Press.
Dasgupta, P. and K.G. Maler (1991), 'The Environment and Emerging Development Issues', Proceedings of the World Bank Annual Conference on Development Issues.
Davies, J. and M. Richards (1999), 'The Use of Economics to Assess Stakeholders' Incentives in Participatory Forest Management: A Review', European Union Tropical Forestry Paper No. 5, Overseas Development Institute, London.
Dayton-Johnson, J. (2000), 'Determinants of Collective Action on the Local Commons: A Model with Evidence from Mexico', *Journal of Development Economics*, **62**(1), 181–208.
Demsetz, H. (1964), 'The Exchange and Enforcement of Property Rights', *Journal of Law and Economics*, **70**(4), 414–30.
Demsetz, H. (1967), 'Towards a Theory of Property Rights', *American Economic Review*, **52**(2), 347–79.
Dove, M.R. (1993), 'A Revisionist View of Tropical Deforestation and Development', *Environment Conservation*, **20**(1), 17–24.
Drake, L., P. Bergstrom and H. Svedsater (1999), 'Farmers' Attitude to and Uptake of Countryside Stewardship Policies', draft final report to the STEWPOL Project, Chapter 5, FAIRI/CT95/0709, University of Uppsala, Sweden.
Drennan, L.G. (2000), 'The Efficient Mix of Performance Controls: A Transaction Cost Economics Perspective', Proceedings of International Symposium of New Institutional Economics, 22–24 September 2000, Tubingen, Germany.
Eggertsson, T. (1990), *Economic Behaviour and Institutions,* Cambridge, UK: Cambridge University Press.
Falconer, K. (2000), 'Farm-level Constraints on Agri-environmental Scheme Participation: A Transactional Perspective', *Journal of Rural Studies*, **16**(3), 379–94.
Feeny, D., F. Berkes, B.J. McCay and J.M. Acheson (1998), 'The Tragedy of the Commons: Twenty-two Years Later', in J. Baden and D. Noonan (eds), *Managing the Commons*, Bloomington, IN: Indiana University Press.
Fenoaltea, S. (1984), 'Slavery and Supervision in Comparative Perspectives: A Model', *Journal of Economic History*, **44**(3), 635–68.
Gibbs, J.N. and D.W. Bromley (1989), 'Institutional Arrangements for Management of Rural Resources: Common Property Regimes', in F. Berkes (ed.), *Common Property Resources: Ecology and Community-based Sustainable Development*, London: Belhaven Press.
Gordon, H.S. (1954), 'The Economic Theory of a Common Pool Resource: The Fishery', *Journal of Political Economy*, **62**(2), 124–42.
Griffin, R.C. (1991), 'The Welfare Analytics of Transaction Costs, Externalities and Institutional Choice', *American Journal of Agricultural Economics*, August, 601–14.
Grima, A.P.L. and F. Berkes (1989), 'Natural Resources: Access, Rights-to-use and Management', in F. Berkes (ed.), *Common Property Resources – Ecology and Community Based Sustainable Development*, London: Belhaven Press.
Guggenheim, S. and J. Spears (1991), 'Sociological and Environmental Dimension of Social Forestry Projects', in M.M. Carnea (ed.), *Putting People First: Sociological Variables in Rural Development*, New York: Oxford University Press, pp. 304–39.
Hanna, S. (1995), 'Efficiencies of User Participation in Natural Resource Management', in S. Hanna and M. Munasinghe (eds), *Property Rights and the Environment: Social and*

Ecological Issues, The Beijer International Institute of Ecological Economics and The World Bank.
Hanna, S. and M. Munasinghe (1995), 'An Introduction to Property Rights and the Environment', in S. Hanna and M. Munasinghe (eds), *Property Rights and the Environment: Social and Ecological Issues*, The Beijer International Institute of Ecological Economics and The World Bank.
Hanna, S., C. Folke and K. Maler (1995), 'Property Rights and Environmental Resources', in S. Hanna and M. Munasinghe (eds), *Property Rights and the Environment: Social and Ecological Issues*, The Beijer International Institute of Ecological Economics and The World Bank.
Hardin, G. (1968), 'The Tragedy of Commons', *Science*, **162**(3859), 1243–8.
Hill, I. and D. Shields (1998), 'Incentives for Joint Forest Management in India: Analytical Methods and Case Studies', World Bank Technical Paper No. 394, Washington, DC: The World Bank.
Johnson, O.E.G. (1972), 'Economic Analysis, the Legal Framework and Land Tenure Systems', *Journal of Law and Economics*, **15**(1), 259–76.
Kanbur, R. (1992), 'Heterogeneity, Distribution and Cooperation in Common Property Resource Management', background paper for the 1992 *World Development Report*, Office of the Vice President, Development Economics, The World Bank, Washington, DC.
Kant, S. (1998), 'Community Management: An Optimal Resource Regime for Tropical Forests', Department of Economics, Faculty of Forestry, University of Toronto, mimeo.
Kant, S. (2000), 'A Dynamic Approach to Forest Regimes in Developing Economies', *Ecological Economics*, **32**(2), 287–300.
Keohane, R.O. and E. Ostrom (eds) (1995), *Local Commons and Global Interdependence, Heterogeneity and Cooperation in Two Domains*, London: Sage.
Kherallah, M. and J. Kirsten (2001), 'The New Institutional Economics: Applications for Agricultural Policy Research in Developing Countries', International Food Policy Research Institute (IFPRI), Washington, DC.
Kumm, K.I. and L. Drake (1998), 'Transaction Costs to Farmers of Environmental Compensation', unpublished report, Department of Economics, SLU, University of Uppsala, Sweden.
Kuperan, K., N. Mustapha, R. Abdullah, R.S. Pomeroy, E. Genio and A. Salamanca (1998), 'Measuring Transaction Costs of Fisheries Co-management', paper presented at the 7th Biennial Conference of the International Association for the Study of Common Property, Vancouver, www.indiana.edu/~iasip/Drafts/kuperan.pdf, accessed 18 August 2009.
Leffler, K. and R. Rucker (1990), 'Transaction Costs and the Efficient Organization of Production: A Study of Timber Harvesting Contracts', *Journal of Political Economy*, **99**(5), 1060–87.
Libecap, G. (1989), *Contracting for Property Rights*, New York: Cambridge University Press.
Lin, J.Y. and J.B. Nugent (1995), 'Institution and Economic Development', in J. Behrman and T.N. Srinivasan (eds), *Handbook of Development Economics*, Vol. III, Amsterdam: Elsevier, pp. 2303–70
McKean, M. (1992a), 'Success on the Commons: A Comparative Examination of Institutions for Common Property Resource Management', *Journal of Theoretical Politics*, **4**(3), 247–81.
McKean, M. (1992b), 'Management of Traditional Common Lands (Iriachi) in Japan', in *National Research Council Proceedings of the Conference on Common Property Resource Management*, Washington, DC: National Academic Press, pp. 533–89.
McKean, M. (2000), 'Common Property: What It Is, What Is It Good for, and What Makes It Work', 'Forests, People and Governance: Some Initial Theoretical Lessons', in C.C. Gibson, M.A. McKean and E. Ostrom (eds), *People and Forests, Communities, Institutions, and Governance*, Cambridge, MA: The MIT Press.
Meinzen-Dick, R. and A. Knox (1999), 'Collective Action, Property Rights and Devolution of Natural Resource Management: A Conceptual Framework', paper presented at the

International Workshop on Collective Action, Property Rights and Devolution of Natural Resource Management, Exchange of Knowledge and Implications for Policy, 21–25 June, the Philippines, http://www.capri.cgiar.org/pdf/capriwp11.pdf, accessed 19 August 2009.

Menard, C. (2000), 'Enforcement Procedures and Governance Structures: What Relationship?', in C. Menard (ed.), *Institutions, Contracts and Organizations: Perspectives from New Institutional Economics*, Cheltenham, UK and Northampton, MA, USA Edward Elgar.

Meshack, C.K., B. Adhikari, N. Doggart and J.C. Lovett (2006) 'Transaction Costs of Community Based Forest Management: Empirical Evidence from Tanzania', *African Journal of Ecology*, **44**(4), 468–77.

Molinas, J.R. (1998), 'The Impact of Inequality, Gender, External Assistance and Social Capital on Local-level Collective Action', *World Development*, **26**(3), 413–31.

Nabil, M.K. and J.B. Nugent (1989), 'The New Institutional Economics and its Applicability to Development', *World Development*, **17**(9), 1333–47.

North, D.C. (1981), *Structure and Change in Economic History*, New York: Norton.

North, D.C. (1984), 'Transaction Costs, Institutions and Economic History', *Journal of Institutional and Theoretical Economics*, **140**, 7–17.

North, D.C. (1990), *Institutions, Institutional Change and Economic Performance*, New York: Cambridge University Press.

North, D.C. (1997), 'Transaction Costs through Time', in C. Menard (ed.), *Transaction Cost Economics: Recent Developments*, Cheltenham, UK and Lyme, NH, USA: Edward Elgar.

North, D.C. and R.P. Thomas (1977), 'The First Economic Revolution', *Economic History Review*, **30**(2), 229–41.

Oakerson, R.J. (1986), 'A Model for the Analysis of Common Property Problems', *Common Property Resource Management*, proceedings of a conference prepared by the Panel on CPRM Office for International Affairs, Washington, DC: National Research Council National Academy Press, pp. 13–30.

Olson, M. (1965), *The Logic of Collective Action: Public Goods and the Theory of Groups*, Cambridge, MA: Harvard University Press.

Olsson, O. (1999), 'A Microeconomic Analysis of Institutions', Working Paper in Economics No. 25, Department of Economics, Gothenburg University.

Ostrom, E. (1987), 'Institutional Arrangements for Resolving the Commons Dilemma: Some Contending Approaches', in B.J. McCay and J.M. Acheson (eds), *The Question of the Commons*, Tucson: University of Arizona Press, pp. 250–65.

Ostrom, E. (1990), *Governing the Commons*, Cambridge, UK: Cambridge University Press.

Ostrom, E. (1998), 'A Behavioural Approach to the Rational Choice Theory of Collective Action', *American Political Science Review*, **92**(1), 1–22.

Ostrom, E. and R. Gardner (1993), 'Coping with Asymmetries in the Commons: Self-governing Irrigation Systems Can Work', *Journal of Economic Perspectives*, **7**(4), 93–112.

Pelletier-Fleury, N., V. Fargeon, J. Lanoe and M. Fardeau (1997), 'Transaction Costs Economics as a Conceptual Framework for the Analysis of Barriers to the Diffusion of Telemedicine', *Health Policy*, **42**, 1–14.

Posner, R.A (1987), 'The Law and Economics Movement', *American Economic Review*, **77**(2), 1–13.

Putnam, R. (1993), *Making Democracy Work: Civic Traditions in Modern Italy*, Princeton, NJ: Princeton University Press.

Richards, M. (1997), 'Common Property Resource Institution and Forest Management in Latin America', *Development and Change*, **28**(1), 95–117.

Richards, M., K. Kanel, M. Maharjan and J. Davies (1999), *Towards Participatory Economic Analysis by Forest User Groups in Nepal*, London: ODI.

Room, J. (1980), 'Assessing the Benefits and Costs of Social Forestry Projects', *The Indian Forester*, **106**(7), 445–55.

Runge, C.F. (1981), 'Common Property Externalities: Isolation, Assurance and Resource Depletion in a Traditional Grazing Context', *American Journal of Agricultural Economics*, **63**(4), 595–606.

Runge, C.F. (1984), 'Institutions and the Free Rider: The Assurance Problem in Collective Action', *Journal of Politics*, **46**, 154–89.

Schlager, E. and W. Blomquist (1998), 'Resolving Common Pool Resource Dilemmas and Heterogeneities Among Resource Users', paper presented at 'Crossing Boundaries', the Seventh Annual Conference of the International Association for the Study of Common Property, 10–14 June, Vancouver, BC, Canada.

Scott, A. (1955), 'The Fishery: The Objectives of Sole Ownership', *Journal of Political Economy*, **63**(2), 116–24.

Seabright, P. (1993), 'Managing Local Commons: Theoretical Issues in Incentive Design', *The Journal of Economic Perspective*, **7**(4), 113–34.

Shelanski, H.A. and P.G. Klein (1995), 'Empirical Research in Transaction Cost Economics: A Review and Assessment', *The Journal of Law, Economics & Organization*, **11**(2), 335–61.

Smith, R.J. (1981), 'Resolving the Tragedy of the Commons by Creating Private Property Rights in Wildlife', *CATO Journal*, **1**(2), 439–68.

Smith, V.L. (1968), 'Economics from Production of Natural Resources', *American Economic Review*, **58**(3), 409–31.

Stevenson, G.G. (1991), *Common Property Economics: A General Theory and Land Use Application*, Cambridge, UK: Cambridge University Press.

Sugden, R. (1984), 'Reciprocity: The Supply of Public Goods through Voluntary Contributions', *Economic Journal*, **94**(376), 772–87.

Tang, S.Y. (1992), *Institutions and Collective Action: The Case of Irrigation Systems*, San Francisco: ICS Press.

Uphoff, N. (1998), 'Community-based Natural Resource Management: Connecting Micro and Macro Processes, and People with their Environment', paper presented at International Workshop in Community-based Natural Resource Management, 10–14 May, Washington, DC: World Bank.

Varughese, J. (1999), 'Villagers, Bureaucrats, and Forests in Nepal: Designing Governance for Complex Resource', unpublished PhD dissertation, Indiana University, USA.

Varughese, J. and E. Ostrom (2001), 'The Contested Role of Heterogeneity in Collective Action: Some Evidence from Community Forestry in Nepal', *World Development*, **29**(5), 747–65.

Vätn, A. (2001), 'Environmental Resources, Property Regimes, and Efficiency', *Environment and Planning: Government and Policy*, **19**(5) 665–80.

Velded, T. (2000), 'Village Politics: Heterogeneity, Leadership and Collective Action', *Journal of Development Studies*, **36**(5), 105–34.

Wallace, M.B. (1981), 'Solving Common-property Resource Problems: Deforestation in Nepal', unpublished PhD Dissertation, Harvard University, Cambridge, MA.

Wang, N. (2001), 'The Coevolution of Institutions, Organizations, and Ideology: The Longlake Experience of Property Rights Transformation', *Politics and Society*, **29**(3), 415–45.

Williamson, O.E. (1979), 'Transaction-cost Economics: The Governance of Contractual Relations', *Journal of Law and Economics*, **22**(2), 233–61.

Williamson, O.E. (1985), *The Economic Institution of Capitalism*, New York: The Free Press.

Williamson, O.E. (1991), 'Comparative Economic Organization: The Analysis of Discrete Structural Alternatives', *Administrative Science Quarterly*, **36**(2), 269–96.

Williamson, O.E. (1999), 'Public and Private Bureaucracies: A Transaction Cost Economics Perspective', *Journal of Law, Economics, and Organizations*, **15**(1), 306–41.

Williamson, O.E. (2000), 'The New Institutional Economics: Taking Stock, Looking Ahead', *Journal of Economic Literature*, **38**(3), September 2000, 595–613.

Zak, P.J. and S. Knack (2001), 'Trust and Growth', *The Economic Journal*, **111**(470), 295–321.

6. Analysing dominant policy perspectives – the role of discourse analysis
David G. Ockwell and Yvonne Rydin

Introduction

The last decade has seen a 'linguistic turn' within policy analysis (Edelman, 1988; Rydin, 1998, 1999; Hastings, 1999) as it becomes increasingly accepted that language use and appeals to different discourses by various actors in the policy-making sphere have a direct influence on the nature of any policy. In this chapter we explore how to undertake a discursive policy analysis. Rather than focus on the theoretical debates on this approach, we address the practical problems and potential for undertaking discourse analysis of environmental policy through a case study of the policy governing anthropogenic fire in Cape York Peninsula, Queensland, Australia. We begin by exploring the rationale for and benefits of using discourse analysis. Then we emphasize the need to find an appropriate 'middle range' theory for application in any specific context. To illustrate our point, two alternative frameworks for undertaking such an analysis are outlined. We then apply these frameworks in detail to our case study and use them to understand why a particular policy perspective has dominated fire policy in Cape York. This demonstrates the nature of the insights that the two approaches facilitate and provides the opportunity for exploring the methodological difficulties and practicalities of such an analysis.

The arguments for a discursive approach to policy analysis

The term 'discourse' is both complex and contested. It has multiple roots in the social sciences and humanities (Hastings, 1999, 2000). Dryzek (1997, p. 8) defines a discourse as 'a shared way of apprehending the world. Embedded in language it enables subscribers to interpret bits of information and put them together into coherent stories or accounts. Each discourse rests on assumptions, judgements and contentions that provide the basic terms for analysis, debates, agreements and disagreements'.

The key point of attending to discourse within environmental policy analysis is to respond to the assumption that policy language is a neutral medium through which ideas and an objective world can be represented and discussed (Darcy, 1999). This assumption overlooks the extent to which policy is contingent on social constructions of reality and the way the expression of policy issues will both be the result of power relations, ideological contestations and political conflict, and actively shape such relations, contestations and conflicts. Advocates of discourse analysis claim that it is crucial to examine and explain how language is used in such contexts in order to reveal aspects of social and political processes that were previously obscured or misunderstood. Furthermore, discourse analysis can serve to illuminate the way in which entrenched policy positions are to some extent sustained by the way in which policy problems are linguistically framed (Scrase and Ockwell, 2009).

More specifically, three distinct benefits of policy discourse analysis can be identified (Rydin, 2005). First, it enables one to understand different policy actors' perspectives and their self-presentation within the policy process. These will be expressed through the language that policy actors use and can help explain how different actors operate within policy contexts. An actor may use specific forms of language that are particularly appropriate and effective in a given policy context; by contrast, the weak situation of community representatives at formal hearings and inquiries is often at least partly due to their lack of command of the appropriate formal language. There are also strong links between the identity of actors and their use of language; identity is constructed through linguistic means. This has implications for how actors are categorized and treated in policy contexts; what it is to be actor X in a certain policy situation is discursively constructed. The argument here is that language is not just a medium of interaction but is also constitutive of actors, their identities and their values. Actors' values, therefore, cannot be seen in terms of their hard-wired preferences but rather as generated through debate, discussion and enunciation of those values (DeLuca, 1999). Furthermore, this is not an individual undertaking but is inherently social, occurring through interactions between actors.

Second, the attention to language allows consideration of how actors' power is at least in part discursive. Interaction between actors then becomes not just a series of encounters in which interests are balanced against or do battle with each other on the basis of their material resources. Rather it points to how the language that actors actively or unconsciously use in their communications with others is involved in persuasion and rationalization and influences the dynamics of policy debates. This is not just a matter of individual actors' capacity in using language. Linkages with prevailing societal discourses will also be important in supporting a particular actor's case. The reliance on various forms of economic rationality are a case in point, where the discursive reliance on a widely used and referenced argument about the importance of economic growth carries weight quite independently from the material power of economic actors and their skill in persuasion (Rydin, 2003).

Another way of considering the inter-relationship between language and actors' interactions with one another is to see language as also constitutive of the incentives facing actors – the costs and benefits (monetary and otherwise) that they take into account in deciding on their behaviour. For example, researchers have suggested that reputation can be a key factor shaping behaviour (Ostrom, 1990; Chong, 1991). Actors may seek to promote and protect a good reputation and avoid behaviour that is going to expose them to public shame and blame. But reputation is a social variable constructed through social communicative interactions. The detailed language of interactions will be central to assessments (by the actor concerned and others) of whether a reputation is being damaged or enhanced.

And third, as well as understanding policy actors and the dynamics of policy processes more fully, the discursive dimension allows the possibility for devising new modes of communication to achieve normatively better policy outcomes. There has been a lot of emphasis on creating more inclusive and deliberative spaces for communication (Burgess et al., 1998; Hillier, 1998; Healey, 1999; Mason, 1999). Much of the literature on these policy innovations has arisen from an engagement with the notion of policy as an inherently discursive process. There has, however, been a tendency towards an overly procedural approach in devising new spaces for communication. Graham Smith is one of the most insightful writers on deliberation but in his discussion of institutional design and deliberation (2003), he concentrates on procedures and decision rules, specifically how to ensure equality of voice, defence of deliberation against strategic action and sensitivity to the scope, scale and complexity of environmental issues. But how do these procedures and decision rules change the communication within deliberation? This can only be understood through considering the detailed language of that communication. The lesson of the linguistic turn is that communication between actors is not just a matter of how that communication is arranged. The language of the interaction also needs to be considered.

Even more restrictively, deliberation is often equated with communication as if this not only characterizes deliberation but distinguishes it from other types of policy intervention. This fails to see the communicative dimension of all policy work (Majone, 1989) and, furthermore, the linguistically mediated nature of all social activity. Language does play a pivotal role in deliberation, but it is also implicated in many other institutional arenas. This can provide insights into how language expresses values and enables or disables agreement, including consensus.

The roots of this lack of attention to the details of language even within those analysts of policy who see policy as essentially communicative, derives from the Habermasian roots of most work on deliberative democracy and its planning applications (Dryzek, 1990, 2000; Healey, 1999). This identified a specific potential within communication between actors – that is, the potential to create consensus. Where actors engage in communication with a performative stance towards mutual understanding, then consensus between parties is immanent in the communication. The major constraint on achieving such a consensus is pinpointed as the absence of an ideal speech situation in which communicative rationality can hold sway. This has led to the emphasis being placed on the circumstances within which communication occurs rather than the nature of the communication in linguistic terms. This has been confirmed in Habermas's most recent work (1996). In our chapter we want instead to turn back towards the language of policy and the detailed analysis of that language in specific situations.

There is one particular problem within environmental policy that this approach seems particularly well suited to address. This concerns the mismatch between complex environmental problems and simple dominant policy responses. This simplicity is at odds with the inherent variability and complexity of the ecosystems whose healthy functioning such policy aims to sustain. It ignores the fact that environmental problems are characterized by a high degree of uncertainty. On one level, uncertainty exists as a result of scientists' incomplete understanding of ecosystem functioning. Additionally, however, environmental problems are invariably linked to issues of resource distribution. As a result, there are always economic, social and political implications of any environmental policy (see, for example, Wheeler and McDonald, 1986; Rees, 1990; see also, for example, Flournoy, 1993). As Dryzek (1997) highlights, when ecological systems interact with economic, social and political systems through the policy process, the level of uncertainty associated with environmental problems is greatly magnified. Hajer and Wagenaar (2003) refer to this as policy-making under conditions of 'radical uncertainty'.

Increasing evidence suggests that such uncertainty is often reflected in inappropriate environmental policies due to the application of standardized management techniques that ignore the spatial, temporal, social, economic and political complexity of environmental problems. Leach and Mearns (1996) give the example of the woodfuel crisis in Africa, which is widely perceived as a classic example of a supply gap where demand for woodfuel exceeds supply. This has been met by a standardized response of mass tree planting by governments, NGOs and inter-governmental organizations. The basic assumptions that define the supply gap, however, ignore more subtle issues such as the fact that most woodfuel comes from clearing wood for agriculture or from lopping branches valued for fruit and shade. From a broader perspective there is not one big problem of energy supply but many smaller problems of command over trees and their products to meet a wide range of basic needs. The range of policy solutions is therefore equally diverse. Why is it then that one policy outcome can emerge and be sustained in the face of conflicting evidence? Discourse analysis offers a valuable approach to answering this question.

Theoretical perspectives on discourse analysis
Just as the term 'discourse' itself is a contested concept used in different ways, so are there a variety of different perspectives on discourse analysis, both methodologically and theoretically. One of the key distinctions that runs through the literature is between those approaches that are derived from a Foucauldian perspective and those derived from a Habermasian one. The two methodologies that we explore in this chapter derive from both sides of this distinction: Maarten Hajer, who developed his framework from an engagement with Foucault's work; and John Dryzek, whose interest in discourse derives from his earlier work on Habermas and normative theories of deliberative democracy. These 'middle-range' theorists are important because they have sought to develop and apply Foucault's and Habermas's broader ideas to environmental issues and specifically to environmental policy. They have done considerable work in qualifying and operationalizing Foucault's and Habermas's theories. In doing so, they have to some extent modified these original theories but retain the essentials of the Foucauldian and Habermasian perspectives. We will explore these two approaches next, before demonstrating how they can be applied to a specific case study.

Hajer
In the development of the study of discourse Foucault's work has been pivotal. Through the study of the history of sexuality, madness and the disciplinary basis of the academy, Foucault, referring to what he called power/knowledge, developed the idea that knowledge, and hence discourse, is a reflection of power within society. As such, language is seen as the operation of power (Foucault, 1980, 1984; Bevir, 1999; Rydin, 1999; Hastings, 2000; Watt and Jacobs, 2000). Hajer has sought to work within a Foucauldian framework, in terms of engaging with the combined concept of power/knowledge, while adjusting it to the problem of understanding environmental policy situations. Hajer's work focuses on the discursive nature of environmental policy-making (Hajer, 1995; Hajer and Wagenaar, 2003). Discourse is seen as constituting both text and practice

with a strong emphasis on social constructivism running throughout (Dryzek, 1995; Hajer, 1995; Bakker, 1999; Keeley and Scoones, 2000; Richardson and Jensen, 2000). In this view discourses are produced both through individual activities and institutional practices that reflect particular types of knowledge. Discourses are therefore actively produced through human agencies that undertake certain practices and describe the world in certain ways. Actors are not, however, seen as acting within a vacuum. Discourses simultaneously have structuring capabilities. They provide parameters within which people act and mould the way actors influence the world around them (Hajer, 1995; Keeley and Scoones, 2000).

In Hajer's view, politics is a struggle for discursive hegemony in which actors struggle to achieve 'discursive closure' by securing support for their definition of reality. There is a significant Foucauldian influence within Hajer's work in terms of the regulatory power of discourses as they act to select appropriate and meaningful utterances and actions within a struggle for hegemony in the policy-making process (Foucault, 1979, 1990; Buttel, 1997; Rydin, 1998; Richardson and Jensen, 2000). The notion of 'story-lines' is brought in to describe the common adoption of narratives through which elements from many different spheres are combined to provide actors with symbolic references that imply a common understanding (Hajer, 1995; Rydin, 1999). Essentially, the assumption is that actors don't draw on a comprehensive discursive system, instead this is evoked through story-lines. By uttering a specific word or phrase, for example, 'global warming', a whole story-line is in effect reinvoked. Story-lines can, in this way, therefore act to define policy problems.

The widespread adoption of a story-line results in the formation of 'discourse coalitions' where groups of actors are drawn to particular story-lines as they represent common interests (Hajer, 1995; Bakker, 1999; Rydin, 1999). These actors might not have ever met and might apply different meanings to a story-line, but in the struggle for discursive hegemony that is assumed to play out within the policy-making process, story-lines form the 'discursive cement' that keeps the discourse coalition together by producing 'discursive affinities'. Particularly strong discursive affinity is referred to as 'discursive contamination'. The Foucauldian basis of Hajer's approach views story-lines as playing an essential role in positioning actors. They add credence to the claims of certain groups and render those of other groups less credible. Story-lines therefore act to create social and moral order within a given domain by serving as devices through which actors are positioned and ideas of blame, responsibility and urgency are ascribed.

The social constructivist approach emphasizes the role of institutional

174   *A handbook of environmental management*

arrangements in structuring discourses, forming routine understandings where complex research is often reduced to visual reproduction or catchy one-liners that ignore uncertainty and entail significant loss of meaning. Routine forms of discourse therefore express a continuous power relationship that is particularly effective in that it avoids confrontation. The use of the term 'sustainable development' in contemporary British public policy arguably provides an example of this.

To shape policy, a new discourse must both dominate public discussion and policy rhetoric and penetrate the routines of policy practice through institutionalization within laws, regulations and routines (Hajer, 1993; Nossiff, 1998; Healey, 1999). In terms of policy change then, promoting a new story-line is a difficult task involving dismantling previous story-lines and confronting the interests of those who were able to achieve prominence for their claims and viewpoint originally (Rydin, 1999). Discourse analysis from Hajer's perspective is a method to shed light on the social and cognitive basis of the way in which policy problems are constructed (Hajer, 1995), with analysis focused on the socio-cognitive processes in which discourse coalitions are established. He puts emphasis on the constitutive role of discourse in political processes, but assigns a central role to discoursing subjects, although in the context of a duality of structure. Social action is seen as stemming from human agency, however, social structures of various sorts both enable and constrain their agency. It is therefore possible for agents to accomplish policy change through discursive interaction within the context of these structures, but this inherently requires deconstructing the discursive hegemony that existing dominant political interests have achieved.

Dryzek
John Dryzek is well known as a normative political theorist. He has sought to apply Habermas's concept of communicative rationality and deliberative democracy to the specific issue of the environment (Dryzek, 2000), a development that has been welcomed as a major advance in political theory. Here, though, the focus is on his work in analysing environmental discourse. In this work he explicitly counters the Foucauldian approach and instead adopts a more agency-centred model. Dryzek sees discourse and power as interconnected in all kinds of ways, whereas Foucault would deny such a distinction between power and discourse as discourse *is* the operation of power (Dryzek, 1997). Furthermore, Dryzek sees constraints on the power of discourse, as powerful actors may override developments at the discursive level by ignoring them in terms of policy. Alternatively, discourses may be absorbed to suit the interests of a firm or government (ecological modernization springs to mind). Another constraint may arise

from the need for capitalist governments to fulfil a number of basic functions irrespective of discourses that may have been captured by government officials, especially continued economic growth. Within his rejection of the Foucauldian discursive approach, Dryzek points to the very existence of authors such as Foucault as evidence that individuals subject to discourses are able step back and make comparative assessments and choices across different discourses. Dryzek also rejects the hegemonic terms used by Foucauldians to describe discourse. He asserts that variety is as likely as hegemony, with the disintegration of the previously hegemonic discourse of industrialism since the 1960s as evidence of just such variety.

Dryzek sets out his approach to discourse analysis with the aim of advancing 'analysis of environmental affairs by promoting critical comparative scrutiny of competing discourses of environmental concern' (Dryzek, 1997, p. 20). He is thus interested in explaining how environmental discourses inform political programmes. A four-fold typology is put forward where ways of thinking about environmentalism are characterized in terms of their departure from the discourse of industrialism (Elliott, 1999). These departures can be reformist or radical, prosaic or imaginative and result in the identification of four main categories of environmental thought. These four strands are categorized as follows. The discourse of environmental 'problem-solving' is prosaic and reformist. It takes the political-economic status quo as given but in need of adjustment to cope with environmental problems, especially via public policy. The discourse of 'survivalism', popularized by the Club of Rome (Meadows et al., 1972) is also prosaic, but radical. It is radical because it seeks wholesale redistribution of power within the industrial political economy and wholesale reorientation away from perpetual economic growth to avoid exhausting natural resources and the assimilative capacity of the environment. It is prosaic because it sees solutions in terms of options set by industrialism, especially greater control of existing systems by administrators, scientists and other responsible elites. The discourse of 'sustainability' is imaginative and reformist, beginning in the 1980s with imaginative attempts to dissolve the conflicts between environmental and economic values that are characteristic of the discourses of problem-solving and survivalism. Finally, the discourse of 'green radicalism' is also imaginative, but radical. It rejects the basic structure of industrial society and the way the environment is conceptualized therein. Due to such radicalization and imagination, it features deep intramural divisions.

In common with the roots of Habermas's work in argumentation theory, Dryzek also adopts a rhetorical method, marrying this with a more basic social constructivist perspective (see Rydin, 2003, ch. 2). This offers a fairly tightly controlled comparative analytical device for identifying the elements through which discourses construct stories. First the basic entities whose existence is recognized or contrasted must be identified. Dryzek refers to this as the 'ontology' of a discourse. Second, assumptions held about natural relationships, such as Darwinian struggle or cooperation, should be explored. This includes hierarchies of gender, expertise, political power, intellect, race, and so on. Propositions about agency and motivation constitute the third area of investigation. Agents may, for example, be seen as benign, public-spirited administrators or selfish bureaucrats. They could include enlightened citizens, rational consumers, ignorant and short-sighted populations, and so on. Finally, the key metaphors and rhetorical devices that a discourse invokes should be scrutinized. This might include, for example, metaphors such as spaceships (Boulding, 1966) that may act as rhetorical devices to convince listeners or readers. It may also include other devices like an appeal to widely accepted institutions or practices such as established rights. A discourse could also accentuate negatives such as horror stories regarding government mistakes.

In order to demonstrate the nature of the insights that Hajer and Dryzek's approaches facilitate and to explore the methodological difficulties and practicalities of such analysis, we now apply them to a case study of environmental policy governing anthropogenic burning in Cape York Peninsula (Cape York), Queensland, Australia.

Discourse analysis in practice – a case study[1]
Cape York is situated at the north-eastern tip of Queensland, Australia (Figure 6.1). Covering an area roughly equivalent in size to England, Cape York has a low population density of just 18 000 people mostly concentrated in a few mining towns and Aboriginal reserves as well as scattered cattle stations. Northern Australia, including Cape York, is thought to have a long history of anthropogenic burning stretching back at least 40 000 years (some estimates date it as far back as 70 000 years), coinciding with the arrival of the first Aborigines (Stocking and Mott, 1981). The idea of 'fire-stick' farming was popularized by Rhys Jones (1969) to describe the practices of indigenous land users where low-intensity, early dry-season burning across small areas was used to drive game into hunting grounds and increase the productivity of resource-rich areas such as monsoon forests.[2]

There has been considerable controversy over the impact of Aboriginal use of fire on the ecology of Australia. Most prominent is the debate around whether, in tropical northern Australia, Aboriginal burning caused the recession of earlier rainforest in favour of savanna or whether the recession of the rainforest was in fact the result of climate change (Flannery, 1994; Rose, 1996; Bowman, 1998, 2000; Hill, 2003). Bowman

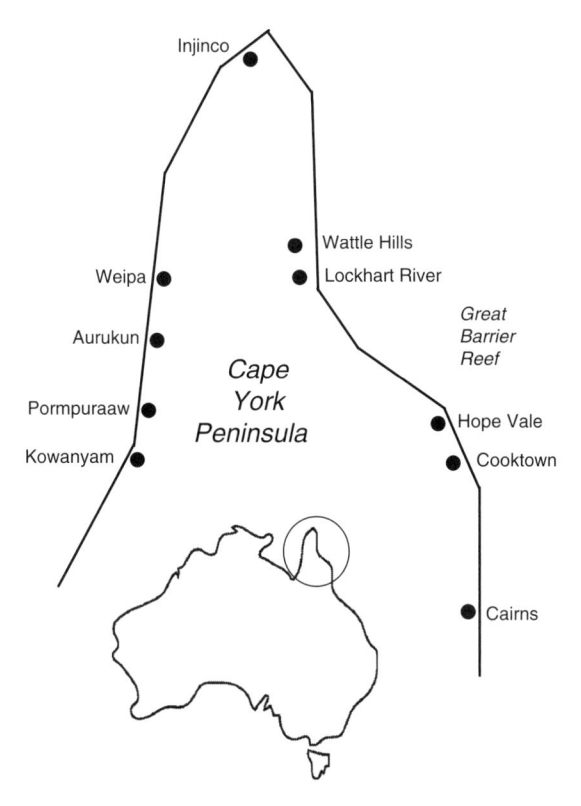

Figure 6.1   Map of Cape York Peninsula

(1998, p. 2) characterizes the lack of scientific consensus surrounding this debate as 'an inherent circular argument concerning the cause and effect of climate change, vegetation change, and burning through the late Quaternary'. It is, however, widely accepted that the pattern of burning in tropical northern Australia has changed in modern times, coinciding with the displacement of Aborigines by European settlers. Late dry-season, high-intensity burns now define anthropogenic burning with increased fuel loads over larger areas. This has reduced fire-sensitive vegetation in some areas. There has also been a lack of fire in other areas, which has enhanced fire-sensitive ecosystems (Gill et al., 1990; Bowman, 1998, 2000; Hill, 2003).

Anthropogenic burning in Cape York provides a typical example of such changes in burning practices. Overall, an estimated 80 per cent of the total area of Cape York currently burns each year (Cape York Peninsula Sustainable Fire Management Programme, personal communication,

2004). Fire-assisted pastoralism forms the dominant land use in Cape York. Pastoralists tend to burn land to promote the growth of green grass for their cattle to feed on. There are also extensive areas of Cape York (approximately 14.2 per cent of the total land area of Cape York in 2004 and increasing annually – Queensland Parks and Wildlife Service, personal communication, 2005) set aside as national park and wildlife reserves under the control of the Queensland Parks and Wildlife Service (QPWS) who also use fire as a significant part of their land management approach. They justify their use of fire through a number of reasons including hazard reduction where, it is argued, burning available ground fuel (dry leaf litter and so on) avoids the spread of wildfires later in the dry season, habitat management, including weed and pest management, and maintaining habitat diversity. There are also several Aboriginal reserves where indigenous communities are free to pursue their own traditional burning practices.

Despite the lack of consensus within the scientific literature regarding the environmental impacts of fire, environmental policy in Cape York tends to be unquestioningly pro the use of fire as a land management tool. This can be seen in both the policies of the Queensland Rural Fire Service who are responsible for policing the use of fire in Cape York (RFS, 2001) and the policies of those government departments with direct jurisdiction over the management of areas of land there (see, for example, Grice and Slatter, 1997; Gill et al., 1999; Marlow, 2000; QPWS, 2000; see also, for example, DNR, 2001, 2004; EPA, 2002).[3]

The Permit to Light Fire system operated by the Queensland Rural Fire Service (Queensland State Government, 1990) provides that landholders can, in theory, be prosecuted if they light a fire outside the terms of a Permit. The problem, however, has always been in proving who lit a fire. Once a fire is lit on Cape York it has the potential, especially at dry times of year, to burn for weeks or even months across hundreds of thousands of hectares of land, therefore affecting areas nowhere near where it was first lit. The introduction of the Cape York Peninsula Sustainable Fire Management Project has had some success in addressing the issue of accountability. The project provides an online service[4] where fires on Cape York can be tracked by satellite and thus, when cloud cover does not interfere, the origin of fires can sometimes be identified. The attitude amongst most stakeholders, however, tends to be very much that if the fire is on your land, it's your problem whether you lit it or not. This is characterized by a traditional saying often used by landowners on Cape York; 'He who owns the fuel owns the fire' (Queensland State Government, 1990; RFS, 2001). The insinuation is that each landholder ought to engage in hazard reduction burning to ensure that there is insufficient ground-fuel build-up (dry leaf litter and so on) on their land to allow a fire to encroach.

As well as its role in coordinating the Permit to Light Fire system, the Rural Fire Service also works with Cape York landholders in implementing a series of 'controlled' burns at the beginning of each dry season. This is done via a series of workshops held across Cape York prior to each dry season where interested landholders can attend and request that the Rural Fire Service carry out burning on their land. The Rural Fire Service then flies a light aircraft along the boundaries of participating properties and drops incendiary bombs that are intended to burn a series of firebreaks between each property. The rationale for this practice is one of hazard reduction. As outlined above, the rationale is that by burning the available ground fuel on property perimeters, the spread of wildfires later in the dry season might be avoided. Some stakeholders, however, are critical of this practice as being too indiscriminate and not accounting for environmental considerations in terms of whether the various affected ecosystems are able to cope with regular, or indeed, any fire. Indeed, as Russell-Smith et al. (2003) highlight, as well as there being no consensus over what fire-oriented landscape management should aim for, neither is it clear whether humans have the tools or resources to implement particular regimes over a large spatial scale.

When applying for a Permit to Light Fire, a landholder is required to disclose to the local Fire Warden any arrangement they have with other government agencies that obliges them to protect some aspect of their property from fire for environmental reasons or otherwise. Such arrangements are rare but do include agreements between landholders and the Queensland Department of Natural Resources, Mines and Energy to undertake burning to kill back weeds such as rubber-vine (DNR, 2001), or, in a few examples, to maintain the habitat of the endangered golden shouldered parrot, *Psephotus chrysopterygius*, which relies on late dry-season fires to create necessary nesting conditions (Crowley et al., 2003). The only instance where such an agreement might limit burning is where a public road runs through a property. There is only one main, unsurfaced dirt road in Cape York so this is rarely an issue. Once a landholder has informed their Fire Warden of any such agreement, it is then up to the Warden to consider this when detailing the conditions of the Permit to Light Fire. Wardens are all local stakeholders themselves chosen by the Rural Fire Service on the basis of assumed local knowledge of their area of responsibility (areas of responsibility range from just the Warden's own land to including theirs and a few of their neighbours' properties). The Rural Fire Service states that 'the local volunteer Fire Warden should not be responsible for policing environmental issues' (RFS, 2001, 1-1) on the basis of them not possessing sufficient knowledge to do so.

In addition to the Rural Fire Service, Queensland's Environmental

Protection Agency also supports the pro-burning policy discourse. This agency is responsible for achieving 'ecologically sustainable development' under the terms of the Environmental Protection Act 1994 (Queensland State Government, 1994). The Queensland National Parks and Wildlife Service is part of the Environmental Protection Agency with direct responsibility for the management of National Parks. As outlined above, the Environmental Protection Agency actively uses fire as a land management tool on National Park land as part of their land management policy. This is also in line with the rationale for the Department of Natural Resources, Mines and Energy's fire-relevant policies (Marlow, 2000; DNR, 2001).

The pro-burning policy stance in Cape York does not, however, have unanimous support from all stakeholders. Whilst many stakeholders are pro-burning, including Aboriginal communities, pastoralists and government scientists whose rationales for burning were summarized above, there are also two key stakeholder groups who are primarily anti-burning. These are the residents of Wattle Hills (a self-sufficiency community pursuing sustainable forestry practices on their 35 650 ha property), and several independent scientists who cite a growing body of scientific and anecdotal evidence that questions the environmental sustainability of the dominant pro-burning policy paradigm in Cape York (see, for example, Ockwell and Lovett, 2005). Cape York stakeholders can therefore be seen as polarized between two opposing discourses of 'pro-' and 'anti-' burning.

In order to demonstrate the value of discourse analysis in understanding the policy dominance of the pro-burning discourse in Cape York, we analyse primary data in the form of the transcript of a seminar hosted by the Cairns and Far North Environment Centre (CAFNEC) in 1992 entitled 'Tropics Under Fire. Fire Management On Cape York Peninsula' (CAFNEC, 1992). The seminar invited stakeholders to come together and give short, 20-minute presentations outlining their views on the use of fire in Cape York. Overall, ten presentations were made by the stakeholders that were present. This seminar was held some time ago and other seminars have been held since, however, this particular seminar has been chosen for analysis for two reasons. First, analysis of later conferences and extensive consultation with Cape York stakeholders has demonstrated that there has been little change in attitude since the 1992 seminar was held. Second, the 1992 seminar constituted the widest and most equally proportioned representation of the various stakeholder groups. The transcript thus provides a useful summary of both the pro- and anti-burning discourses from the perspectives of all the key stakeholders and interest groups.

Double close-reading of the transcript of the seminar by both authors forms the basis of the analysis presented here. Both Hajer's Foucauldian and Dryzek's Habermasian frameworks for undertaking discourse analysis were then applied. As the analysis demonstrates, the discursive construction of the burning issue is more complex and nuanced than the two broadly opposing pro- and anti-burning positions imply. We begin our discourse analysis by applying Hajer's Foucauldian-inspired framework before moving on to examine what Dryzek's Habermasian framework can add to the analysis.

Applying Hajer's analytic framework[5]
In keeping with Foucault's own work, Hajer has applied his ideas in detailed case studies. This allows more extended engagement with the ideas he has developed than is possible here, in the context of a discourse analysis of a specific text. Hence, the emphasis of the analysis will be on the particular concept of story-lines. It must therefore be acknowledged that some aspects of the Foucauldian approach are underplayed. The idea that a discourse comprises a set of practices as well as a set of representations is not given full weight by a text-based methodology. The emphasis on practices within a Foucauldian approach should not, however, be taken to exclude a concern with the text and the words it comprises. The use of the Cape York seminar transcript provides an opportunity for considering the story-lines embedded in these words and the potential for discourse coalitions thereby created.

The main story-lines adopted by the key actors are set out in Table 6.1. There is the clear distinction between two opposing discourses, constructed as pro- and anti-burning. The strongly drawn distinction sets the context or frame for the policy discussion and, as such, it can support the argument for two discourse coalitions coalescing around these two opposition perspectives on the use of fire. In the 'pro' coalition are the Aborigines, the government scientists and the pastoralists, while the residents of Wattle Hills and the independent scientists make up the 'anti' coalition.

It might be argued that the anti-burning discourse is placed in the position of challenging the established pro discourse, a task that is bound to require additional discursive resources as the existing policy scenario is well established. Those stakeholders who are anti-burning have to promote a new story-line of 'fire is undesirable' and actively dismantle the 'fire is desirable' story-line. This suggestion of an embedded bias towards the established 'pro' story-line is, however, somewhat undermined by the framing of the seminar itself. The title of the seminar 'Tropics Under Fire' and the illustration on the front cover of the conference transcript of a bird and other iconic animals (such as kangaroos) fleeing smoke and flames lit by a giant, match-wielding human hand suggests that, at least within CAFNEC (the environment centre that set up the seminar) there might be a bias towards the 'fire is undesirable' story-line. Instead of relying on the

182   *A handbook of environmental management*

Table 6.1   Applying Hajer's story-lines

	Link to Dominant Oppositional Discourses	Story-lines
Aborigines	Pro-burning	Fire as a cultural practice of traditional owners and custodians of land, embodying local indigenous knowledge *Linked to* History of repression, suppression and removal of Aborigines leading to loss of knowledge
Government scientists	Pro-burning	Human intervention in natural systems needed to ensure conservation of habitats and ecosystems *Elaborated as* Burning as promoting habitat diversity; *and* Burning as 'fuel reduction' preventing larger-scale, 'natural' fires
Pastoralists	Pro-burning	Burning is environmentally beneficial (through stabilizing ecosystems) but also economically beneficial (through providing fodder, safe mustering of cattle and promoting tourism)
Wattle Hills	Anti-burning	Letting 'nature take its course' will promote environmental quality and economic benefits
Independent scientists	Anti-burning	Scientific knowledge does not support burning; too much uncertainty about impacts of burning *Linked to* Global stories of biodiversity maintenance, rainforest protection and endangered species

simple construction of the two opposition discourses, Hajer's approach highlights that any construction of a broad 'anti' or 'pro' discourse coalition depends on how much the individual story-lines being used by actors support or undermine each other. It is notable here, therefore, that the three pro-burning story-lines all represent to some extent a different take on the same underlying story about human engagement with nature.

There is a strong emphasis in the Aboriginal story-line on the historic knowledge of how to use fire gained through millennia of active engagement between Aborigines and their local environment. This is contextualized with a moral claim that the Aborigines are the 'custodians' (p. 6) of the land together with the anti-colonial discourse that emphasizes the European repression of Aborigines. In this way, the Aboriginal story can rely on strong affinities with the established cause of native title claims, which has gained considerable political credibility since the early 1970s. This reflects Hajer's idea of how moral orders are established through discourse in that blame and responsibility are attributed to European descendants in order to justify prioritizing the Aboriginal representatives' views on land management. In terms of social practice, it is significant that an Aboriginal representative was, out of respect, asked to open the seminar. But in terms of the discourse coalition analysis, the important point is the existence of this clear narrative thread about the need for active human–nature engagement; as such, this is available for connecting with the other actors' discourses.

Although couched in very different terminology, government scientists' discourse follows the same narrative thread as that of the Aborigines. Here, human engagement with ecosystems is promoted as necessary to ensure conservation. Two subsidiary discourses are discernible: a story-line of 'habitat diversity' (p. 36), preservation through burning and a story-line of 'fuel reduction' (p. 37) where anthropogenic fire is endorsed as a way of avoiding the catastrophic ecological consequences of naturally occurring fires by reducing the volume of standing fuel. The basis for such engagement is scientific knowledge as opposed to traditional knowledge but the story is the same. By advocating 'patch' or 'mosaic burning' (p. 31), which is thought to have been traditionally pursued by Aborigines (p. 27), and claiming that this leads to ecologically desirable 'stable' vegetation patterns, discursive affinity with the Aboriginal discourse is also exhibited. The connection is also facilitated by constructing Aboriginal practices in scientific terms, with reference to scientific papers on Aboriginal land management and even the inclusion of one such paper in the transcripts of the conference.

This narrative thread is reinforced again by the discourse of the pastoralists. Here, everyday economic practices are portrayed as supporting the practice of burning in both environmental and economic terms. Again, active engagement between humanity and nature (through burning) is seen as beneficial in the long term. The pastoralist representatives maintain the story-line of fire being environmentally beneficial at the same time as emphasizing its economic desirability. By describing pastures as 'botanical communities' (p. 29) they claim that fire can maintain 'botanical stability' (p. 32) and therefore preserve environmental 'integrity'. This provides a clear example of discursive affinity with the government scientists, which is also evident in the pastoralists' adherence to the story-line of burning as desirable in terms of reducing wildfire risks. Discursive affinity with Aborigines is also achieved by asserting that the economic benefits derived by pastoralists from cattle fodder through post-fire new growth and safe mustering of their herds are the same as those traditionally derived by Aborigines. As well as the direct economic benefits from pastoralism, burning is also endorsed as economically desirable for encouraging tourism by clearing the ground for hiking and attracting charismatic species such as wallabies and kangaroos to the fresh new after-growth.

So the story-lines analysis provides some basis for understanding the strength of the pro-burning discourse coalition. It can also help understand the weakness of the anti-burning coalition. The story-lines of this coalition focus on opposing the dominant discourse instead of building links within the coalition. The aim of the proponents of the anti-fire discourse tends to focus chiefly on discrediting the government scientist story-line of fire being ecologically desirable. The national parks agency, the QPWS, tends to constitute the main focus for attack. The Wattle Hills representative promotes a story-line of 'fire as destructive to life', which is diametrically opposite to the government scientists' story-line. The government 'fire management' line is directly confronted and instead a non-interventionist, 'let nature take its course' line is promoted. Both independent scientists advance a story-line of fire as environmentally damaging. Again, a direct attempt is made to deconstruct the government scientists' story-line of fire as ecologically desirable by highlighting the degree of uncertainty surrounding current scientific knowledge of the impacts of fire. Policies that involve using fire are described as 'stabbing in the dark' (p. 43). The 'fuel reduction' story-line is also attacked on the basis of negative impacts on soil fertility as argued by the Wattle Hills representative. The argument is that without fire, dead matter has a chance to decompose, resulting in increased soil fertility, as opposed to being burnt and therefore losing biomass through combustion.

Little discursive connection is evident between the 'let nature take its own course' Wattle Hills discourse and the strong scientific rationality of the independent scientists, with their attempts to link the anti-burning story-line to other global environmental story-lines. While the Wattle Hills residents highlight their own credentials as self-sufficiency pioneers, the independent scientists have significant academic credentials and a strong professional involvement with Cape York including, in one case, having worked on television documentaries on bird life in the Cape. The rational, scientific tone of their representations reflects their scientific background with extensive intertextual reference made to scientific papers in support of their arguments. Several popular global story-lines are invoked to support avoiding the use of fire, including 'biodiversity maintenance', protection of 'endangered species', expansion of 'valuable rainforests' and maintenance of 'sustainable populations' of wildlife. Little attempt is made to bridge the 'global' of the scientists with the 'local' of the Wattle Hills residents and yet, without this, the 'anti' discourse coalition remains discursively weak.

Applying Dryzek's analytic framework
The first part of the Dryzek framework concerns the identification of actors' discourses with his typology of four societal discourses on environmental issues. Table 6.2 represents an attempt to classify the discourses and sub-discourses present in the transcript in terms of Dryzek's categories. This displays some difficulties. It was not possible to assign actors' discourses to Dryzek's categories in a simple and non-contestable way.

For example, the Aboriginal discourse can be considered to fall into both the survivalism and green radicalism categories. There is a clear radical strand to their discourse; the emphasis on the colonial history of European settlement and repression marks it as such. While the status quo of the established political economy is not challenged directly, it is implied that Aboriginal rights should have priority over other concerns, including any economic imperative. There is a tension, however, between much of the tone of the Aboriginal presentations, which is distinctly prosaic, and the sub-text, which emphasizes the distinctive spiritual claims of the Aborigines. The latter is clearly imaginative not prosaic. The emphasis on the prosaic in this particular context may be a strategic discursive decision

Table 6.2 *Applying Dryzek's discourses*

	Prosaic	Imaginative
Reformist	*Problem-solving discourse* Government scientists Independent scientists (less critical discourse) Pastoralists	*Sustainability discourse* Pastoralists
Radical	*Survivalism discourse* Aborigines (everyday management discourse) Independent scientists (more critical discourse) Wattle Hills (everyday management discourse)	*Green radicalism discourse* Aborigines (underlying spiritual discourse) Wattle Hills (self-sufficiency discourse)

and/or may relate to the way that Aboriginal voices were largely represented by other semi-professionalized voices. In either case, it suggests that the categories are not very helpful in furthering the analysis where the Aboriginal discourse is concerned.

A similar point can be made about the discourses of the independent scientists and the residents of Wattle Hills. The independent scientists adopted a strong critique of contemporary industrialism. This very much put them on the borders between a reformist and a radical discourse. It certainly distanced them from the more firmly pro-status-quo stance of the government scientists. And while, like the government scientists, they remained prosaic rather than imaginative in their discourse, it did make it difficult to simply assign the independent scientists to either the problem-solving or survivalism category. The Wattle Hills story-line emphasized the need to 'let nature take its course'. Policies promoting fire were seen as misleading and neglectful of life, reverence for life being of primary importance. This suggests a radical, imaginative departure from industrialism that fits within Dryzek's definition of 'green radicalism' (Dryzek, 1997). In much of their more specific discussion of management practices, however, the discourse of the Wattle Hills residents is distinctly prosaic, suggesting a place in the survivalism category instead.

Indeed, only the government scientists (clearly a prosaic reformist discourse of problem-solving) and the pastoralists (an imaginative but reformist discourse of sustainability) fall clearly into one box of Dryzek's typology. This can be interpreted in two ways. Either the typology fails the 'ideal type' test and is not a useful analytic tool. Or the lack of discursive clarity of some of the actors' discourses may be a significant factor in explaining the pattern of discursive coalition formation and the success or failure of individual actors to achieve discursive influence. Before concluding on this, however, the second part of Dryzek's framework deserves consideration.

This concerns the detailed social constructivist/rhetorical analysis of the discourses. Some, but by no means all of this analysis is summarized in Table 6.3. What is immediately evident here is the wealth of detail that such an analysis offers. Furthermore, such an analysis is able to incorporate more of the emotional impact of the different discourses. The emphasis on story-lines, while couched in terms of narrative rather than logic, still privileges an account of connections and makes little reference to the language in which the narrative is delivered. Much of the impact of language is not just to be found in the plot of the story being told but in how that story is told. It is here that rhetorical analysis in particular demonstrates its strengths.

Starting with the Aboriginal discourse, the key entities here are

Table 6.3 Applying Dryzek's rhetorical analysis

	Basic Entities	Assumptions re Relationships	Agency and Motivation	Metaphors and Rhetorical Devices
Aborigines	Aborigines with moral responsibilities; 'nature'; Europeans	Aborigines in tune with nature; Europeans destructive of nature	Europeans in pursuit of profit; Aborigines seek to preserve land	Very flat as largely spoken for, that is, mediated voice
Government scientists	Responsible, knowledgeable and realistic scientists, ecosystems	Humans can work with nature within a management/ scientific framework; scientific justification of traditional practices	Ethos of scientific rationality or scientifically informed managers	Fire as 'management tool'; described as 'patch burning', a 'traditional' practice; 'rejuvenation' through burning; having an 'evolutionary' role; 'holocaust' references
Pastoralists	Pastures as 'botanical communities'; responsible and irresponsible individuals	Emphasis on win-win scenarios of sustainable development	Profit can promote environmental protection	'Stability', 'integrity'; religious and emotive imagery
Wattle Hills	Humans as part of nature	Nature knows best	Non-interventionist; self-sufficiency as route to sustainability; nature also has agency	Farmland as 'priceless resource'; emotional tone; use of Haiku poetry; some scientific rhetoric
Independent scientists	Humans and nature as separate	Scientific knowledge reveals relationships	Scientific rationality legitimately dominates; scientists as responsible; humanity could destroy nature	Scientific terminology *plus* emotional language

constructed as Aborigines with an innate set of responsibilities and rights in relation to the land, nature as a discrete entity, and Europeans, constructed as bearing guilt and blame. While Aborigines are viewed as being in tune with the 'natural rhythms of life' and intrinsically seeking to preserve the land, Europeans are constructed as destructive towards nature and operating in pursuit of economic profit. These constructions support the story-line gleaned from a Hajerian analysis of the text. The Dryzekian analysis adds little in this case largely because there is little use of metaphors or other rhetorical devices. This in turn is because the Aboriginal voices are largely reported or represented. Indeed, both the Aboriginal representations at the seminar were made by white academics. There is little active voice in this discourse and the resulting tone is rather 'flat'. The only discursive colour can be found in the references to caring for the land by referring to burning as 'cleaning up' land that would otherwise be 'neglected' (p. 23). Love for and care of the land is equated with burning it; a connection that is also made by the government scientists (p. 34).

By contrast, the analysis of the government scientists' discourse from this perspective is very revealing. The main constructed entities are scientists – constructed as responsible, knowledgeable but also realistic – and ecosystems, which are a scientific category. This combination of constructions gives considerable authority to the government scientists. They constitute the main means of accessing knowledge about ecosystems. 'Experience of over 50 years of research' (p. 37) is referenced to support government pro-burning policy. It is stated by one representative that the government position on the use of fire is 'so clearly established factually' that anyone disputing it should 'go and read the literature' (p. 35).

But the government scientists also present themselves in a range of moral terms: they are responsible: they recognize the limitations of scientific knowledge in terms of uncertainty – 'we are never going to have perfect knowledge' (pp. 12 and 14). They also refer to the importance of local knowledge and of cultural heritage, therefore enabling the link between their discourse and that of the Aborigines. This is taken further in how the key relationships are constructed. Scientific knowledge is seen as justifying the kind of traditional practices undertaken by Aborigines. Intertextual reference is made to carbon dating evidence that suggests a 40 000-year history of the use of fire in Australia to justify an accusation of 'supreme arrogance' on behalf of those opposed to burning as they are 'denying the ancient order' (p. 34). Science can work with tradition within the context of an overall assumption of the possibility of positive human engagement with nature that enables management. Furthermore, such human management is inevitable and has always happened: 'There is no such thing as natural management' (p. 13). This places modern scientific management on a par with traditional Aboriginal management. The ethos of scientific rationality that is relied on here, therefore, does not undermine the potential for an alliance with the Aboriginal discourse.

As well as adding such detail to the workings of a discourse coalition between the government scientists and the Aborigines, the rhetorical analysis provides an insight into the emotional appeal of such a discourse. The government scientists' discourse is a rhetorically rich discourse. In particular, there is very active rhetorical engagement with the key term of 'fire'. As they say, colloquially, 'Fire ain't fire' (p. 13). Instead the discourse describes fire in terms of a 'management tool', a 'traditional practice', a form of 'rejuvenation' (p. 12), and as having an 'evolutionary' role (p. 15). Fire is not a negative thing when discussed in such language. It makes it possible to combine positive reference to 'fire' in a discourse that also talks of having 'love' for the land (p. 34).

And yet the negative effects of using fire are acknowledged. There is reference to individual birds and animals killed by fire. This is, however, set against a synecdochical account of how burning can save the habitat of two key species – the malleefowl and ground parrots (pp. 12–13). The extent of the intended burning is also firmly set in context by invoking the extremely emotive term 'the holocaust'. What the government scientists are doing is making 'choices' (p. 14) or 'playing God' (p. 16) in order to avoid 'policies of inaction' that would result in a 'holocaust of extinction' (p. 38). The fire of government scientists prevents the fire of complete annihilation.

The third party to the dominant discourse coalition, the pastoralists, has a similar discourse in the sense of mixing scientific and emotive rhetoric. The pastoralists' discourse again constructs the key entities in terms of, on the one hand, nature described in scientific terms – botanical communities with scientific names for species – and, on the other hand, humans as individuals who are able to act responsibly (although this is acknowledged as an individual choice, not as inevitable). The relationship assumed between nature and humans is seen in terms of the kind of win-win scenarios that are typical of the broader sustainability discourse. The value that this discourse adds over the scientific/traditional management discourses of the government scientists and the Aborigines is its suggestion that economic profit can also be harnessed to the goal of conservation. As a result, the combination of the three discourses is rhetorically very strong indeed.

Similar to the government scientists' discourse, there is also a rich use of emotive rhetoric within the pastoralists' discourse, combining scientific terminology with moral imagery. Pastures are portrayed as biotic communities with 'stability' and 'integrity', both moral and eco-scientific terms.

There is also use of quasi-religious rhetoric with reference to fire. A more primal type of religiosity is invoked:

> we are always going to have trouble with fire, as long as some of us feel a thrill, a quickening of the pulse, as we light up the edge of a road, or feel a grim satisfaction as we watch the flames leap up the hillside, because since mankind first learned to use it, everyone loves a fire. (p. 42)

In these terms, the pastoralists also see fire as a strong cultural reference point to support the management claims for their burning practices: 'mankind has held fire in both a revered and feared position . . . Fire has been given God qualities and worshipped' (p. 29). The symbolism of fire is related to scientific claims for fire as rejuvenating habitats with 'old trees sacrificed for new seedlings' (p. 12).

Conversion, another religious theme, features too. The representative from the Cape York Peninsula Development Association (CYPDA) employs a persuasive rhetorical technique, emphasized by Dryzek, in stating his original affinity with the anti-burning discourse, which changed over time as he realized that his views were factually misguided (pp. 39–42). Fire is portrayed as an 'emotional issue clouded by folklore', which contrasts to the rational scientific reality of its being merely 'a fast form of oxidation'. The 'community disharmony' that exists over the use of fire is put down to a simple 'lack of understanding' on behalf of those opposed to burning. This establishes further discursive affinities with the government scientists' argument.

Turning to the Wattle Hills residents, this is the only discourse to construct humans and nature as part of the same entity; effectively this places humans as part of nature. This undermines the notion of human agency. Instead, the appropriate role becomes one of non-intervention and self-sufficiency. By contrast, nature is credited with substantial agency as well as superior knowledge: 'nature knows best'. This detailed construction of the residents' discourse puts them discursively at odds with the previous three discourses; it therefore reinforces the structure of the discourse coalitions discussed above. In addition, the Wattle Hills discourse is the most emotive of the five presented in the workshop. It uses Haiku poetry and highly charged language. It refers to Australia as a 'fire-shaped continent' (p. 10) and anthropomorphizes the larger animals, referring also to the 'tender growth' where birds nest (p. 9). Reference is made to 'QPWS arsonists' alongside 'pyromaniacs' and 'vandalism' (pp. 9 and 17). The QPWS policy, which implies 'nature can no longer take its course', is dismissed as being 'spawned' by some 'dark philosophy' (p. 9). A strong line is taken in establishing the destructiveness of fire through the use of emotive phrases such as 'bushfires kill and destroy', they bring 'death and destruction'

(p. 8) resulting in a 'smoking, blackened landscape' (p. 10). The impact of 'The Almighty Match' (CAFNEC, 1992) on wildlife, especially birds, provides the main focus for their arguments with reference made at one point to one of the representatives having observed 'young birds' being 'burnt alive' and 'totally cooked' (p. 44).

The following closing statement is typical of the non-conformist nature of the representation: 'Let us aim for perfection in all things but please, now, stop diminishing our reservoirs of nature and spirit.' The failure to engage with the conventions of non-emotional presentation is explicitly recognized by the representative himself who makes 'no apology for introducing some emotion to the debate'. Such non-conformity, however, may well be interpreted as making it difficult to gain credibility for the argument by positioning it outside the institutional conventions of the other representatives. The Wattle Hills discourse does use a limited amount of scientific rhetoric (p. 9) but has no rhetorical means of combining the scientific and the emotive and this leaves the discourse as predominantly emotional, a position that is bound to reduce its standing in any policy debate.

It does, however, play a role in constituting the identity of Wattle Hills. The residents of Wattle Hills pursue a very different environmental management policy from most other landholders in the Cape and are well known for their alternative lifestyle centred on self-sufficiency and natural regeneration. This is reflected in their presentation's overall departure from the rational, scientific approach adopted by the rest of the representatives at the seminar. Here, the discourse constitutes Wattle Hills as a distinctive community. This, however, undermines their attempt to take a central place within the policy debates. There is an attempt at an appeal to the 'economic advantage' to Cape York landholders of a fire-free management regime, which, they claim, will 'halt the decline in soil fertility', healthy farmland being a 'priceless resource' (p. 11). This could be interpreted as attempting discursive affinity by engaging with those for whom economic gain is a priority. This is not, however, going to be a winning trope in a debate framed around environmental protection and where scientific rationality plays such a key role.

Finally, there is the discourse of the independent scientists. As might be expected this has parallels with the government scientists' discourse in terms of the pattern of social construction and rhetoric. Humans and nature are viewed as distinct with scientific knowledge legitimizing certain practices, through revealing the key relationships affecting natural systems. Scientists are again seen as responsible and scientific terminology is again combined with emotional language. The difference here, however, is the mainly negative ethos and loose use of apocalyptic rhetoric: 'destroyed', 'disaster' and 'catastrophe'. There is also the use of tropes that imply nature would be better off unmanaged. Burning is equated to a violation (p. 17), suggesting a virgin state would be preferable. And wilderness, presented as unmanaged land, is compared favourably with managed landscapes (p. 20), in particular as wilderness is seen as the source of sublime romantic encounters: 'a place where we can stand with our senses steeped in nature' (p. 20).

The rhetoric used drives the independent scientists' argument towards the conclusion that humans will inherently destroy rather than conserve nature and that management conflicts with the natural state of the land. The government scientists, by contrast, managed the ethos to imply a more positive message. In policy debates this may well carry greater weight than a purely negative and oppositional discourse. While negative rhetoric such as this is highly influential within environmentalism and can assist in building coalitions among environmentalist groups, it is less effective within governmental policy settings.

Concluding on discourse analysis of a dominant policy perspective
What we hope to have shown in this application of two discourse analysis approaches to a specific environmental policy case study is the kind of insights that can be achieved through an attention to language and discourse. For example, throughout the analysis, Hajer's story-line concept provided a powerful heuristic device enabling the conflicting claims of the pro- and anti- burning discourses to be clearly illustrated. Two separate discourse coalitions are discernible throughout, subscribing either to a story-line of 'fire is desirable' or 'fire is undesirable'. Precisely as Hajer posits, the members of these coalitions subscribe to the same story-line but tend to apply different meanings to it. For example, the Aboriginal representatives subscribe to the 'fire is desirable' story-line on the grounds of cultural tradition, whereas the government scientists subscribe on the grounds of ecological desirability.

Discursive affinities are easily discernible within representations of members of the same discourse coalition, such as the promotion of the fuel reduction story-line by both government scientists and pastoralists. Obvious attempts to discredit the storylines of opposite discourses are also observed throughout the representations. From the perspective of Dryzek's work, his typology of societal discourses on the environment proves a rough-and-ready tool for analysing the construction of the discourses. By describing basic entities, assumptions and motivations within different representatives' story-lines, Dryzek's approach is useful in enabling differentiation between those subscribing to the same overall story-line. It also provides an access point for the emotive use of language, which plays a central role in the effectiveness of discourses. This is further enriched by the detailed analysis that rhetoric affords. From the perspective of the two approaches, the sustained dominance of the pro-burning discourse within government policy can be explained by a combination of the failure of the anti-burning discourse to achieve sufficient discursive affinity to effectively challenge the dominant pro discourse and the rhetorical strength of the various proponents of the pro discourse.

But it should be recognized that these are just two possible perspectives on discourse analysis. For example, they ignore the issue of 'pervasive power' accounts of dominant policy perspectives more commonly found in mainstream Foucauldian approaches that do not make the adjustments that Hajer considers necessary. This emphasizes the necessity of making theoretical choices (with methodological implications) in undertaking any discourse analysis. An emphasis on discourse is not sufficient; the analyst needs to adopt a specific take on how discourse operates socially and within policy contexts. Dryzek and Hajer offer two such possibilities. They are able to expose the ontological and epistemological assumptions and constructions that underlie the dominant policy perspective on anthropogenic burning in Cape York as well as challenging the stance of objectivity that is often assumed by policy-makers in the face of conflicting discourses. As environmental decision-making inherently takes place under conditions of radical uncertainty, the significance of attending to such exposure should not be underestimated.

We end with some reflection on the experience of undertaking discourse analysis as a methodological approach. First of all, we would endorse Lees' point that any discourse analysis must be undertaken with rigour, just as with any methodological approach (Lees, 2004). In our view this involves having a clear theoretical framework that ties in closely to the form of attention to discourse that is adopted. There is a need to specify the concept and terms that will be used to analyse the discourse and even to collect the data that constitutes evidence of the discourse. We have found Hajer and Dryzek's approaches helpful here in identifying such 'middle-range' concepts and terms. A detailed attention to the words and language of texts has been invaluable in providing a rigorous form of discourse analysis that is transparent and justifiable. Collaborative work at the level of close-reading has proved an essential element of maintaining rigour while also introducing the creativity needed for achieving insightful discourse analysis. We would see such analysis as inevitably a creative process. It is also a reflective process, drawing on the everyday skills and knowledge that any researcher has as a language user but in a way that builds a link between theoretical understanding and linguistic practice (Myerson and Rydin, 1996). These links between theory, careful empirical work and a balance between creativity and reflection on the part of the researcher are all elements that make for successful and effective discourse analysis.

Finally, however, a note of caution must be raised in the context of the pervasive power accounts of the influence of discourse discussed above. These arguably suggest an implicit responsibility for us as analysts to be aware of our own ontological and epistemological beliefs and understandings. This implies a need to adopt a critical and fundamentally reflexive approach where we explicitly consider how our own interests may wittingly, or unwittingly, be reflected in the production of knowledge that might be inherently perspective-bound (Hastings, 2000; Heller, 2001). This is of particular importance given the degree of legitimacy afforded to academic perspectives (Hastings, 2000). The question must always be asked: would another author analysing the same text reach the same conclusions? Discourse analysis carried out in the spirit of such critical and reflexive enquiry can then fulfil its potential as a heuristically powerful and potentially emancipatory tool for policy analysis.

Notes

1. For further detailed explorations of this case study see Ockwell and Rydin (2006) and Ockwell (2008).
2. This euro-anthropocentric interpretation of Aboriginal burning is contested by contemporary anthropologists and many Aboriginals – this is outlined further below.
3. See Ockwell and Rydin (2006) for a more in-depth discussion of the dominant pro-burning policy paradigm in Cape York.
4. See www.firenorth.org.au/nafi2, accessed 19 August 2009.
5. Note: page references in parentheses from hereon refer to the 'Tropics Under Fire' seminar transcript under analysis.

References

Bakker, K. (1999), 'Deconstructing discourses of drought', *Transactions of the Institute of British Geographers*, **24**(3), 367–78.
Bevir, M. (1999), 'Foucault, power, and institutions', *Political Studies*, **47**(2), 345–59.
Boulding, K.R. (1966), *The Economics of the Coming Spaceship Earth*, Baltimore, MD: Johns Hopkins University Press.
Bowman, D.M.J.S. (1998), 'Tansley Review No. 101 – The impact of Aboriginal landscape burning on the Australian biota', *New Phytologist*, **140**, 385–410.
Bowman, D.M.J.S. (2000), *Australian Rainforests: Islands of Green in a Land of Fire*, Cambridge, UK and New York: Cambridge University Press.
Burgess, J., C.M. Harrison and P. Filius (1998), 'Environmental communication and the cultural politics of environmental citizenship', *Environment and Planning A*, **30**(8), 1445–60.
Buttel, F.H. (1997), 'The politics of environmental discourse: ecological modernization and the policy process', *Social Forces*, **75**(3), 1138–40.
CAFNEC (1992), 'Tropics under fire: fire management on Cape York Peninsula', Public Seminar, Cairns and Far North Environment Centre, Australia.
Chong, D. (1991), *Collective Action and the Civil Rights Movement*, Chicago: The University of Chicago Press.
Crowley, G.M., S. Garnett and S. Shephard (2003), 'Management guidelines for golden-

shouldered parrot conservation', Brisbane: Queensland Environmental Protection Agency.
Darcy, M. (1999), 'The discourse of "community" and the reinvention of social housing policy in Australia', *Urban Studies*, **36**(1), 13–26.
DeLuca, K. (1999), *Image Politics: The New Rhetoric of Environmental Activism*, London: The Guilford Press.
DNR (2001), *2000/01 Technical Highlight*, Queensland: Queensland Department of Natural Resources, Mines and Energy.
DNR (2004), *Vegetation Management Code for Ongoing Clearances. Cape York Peninsula Bioregion*, Queensland: Queensland Department of Natural Resources, Mines and Energy.
Dryzek, J.S. (1990), *Discursive Democracy: Politics, Policy and Political Science*, New York: Cambridge University Press.
Dryzek, J.S. (1995), 'Review of "Risk Society – Towards a New Modernity" – Beck, U.', *Policy Sciences*, **28**(1), 231–42.
Dryzek, J.S. (1997), *The Politics of the Earth. Environmental Discourses*, Oxford, UK: Oxford University Press.
Dryzek, J.S. (2000), *Deliberative Democracy and Beyond: Liberals, Critics, Contestations*, Oxford, UK: Oxford University Press.
Edelman, M. (1988), *Constructing the Political Spectacle*, Chicago: University of Chicago Press.
Elliott, L. (1999), 'Review of "The politics of the earth: environmental discourses"', *Australian Journal of Political Science*, **34**(1), 130–31.
EPA (2002), *Fire Management*, Brisbane, Queensland: Queensland Environmental Protection Agency and Queensland National Parks and Wildlife.
Flannery, T. (1994), *The Future Eaters*, Sydney: New Holland Publishers.
Flournoy, A.C. (1993), 'Beyond the spotted owl problem – learning from the old-growth controversy', *Harvard Environmental Law Review*, **17**(2), 261–332.
Foucault, M. (1979), *Discipline and Punish: the Birth of the Prison*, New York: Vintage.
Foucault, M. (1980), *Power/Knowledge. Selected Interviews and Other Writings 1972–1977*, Brighton: Harvester.
Foucault, M. (1984), *The Care of the Self: The History of Sexuality Volume 3*, Harmondsworth: Penguin.
Foucault, M. (1990), *The History of Sexuality Volume 1: An Introduction*, London: Penguin.
Gill, A.M., J.R.L. Hoare and N.P. Cheny (1990), 'Fires and their effects in the wet-dry tropics of Australia', in J.G. Goldammer (ed.), *Fire in the Tropical Biota*, Berlin: Springer-Verlag, pp. 159–78.
Gill, A.M., J.C.Z. Woinarski and A. York (1999), *Australia's Biodiversity – Responses to Fire. Plants, Birds and Invertebrates*, Department of the Environment and Heritage, Canberra, ACT.
Grice, T.C. and S.M. Slatter (eds) (1997), *Fire in the Management of Northern Australian Pastoral Lands*, St Lucia, Queensland: The Tropical Grassland Society of Australia, St Lucia.
Habermas, J. (1996), *Between Facts and Norms: Contributions to a Discourse Theory of Law and Democracy*, Cambridge, UK: Polity Press.
Hajer, M.A. (1993), 'Discourse coalitions and the institutionalization of practice', in F. Fischer and J. Forester (eds), *The Argumentative Turn in Policy Analysis and Planning*, Durham, NC: Duke University Press, pp. 43–71.
Hajer, M.A. (1995), *The Politics of Environmental Discourse. Ecological Modernization and the Policy Process*, Oxford, UK: Oxford University Press.
Hajer, M.A. and H. Wagenaar (2003), *Deliberative Policy Analysis. Understanding Governance in the Network Society*, Cambridge, UK: Cambridge University Press.
Hastings, A. (1999), 'Discourse and urban change: introduction to the special issue', *Urban Studies*, **36**(1), 7–12.
Hastings, A. (2000), 'Discourse analysis: what does it offer housing studies?', *Housing Theory and Society*, **17**(3), 131–9.

Healey, P. (1999), 'Sites, jobs and portfolios: economic development discourses in the planning system', *Urban Studies*, **36**(1), 27–42.
Heller, M. (2001), 'Critique and sociolinguistic analysis of discourse', *Critique of Anthropology*, **21**(2), 117–41.
Hill, R. (2003), 'Frameworks to support indigenous managers: the key to fire futures', in G. Cary, D. Lindenmayer and S. Dovers (eds), *Australia Burning: Fire Ecology, Policy and Management Issues*, Collingwood, Australia: CSIRO Publishing, pp. 175–86.
Hillier, J. (1998), 'Beyond confused noise: ideas towards communicative procedural justice', *Journal of Planning Education and Research*, **18**(1), 14–24.
Jones, R. (1969), 'Fire-stick farming', *Australian Natural History*, **16**(7), 224–8.
Keeley, J. and I. Scoones (2000), 'Knowledge, power and politics: the environmental policy-making process in Ethiopia', *Journal of Modern African Studies*, **38**(1), 89–120.
Leach, M. and R. Mearns (1996), *The Lie of the Land. Challenging Received Wisdom on the African Environment*, Oxford, UK: James Currey.
Lees, L. (2004), 'Urban geography: discourse analysis and urban research', *Progress in Human Geography*, **28**(1), 101–7.
Majone, G. (1989), *Evidence, Argument and Persuasion in the Policy Process*, New Haven, NJ: Yale University Press.
Marlow, D. (2000), *A Guide to Fire Management in Queensland (incorporating fire management theory and departmental practice)*, Queensland: Queensland Department of Natural Resources, Mines and Energy.
Mason, M. (1999), *Environmental Democracy*, London: Earthscan.
Meadows, D.H., D.L. Meadows, J. Randers and W.W. Behrens (1972), *The Limits to Growth* (a report for the Club of Rome's project on the predicament of mankind), New York: Universal.
Myerson, G. and Y. Rydin (1996), *The Language of Environment: A New Rhetoric*, London: UCL Press.
Nossiff, R. (1998), 'Discourse, party, and policy: the case of abortion, 1965–1972', *Policy Studies Journal*, **26**(2), 244–56.
Ockwell, D. (2008), '"Opening up" policy to reflexive appraisal: a role for Q Methodology? A case study of fire management in Cape York, Australia', *Policy Sciences*, **41**(4), 263–92.
Ockwell, D. and J.C. Lovett (2005), 'Fire assisted pastoralism vs. sustainable forestry – the implications of missing markets for carbon in determining optimal land use in the wet-dry tropics of Australia', *Journal of Environmental Management*, **75**(1), 1–9.
Ockwell, D. and Y. Rydin (2006), 'Conflicting discourses of knowledge: understanding the policy adoption of pro-burning knowledge claims in Cape York Peninsula, Australia', *Environmental Politics*, **15**(3), 379–98.
Ostrom, E. (1990), *Governing the Commons: The Evolution of Institutions for Collective Action*, New York: Cambridge University Press.
QPWS (2000), *Fire Management System*, Brisbane: Queensland Parks and Wildlife Service.
Queensland State Government (1990), Fire and Rescue Service Act 1990.
Queensland State Government (1994), Environmental Protection Act.
Rees, J. (1990), *Natural Resources. Allocation, Economics and Policy*, London: Routledge.
RFS (2001), *Fire in the Landscape and Beyond 2000*, Queensland: Rural Fire Service.
Richardson, T. and O.B. Jensen (2000), 'Discourses of mobility and polycentric development: a contested view of European spatial planning', *European Planning Studies*, **8**(4), 503–20.
Rose, D.B. (1996), *Nourishing Terrains: Australian Aboriginal Views of Landscape and Wilderness*, Canberra: Australian Heritage Commission.
Russell-Smith, J., P.J. Whitehead, R.J. Williams and M. Flannigan (2003), 'Fire and savanna landscapes in northern Australia: regional lessons and global challenges – preface', *International Journal of Wildland Fire*, **12**(3/4), v–ix.
Rydin, Y. (1998), 'The enabling local state and urban development: resources, rhetoric and planning in East London', *Urban Studies*, **35**(2), 175–91.
Rydin, Y. (1999), 'Can we talk ourselves into sustainability? The role of discourse in the environmental policy process', *Environmental Values*, **8**(4), 467–84.

Rydin, Y. (2003), *Conflict, Consensus and Rationality in Environmental Planning: An Institutional Discourse Approach*, Oxford, UK: Oxford University Press.

Rydin, Y. (2005), 'Geographical knowledge and policy: the positive contribution of discourse studies', *Area*, **37**(1), 73–8.

Scrase, I. and D. Ockwell (2009), 'Energy issues: framing and policy change', in Ivan Scrase and Gordon MacKerron (eds) *Energy for the Future, A New Agenda*, Basingstoke: Palgrave Macmillan.

Smith, G. (2003), *Deliberative Democracy and the Environment*, London: Routledge.

Stocking, G.C. and J.J. Mott (1981), 'Fire in the tropical forests and woodlands of northern Australia', in A.M. Gill, R.H. Groves and I.R. Noble (eds), *Fire and the Australian Biota*, Canberra: Australian Academy of Science.

Watt, P. and K. Jacobs (2000), 'Discourses of social exclusion – an analysis of "Bringing Britain Together: A National Strategy for Neighbourhood Renewal"', *Housing Theory and Society*, **17**(1), 14–26.

Wheeler, W.B. and M.J. McDonald (1986), *TVA and the Tellico Dam 1936–1979. A Bureaucratic Crisis in Post-industrial America*, Knoxville: University of Tennessee Press.

7. Theoretical perspectives on international environmental regime effectiveness: a case study of the Mediterranean Action Plan
Sofia Frantzi[1]

Introduction

Many modern environmental problems are not occasional random events that suddenly arise, but are rather the result of long-term processes requiring effective management through time instead of instant solutions. Their causes and effects are complex issues, strongly interlinked with other aspects of social, political and economic realities. When these problems are of a transboundary, or global nature, then their management must be attempted through regional (bilateral or multilateral) or international agreements.

Traditionally the focus of academic research has been on issues associated with the challenge of achieving international cooperation, in other words on regime formation, but recently there has been an increasing interest in implementation issues, that is, regime effectiveness. This chapter aims to discuss the concept of effectiveness of international environmental agreements as debated within the academic literature. In the first section the major theoretical perspectives on international relations are presented as the context for understanding different explanations given to international cooperation. Different approaches to defining and measuring effectiveness of the agreements are then described in more detail. In the second section there is specific reference to a particular example of an environmental agreement. The Mediterranean Action Plan was chosen for this purpose since it has not been studied extensively and in addition its effectiveness is ambiguous according to different viewpoints. Finally in the last section, a new definition of effectiveness is given, drawing insights from the aforementioned literature, suggesting that for a regime to be effective it has to use a holistic approach, to have a pragmatic vision and to be of a dynamic nature. This perspective attempts to provide a new approach to the future study of international environmental agreements.

International environmental regimes
Modern environmental problems are often so extensive that they do not respect national boundaries and cannot be managed by one country acting alone. The need for international cooperation was at the forefront of concern about the environment in the 1970s, and since the 1972 United Nations Conference on the Human Environment, international environmental institutions have proliferated and over 60 multilateral environmental treaties have been signed. For example, new treaties have been established for the protection of stratospheric ozone, the protection of many regional seas from pollution, the control of European acid rain and the conservation of biodiversity, amongst many others (Sands, 2003).

Extensive research has been devoted to the 'high politics' surrounding the negotiations of these international agreements. However, little attention has been paid to the actual effectiveness of implementation after these treaties come into force. The main question that has puzzled researchers is: 'Do regimes matter?' Generally the sequence of events is that scientists recognize an environmental problem, an international agreement is negotiated, a regime is established and operates for some time, but does the regime really make any difference? Some scholars argue that the environmental impact of agreements might be negligible. Others answer that it is the political benefits that are of significance and this diplomatic activity counterbalances any weakness in tackling the actual environmental problem. It can be argued that it is the combination or trade-off of benefits, in both environmental and political terms, that is the key to a regime's success. However, it can also be argued that regimes make a difference irrespective of whether this difference is in the environmental or political field. Below, the main academic and research viewpoints considering effectiveness are described in more detail.

International relations and regime theory
The study of international environmental agreements has become an increasingly important issue in the literature of international relations. Historically the study of these agreements is based in realism, neorealism and neoliberal institutionalism, evolving into what is now called regime theory. In addition, international political economy approaches based on historical materialism have often been used to study cooperation on environmental problems. Some basic observations about these different theoretical perspectives are given below in order to demonstrate the background to explanations used for international environmental cooperation.

Realism The realist approach has descended from traditional texts such as Thucydides' *History of the Peloponnesian War*, Machiavelli's *The*

Prince and Hobbes's *Leviathan* and has been mainly concerned with state security (Haas, 1990, p. 35). Emphasizing the political sphere, the realist approach analyses relationships among states only according to issues of power and self-interest (Kütting, 2000a, p. 12). It assumes that states are only guided by national interest and that their purpose must be to maximize power, a process that ultimately leads to war as states compete amongst themselves. According to realists the actors (states) act rationally and prefer those options that best suit their interests, under the assumption that they have full awareness of world events and thus can estimate both costs and benefits of alternative solutions. Those solutions chosen concern the acquisition of power (Haas, 1990, p. 35). Hence only when the effectiveness of an international environmental agreement coincides with the interests of the states, can the agreement be effective (Kütting 2000a, p. 12). However, since international discussions about environmental problems are often concerned with common threats to livelihoods and not about power, there is a difference in focus between realist thought about war and power on one hand, and concerns about environmental degradation on the other. Moreover, Haas (1990, p. 36) notes that there has been substantial criticism about realism not being an appropriate model for the analysis of environmental cooperation because of the importance it places on matters of security, which are generally not salient features of environmental agreements. However, if security could be extended to matters of public health or security of borders then it could be included as a theme when studying international environmental agreements.

Neorealism Neorealism is the most recent version of classical realism in international relations and is also known as structural realism. With Kenneth Waltz (1979) as its main representative (Keohane, 1986), this approach describes and studies international relations according to the system's structure. Neorealists take methods from game theory and microeconomics in order to explain how states behave under anarchy, and how they negotiate among themselves, resulting in hypotheses about their motives and the results of this negotiation (Haas, 1990, p. 37). However, realism and neorealism share some basic principles such as the international system still operating under anarchy and the states still being the main actors within it. Neorealism, however, allows for some kind of cooperation among states so as to reach a shared goal as, for example, tackling a common environmental problem, since its centre of attention has shifted from war (Kütting, 2000a, p. 13). This form of cooperation can be explained in two different ways, first through hegemonic stability theory and second through game-theoretic approaches (Haas, 1990; Paterson, 1996).

Hegemonic stability: hegemonic stability theory suggests that cooperation is most likely to occur when it is imposed by a dominant state or a 'hegemon' within a system (Haas, 1990, p. 40). The difference between the states that just dominate and the hegemons is that the latter already have their power and leading role legitimately approved by the other states (Paterson, 1996, p. 94; Kütting, 2000a, p. 13). However, according to Kütting (2000a, p. 14) this theory can only explain the existence of cooperation among states but not the quality of that cooperation, because the latter is out of its remit and therefore doesn't have the appropriate methods. For this reason it is not appropriate for studying the effectiveness of international environmental regimes.

Cooperation under anarchy (rational choice and game theory): the 'cooperation under anarchy' tradition is another school within neorealism, which suggests that even in the absence of a hegemon cooperation is still possible. As Paterson (1996, p. 101) observes, scholars of this tradition, influenced largely by game theory, believe that cooperation is indeed possible under conditions of anarchy without, however, suggesting generally that this cooperation could change the primarily anarchic character of the international political order. Rational choice and game theory study and foresee the behaviour of the actors by calculating the best possible decision, under rational terms, for any actor under a particular state of affairs (Kütting, 2000a, p. 14). This school looks at game-theoretic work focusing primarily on repeated game situations such as the Prisoner's Dilemma, the Chicken Game and Stag Hunt. One of the best-known options in empirical research for measuring regime effectiveness by using rational choice and game-theoretic approaches is the so-called Oslo-Potsdam solution, for which further details are given later in this chapter.

A difference with the hegemonic stability school is that cooperation under anarchy suggests that various factors can cause the maintenance of the agreements by states after the decline of a hegemonic power that was initially necessary for the creation of these agreements. Moreover, the supporters of this school, in contrast to the realists, assume imperfect information, variable interest and choices of the actors, and only limited effort at seeking alternative solutions to the problem (Haas, 1990, p. 44). However, according to some authors (Paterson, 1996; Kütting, 2000a) rational choice, game-theoretic approaches and neorealist approaches in general, do not offer a major contribution to the study of the effectiveness of international environmental agreements for various reasons. First, they focus on the behaviour of units (states) and do not really include the object of cooperation (the environmental problem) in their analysis in the sense of dealing with the environmental degradation per se (Kütting, 2000a, p. 15). Second, their main assumption is that states can be treated as actors with given interests on a particular matter, generated by their position in the international system, whereas on environmental issues the interests of states can vary according to their internal structure, for example, the interests of states in the climate change debate (Paterson, 1996, p. 108). Third, according to Young (2001, 2003), while specifically criticizing the Oslo-Potsdam solution, these approaches encounter many analytical and empirical problems that are largely to do with neglecting important factors when accounting for the hypothetical situation in the absence of the regime, and for the collective optimal solution.

Historical materialism and international political economy Another approach often used for assessing international cooperation is an international political economy approach based largely on historical materialism (Paterson, 1996, ch. 8). Historical materialism is mainly concerned with the distribution of economic resources and international equality, often expressed as the North–South divide. Historical materialists explain cooperation in terms of the control of powerful capitalist states (for example, North American and European countries) over weaker ones (for example, developing or Third World countries). According to them the world is broadly divided into three categories on the basis of the division of labour internationally. These are the highly industrialized Western countries, the industrializing countries and finally the developing countries (Haas, 1990, p. 47). Historical materialists identify a much less democratic and equitable structure of international relations (both economic and political) than the neorealists, by suggesting that in cases where effective cooperation does take place it always repeats the principles of capitalism, that is, reproducing the structures where the North takes advantage of the South (Haas, 1990, p. 47). Some authors have found the international political economy approach appropriate for understanding the complex patterns of cooperation with regard to international environmental agreements. For example, according to Paterson (1996, ch. 8) it has been useful in assessing the difficult negotiations among countries over global warming and the UN Framework Convention on Climate Change. However, economic globalization gives rise to complex relations between environment on one hand and global trade and investment on the other, and so raises debates (Stevis and Assetto, 2001; Clapp and Dauvergne, 2005, ch. 5). According to Clapp (2006) there are three different views within this debate. The first one can see positive effects for the environment from international growth and even in cases where some negative side-effects appear, then environmental issues can find ways around them without restricting economic relations. The second view is primarily negative, suggesting that international economic growth can only harm the environment, hence requiring environmental agreements to restrict international economic relations. Finally, the last view is somewhere in between, admitting the potential for both advantages and disadvantages, arguing though that proper management of the global economy can generate benefits for both sides, environment and growth (Clapp, 2006). In this sphere of 'global governance' some writers suggest that in order for this link between trade and environment to work beneficially, the creation of a World Environment Organization (Biermann, 2000, 2006) might balance the negotiating power of the World Trade Organization. To conclude, according to the new perspective on the relationship between international political economy and the environment, the former could potentially offer some explanation of international environmental cooperation that differs significantly from historical materialism.

Neoliberal institutionalism Neoliberal institutionalism has dominated the study of international environmental agreements (Paterson, 2000, p. 12) and centres on the work of regime theorists such as Keohane, Young, Levy and others. This theory evolved from the development of traditions as old as those of Grotius and Kant (Paterson, 1996, p. 115; Kütting, 2000a, p. 15). In spite of the establishment of the United Nations after the Second World War, institutionalism faded mainly because it was considered to have failed in preventing international violence during the inter-war period (Paterson, 1996, p. 115). However, the strengthening of international reliance and collaboration and the emergence of regional integration in the 1950s and 1960s (in particular the European Community) led to its recurrence in an advanced form and its subsequent significance in the 1990s (Paterson, 1996, p. 115; Kütting, 2000a, p. 15). Neoliberal institutionalism, when studying the effectiveness of international environmental agreements, is closely interlinked with regime theory. Regime theory and a different approach within it, of great influence in the past decade, that of Haas's 'epistemic communities', will be discussed in detail below.

Regime theory: regime theory or neoliberal institutionalism evolved out of general developments in the international relations sphere and specifically out of neorealism, thus producing a whole new range of views about the role and importance of international institutions (Paterson, 1996, p. 116). According to Krasner (1983, p. 358), who was one of the first and more important authors on the subject, 'once regimes are established they assume a life of their own'. He suggests that there are many ways in which international institutions affect outcomes by influencing state behaviour. They can alter the capabilities of actors, including states', they can alter states' interests, they can be a source of power that states can appeal to and they may alter the calculations of states concerning the maximization of their self-interest (Krasner, 1983, p. 361). So regime theory could in many cases be seen as synonymous with institutionalism as already described since both focus on the effect of the processes held to influence states' behaviour, and within which sovereign states are caught (Paterson, 1996, p. 117).

The best-known and cited definition of regimes was given by Krasner (1983, p. 2) who stated that:

> Regimes can be defined as sets of implicit or explicit principles, norms, rules and decision-making procedures around which actors' expectations converge in a given area of international relations. Principles are beliefs of fact, causation and rectitude. Norms are standards of behaviour defined in terms of rights and obligations. Rules are specific prescriptions or proscriptions for action. Decision-making procedures are prevailing practices for making and implementing collective choice.

This definition is closely related to Young's and Keohane's definitions of institutions. Young (1989, p. 32) defines institutions as 'social practices consisting of easily recognised roles coupled with clusters of rules or conventions governing relations between occupants of these roles'. Keohane (1989, p. 3) gives another definition as 'persistent and connected sets of rules (formal and informal) that prescribe behavioural roles, constrain activity, and shape expectations'. Moreover, Keohane et al. (1993, p. 5) extend the definition of institutions by adding that 'they may take the form of bureaucratic organisations, regimes (rule-structures that do not necessarily have organisations attached), or conventions (informal practices)'. Later Levy et al. (1994, 1995, p. 274) in their work 'The study of international regimes' (1995) define international regimes as 'social institutions consisting of agreed upon principles, norms, rules, procedures and programs that govern the interactions of actors in specific issue-areas'.

The above definitions, differing slightly one from another, all allow for the study of international agreements regarding them as regimes and explaining their attributes according to them. For the purposes of this study Krasner's definition will be the point of reference.

According to Krasner (1983, pp. 6–10) there are three orientations of regime theory. The realist/structuralist view sees the states as actors in the international system that want to maximize their power, thus they use regimes only as means to establish rules expressing their interests. It does not allow for regimes to have an independent impact on behaviour, so it views the regime concept as useless. The modified realist/structuralist view sees regimes as the outcome of negotiations and bargaining, often analysed by rational choice and game theory, and includes other factors of international cooperation such as social or technological, hence moving away from the pure politics of maximization of interest. This view suggests that regimes may matter but only under fairly restrictive conditions, for instance when independent decision-making leads to unwanted outcomes. Finally, the Grotian view lays emphasis on social factors, and even though it sees the states as still the main actors in the international sphere, it assumes that these actors are necessarily bound by specific norms and rules. This last orientation considers regimes as much more persistent and accepts them as a fundamental part of all patterned human interaction, including interaction in the international system (Krasner, 1983, pp. 6–10).

Nevertheless, the distinction between the above three orientations does not really play an important role. Regimes cannot always be irrelevant, and they cannot always be necessary. So the view that regimes may matter under certain conditions, is the most appropriate. Their effectiveness is of great importance, since only effective regimes may make a difference. More details will be given below on the way that regime theory is applied to the study of international environmental regimes when discussing how regime theorists define and measure environmental regime effectiveness.

Epistemic communities: a popular tradition within environmental international regime theory is that of 'epistemic communities' (Haas, 1989). This theory highlights the role of knowledge-based 'epistemic communities' consisting of specialists responsible for articulating policies and identifying the national interest. Initially, the term 'epistemic community' was used in literature on the sociology of knowledge. It was later borrowed by international relations specialists and adapted to describe a specific community of experts. This community 'shares a belief in a common set of cause-and-effect relationships as well as common values to which policies governing these relationships will be applied' (Haas, 1989, p. 384). The community, even though originating from various disciplines, operates within a common network where there is an exchange of ideas, concerns, results and solutions, aiming at the same political objectives (Haas, 1990, p. 55). This approach focuses on the groups of people who initiate cooperation rather than on which states are the leading actors who start the process. However, supporters of this theory do not suggest it should replace the older international relations theories, but rather complement them. For instance, as will be described below, Haas (1990) in his study about the Mediterranean Action Plan explains the cooperation by referring to 'epistemic communities', but he also offers other explanations from the perspectives of realism/neorealism and historical materialism.

Defining and measuring regime effectiveness
Within regime theory there have been many efforts from researchers to rigorously study international environmental regimes and try to identify

not only how these agreements were formed, but also if they were effective afterwards. There is a growing interest in the effectiveness aspect of regimes, but it is a matter of debate because quite different definitions are used, resulting in different ways of estimating effectiveness. As Kütting (2000a, p. 30) observes 'Within the effectiveness debate in regime theory. . . on one level effectiveness is seen in terms of institutional workings through good institutional structures. . . on another level effectiveness is measured on the basis of environmental impact'.

Usually regime theorists look at effectiveness as institutional performance and not as environmental improvement. Even though some of them recognize the need to look at the environmental impact, only a few actually try to measure it. For example, some of the Norwegian regime theorists (Wettestad and Andresen, 1991; Underdal, 1992) have considered the environmental problem but still focus on the institutional performance of a regime. Also, Haas et al. (1993, p. 7) ask the question whether the quality of the environment is better because of the regime but they do not indicate how such change could be measured and how much of it could be assigned to the regime itself, rather than to other external factors. Nevertheless, change itself is not a sufficient measurement of effectiveness (Kütting, 2000b). However, recently there has been an attempt by Kütting (2000a, 2000b) to introduce the concept of environmental effectiveness when studying environmental regimes by distinguishing the concept of effectiveness as seen in institutional terms from that of accounting for improved environmental quality, though still having a regime theory perspective.

Furthermore, the attempts to measure effectiveness have been mainly qualitative. These qualitative methods vary in whether their view is descriptive (trying to explain what did happen), predictive (trying to estimate what will happen), normative (looking at what should ideally happen) or explanatory (trying to explain the reasons why something happened) (Mitchell and Bernauer, 2002, p. 2). However, a small but increasing number of researchers have approached the subject quantitatively, recognizing the need for these methods to complement each other in order to produce more reliable results. A brief discussion of some of these methods is provided below.

Qualitative approaches In order to estimate whether international environmental institutions are effective, Haas et al. (1993) refer to certain conditions known as the three Cs. They measure the impact of international institutions on three conditions essential for effective action in environmental problems: high levels of governmental *concern*, a hospitable *contractual* environment in which agreements can be made and kept, and sufficient political and administrative *capacity* in national governments. In each regime they examine three phases of activity; agenda-setting, international policies and national policy responses, which are referring to each of the three conditions respectively. Thus, a regime is deemed effective if it increases governmental concern, enhances the contractual environment and builds national capacity. They ask the question 'Is the quality of the environment or resource better because of the institution?', but due to lack of available data concerning changes to the state of the biophysical environment that can be actually assigned to the institution, they decide to focus on 'observable political effects rather than directly on environmental impact' (Haas et al., 1993, p. 7).

Young (1999) looks at causal connections and behavioural mechanisms. A regime is considered effective based on the extent it ameliorates the problem that led to the regime's creation in the first place. However, he admits that this approach is practically difficult to analyse since complex social and natural systems within which regimes operate do not allow for the observed changes to be assigned to the regime itself. According to the legal approach, the regime is effective to the extent it is followed by legal compliance, and in the economic approach if it incorporates the legal definition and adds a cost-efficiency criterion. In the normative approach, effectiveness equals achievement of values such as fairness or justice, stewardship and participation, whereas in the political approach a regime is effective if it causes changes to the behaviour of actors, in the interests of actors, or in the policies and performance of institutions in ways that contribute to positive management of the targeted problem. Moreover, Young differentiates the effects of environmental regimes in three dimensions. First, he divides them into internal and external to the behavioural complex, which is the group of actors, interests and interactions on a specific issue area. Second, he separates them into direct and indirect effects. Finally, he divides them into good or bad according to the impact on the problem, in other words if they ameliorate or worsen it (Young, 1999).

Another approach to the measurement of effectiveness focuses on institutional factors and addresses a series of related questions based on the identification of problem structure, institutions and institutional fit and the analysis of legal and organizational issues that arise from this approach (von Moltke, 2000). This research strategy begins with consideration of a problem's structure. It then proceeds to identify the institutions that may be needed – and those that have been employed – to address the issue in light of its problem structure. Von Moltke's underlying hypothesis is that it is more likely for a regime to be effective when it achieves a good fit between problem structure and institutional characteristics, and that it is the desirable fit between problem structure and institutions that is a primary reason for its effectiveness. Moreover, he stresses the importance of science assessment (the interpretation of the research for policy purposes), and the need for transparency and participation. He goes on to discuss the issue of dispute settlement mechanisms, without considering them necessary for environmental regimes since they pursue effectiveness and implementation in entirely different ways (von Moltke, 2000).

As mentioned earlier, one qualitative approach that is different from the others in the sense of introducing the concept of environmental effectiveness, is that of Kütting (2000a, 2000b). She suggests a distinction between institutional and environmental effectiveness, since most regime theories are interested in the structure of the institution and the behaviour of the actors in it, judging its effectiveness by the occurrence of change in this behaviour, which it is assumed would eventually lead to a positive environmental result. However, the change in actors' behaviour might not actually result in environmental improvement, and even if it does, this improvement might not be sufficient to solve the problem. In addition, the assessment of the state of the environment before and after the regime and how much of a change can be actually assigned to the regime itself poses another methodological problem. For this reason, Kütting regards the distinction between institutional and environmental effectiveness as necessary, stressing, however, that a good definition should incorporate both these dimensions since these are 'two sides of the same coin' (Kütting, 2000a, pp. 30–34). Her approach looks at four areas of environmental effectiveness, which describe the relation of an environmental problem to the particular regime established for its abatement, and the social structures within which they are found. These four determinants are economic structures, time, science and regulatory structures.

Economic structures include not only the structures concerned directly with the agreement but also refer to the economic organization of the society. Environmental problems can occur through the economic organization of the society but they can also be avoided through the same structures. Time is crucial when damage may be irreversible and this is frequently the case in environmental problems so the time plan of the environmental regimes has to account for that pressure. Science is necessary in policy-making in order to define the roots and the solutions to the problems, but according to Kütting its importance should not only be limited to being an input in the creation of the regime, but it should also be regarded as a social activity consistent with other social processes, emphasizing the constant interaction between science and policy. Finally, regulatory structures are mainly concerned with institutional design and effectiveness, referring not only to the structure of the agreement but also to the other bureaucratic structures within which the regime operates, and they are important because regime design matters (Kütting, 2000a, ch. 4).

Generally, when specific cases are studied in qualitative research, there is a problem about generalizing the results and assuming they will apply in all cases. Even though results may be reliable for a particular case, they cannot always be extended to others. Moreover, no matter how well a study of effectiveness is designed and carried out, its relative outcome depends heavily on the initial definition of effectiveness, and the criteria used to assess it.

Quantitative approaches A discussion of the main quantitative approaches in the study of environmental regime effectiveness is given below. Some of them are described briefly, whereas others are given in more detail due to their complex statistical nature. One of the most well-known options in empirical research for measuring regime effectiveness is the so-called Oslo–Potsdam solution. This is an 'umbrella term' referring to two closely interlinked approaches, that of Underdal (1992, 2002) and that of Helm and Sprinz (2000).

Underdal (1992, 2002, pp. 5–6) focuses on the relationship between the regime's output – the institution established as a new set of rules and regulations; its outcome – the change in the behaviour of states; and its impact – the actual change in the state of the biophysical environment. He suggests that regime effectiveness has two components: changes in human behaviour and changes in the state of the biophysical environment itself. Moreover, he asks some critical questions. First, what is the object to be evaluated, because it makes a vast difference whether the evaluation concerns only the regime, or whether it concerns the whole problem-solving effort that might include various kinds of costs or positive side-effects associated with the process of its establishment and maintenance. Second, he discusses the standard against which this should be evaluated, stressing however, that effectiveness is only a relative term and should be defined in each regime independently. The issue he raises about standards is important since environmental scientists and activists on one hand and regime theorists on the other, could have diverse opinions about the nature of standards against which they measure effectiveness. Third, he raises the issue of methodology in order to measure the object of evaluation against the standard. Methodologically, Miles and Underdal use counterfactual analysis against certain behavioural and technical optima by comparing the actual regime versus no-regime and the regime versus the collective optimum (Miles et al., 2002, ch. 2). They use qualitative case studies (for example, the Vienna Convention and Montreal Protocol, the International Whaling Commission, inter alia) to assess effectiveness on a 0–4 scale for behavioural change and on a 1–3 scale for environmental improvement. They then normalize the scales to range from 0 to 1 in order to make comparisons between them. A weakness of this approach is the difficulty of estimating the counterfactual by assuming hypothetical conditions in the absence of the regime. This is largely true since assessing the current state of the environment is difficult in itself. Even more difficult, if not impossible, is the idea of estimating how the state of the environment would be today if the regime in question did not exist in the first place. Moreover, the basis of this technique is still qualitative since environmental improvement and behavioural change are still assessed through qualitative case studies.

Helm and Sprinz (2000) also use counterfactual analysis based largely on the questions Underdal posed about the object of evaluation, the standard against which it should be evaluated and the methodological approach used. According to them regime effects are improvements in the object of evaluation, measured by application of policy instruments leading to changes such as emission reductions. A lower bound is determined by the no-regime counterfactual (NR), which is the degree of policy-instrument application that would have occurred in the absence of the regime. An upper bound is established by the collective optimum (CO), the degree of application that would have been obtained by a perfect regime. Accordingly, the regime potential is expressed in units of policy-instrument use and is the difference between the no-regime counterfactual and the collective optimum. The actual policies executed by countries (AP) usually fall into this interval. Thus, the effectiveness of a regime can be measured as the percentage of the regime potential that has been achieved, where this score falls into the interval of 0–1 (Figure 7.1).

They estimate scores by using a combination of methods such as game theory, optimization or experts' judgements. However, their approach has been criticized. Young (2001, pp. 110–14) points out that use of the Nash equilibrium leaves no room for cooperation, since it assumes that all actors try not to be taken advantage of, and it might also produce results that are worse for everyone, compared with those that could be achieved through other potential ways of cooperation. Moreover, he argues that the interactive decision-making used to calculate the no-regime counterfactual

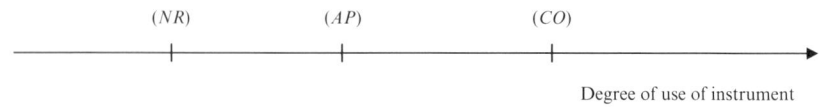

Effectiveness score $ES = (AP - NR)/(CO - NR)$
where (NR) = no-regime counterfactual, (CO) = collective optimum, (AP) = actual performance

Source: Helm and Sprinz (2000).

Figure 7.1   General concept for measuring regime effectiveness

leaves out many important factors such as political, technological, demographic and social factors. He has similar concerns about the collective optimum, pointing out that it neglects important side-effects of regimes when accounting for regime consequences. Empirically, Young suggests that the use of the counterfactual poses the same methodological problems discussed before since the use of expert judgements to estimate it are insufficient especially when they do not account for social or technological factors (Young, 2001, pp. 110–14). This critique led to a fruitful debate on the issue and on potential ways to improve these approaches (Hovi et al., 2003a, 2003b; Young, 2003).

Another approach to measuring effectiveness is given by Mitchell (2004) who, in order to evaluate international environmental regimes, uses regression analysis on panel data. He proposes a quantitative approach by developing a model for a single regime's effects. In this model he uses time-series data for one country at a time for the 1985 Sulphur Protocol of the European Convention on Long-range Transboundary Air Pollution (LRTAP). He specifies the following model to estimate national sulphur emissions for the LRTAP case (Mitchell, 2004, p. 127):

$$EMISS = \alpha + \beta_1 * MEMBER + \beta_2 * INCOME + \beta_3 * POP + \beta_4 * COAL + \beta_5 * EFFIC + \ldots + \beta_N * OTHER + \varepsilon$$

where *EMISS* is annual emissions of sulphur dioxide and *MEMBER* is coded as 0 in years of non-membership of the country to the regime and as 1 in years of membership. Generic drivers of emissions of most pollutants are also included such as per capita income (*INCOME*) and population (*POP*). Emission-specific drivers are included, such as the country's coal power plants (*COAL*) and their average efficiency (*EFFIC*). The model estimates difference in sulphur emissions and how these are explained by the different variables. For instance β_1 represents the expected difference in emissions that would arise from a country becoming a regime member, holding all other variables constant. The coefficients of the other independent variables β_2 through β_N correspond to the estimated increase in emissions that would arise from a one-unit increase in that variable. The *t*-statistic on the coefficients shows the statistical significance of the independent variables, whereas the goodness-of-fit (R^2) of the model equation as a whole provides an estimate of how completely the analyst has modelled the dependent variable.

Mitchell (2004, p. 129) advances his method by developing another model that allows comparison by combining data from different regimes. He uses time-series data and data across regimes. As an example he develops a model to assess the simple claim that sanctions are necessary for a regime to significantly influence behaviour. An extension of this model could be used to evaluate how much a regime's effectiveness depends on contextual factors. For example, international conferences and reports might raise the importance of an environmental issue for a few years, and therefore lead to increased levels of implementation and compliance (Brown Weiss and Jacobson, 1998).

An advantage of this technique, and also of other quantitative methods, is that its conclusions can hold reasonably well across many cases even though they cannot completely explain any specific regime (Mitchell, 2004, p. 122). However, it is important to avoid confusion between the notions of statistical significance and policy significance of the independent variables (Mitchell, 2004, p. 128). For instance, a study might show that an independent variable is statistically significant, which means that it can definitely explain the variation in the dependent variable. Despite that, the change in the variation might be so small as to be environmentally meaningless.

Mitchell's approach is a promising new angle to assess effectiveness with the use of econometrics and by using actual scientific measurement of the environmental problem (for example, emissions). However, it largely depends on availability of similar data for other regimes. For instance, when measuring marine pollution, it is almost impossible to keep a long time-series record of pollutants released into the sea, which is necessary for this type of analysis. Methodological problems would include which pollutants to measure, at what locations (since pollution may be a localized phenomenon), and how to connect these releases directly to the regime's regulations. Moreover, the high costs of marine monitoring deter countries from keeping regular data. So this approach may prove innovative and useful in certain cases, but its applicability in others remains in question.

All the above quantitative techniques have many advantages, as they can be based on actual measurements and their conclusions can be valid for many cases. They counterbalance the problem of generalization of results that qualitative techniques face. However, they might ignore aspects that are difficult to measure numerically (for example, political benefits) and might not completely explain particular cases. In that respect, quantitative analysis should not replace qualitative approaches, but instead a combination of the two can enable an integrated study of regime effectiveness.

Other issues related to the study of regime effectiveness In addition to the definition and measurement of regime effectiveness some other issues related to the study of environmental regimes are worth mentioning, notably institutional economics, compliance and verification, transparency, openness and participation, and environment and security. These issues can directly or indirectly affect the effectiveness of regimes, therefore they should be taken into account when studying a particular regime.

Institutional economics: within the framework of effectiveness of international environmental regimes, and since they belong to the broader category of institutions, an issue that is certainly worth looking at is economic efficiency or cost-effectiveness. This is the extent to which the production of the best economic outcome is produced by means of the least-cost combination of inputs. As North (1990) observes, transaction costs are the measure of economic efficiency of institutions. He stresses the message from Coase's theorem that when it is costly to transact, then institutions matter (Coase, 1960). North's theory of institutions combines human behaviour with the costs of transacting. The key to the costs of transacting is the costliness of information. This is because transaction costs include the price of what is being exchanged, and the costs of protecting rights and policing and enforcing agreements. He also argues that it needs resources not only to protect property rights and to enforce agreements but also to define these rights and agreement rules beforehand. Environmental regimes must perform certain functions such as limiting use, coordinating users and responding to changing environmental conditions, which include the transaction costs of coordination, information-gathering, monitoring and enforcement. It is easily possible to create a regime so costly to implement that it overcomes the benefits to be gained from its existence. Therefore, when examining the effectiveness of international environmental regimes, researchers should also take into account economic efficiency and transaction costs. No matter how effective a regime is in the amelioration of the problem it was designed for, it could not perform in the long term if it costs the countries too much.

Compliance and verification: when studying international environmental agreements and their effectiveness, Ausubel and Victor (1992) introduce the importance of verification of compliance. They suggest that verifiable international environmental agreements have more chances to have successful negotiation procedures and thereafter are more likely to be implemented properly by the participants. They define verification as 'the process determining whether or not a party is in compliance' (p. 4) and note that it has not been regarded as a significant aspect of most international environmental issues to date. In order to fulfil this criterion the creation of large costly new international or national organizational infrastructures is necessary, which in most cases has not been done, so most of the formal information under the regimes is collected, if indeed it is, by national organizations already existing before the regime was established. In many cases other actors such as NGOs are involved in this process. However, verification is still mainly dependent on national reports, which might be unreliable or even false especially when national interests are at stake. Hence, practically, it could be the case that compliance is not achieved even if reporting indicates to the contrary. Furthermore, it is crucial to properly set the standards against which compliance will be measured so as to be meaningful (Ausubel and Victor, 1992). Only recently have studies paid attention to the issue of compliance in international climate regimes (Barrett and Stavins, 2003; Victor, 2003) noting that successful implementation means high levels of participation and compliance. Barrett and Stavins (2003), commenting on the Kyoto Protocol, find that it does not induce significant participation and compliance and propose different approaches to improve it by offering positive or negative incentives. In the Montreal Protocol, for instance, a threat of restrictions on trade of CFCs or products containing CFCs between the countries participating in the agreement and those not participating proved successful in motivating more countries to participate. However, it is commonly acknowledged, especially in the case of the Kyoto Protocol, that compliance alone (even if fully achieved) cannot always mitigate the environmental problem, since some of the heavier polluters might not choose to participate in the agreement at all.

Transparency, openness and participation: one of the issues requiring attention from international environmental regime practitioners and scholars is transparency and openness. According to Ausubel and Victor (1992) transparency refers to the clear presentation of the regime's activities and information collected, whereas openness means access of actors to the negotiating process and information, irrespective of whether these actors come from within the government or not. They also note that successful environmental regimes should provide for these conditions, since in their case studies of arms control regimes the latter proved unsuccessful partly due to concealment and restricted participation (Ausubel and Victor, 1992). Moreover, von Moltke (2000) also stresses the importance of transparency and participation in environmental affairs in general, though remarkably few formal rules have been adopted in international environmental agreements to address these needs. A first step in this direction was the adoption in 1998 of the Aarhus Convention on Access to Information, Public Participation in Decision-making and Access to Justice in Environmental Matters (von Moltke, 2000).

Domestic politics: another important factor that affects international environmental cooperation is change in the patterns of domestic politics. According to Weale (1992, p. 200), domestic public policy can naturally be affected by actors and procedures in the international sphere. However, with the internationalization of political life, domestic actors and procedures may similarly affect and shape foreign policy-making. Active pressure groups may play a crucial role by shedding light on important issues and attracting media attention. This extra power can prove very useful with regard to international environmental agreements since it can be used to push governments into participating and complying with them. Moreover, as Carter (2001, p. 239) observes, domestic political pressure can originate from environmental groups, from the media, public opinion or political parties (especially the Greens). This pressure can persuade a government to change its position in the negotiations surrounding an international environmental regime, often resulting in that country becoming a 'lead state' with a key role in persuading or forcing other states to join efforts to form the regime. Carter gives as examples the swing of the West German government in the 1980s from veto to lead state on acid rain as a response to the Green Party becoming an electoral force and the decision of the Australian Labor Party to reject the Antarctic Minerals Treaty as a result of its pro-green position at the 1987 election, which aimed to win the support of environmentally conscious voters (Carter, 2001, p. 239). Finally, Haas et al. (1993, p. 17) argue that 'lead states' are subject to more intense domestic political pressure than other countries, something that led to US leadership on marine oil pollution in the 1970s and on ozone in the 1980s. This pressure, together with the frequently greater damage to the country from the environmental problem, and the advanced policies for that problem, increase governmental concern and capacity, resulting in promotion of institutional solutions to the problems by the 'lead states'.

Environment and security: traditionally, in political science, security has been considered as protection of a sovereign state from other sovereign states that might threaten it by means of military power (Morgenthau, 1978). Nowadays there is an increasing concern that environmental problems can threaten security by leading to violent conflict. According to Swatuk (2006) in the 1990s two different debates arose within the academic community. One is concerned with the redefinition of security in order to include environmental concerns, whereas the other focuses on the ways and extent that environmental issues may threaten security in the first place. The two sides have failed to reach a consensus. The first tries to interlink environmental change with the causes of conflict, identifying ways in which this might happen (Homer-Dixon, 1999). The other group, by contrast, suggests that whilst the high degree of global interdependence might result in environmental problems producing complex situations, it is rather unlikely to lead to violent conflict (Deudney, 1990).

Some argue that many environmental problems may present significant threats to human health and welfare, which in turn would affect the well-being of nations themselves and therefore these problems should be taken into account when considering issues of national and regional security (Kullenberg, 2002). This necessity for combining security matters with environmental issues is most appropriate in many cases of environmental problems, and especially so when addressing maritime affairs (Kullenberg, 2002). Moreover, Carter (2001) observes that according to realists the environment could be considered a security issue in cases where global commons problems might cause conflicts among countries. Such cases where military conflict arose straight from disputes over environmental problems are rare. However, a significant emerging issue is the rising number of environmental refugees who, while trying to escape from natural disasters such as drought, famine, degraded land and deforestation, seek a more secure future by crossing national borders (Carter, 2001, p. 227).

As Paterson (2000, pp. 18–23) puts it, again according to a realist view, there are two senses in which environmental change can threaten security. It may lead to interstate war especially over shared renewable resources, traditionally water, although this is an unlikely possibility. It is likely to cause internal instability of states, especially when combined with or caused by population growth. In that case environmental change may lead to a complete collapse of the social structure by unplanned urbanization, spreading disease and ecological marginalization of poor people. Haas (1990, p. 36) also recognizes the threat that environmental degradation poses to international security. He admits that the realist view of security has received criticism concerning whether it is appropriate for environmental issues and he finds it ambiguous. If the idea of security includes public health, security of borders, social and economic stability, then cooperative solutions would be more easily achieved for environmental problems. Haas argues that countries might still underestimate the environmental issues when matters of national power are involved, describing, for instance, the political tension that persisted in the negotiations of the Mediterranean Action Plan, resulting from a Greek–Turkish diplomatic incident at sea. No matter which side of the debate one takes, environmental degradation does seem to be associated with the security of nations, even if only in the sense of internal stability and social integrity, hence environmental regimes should also be assessed as an aspect of security.

The Mediterranean Action Plan and the Barcelona Convention
Having reviewed the literature concerning effectiveness of international environmental regimes in general, in this part of the chapter the discussion will focus on a particular environmental regime, the Mediterranean Action Plan. As mentioned earlier, it was chosen as a case because it has not been studied extensively (but see Haas, 1989, 1990; Skjaerseth, 1996,

2002; Kütting, 2000a, 2000b), and also its effectiveness is ambiguous according to different viewpoints.

The Mediterranean Action Plan (MAP) was created in 1975, under the auspices of the United Nations Environment Programme (UNEP), only three years after the Stockholm Ministerial Conference set up the latter programme. MAP was adopted as a Regional Seas Programme under UNEP's aegis. The UNEP Regional Seas Programme is a promising attempt to develop treaties and soft rules and standards at the regional level taking into consideration the different characteristics – both needs and capabilities – of the different regions (Sands, 2003). MAP was the first plan adopted and has worked since then as a model for designing the other plans.

The Barcelona Convention was signed in 1976 and forms the legal part of MAP, in force since 1978 and amended in 1995. It includes six protocols, namely, the Dumping Protocol, the Prevention and Emergency Protocol, the LBS (Land-Based Sources) Protocol, the SPA (Specially Protected Areas) and Biodiversity Protocol, the Offshore Protocol and the Hazardous Wastes Protocol. The Barcelona Convention is complemented by a research component (MED POL), policy-planning programmes (Blue Plan and Priority Actions Programme) and financial/institutional arrangements.

The Mediterranean Action Plan (UNEP/MAP) involves 21 countries bordering the Mediterranean Sea, as well as the European Union, which are Contracting Parties to the Barcelona Convention and its protocols.

MAP's main objectives (UNEP, 1995b: Annex IX) are:

- to ensure sustainable management of natural marine and land resources and to integrate the environment in social and economic development, and land use policies;
- to protect the marine environment and coastal zones through prevention of pollution, and by reduction and, as far as possible, elimination of pollutant inputs, whether chronic or accidental;
- to protect nature, and protect and enhance sites and landscapes of ecological or cultural value;
- to strengthen solidarity among Mediterranean coastal states in managing their common heritage and resources for the benefit of present and future generations; and
- to contribute to improvement of the quality of life.

Origins, negotiations and formation of the Mediterranean Action Plan
Haas (1990, ch. 3) gives a detailed overview of the history and negotiations up to the adoption of MAP, which is summarized below. Early worries

218 *A handbook of environmental management*

about Mediterranean Sea pollution arose between the late 1960s and 1974 when some Mediterranean officials expressed for the first time a need for action and governments sought ways to obtain information on the extent of marine pollution by identifying sources and types of pollutants and on possible ways to deal with the situation. Since adequate information was not yet available the attention focused on oil pollution resulting from maritime traffic and accidental spills, as this was the most visible form. Afterwards, however, several scientific meetings and conferences revealed a variety of pollutants and their sources, with the most important being the land-based, so in 1974 a first draft of a treaty was prepared by the Food and Agriculture Organization. However, later the same year, Mediterranean governments approached another United Nations organization, the United Nations Environment Programme (UNEP) to guide and support this regional effort, which in turn, with the help of 40 Mediterranean marine experts, developed a comprehensive plan. Finally, in 1975 the Mediterranean Action Plan was adopted including seven monitoring and research projects, for an entire set of pollutant types and sources, and several pilot demonstration projects (Haas, 1990, ch. 3). Thereafter MAP gradually widened its scope through creation of protocols covering land-based sources of pollution, marine dumping, tanker oil pollution, as well as pollution transported by rivers and in the atmosphere and by extending the lists to include more pollutants. The environmental assessment component of MAP also evolved as the research and monitoring projects increased from seven to 12 and some interim standards were developed (Haas, 1990, ch. 4).

However, following the 1992 UN Conference on Environment and Development 'Earth Summit' in Rio and the requirements of the Rio Declaration on Environment and Development (Agenda 21), MAP attempted to translate the results of the summit onto the regional Mediterranean level, and adapted Agenda 21 to the Mediterranean context by setting up Agenda MED 21. This led to adoption of the Action Plan for the Protection of the Marine Environment and Sustainable Development of the Coastal Areas of the Mediterranean (MAP II) on 10 June 1995 (UNEP, 1995b). MAP II reflected both increasing concern for the pressures exerted on the Mediterranean environment and commitment of Mediterranean states to the ideal of sustainable development.

International environmental cooperation and the creation of the Mediterranean Action Plan
Regional cooperation was necessary to create a treaty aimed at protection of the Mediterranean against pollution. Environmental cooperation, as with any other international relations procedure, requires different actors

or states to coordinate decisions and actions with the other actors involved. Reaching an international political agreement is difficult and there are different explanations about the conditions under which cooperation in the Mediterranean Basin was achieved through the framework of MAP and the Barcelona Convention. Haas (1990) summarizes these different interpretations into the main categories described below, explaining the causes of cooperation, its effects and its forms in each one of the views.

Realism and neorealism are concerned mainly with the relationship between state power and order in security affairs and the political economy of advanced industrialized societies. Realists and neorealists would relate cooperation to the distribution of power between the Mediterranean states. Under this perspective the regional hegemonic leadership of France would play a key role in developing cooperation under conditions of international anarchy. This hegemony would dictate that the scope of the agreements would mainly cover pollutants of interest to France but also extend to other issues of national French interest. The strength of cooperation – how weak or binding it is – would be dependent on French power and might also depend on information available. Under a realistic view the duration of the cooperation – how persistent it is – would also vary with the two previous factors and the effects of the cooperation would be to strengthen the influence of France in the region and achieve common benefits for all the Parties. However, this explanation did not prove adequate when, after the decline of the regional French hegemony, MAP continued to exist and to receive increased support both from the hegemon and also from weaker states, showing that it is difficult to predict potential change in the motives of the states (Haas, 1990, ch. 6).

Historical materialism, as discussed earlier, is basically concerned with distribution of economic resources and international equality, very often expressed as the North–South divide. Historical materialists explain cooperation in terms of the control of powerful capitalist states (that is, European countries in the case of the Mediterranean region) over weaker less-developed ones (that is, North African and/or Middle East countries in the same case). According to them the imperialism of European states would lead to cooperation under conditions of capitalism. The scope of the cooperation would not be clear but it would strengthen areas where European states have interests. Both strength and duration of the cooperation would vary with European dominance and effects of cooperation would be imposition of unwanted forms of development on less-developed countries, excluding alternatives, and the provision of relatively more benefits to European states, thus increasing commercial dependence of the less-developed countries on them. So, in the context of MAP, under a historical materialist interpretation, northern

220 *A handbook of environmental management*

Mediterranean countries would try to impose capitalist policies on the southern Mediterranean developing countries. However, the negotiations proved to be a compromise where both sides' interests were equally represented, indicating that historical materialism was not able to provide a satisfactory explanation of cooperation (Haas, 1990, ch. 7).

A third set of explanations introduced by Haas (1989, 1990) involves the 'epistemic communities' theory. This theory highlights the role of specialist knowledge-based 'epistemic communities' in formulating government policy and altering national interests and finally leading to international cooperation. The 'epistemic community' approach gives a more flexible character to the cooperation, having a broader scope than the other explanations. According to this approach the acquisition of new information and the negotiations between the states would lead to cooperation under conditions of scientific uncertainty. The scope of the cooperation would be broad and specifically outlined by the 'epistemic community' and the strength and duration of the cooperation would vary with the extent of the involvement of the 'epistemic community' and coalitions within the states. This cooperation would lead to adoption of convergent pollution control policies, and would eventually inspire Mediterranean governments to design and implement new models of comprehensive environmental policy. Indeed, the countries where scientific experts were strong had deeper involvement in MAP and became its strongest proponents, and vice versa. The 'epistemic communities' explanation complements the previous two theories, since it accounts for variability in the preferences of the states through time, an aspect missing from other explanations (Haas, 1989, 1990, ch. 8). However, even though this theory has been useful in explaining the negotiations and creation of MAP, it is open to question whether the current operation of the regime is based on 'epistemic communities'. Moreover, the generalizability of the theory to explain other regimes is not yet proven.

Structure of the Mediterranean Action Plan and its components
According to Raftopoulos (1993) regional action plans usually consist of five components: the assessment component, the management component, the legal component, the institutional component and the financial component. The basic characteristics of each MAP component are described below.

The legal component of MAP MAP seeks to achieve all its objectives through its legal component, the Barcelona Convention and related protocols. The Convention for the Protection of the Mediterranean Sea against Pollution was signed in 1976, and has been in force since 1978. In 1995 it was replaced by an amended version taking into account recommendations of the 1992 Rio Conference on Environment and Development and it was recorded as the Convention for the Protection of the Marine Environment and the Coastal Region of the Mediterranean, being in force since 2004. The amended version of the Barcelona Convention introduces new principles such as Environmental Impact Assessment (EIA), the polluter pays principle and the precautionary principle and also suggests time limits for environmental regulations (UNEP/MAP, 2005a). The 22 Contracting Parties to the Barcelona Convention are: Albania, Algeria, Bosnia and Herzegovina, Croatia, Cyprus, Egypt, the European Community, France, Greece, Israel, Italy, Lebanon, Libya, Malta, Monaco, Morocco, Serbia and Montenegro, Slovenia, Spain, Syria, Tunisia and Turkey.

As described in Article 1.1 of the Convention (UNEP/MAP, 2005a), geographically it covers:

> the maritime waters of the Mediterranean Sea proper, including its gulfs and seas, bounded to the west by the meridian passing through Cape Spartel lighthouse, at the entrance of the Straits of Gibraltar, and to the east by the southern limits of the Straits of the Dardanelles between Mehmetcik and Kumkale lighthouses.

As is obvious from the above definition, the internal waters of the Contracting Parties are excluded in the provisions, as are the Black Sea, the Sea of Marmara and the Bosphorus, since the 'demarcation line' is the southern limit of the Straits of the Dardanelles. In the following provisions, the Convention may be extended to include coastal areas as defined by each Contracting Party within its own territory, and also any Protocol to the Convention may extend geographical coverage to that which the particular Protocol applies.

In Article 2(a) pollution is defined and described as:

> the introduction by man, directly or indirectly, of substances or energy into the marine environment, including estuaries, which results, or is likely to result, in such deleterious effects such as harm to living resources and marine life, hazards to human health, hindrance to marine activities, including fishing and other legitimate uses of the sea, impairment of quality for use of seawater and reduction of amenities.

The protocols to the Barcelona Convention, also summarized in Table 7.1, are the following:

- *Dumping Protocol.* The full title is 'Protocol for the Prevention of Pollution in the Mediterranean Sea by Dumping from Ships and Aircraft'. It was signed in 1976 and has been in force since 1978.

Table 7.1 MAP Protocols

Protocol	Entry into Force	Description
Dumping Protocol Protocol for the Prevention of Pollution in the Mediterranean Sea by Dumping from Ships and Aircraft	Adoption: 1976 Entry into force: 1978 Amendments: 1995 but oldest version in force	Aims at prohibiting discharge of wastes and other materials by committing states to ban dumping of certain substances – the 'black list' – and issue permits for the dumping of less hazardous substances – the 'grey list'
Prevention and Emergency Protocol Protocol Concerning Cooperation in Preventing Pollution from Ships, and, in Cases of Emergency, Combating Pollution of the Mediterranean Sea	Adoption: 2002 Entry into force: 2004 Replaced the oldest version in force since 1976	Focuses on promoting means of combating oil pollution through multilateral cooperation, by committing states to notify each other in case of an oil spill and to cooperate in the clean-up
LBS (Land-Based Sources) Protocol Protocol for the Protection of the Mediterranean Sea against Pollution from Land-Based Sources	Adoption: 1980 Entry into force: 1983 Amendments: 1995 but oldest version in force	Focuses on eliminating persistent toxic substances by committing states to ban or strictly limit a number of compounds such as organohalogen, organophosphorus and organotin compounds, heavy metals and chlorinated hydrocarbons, inter alia
SPA (Specially Protected Areas) and Biodiversity Protocol Protocol Concerning Specially Protected Areas and Biological Diversity in the Mediterranean	Adoption: 1995 Entry into force: 1999 Replaced the oldest version in force since 1982	Encourages creation and development of marine parks to safeguard representative types of coastal and marine ecosystems and their biodiversity, endangered habitats and species and sites of aesthetic or cultural importance

A case study of the Mediterranean Action Plan 223

Table 7.1 (continued)

Protocol	Entry into Force	Description
Offshore Protocol Protocol for the Protection of the Mediterranean Sea against Pollution Resulting from Exploration and Exploitation of the Continental Shelf and the Seabed and its Subsoil	Adoption: 1994 Not yet in force	This Protocol requires authorization by national authorities for any offshore activity, which should be granted only after the examination of a study of the activity's potential effects on the environment (Environmental Impact Assessment)
Hazardous Wastes Protocol Protocol on the Prevention of Pollution of the Mediterranean Sea by Transboundary Movements of Hazardous Wastes and their Disposal	Adoption: 1996 Not yet in force	The Protocol requires Parties to take all appropriate measures to eliminate pollution resulting from the transboundary movement and disposal of hazardous wastes to the fullest possible extent and to eliminate such movements if possible

This Protocol was amended and recorded as the 'Protocol for the Prevention and Elimination of Pollution in the Mediterranean Sea by Dumping from Ships and Aircraft or Incineration at Sea'. It was signed in 1995 but still awaits entry into force. The Dumping Protocol commits states to banning dumping of certain substances – the 'black list' – and issue permits for dumping of less hazardous substances – the 'grey list'. Factors to be considered when establishing criteria governing issue of permits include characteristics and composition of wastes or other matter, features of the dumping site and method of deposit of matter to the site. An exception to the Protocol's provisions is the case of force majeure due to stress of weather or any other cause when human life or the safety of a ship or aircraft is threatened.

- *Prevention and Emergency Protocol.* The full title is 'Protocol Concerning Cooperation in Preventing Pollution from Ships, and, in Cases of Emergency, Combating Pollution of the Mediterranean

Sea'. This Protocol was signed in 2002, and has been in force since 2004, replacing the existing 'Protocol Concerning Cooperation in Combating Pollution of the Mediterranean Sea by Oil and Other Harmful Substances in Cases of Emergency', which was in force from 1976. The Prevention and Emergency Protocol commits states to notify each other in case of an oil spill and to cooperate in the clean-up. In the event of an oil spill or other emergencies UNEP and also any other state likely to be affected must be informed. Moreover, in the framework of this Protocol, a regional activity centre (REMPEC – Regional Marine Pollution Emergency Response Centre for the Mediterranean Sea) has been established in Malta, administered by the International Maritime Organization and the United Nations Environment Programme to deal with the implementation of this Protocol. Cooperation in the clean-up includes salvage or recovery of packages containing hazardous or noxious substances released or lost overboard. The Protocol also provides for other actions such as dissemination of reports and information. The article about assistance allows for it to be asked for and given by the regional activity centre or by any other signatory state in the form of equipment, products and facilities, expert advice and the costs of any action shall be borne by the requesting Party.

- *LBS (Land-Based Sources) Protocol.* The full title is 'Protocol for the Protection of the Mediterranean Sea against Pollution from Land-Based Sources'. It was signed in 1980 and has been in force since 1983. This Protocol was amended as the 'Protocol for the Protection of the Mediterranean Sea against Pollution from Land-Based Sources and Activities'. The amendment was signed in 1995 but still awaits entry into force. The LBS Protocol covers some sectors of activity, including heavy metal industries, agriculture, energy production and waste treatment, binding the countries to adopt new industrial, agricultural and waste treatment practices. It also commits states to ban or strictly limit a number of compounds such as organohalogens, organophosphorus compounds, organotins, heavy metals, chlorinated hydrocarbons, radioactive substances and thermal discharges inter alia. The Protocol in itself does not define specific emission or time limits; however, it provides that states should progressively adopt such guidelines and measures. Following this, in 1997 the MED POL programme assisted countries to design and adopt the 'Strategic Action Programme to Address Pollution of the Mediterranean Sea from Land-Based Activities (SAP)', which entails more specific emission and time limits for pollution reduction.

- *SPA (Specially Protected Areas) and Biodiversity Protocol.* The full title is 'Protocol Concerning Specially Protected Areas and Biological Diversity in the Mediterranean'. This Protocol was signed in 1995, and came into force in 1999, replacing the existing 'Protocol Concerning Mediterranean Specially Protected Areas', which came into force in 1982. The SPA Protocol was outside the scope of the programme as this was initially anticipated in the Barcelona Convention and MAP and this is why it is considered different from the other protocols, which were provided for in the Convention. It encourages creation and development of marine parks to safeguard representative types of coastal and marine ecosystems and their biodiversity, endangered habitats, and habitats critical to the survival of endangered species. The Protocol also provides for protection of sites of particular importance because of their scientific, aesthetic, cultural or educational interest. It suggests the regulation of certain activities such as fishing, hunting and trading animals, and the passage, stopping or anchoring of ships. Moreover, it suggests the establishment of a 'List of Specially Protected Areas of Mediterranean Importance' or 'SPAMI List'. A regional activity centre has been established in Tunis (SPA/RAC) to deal with issues of protected areas. However, the Protocol only encourages development of specially protected areas and does not oblige the signatory states to take any form of action, so the question of whether this issue should be treated in the form of a Protocol remains unanswered.
- *Offshore Protocol.* The full title is 'Protocol for the Protection of the Mediterranean Sea against Pollution Resulting from Exploration and Exploitation of the Continental Shelf and the Seabed and its Subsoil'. It was signed in 1994 but still awaits entry into force. This Protocol requires authorization by national authorities for any offshore activity, which should be granted only after study of the activity's potential environmental effects. It includes lists of harmful or noxious materials and substances, the disposal of which is either prohibited or requires a special permit, and provides for monitoring of planned installations for environmental and safety effects. In addition to this the Protocol provides that each Party shall prescribe sanctions to be imposed for breach of obligations and that as soon as possible appropriate rules and procedures for the determination of liability and compensation for damage resulting from relevant activities should be formulated and established. Delay in adoption and ratification of this Protocol is attributed to involvement of offshore industries, especially the oil industry, in the decision-making of the governments.

226 *A handbook of environmental management*

- *Hazardous Wastes Protocol.* The full title is 'Protocol on the Prevention of Pollution of the Mediterranean Sea by Transboundary Movements of Hazardous Wastes and their Disposal'. It was signed in 1996 but still awaits entry into force. The Protocol requires Parties to take all appropriate measures to eliminate pollution resulting from the transboundary movement and disposal of hazardous wastes to the fullest possible extent and to eliminate such movements if possible. Contracting Parties are obliged to generally prohibit the export and transit of hazardous wastes to developing countries and the Parties that are non-EU members should prohibit all imports and transits. Moreover, the countries directly or with the help of competent authorities should implement programmes of financial and technical assistance to developing countries for the implementation of this Protocol. Lists of hazardous wastes and hazardous characteristics of substances are also described, and provisions for liability and compensation for damage resulting from the transboundary movement of hazardous wastes are also included in the Protocol. The delay in the adoption and ratification of this Protocol is also considered to be occurring for the same reasons as for the Offshore Protocol, that is, due to conflicting interests with the oil industry.

In addition there is a seventh Protocol under preparation concerning Integrated Coastal Zone Management (ICZM). In most cases the protocols have been revised and supplemented. Most of the amendments, including the new Barcelona Convention, are still in the process of ratification as summarized in Table 7.1.

The Barcelona Convention and protocols raise the issue of dealing with a legally and institutionally complex scheme, because it concerns an international environmental order, which develops 'diachronically rather than synchronically and contextually rather than in isolation from its relational foundation' (Raftopoulos, 1993, p.42). The legal component of MAP is divided in two broad categories, the common environmental norms and rules and the community membership norms and rules. The former relate to specific environmental provisions, whereas the latter give standard 'membership' powers and duties to each 'Contracting Party' (Raftopoulos, 1993).

The institutional component of MAP The institutional component of MAP, as defined within the framework of the Barcelona Convention, is structured in such a way as to give authority to two organs: the Meetings of the Contracting Parties and the Secretariat. The highest authority in decision-making is given to the Meetings of the Contracting Parties, which occur every two years, and reflect shared interests of all the Parties. They also make sure that current legal obligations are met, and oversee formation of new rules. The second authority is the Secretariat of MAP, based in Athens, which supports its operation by carrying out all the administrative tasks that secure its smooth implementation, but which also helps to integrate stakeholder interests into the legislative goals (Raftopoulos, 1993, p. 73).

Moreover, following launch of the MAP II process and a shift towards a 'sustainable development' orientation, the Mediterranean Commission on Sustainable Development (MCSD) was set up as an advisory body to MAP in 1996 as a think-tank on policies for promoting sustainable development in the Mediterranean Basin. Moreover, the operation of MAP is supported through six Regional Activity Centres (RACs) in six Mediterranean cities, which help operations in a more decentralized way under supervision of the Secretariat, each offering expertise in specific fields of action for facilitating the operation of MAP, as shown in Table 7.2.

The environmental assessment component of MAP The environmental assessment component of MAP, stated in the official text of UNEP (1978) as the 'Co-ordinated Pollution Monitoring and Research Programme in the Mediterranean' is widely known as MED POL. It is the most straightforward technical aspect of MAP and has played 'an important cohesive role for the development of a concrete, scientifically based, regional approach to the problems of the Mediterranean pollution' (Raftopoulos, 1993, p. 5). MED POL operates in phases. Its first phase, MED POL – Phase I, lasted from 1975 until 1980. At that time there was not enough scientific expertise either in the number of trained scientists or in terms of facilities established, therefore it was constructed upon pilot projects. This was considered a necessary condition, bearing in mind that full-scale regional assessments require identified pollution problems common to all the participating states (Raftopoulos, 1993). Initially there were seven pilot projects approved in 1975 followed by several others to support the programme. States had designated national research centres to participate in the pilot projects, and the planning and carrying out of necessary actions was a collaborative effort of UNEP with several international organizations (ECE, UNIDO, FAO, WHO, WMO, UNESCO, IAEA and IOC of UNESCO). According to Raftopoulos (1993, pp. 8–9) MED POL – Phase I proved largely successful in transferring technology and scientific expertise to many Mediterranean states, especially in less-developed countries since UNEP at the time followed a policy of allowing most of the resources to those needing them most.

Table 7.2   Regional Activity Centres (RACs)

Regional Activity Centre	Establishment	Description
REMPEC Regional Marine Pollution Emergency Response Centre for the Mediterranean Sea	Year: 1976 Place: Manoel Island, Malta Status: Centre under IMO/UNEP agreement, administrated by IMO	Aims at preventing and combating pollution from oil and other harmful substances by helping Mediterranean coastal states to be prepared for major marine pollution incidents and to cooperate in the clean-up
BP/RAC Blue Plan Regional Activity Centre	Year: 1977 Place: Sophia Antipolis, France Status: National Centre, with an NGO status, with regional function	Adopts a systemic and prospective approach to Mediterranean environment and development issues using observation and evaluation tools, generating indicators and publishing several studies accordingly
PAP/RAC Priority Actions Programme Regional Activity Centre	Year: 1980 Place: Split, Croatia Status: National Centre with regional function	Aims to improve the Mediterranean environmental situation by addressing priority actions. It is mainly concerned with integrated coastal area management to lessen development problems in built up coastal areas
SPA/RAC Specially Protected Areas Regional Activity Centre	Year: 1994 Place: Tunis, Tunisia Status: National Centre with regional function	Focuses on biodiversity issues and is involved in the protection of Mediterranean species, their habitats and ecosystems by producing, inter alia, strategies for biodiversity conservation
CP/RAC Cleaner Production Regional Activity Centre	Year: 1995 Place: Barcelona, Spain Status: Public company put at the disposal of MAP	Focuses on promoting and disseminating cleaner production technologies for industrial sector in order to reduce industrial waste at source of the Mediterranean industrial sector

Table 7.2 (continued)

Regional Activity Centre	Establishment	Description
INFO/RAC Information and Communication Regional Activity Centre previously *ERS/RAC* Environment Remote Sensing Regional Activity Centre	Year: 2005 INFO/ RAC Year: 1993 ERS/ RAC Place: Rome and Palermo, Italy Status: Public body put at the disposal of MAP	Aims to provide information and communication services and technical support to MAP, also by enhancing public awareness (Initially ERS/RAC would promote and introduce remote sensing and GIS for environmental monitoring and sustainable development)

Phase II of MED POL lasted from 1981 until 1990 and was named the 'Long-term Pollution Monitoring and Research Programme'. For effective implementation of its specific objectives it was divided into four distinct components: monitoring, research and study topics, data quality assurance, and assistance. Overall coordination of Phase II was in the hands of the Mediterranean Action Plan Coordinating Unit (the Secretariat of the Barcelona Convention) acting on behalf of UNEP, even though the countries were fully responsible for monitoring activities as stated in Article 12 of the Barcelona Convention and in Article 8 of the Land-Based Sources Protocol.

MED POL has recently finished its Phase III, which started in 1996 and lasted until 2005. Just before the end of Phase II important events at both international and regional levels took place, which guided MED POL to change its direction. These events were the adoption of Agenda 21 in Rio 1992 and the Global Plan of Action (GPA) in 1995 in Washington (UNEP, 1995a) to address pollution from land-based sources and activities, and creation of the Mediterranean Commission for Sustainable Development (MCSD) together with the amended LBS Protocol at regional level. Hence there was a slow change from pollution assessment to pollution control, with MED POL becoming a tool for the countries to properly manage their marine and coastal areas. MED POL Phase III, adopted in 1995 and called the 'Programme for the Assessment and Control of Pollution in the Mediterranean Region', was directly concerned with implementation of the two relevant protocols (Dumping and LBS), since it focused more on management of pollution control (UNEP, 1999). It included activities such as pollutant trend monitoring and assessing effects of contaminants on living organisms as well as inventory of pollution sources and loads and finally the setting up of a database. Regarding control, compliance of the countries is monitored by an annual report discussing the country's existing action plans, programmes and measures for pollution control and how well these comply with national, regional or international legislation. All the above activities have to be described in agreements between each country and MED POL.

From 2005 until 2013, a new phase of MED POL has come into operation as put forward in the 13th Ordinary Meeting of the Contracting Parties to the Barcelona Convention (UNEP/MAP, 2003). However, the starting points for its objectives and goals are those set out in Phase III, which was considered adequate for supporting the overall objectives of the Convention and the protocols. In that respect it will continue to operate with the same tools (monitoring, compliance monitoring, assessments, capacity building, and so on). However, taking into account recommendations of the evaluation of Phase III (UNEP/MAP, 2005b), it focuses more on some aspects of Control and Assessment and Public Participation and it tries to use the Ecosystem Approach more widely in all its aspects (UNEP/MAP, 2005c).

The environmental management component of MAP The MAP environmental management component is called 'Integrated Planning of the Development and Management of the Resources of the Mediterranean Basin' (UNEP, 1978) and was the first of the main aspects of MAP to be implemented. Its aim is also to protect the Mediterranean marine environment but instead of focusing only on pollution sources, it integrates development issues of the region in the sense of environmental management. From the beginning it was divided into a long-term research and study programme, the Blue Plan and a more straightforward and immediate programme aiming at performing specific actions, the Priority Actions Programme (Raftopoulos, 1993).

To assist implementation of the Blue Plan, a Regional Activity Centre was established in France, namely the BP/RAC. Initially the Blue Plan performed 12 investigative thematic studies with the help of experts from both North and South Mediterranean in each study. Later on a more thorough and complete scientific study was performed in order to examine the potential for integrating social and economic development in the region to enhance environmental protection. A synthesis and presentation phase was also planned in order to guarantee dissemination of the results of the above studies, nevertheless the Blue Plan was criticized for not being able to achieve that goal (Raftopoulos, 1993, p. 27). According to Raftopoulos, it has not succeeded in getting through to the non-expert Mediterranean community such as stakeholders, or the wider public, mainly due to a poor communication network.

For the Priority Actions Programme (PAP), another Regional Activity Centre was established, namely the PAP/RAC. Contrary to the Blue Plan, it involved particular actions to be taken on issues considered as priorities at the time. Following the example of MED POL it was designed to be implemented through demonstration and pilot projects. At that time there was inadequate scientific awareness on the integration of environment and development for the purposes of environmental management so this approach was the only solution (Raftopoulos, 1993).

However, according to Raftopoulos (1993, p. 32), the environmental management component, consisting of the Blue Plan and the Priorities Action Programme, even though a rather large and important aspect of MAP, was clearly not covered in the Barcelona Convention. This means its ideas and findings were not translated into legal provisions, so to a large extent integration of environment and development was only in the form of words and not action.

The financial component of MAP   Finally, the financial component of the Mediterranean Action Plan is mainly covered by the Mediterranean Trust Fund. This is a fund that all the Contracting Parties to the Convention contribute to, according to their respective national wealth. The Contracting Parties may also contribute to the operations of MAP through in-kind contributions (for example, through participation of their national institutes in the MED POL programme especially in MED POL Phase II). Additionally some Contracting Parties may provide extra voluntary contributions to the Mediterranean Trust Fund, even on a regular basis such as, for instance, the European Union. The financial arrangements of MAP are also supported on certain occasions by UNEP through project funding, and this was the case especially in the first years of MAP's operation.

Effectiveness of the Mediterranean Action Plan
Effectiveness of the Mediterranean Action Plan has not been extensively studied by international relations academics. A few exceptions include Haas, who brought MAP to the attention of the academic community by praising it as a success, and some others like Skjaerseth and Kütting who were more critical. Other types of studies carried out discussed certain aspects of MAP or tried to assess specific features (for example, legal perspectives) of its operation (Boxer, 1978; Raftopoulos, 1993, 1997; Jeftic, 1996; Pavasovic, 1996; Vallega, 1996; Massoud et al., 2003; Raftopoulos and McConnell, 2004, inter alia).

Haas's study of 'epistemic communities' (Haas, 1989, 1990) did much to bring the Mediterranean Action Plan to the attention of the academic community. He suggests that many studies focused on regime negotiations and their creation but few attempts have been made to investigate their real and practical significance and their direct impact on the behaviour of actors (states). He proposes that MAP derives its effectiveness from the influence of 'epistemic communities'. He considers it a success because it 'altered the balance of power within the Mediterranean governments by empowering a group of experts who then contributed to the development of convergent state policies in compliance with the regime' (Haas, 1989, p. 377). He concludes that MAP may signal the emergence of an entirely new international political order for the environment and he stresses the role of 'epistemic communities' in promoting stronger national pollution controls (Haas, 1990). Nevertheless, more than 15 years after Haas's study, this enthusiasm is missing from other researchers of MAP.

Skjaerseth (1996, 2002) also studied MAP but he was not convinced about its success. He notes that the reasons for signing up to the Barcelona Convention did not always have much to do with environmental concern. For the less developed countries it was an opportunity to receive training and equipment for monitoring pollution, since the financial burden, at least until 1979, was carried by UNEP. Also it was a diplomatic opportunity to establish political/diplomatic ties between countries traditionally in conflict. Therefore the states probably had mixed motives that were not necessarily entirely environmental. Moreover, the Barcelona Convention goals were vague, and even though a main goal of MAP in its second phase was to produce specific targets with specific deadlines for the Parties to the Convention, it failed to do so. In addition, the states have not been very willing to provide adequate reporting on the national implementation of their commitments. Therefore, due to the lack of clear targets and the inadequate state reporting, it is difficult to estimate whether there has been behavioural change among target groups. Skjaerseth also considers the MAP budget to be very limited compared with the wide scope of its demands. It is even more difficult to assess the impact of MAP on the state of the marine environment since there is a lack of reliable and continuous pollution and water quality data. It has to be noted though that the collection of these scarce data is largely a result of MAP's establishment. However, even if there is an improvement in the marine environment it is rather difficult to attribute it all to the regime, since other factors such as general socioeconomic and technological change or natural environmental variation have to be taken into account. Moreover, for many countries, much environmental national legislation was also required by other organizations such as the European Union. Skjaerseth concludes that MAP is considered a collaborative political success since it produced a complete plan for de-polluting the Mediterranean Sea and furthermore because it increased the general environmental awareness and preparedness through regional cooperation and transfer of knowledge. However, its impact on behavioural change among target groups is not so clear (Skjaerseth, 1996, 2002).

Kütting (2000a, ch. 5) is also critical of MAP. Even though admitting that the regime has been successful in starting and maintaining a cooperation process for a significant period of time in a region traditionally characterized by many political conflicts, she finds that overall it can not be considered as successful in terms of either institutional or environmental effectiveness. Moreover, she argues that basically MAP faced the typical North–South divide that underpins so many global environmental problems, although in this case at a regional scale. She also considers that MAP has been rather disregarded by the international relations academic community because traditional international relations research focuses on matters of national economic interest when examining international agreements and this was not the case in MAP, as it was formed due to environmental concern. She even asks the question why MAP 'exists at all since there is an apparent lack of motivation?' (Kütting, 2000a, p. 7). Overall, Kütting suggests that MAP may have been a political success but in terms of amelioration of the environmental problem, it has not offered a lot (Kütting, 2000b).

MAP may have succeeded in completing some activities but it is not clear how much can really be assigned directly to it. As mentioned earlier, the most important part of the Mediterranean Action Plan is that which deals with combating pollution from land-based sources, since these are the main polluters of the Mediterranean marine environment. More than 20 years after the LBS Protocol's entry into force, its effectiveness cannot be clearly estimated. There have been several noteworthy actions, such as construction of sewage treatment plants in many Mediterranean cities, nevertheless, it is quite likely that some of these actions would have been taken anyway.

In conclusion, the focus should be on areas where the Mediterranean Action Plan has undoubtedly been successful. Even if it has not achieved an enormous change in the state of the biophysical environment, it has certainly enhanced cooperation, stability and security in a traditionally unstable and politically heterogeneous region. Moreover, MAP has promoted environmental awareness and capacity-building especially in the less developed countries of the southern Mediterranean. In some ways the political, rather than scientific, success of MAP is ironic as it was the expert scientific 'epistemic community' that first created the international collaboration responsible for launching the Barcelona Convention. But the legacy is diplomatic rather than scientific.

A new approach to defining and measuring effectiveness
In the first part of this chapter various theories about international environmental regime effectiveness and a range of efforts to define and measure this effectiveness in applied cases were reviewed and examined. What is evident is that there is no one way to define and measure such a concept, especially when dealing with complex interactive systems consisting of socioeconomic factors, policy and politics and the global environment. The second part presented an overview of an international regime and a critical discussion of its effectiveness as a case study. The assessments of MAP made by different academics and practitioners largely varied according to which criteria they used in assessing the regime.

The theories of realism and neorealism are primarily concerned with state security and national interest. They do not include environmental concerns in their analysis, and assume that states have given interests, which is not the case in environmental issues. Looking at the formation of MAP, as Haas noted, an explanation through the hegemonic stability strand failed once France declined as the hegemon. MAP continued to exist and be supported by both the lead and weaker states. Historical materialism and international political economy, especially with the dominance of economic globalization, can in some cases explain environmental cooperation better, but in the case of MAP these theories also failed since interests of both sides, developed and less-developed states, are represented equally in the regime. Therefore, neoliberal institutionalism and strand regime theory are the most suitable traditions to explain international environmental cooperation. The distinction between Krasner's different orientations is not important because regimes may matter under certain conditions, meaning that effective regimes do matter.

Concerning the different approaches used when defining and measuring effectiveness, most of the regime theorists focus on institutional performance of a regime. Even those that consider the environmental problem do not clearly define how this aspect can be assessed. A different approach by Mitchell gives an example of such an assessment, but it leaves out of the calculation factors that cannot be easily measured by numbers, such as the political benefits of cooperation. Kütting makes clear the need for a distinction between institutional and environmental performance, although looked at from a regime theory perspective. As far as qualitative and quantitative techniques used in the study of regime effectiveness are concerned, the former usually explains a case well, since time and effort are spent in researching that particular case, however generalizability poses problems. On the other hand, quantitative approaches can be valid for many situations, but they might miss important case-specific factors. For instance, in counterfactual analysis it is difficult to estimate the hypothetical situation of the absence of a regime, and Mitchell's econometric approach depends heavily on the availability of data, which renders it difficult to apply. Therefore a new approach, which would take into account both institutional and environmental parameters using complementary qualitative and quantitative techniques would be ideal to assess the effectiveness of regimes.

Looking at the case of the Mediterranean Action Plan, the handful of important studies on its effectiveness show a varied set of opinions, demonstrating that assessing effectiveness depends primarily on defining the criteria used for this process. Haas's prominent study on 'epistemic communities' found the regime successful and argued enthusiastically that it would introduce a new concept in international environmental cooperation. His theory can provide a satisfactory explanation for the role of scientific groups in the creation of MAP, but its continued success was mostly political. It remains highly questionable whether these scientific groups are the power behind its implementation, or if these groups are instigating such processes in other international environmental regimes as well. On the contrary, the study of Skjaerseth is more critical about the achievements of MAP. Even though he recognizes its political contributions to cooperation and its overall enhancement of general environmental awareness, he notes that the desired change in behaviour of the actors is not very evident. Finally, Kütting, distinguishing between institutional and environmental effectiveness, concludes that unfortunately MAP was not successful in the long term in either of the two aspects. According to her criteria, its only real achievement is the instigation and continuation of a cooperation effort in a politically very difficult region of the world. Drawing from the previous three studies, and their different outcomes on the same case, it is essential to define effectiveness before any attempt to assess it.

Undoubtedly, for a scientist, only improvement of the environment is the raison d'être of an environmental regime. However, the regime's institutional performance is equally important as an indirect way to achieve this as a means to an end and not as an end per se. Hence, the institutional and environmental aspects of effectiveness do not need to be separated, but rather integrated in order to provide a holistic view. The need for an interdisciplinary approach is the first and foremost rule in that respect. So far academics that study environmental regime effectiveness have come mainly from a political science background rather than a scientific one. On the other hand scientists might not rigorously research international relations issues. 'Epistemic communities' drawing expertise from all disciplines and using both qualitative and quantitative methods of analysis might prove useful in order to design and implement these regimes. For instance, in the Mediterranean Action Plan the first question would be: is the Mediterranean cleaner than before? Or at least cleaner than it would be without MAP? Then methodological problems such as how to measure cleanliness would arise, which could only be superseded by proper design of long-term environmental assessment and more importantly by a proper feedback mechanism between science and policy. In the absence of a clear scientific answer the question might be asked, how well is MAP performing? Then the political aspects would come into play, combining all the relevant issues, whether the regime enhances international cooperation and security, creates structures, changes the behaviour of the actors, allows for multi-stakeholder participation and so on. Such a holistic approach could be the first rule for effective international environmental agreements.

Furthermore, practice has frequently deviated from theory. High expectations, ambitious plans and disregard of social and economic considerations have sometimes led to the establishment of regimes that are difficult to implement. A general drawback of international law is its voluntary nature, as it cannot legally bind any state, apart from those willingly participating in the regimes. For this reason a regime should provide incentives to its members for participation, and also for compliance in the long term, irrespective of whether these incentives would be of a political or economic nature. Even the imposition of rules such as sanctions might deter countries from agreeing, thus achieving even poorer results. Economic considerations should also be taken into account in terms of financial resources for all the parties to implement the provisions of the agreement, but also in terms of fair social policy. It may be that the environment is the object of protection, but in no way should this happen at the expense of human needs. People in developing countries need bread to eat before saving the earth and the sea, and even in developed ones governments might not accept strict agreements requiring, for instance, the closure of polluting industries, for fear of unemployment. Hence the environmental and time limits of an agreement should be specific but at the same time realistic. Only regimes with a pragmatic vision have more chance to succeed in the long term.

Ultimately, so far the discussion has focused on the criteria used when assessing effectiveness of environmental regimes. Various scholars define various criteria accordingly. Hence they examine each case by using this set of criteria and how the regime performs in each one of them at a given moment in time. Nevertheless, times change and with them whole new concepts in the environmental and political sphere arise. Some regimes have a life of more than 30 years such as the Mediterranean Action Plan. Which leads to the logical question: how can the effectiveness of MAP be assessed today, since other criteria were used for its design 30 years ago? Even concepts such as marine pollution had a different meaning before the introduction of ideas such as habitat degradation or coastal zone management. In that respect a regime should always be ready to adapt properly and quickly to new needs, new definitions and new realities. It should have such an institutional structure that would allow for right and rapid amendments, and would eradicate any trace of bureaucracy. It should influence other international or national policies and politics and be open to be influenced by them. Effective regimes are the living ones, which can move through time being older and wiser, not older and weaker. Hence regime effectiveness could not be assessed by static criteria, the only exception to this being the criterion of the regime's dynamic nature.

This new perspective on effectiveness would require a regime to use a holistic approach based on science, policy and their interaction, have a pragmatic vision for its ultimate goals and be of a dynamic nature to evolve through time.

Meeting all the above conditions is hard but perhaps it might prove successful in the quest for effective international environmental regimes. Bearing in mind that the above definition presents a very broad approach, it will be further developed in forthcoming studies and particularly applied in a specific case study, that of the Mediterranean Action Plan.

Conclusion

Environmental problems, instead of a solution per se, demand an effective management through time. Since this management especially in the case of global or transboundary environmental problems is very often in the hands of international environmental regimes, special attention should be paid to the design and implementation of these regimes, as well as to their assessment. The new perspective on international environmental regime effectiveness might perhaps prove a helpful tool towards this direction.

Nevertheless, further research is needed in order to specify new ways that would bridge the gap between science and policy, which would provide realistic solutions reconciling conflicting interests and that would give life to human-made institutions.

Note

1. The author is very grateful to Neil Carter for his valuable comments on earlier drafts of this study. Special thanks go to Evangelos Raftopoulos for his inspiring ideas during several discussions on the research. Thanks also go to Jon Lovett and Gabriela Kütting for their overall support. The Firos Foundation is gratefully acknowledged for funding the author's doctoral studies.

References

Ausubel, J.H. and D.G. Victor (1992), 'Verification of International Environmental Agreements', *Annual Review of Energy and Environment*, **17**(1), 1–43.

Barrett, S. and R. Stavins (2003), 'Increasing Participation and Compliance in International Climate Change Agreements', *International Environmental Agreements: Politics, Law and Economics*, **3**(4), 349–76.
Biermann, F. (2000), 'The Case for a World Environment Organization', *Environment*, **42**(9), 22–31.
Biermann, F. (2006), 'Global Governance and the Environment', in M.M. Betsill, K. Hochstetler and D. Stevis (eds), *International Environmental Politics*, Basingstoke: Palgrave.
Boxer, B. (1978), 'Mediterranean Action Plan: An Interim Evaluation', *Science*, **202**(4368), 585–590.
Brown Weiss, E. and H.K. Jacobson (1998), *Engaging Countries: Strengthening Compliance with International Environmental Accords*, Cambridge, MA: MIT Press.
Carter, N.T. (2001), *The Politics of the Environment: Ideas, Activism, Policy*, New York and Cambridge, UK: Cambridge University Press.
Clapp, J. (2006), 'International Political Economy and the Environment', in M.M. Betsill, K. Hochstetler and D. Stevis (eds), *International Environmental Politics*, Basingstoke: Palgrave.
Clapp, J. and P. Dauvergne (2005), *Paths to a Green World: The Political Economy of the Global Environment*, Cambridge, MA: MIT Press.
Coase, R.H. (1960), 'The Problem of Social Cost', *Journal of Law and Economics*, **3**(2), 1–44.
Deudney, D. (1990), 'The Case against Linking Environmental Degradation and National Security', *Millennium*, **19**(3), 461–76.
Haas, P.M. (1989), 'Do Regimes Matter? Epistemic Communities and Mediterranean Pollution Control', *International Organization*, **43**(3), 337–403.
Haas, P.M. (1990), *Saving the Mediterranean: The Politics of International Environmental Cooperation*, New York: Columbia University Press.
Haas, P.M., R.O. Keohane and M.A. Levy (1993), *Institutions for the Earth: Sources for Effective International Environmental Protection*, Cambridge, MA: MIT Press.
Helm, C. and D.F. Sprinz (2000), 'Measuring the Effectiveness of International Environmental Regimes', *Journal of Conflict Resolution*, **44**(5), 630–52.
Homer-Dixon, T. (1999), *The Environment, Scarcity and Violence*, Princeton: Princeton University Press.
Hovi, J., D.F. Sprinz and A. Underdal (2003a), 'The Oslo–Potsdam Solution to Measuring Regime Effectiveness: Critique, Response, and the Road Ahead', *Global Environmental Politics*, **3**(3), 74–96.
Hovi, J., D.F. Sprinz and A. Underdal (2003b), 'Regime Effectiveness and the Oslo–Potsdam Solution: A Rejoinder to Oran Young', *Global Environmental Politics*, **3**(3), 105–7.
Jeftic, L. (1996), 'Integrated Coastal and Marine Areas Management (ICAM) in the Mediterranean Action Plan of UNEP', *Ocean & Coastal Management*, **30**(2–3), 89–113.
Keohane, R.O. (1986), *Neorealism and its Critics*, New York: Columbia University Press.
Keohane, R.O. (1989), *International Institutions and State Power: Essays in International Relations Theory*, Boulder, CO: Westview Press.
Keohane, R.O., P.M. Haas and M.A. Levy (1993), 'The Effectiveness of International Environmental Institutions', in P.M. Haas, R.O. Keohane and M.A. Levy (eds), *Institutions for the Earth: Sources for Effective International Environmental Protection*, Cambridge, MA: MIT Press.
Krasner, S.D. (1983), *International Regimes*, Ithaca, NY: Cornell University Press.
Kullenberg, G. (2002), 'Regional Co-development and Security: A Comprehensive Approach', *Ocean and Coastal Management*, **45**(11–12), 761–76.
Kütting, G. (2000a), *Environment, Society and International Relations: Towards more Effective International Environmental Agreements*, New York: Routledge.
Kütting, G. (2000b), 'Distinguishing between Institutional and Environmental Effectiveness in International Environmental Agreements: The Case of the Mediterranean Action Plan', *The International Journal of Peace Studies*, **5**(1), 15–33.

Levy, M.A., O.R. Young and M. Zürn (1994), 'The Study of International Regimes', Working Paper No. 94-113, Laxenburg, Austria: International Institute for Applied Systems Analysis.
Levy, M.A., O.R. Young and M. Zürn (1995), 'The Study of International Regimes', *European Journal of International Relations*, **1**(3), 267–330.
Massoud, M.A., M.D. Scrimshaw and J.N. Lester (2003), 'Qualitative Assessment of the Effectiveness of the Mediterranean Action Plan: Wastewater Management in the Mediterranean Region', *Ocean & Coastal Management*, **46**(9–10), 875–99.
Miles, E.L., A. Underdal, S. Andresen, J. Wettestad, J.B. Skjaerseth and E.M. Carlin (2002), *Environmental Regime Effectiveness: Confronting Theory with Evidence*, Cambridge, MA: MIT Press.
Mitchell, R.B. (2004), 'A Quantitative Approach to Evaluating International Environmental Regimes', in A. Underdal and O.R. Young (eds), *Regime Consequences: Methodological Challenges and Research Strategies*, Dordrecht: Kluwer Academic.
Mitchell, R.B. and T. Bernauer (2002), 'Qualitative Research Design in International Environmental Policy', Working Paper No. 2-2002. Zurich: Swiss Federal Institute of Technology.
Morgenthau, H. (1978), *Politics Among Nations*, 5th edition, New York: Knopf.
North, D.C. (1990), *Institutions, Institutional Change and Economic Performance*, New York and Cambridge, UK: Cambridge University Press.
Paterson, M. (1996), *Global Warming and Global Politics*, London: Routledge.
Paterson, M. (2000), *Understanding Global Environmental Politics: Domination, Accumulation, Resistance*, London: Macmillan.
Pavasovic, A. (1996), 'The Mediterranean Action Plan Phase II and the Revised Barcelona Convention: New Prospective for Integrated Coastal Management in the Mediterranean Region', *Ocean & Coastal Management*, **31**(2–3), 133–82.
Raftopoulos, E. (1993), *The Barcelona Convention and Protocols. The Mediterranean Action Plan Regime*, London: Simmonds and Hill.
Raftopoulos, E. (1997), *Studies on the Implementation of the Barcelona Convention: the Development of an International Trust Regime*, Athens: Ant. N. Sakkoulas.
Raftopoulos, E. and M.L. McConnell (2004), *Contributions to International Environmental Negotiation in the Mediterranean Context*, Athens and Brussels: Ant. N. Sakkoulas, Bruylant.
Sands, P. (2003), *Principles of International Environmental Law*, 2nd edition, Cambridge, UK: Cambridge University Press.
Skjaerseth, J.B. (1996), 'The 20th Anniversary of the Mediterranean Action Plan: Reason to Celebrate?', in H.O. Bergesen and G. Parmann (eds), *Green Globe Yearbook 1996*, Oxford, UK: Oxford University Press.
Skjaerseth, J.B. (2002), 'The Effectiveness of the Mediterranean Action Plan', in E.L. Miles, A. Underdal, S. Andresen, J. Wettestad, J.B. Skjaerseth and E.M. Carlin (eds), *Environmental Regime Effectiveness: Confronting Theory with Evidence*, Cambridge, MA: MIT Press.
Stevis, D. and V.J. Assetto (2001), *The International Political Economy of the Environment: Critical Perspectives*, Boulder, CO: Lynne Rienner.
Swatuk, L.A. (2006), 'Environmental Security', in M.M. Betsill, K. Hochstetler and D. Stevis (eds), *International Environmental Politics*, Basingstoke: Palgrave.
Underdal, A. (1992), 'The Concept of Regime Effectiveness', *Cooperation and Conflict*, **27**(3), 227–40.
Underdal, A. (2002), 'One Question, Two Answers', in E.L. Miles, A. Underdal, S. Andresen, J. Wettestad, J.B. Skjaerseth and E.M. Carlin (eds), *Environmental Regime Effectiveness: Confronting Theory with Evidence*, Cambridge, MA: MIT Press.
UNEP (1978), *Mediterranean Action Plan and the Final Act of the Conference of Plenipotentiaries of the Coastal States of the Mediterranean Region for the Protection of the Mediterranean Sea*, New York: United Nations.
UNEP (1995a), *Report of the Ninth Ordinary Meeting of the Contracting Parties to the*

Convention for the Protection of the Mediterranean Sea against Pollution and its Protocols, UNEP(OCA)/MED IG.5/16, Athens: UNEP.

UNEP (1995b), *Global Programme of Action for the Protection of the Marine Environment from Land-Based Activities*, UNEP(OCA)/LBA/IG.2/7, Washington, DC: UNEP.

UNEP (1999), *MED POL – Phase III, Programme for the Assessment and Control of Pollution in the Mediterranean Region*, MAP Technical Reports Series No. 120, Athens: UNEP.

UNEP/MAP (2003), *Report of the Thirteenth Ordinary Meeting of the Contracting Parties to the Convention for the Protection of the Mediterranean Sea against Pollution and its Protocols*, UNEP(DEC)/MED IG.15/11, Athens: UNEP/MAP.

UNEP/MAP (2005a), *Convention for the Protection of the Marine Environment and the Coastal Region of the Mediterranean and its Protocols*, Athens: UNEP/MAP.

UNEP/MAP (2005b), *Evaluation of MED POL Phase III Programme (1996–2005)*, UNEP(DEC)/MED WG.270/Inf.10, Athens: UNEP/MAP.

UNEP/MAP (2005c), *Recommendations for 2006–2007*, UNEP(DEC)/MED IG.16/5, Athens: UNEP/MAP.

Vallega, A. (1996), 'Geographical Coverage and Effectiveness of the UNEP Convention on the Mediterranean', *Ocean & Coastal Management*, **31**(2–3), 199–218.

Victor, D.G. (2003), 'International Agreements and the Struggle to Tame Carbon', in J.M. Griffin (ed.), *Global Climate Change: The Science, Economics and Politics*, Cheltenham, UK and Northampton, MA, USA: Edward Elgar.

von Moltke, K. (2000), 'Research on the Effectiveness of International Environmental Agreements: Lessons for Policy Makers?', in *Final Conference within the EU-financed Concerted Action Programme on the Effectiveness of International Environmental Agreements and EU Legislation held in Barcelona 9–11 November 2000*, Lysaker, Norway: The Fridtjof Nansen Institute.

Waltz, K.N. (1979), *Theory of International Politics*, Reading, MA: Addison-Wesley.

Weale, A. (1992), *The New Politics of Pollution*, Manchester: Manchester University Press.

Wettestad, J. and S. Andresen (1991), *The Effectiveness of International Resource Cooperation: Some Preliminary Findings*, Lysaker, Norway: The Fridtjof Nasnen Institute.

Young, O.R. (1989), *International Cooperation: Building Regimes for Natural Resources and the Environment*, Ithaca, NY: Cornell University Press.

Young, O.R. (1999), *The Effectiveness of International Environmental Regimes: Causal Connections and Behavioral Mechanisms*, Cambridge, MA: MIT Press.

Young, O.R. (2001), 'Inferences and Indices: Evaluating the Effectiveness of International Environmental Regimes', *Global Environmental Politics*, **1**(1), 99–121.

Young, O.R. (2003), 'Determining Regime Effectiveness: A Commentary on the Oslo–Potsdam Solution', *Global Environmental Politics*, **3**(3), 97–104.

8. The price of fish and the value of seagrass beds: socioeconomic aspects of the seagrass fishery on Quirimba Island, Mozambique[1]
Fiona R. Gell

Introduction

Marine resource use is an important component of the local economy in many tropical coastal areas. The importance of marine resources to a community depends on the geographic and economic situation of the area: the level of development, the role of tourism and the availability of alternative sources of income (Ruddle, 1996a). In areas near international airports and large cities, marine resources can be exported to an international market or sold to tourists and can fetch high prices. In more isolated places without a developed transport infrastructure, marine resources may only be used on a subsistence level in the immediate local area and thus have a much lower economic value (White et al., 1994; Lindén and Lundin, 1996; Birkeland, 1997). However, in such isolated places the local value of marine resources is high because they are often the most important source of income and animal protein.

To manage marine resources it is important to assess the ecological status of the habitats and organisms that are being exploited. It is also important to assess how local people use marine resources and their role in the local economy. As the socioeconomic structures of coastal communities develop and change, the intensity with which marine resources are exploited also changes. In the past this generally happened on a local or perhaps national level. However, at the beginning of the twenty-first century, the phenomenon of 'globalization' means that few places really are remote or inaccessible any longer. Economic growth and social change in one place can affect the extent to which marine resources are exploited in other parts of the same country, in other countries and even in other continents. For instance, rapid economic growth in Asia has had a direct impact on the sustainable use of marine resources throughout the Pacific, from Taiwan to the Galapagos, with large Asian companies buying live reef fish, sea cucumbers and other marine products to supply the growing Far Eastern markets (Birkeland, 1997).

Socioeconomic investigations are increasingly being incorporated in biological studies of tropical coastal fisheries (Sri Lanka – Dayaratne et al., 1995; Fiji – Jennings and Polunin, 1996; Kenya – Juma, 1998), as it has become apparent that ecological and socioeconomic aspects of resource use must be considered together. Increasing emphasis is being put on the role of user participation and the use of local knowledge in the management and development of natural resource use in developing countries (Chambers, 1997; Sillitoe, 1998). However, including an element of 'indigenous knowledge' in natural resource projects is rare, with just 1.1 per cent of all projects funded by the UK Department for International Development (DFID) in the past including such research (Sillitoe, 1998)

Fishing and other types of marine resource use are typically last resort sources of food and employment for people who have no alternative (McManus, 1993). In Mozambique half the population's protein intake is estimated to be from fish (Van der Elst et al., 2005). This is one of the reasons that marine resources and particularly those that are easily accessible on foot are put under such severe pressure in very poor, often highly populous areas. Accessible sheltered areas such as reef flats and seagrass beds often support large numbers of very poor fishers who cannot afford boats (McManus, 1993). Seagrasses are therefore often more likely to be intensely exploited than other habitats.

Study area
This study was conducted as part of the Frontier-Moçambique Quirimba Archipelago Marine Research Programme. Much of the socioeconomic data presented here were collected on Quirimba Island, one of the two most densely populated islands in the Archipelago, during 13 months' fieldwork over a period of two years (1996 and 1997). Quirimba Island is 6km long by 2km wide. It is situated within a few kilometres of the Mozambican mainland and is part of Ibo District in the province of Cabo Delgado. At low spring tides it is possible to walk between the island and the mainland and also to Ibo Island to the north and Sencar Island to the south. Quirimba has 3000 inhabitants, most of whom live in Quirimba village at the northern tip of the island. A few hundred people live outside the village, the majority in the Kumilamba area in the south of the island. The seagrass beds of the island have a high diversity of fish with 249 species identified (Gell and Whittington, 2002), which is more than 10 per cent of the total number of fish species in the Western Indian Ocean.

The history of the Quirimbas
The Quirimba Archipelago has a long and colourful history, but there are few existing historical records and the islands have been little studied

(Boxer, 1963). This is surprising because they were a key part of the Arab and Portuguese Indian Ocean empires. Davidson (1961 in Boxer, 1963) suggests that research on the Quirimbas would enable new insights into the wide-ranging scope of the Indian Ocean trade. Before the Portuguese arrived in the sixteenth century, the islands were important as prosperous Arab trading posts for ivory and slaves. In the seventeenth century they suffered in the wars between the Portuguese and the Omani Arabs, and at the end of the seventeenth century the Omanis destroyed most of the buildings in the stone towns on the islands. In the early nineteenth century the islands were devastated by Madagascan raiders (Sousa, 1960; Boxer, 1963). Unlike Zanzibar, Mombasa, Kilwa or Mozambique Island, the Quirimbas did not continue to exist as important Indian Ocean trading ports or become known for their long histories. The Catholic church in Quirimba village (now used as the school) dates back to 1894 and there is also a ruined church, Nossa Senhora do Rozario, that was built around 1580 (Sousa, 1960). Santa Maria beach, where the Frontier-Moçambique project was based, was said to be where some of the first Portuguese settlers in the Quirimba Archipelago had lived (J. Gessner pers. comm. – from C.R. Boxer, the historian who visited Quirimba in the 1950s).

Historically the islands seem to have been very productive agriculturally and able to support residents and visiting traders. They were well placed for trading goods from the interior such as ivory and gold. Ibo and Quirimba were also important centres for the Indian Ocean slave trade. Records from early visitors (mainly Portuguese and British) describe the islands as productive in terms of agricultural produce, goats and, of course, fish, and a good place to stop for supplies (Sousa, 1960; Boxer, 1963).

For over 400 years there were Portuguese inhabitants on the islands and a strong Catholic presence, with parish priests on Ibo and Quirimba. Now the residents are virtually all of African or mixed African, European and Arab origin. The coastal people traditionally associated with the coast of Cabo Delgado are the Mwani who speak a dialect of Kiswahili called Kimwani, and were traditionally traders and fishers. The coastal Mwani people are mainly Muslim. They are not as strictly Muslim as many coastal Tanzanian communities and there is still a strong system of traditional pre-Islamic beliefs. Witch-doctors or *curandeiros* are important figures in the community and are consulted for a huge variety of problems. The other main group on the island are people of Makua origin, the biggest ethnic group in Mozambique (West, 1998) living throughout the northern provinces of the country. A third ethnic group present in small numbers are the Makonde, from the area around the Tanzania–Mozambique border. The Makonde are known for their skilled wood carving.

Until 1975 Mozambique was under Portuguese colonial rule, with many colonial plantations and factories in the rural north. The Portuguese were strict enforcers of fisheries regulations such as mesh size regulations and in colonial times there were regular incinerations of illegal nets on Quirimba (J. Gessner pers. comm.). In 1975 Mozambique gained independence and the 17-year civil war started. The civil war was fought predominantly in the ordinary villages of rural Mozambique and the ordinary people of Cabo Delgado province suffered brutal guerrilla warfare. Tens of thousands of people became *deslocados*, refugees within their own country, fleeing villages that were being destroyed. Many of the *deslocados* fled to the coast, and in particular the islands, because the guerrillas would not cross water and so would not pursue refugees to the islands. This influx of people from the mainland led to an increase in population on the coast (Massinga and Hatton, 1997) and on the islands (J. Gessner pers. comm.) and a greater mix of ethnic groups. Many of the refugees were Makuas and some were Makondes from the northern Mozambican interior.

In the years following the end of the civil war in 1992, Mozambique gained a level of stability with the government rebuilding the political and economic structure (Macia and Hernroth, 1995). The Human Development Index of the country as determined by UNDP has risen over recent years, but Mozambique is still the world's tenth poorest country (UNDP, 2006). Development is now happening very quickly in all sectors of Mozambican life, from tourism to heavy industry. However, the Quirimbas have been somewhat neglected. They are over 2000km from the national capital Maputo and difficult to reach and are therefore rarely visited by politicians and other decision-makers. Infrastructure is poor and there are few opportunities for the people of the Quirimbas to be represented or heard at a national level.

Methods
Socioeconomic data were collected on Quirimba using several methods:

- informal interviews with fishers on fishing trips, at landing sites and in the village;
- formal interviews with boat owners;
- accompanying fishers on fishing trips that employed all major methods;
- Visiting fishers' households, talking to families and observing daily routines;
- informal workshops in the village with fishers.

A number of Quirimba fishers and other residents were key advisers for this study. They were interviewed at length on numerous occasions, suggested other people to talk to, took me to their houses and the houses of family and friends to learn more about life on Quirimba and showed me their allotments (*machambas*) and their other daily work. The information presented in the results is therefore assembled from a large number of mainly informal interviews with a wide spectrum of people in the community. The key informants are acknowledged at the end of this chapter.

Some biological information about the women's invertebrate fishery was collected in the first year of the project and has been presented in Barnes et al. (1998). Additional social and economic information about these fisheries was collected through informal discussion with women in the village, by accompanying them fishing and through two workshops. Workshops were organized with the local OMM (Organizaçao da Mulher Moçambicana – Mozambican Women's Organization) representative in the village. The workshops were conducted in Portuguese, Makua and Kimwani. Most women in Quirimba were illiterate and many did not speak Portuguese so it was important to use visual methods that were accessible to all participants.

Results and discussion

Quirimba during the study period (1996–97)
Quirimba village has a good system of water pumps from wells throughout the village. There is a medical clinic staffed by a trained nurse, but no doctor, a small school, a market selling a very limited range of goods, a few small shops and a bakery. The houses in the village are fairly large (two to four rooms), and well built with a mangrove structure filled out with mud and rock. Most have traditional *mecute* (woven coconut palm) roofs but there are some stone or cement houses with corrugated iron roofs. The houses have large yards where cooking is done and also small toilet and washing shelters. There is no sewerage system in the village. A few houses have shelters with a 'long-drop'-style toilet but many people still use the beach or mangrove area. There is no electricity supply to the village. A few residents have oil-powered generators for their homes, used for a few hours each day. There is a mosque in the village for the predominantly Muslim community.

In Kumilamba in the south of the island, the residents have *machambas*, small allotments used for mainly subsistence agriculture. Houses in Kumilamba are of a less permanent nature than those in the village, and generally smaller (typically just one room). The people who live in the *machambas* have easy access to their agricultural land but they are a 6km walk from the village where they have to go to get drinking water. Also, many children who live in the *machambas* do not go to school because the nearest school is in Quirimba village. Most households on the island have a *chamba*, a piece of land, where they grow sweet potatoes, cassava, papaya, corn and beans. The land tenure system was informal and slightly vague at the time of study and appears to be entering a process of modernization and formalization. Many families have plots that their family had farmed for years, which they think of as their own but have no legal documentation. Others have cleared bush and scrub by slash and burn to claim the newly cleared land as their own.

A large area of the middle section of the island is covered in a coconut plantation owned by farmers of German origin whose family have had a plantation on the island since the beginning of the century. The plantation provides jobs for about 80 people as labourers, guards and processing workers. The coconuts are produced for the copra market. The plantation owners rear cattle to sell in Pemba. There is also a local herd of cattle owned by a number of people in the village. The cattle graze in the coconut plantation.

Although people on Quirimba are fairly self-sufficient in terms of everyday requirements such as food, water and building materials, many people make the trip to Pemba, the provincial capital, to sell produce for a small cash income, to pay for medical treatment and to buy household goods such as cooking pots and clothes. Three motor boats sporadically provide transport for the day-long journey from Quirimba to Pemba. It is also possible to go to Quissanga on the mainland on the other side of the Montpuez Bay by dhow then go by road to Pemba, but this often takes even longer and is more expensive.

Types of marine resource use
The main fishing methods on Quirimba Island are seine netting and *marema* trapping (woven bamboo fish traps) in the subtidal seagrass beds, the collection of invertebrates from the intertidal seagrass beds and a variety of smaller-scale fisheries and resource collection on the coral reefs and in the mangroves. The marine resources captured by these methods can be put into three categories based on their use (see also Table 8.1):

Locally consumed resources   The majority of fish caught locally are consumed locally by fishers and their families or by other local people who trade goods or services for them.

Locally sold resources   Some fresh fish and shellfish are sold locally. There is no fish or seafood market in the village as such – buyers usually go to the beach where fish or shellfish that have been collected that day are landed. Dried produce is occasionally sold between Quirimba residents on a casual basis.

Table 8.1 Marine species exploited on Quirimba Island with method of capture, user group, habitat captured from, market and use for the species

Scientific Name	Common Name	Kimwani	Method of Capture	User Group	Habitat	Market	Use
Fish							
Siganus sutor	African whitespotted rabbitfish	Safi	Seine, gill net and marema trap	Local men	Seagrass and reef flat	Local Mainland/Pemba	Eaten fresh Dried
Leptoscarus vaigiensis	Seagrass parrotfish	Bonju	Marema trap and seine	Local men	Seagrass	Local Mainland/Pemba	Eaten fresh Dried
Lethrinus variegatus	Variegated emperor	Sololo	Suri trap and seine net	Local men	Seagrass	Local Mainland/Pemba	Eaten fresh Dried
Gerres oyena	Blacktip mojarra	Sala	Seine net and gill net	Local men	Seagrass and reef flat	Local Mainland/Pemba	Eaten fresh Dried
Lethrinus lentjan	Pink ear emperor	Njana	Seine net	Local men	Seagrass and coral	Local Mainland/Pemba	Eaten fresh Dried

247

Table 8.1 (continued)

Scientific Name	Common Name	Kimwani	Method of Capture	User Group	Habitat	Market	Use
Large reef fish e.g. acanthurids, scarids			Gill net, spearfishing	Itinerant fishers	Coral reefs	Mainland	Dried
Carangidae	Jacks	*Njolwe*	Fence trap, line fishing	Local men, itinerants	Open water, seagrass, reef	Local	Eaten fresh
Molluscs							
Octopus vulgaris	Octopus	*Mweza*	Stick, hook	Local women	Reef flat	Local	Eaten fresh and dried
Barbatia fusca	Ark shell	*Ombay*	Hand	Local women	Intertidal sand and seagrass	Mainly mainland/Pemba	Dried
Pinna muricata	Pinna shell	*Macaza*	Hand	Local women	Intertidal sand and seagrass	Mainly mainland/Pemba	Dried
Pinctada nigra	Oysters	*Mbari*	Hand	Local women	Intertidal seagrass	Mainly mainland/Pemba	Dried
Pleuroploca trapezium	Tulip shell	*Kome nlume*	Hand/collected on net or trap fishing trips	Men	Seagrass	Operculum sold in Tanzania	Animal eaten fresh

Chicoreus ramosus	Murex	*Makome*	Hand/collected on net or trap fishing trips	Men	Seagrass	Operculum sold in Tanzania	Animal eaten fresh
Terebralia palustris	Mangrove whelk		Hand	Men/children	Mangrove/ intertidal flat	None	Used for baiting traps
Echinoderms							
Holo-thuriidae	Sea cucumber	*Magajojo*	Hand	Local finfishers and itinerants	Seagrass and intertidal	Export	Dried for food and medicinal uses
			Mask and canoe	Local finfishers and itinerants	Seagrass and shallow reef	Export	
			SCUBA	Itinerant fishers	Reefs	Export	
Crustacea							
Panulirus spp.	Crayfish		Hand/spear/ hook	Local fishery/ some itinerant/ tourist	Reef	Pemba	Restaurants
Scylla serrata	Mangrove crabs		Hand	Local fishers	Mangrove	Local European	Eaten fresh
Portunus pelagicus	Swimming crab		Hand/bycatch in traps	Children/trap fishers	Intertidal flats, shallow water	Local	Eaten fresh
Reptiles							
Chelonidae	Turtle	*Assa*	Incidental catch in net	Local and itinerant	Reef and seagrass	Local	Eaten fresh

Source: From personal observation and interviews and workshops with fishers.

Resources sold outside Quirimba Large quantities of dried fish are sold off the island in two main ways. First, island-based traders buy fish from fishers, particularly trap fishers, accumulate a large quantity of dried fish then sell it in the market in Pemba or in markets elsewhere in Cabo Delgado province. They buy fish, often a lot of *Lethrinus variegatus* and *Leptoscarus vaigiensis*, for 3000MZN (meticais) per kilo wet weight, and sell it for 15000–20000MZN dried (12000MZN was equal to approximately US$1 during the study period, so traders paid US$0.25 per kilo for fish and sold it for US$1.25–1.67). This seems like a large profit but in fact the dried fish will have lost up to three-quarters of its wet weight through drying, gutting and cleaning. Net fishing boat owners also buy fish to sell on the mainland. They have an arrangement with their boat's crew whereby they either buy the majority of fish caught at a reduced price or they pay their crew a certain wage and receive the majority of the fish as their share of the operation. They dry this fish and accumulate it in storage until they have enough to make the trip to Pemba to sell it.

The other main route to sale for Quirimban fish is through fishers from the mainland who come to Quirimba and the other islands in the Archipelago to fish. A small proportion of these fishers are from villages on the nearby mainland such as Mahate and as far south as Pemba. Many of them have been coming to the Quirimbas to fish for many years. Most are farmers in the wet season and fish the Quirimbas in the dry season. The dried fish supplements their family's diet of maize and cassava, and can also be sold for cash to buy any goods that the family cannot produce themselves, such as clothes, cooking oil and medicines.

The majority of the visiting fishers are from the next province south, Nampula, and have only been fishing the Quirimbas in large numbers for a few years (from interviews with Nampula fishers and J. Gessner pers. comm.). These fishers are in a similar situation to those described above – they have agricultural work in the wet season and come to the Quirimbas in the dry season to catch fish to dry and take back to eat over the wet season and to sell. Traders also come to Quirimba to buy dried fish, which they said is 'better quality' than the fish they can buy in Nampula. These traders deal in hundreds of kilos of fish, which they drive down to Nampula.

Cash commodities for export The only really lucrative commercially driven fishery on Quirimba is the sea cucumber fishery. In the mid-1990s a large commercial sea cucumber fishing operation based in Pemba was active in the whole of the Quirimba Archipelago. The operation was run by a Chinese export company who processed the sea cucumbers and exported them to the Far East along with other commercially valuable marine resources caught locally, such as shark fin. At the end of the study period this fishery had temporarily ceased. At least one Tanzanian sea cucumber fishing operation was active in the Archipelago, using SCUBA gear. Traditionally, local fishers collect sea cucumbers by snorkelling, thus limiting the depth range over which they are vulnerable. Ordinary net or trap fishers who collect sea cucumbers while they are fishing can sell them individually to traders for up to 1000MZN (less than US$0.10) each. Other incidental catches that can be sold to traders include seahorses for which a fisher can get 5000MZN (US$0.40) for a 7cm individual and the opercula of some shells, which can be sold by the kilo for 150000MZN (US$12.00) or individually for a few hundred meticais. These traders sell these resources on in bulk across the border in Tanzania to be sold on to the Far East.

The prices for fish caught in the seagrass and coral reef fisheries of Quirimba were remarkably consistent. Fresh fish were sold for a set price of between 4000 and 5000MZN per kilo. Occasionally large, good quality lethrinids caught in baited traps or by hook and line were sold for a higher price (reportedly up to 10000MZN per kilo) because of the good quality of lethrinid flesh. Otherwise, everything from a kilo of 7–10cm *Lethrinus variegatus* or other small fish, to a kilo of a large jack (*Carangidae*) or snapper (*Lutjanidae*) caught in a fence trap was sold for around 4000MZN. 5000MZN per kilo of fish was the standard price for fish throughout rural coastal Cabo Delgado. The same price was charged in the rural Mecufi district south of Pemba, where 5000MZN was quoted as the lowest price of fish in Mozambique (Loureiro, 1998). Set prices are common for a variety of commodities in northern Mozambique, from shellfish to sweet potatoes.

The only way to preserve fish on Quirimba is to dry it in the sun. This is done on drying racks outside houses in the village. Dried fish was sold on the mainland for between 15000 and 20000MZN per kilo. Small fish are dried closed (*fechado*) and were sold for 15000MZN per kilo and larger fish are dried open (*aberto* – like a kipper) and sold for the slightly higher price of 16000MZN per kilo. Dried fish often have a slightly 'off' taste, which does not prevent their sale or consumption. Occasionally seagrass parrot-fish (*Leptoscarus vaigiensis*) die in the *marema* traps, and are retrieved in a partially rotted state. These fish are not discarded and are dried as normal and sold for consumption. The highest prices, some exceeding 25000MZN per kilo were given for dried fish sold in villages a long way inland. The prices of fish and marine invertebrates on Quirimba are shown in Table 8.2, along with a selection of other local prices for comparison. Note the high prices per kilo of fresh invertebrates (10000–15000MZN) compared with that of fish (3000–5000MZN), and also the high price of imported goods such as sugar (20000MZN per kilo) and onions (25000 per kilo).

Table 8.2 Prices paid on Quirimba Island for fish, agricultural produce and other goods, their source, mode of production and how they are made available to the community

Item	Cost on Quirimba Meticais	Cost on Quirimba US$	Source	Mode of Production/ Exploitation	Market
Fresh fish/kg	3000 to 5000	0.25–0.40	Local	Caught by men	Sold direct at landing site
Dried fish/kg	12 000 to 15 000	1.00–1.25	Local	Caught by men	Sold from home
Squid/kg	3000	0.25	Local	Caught by men	Sold direct at landing site
Octopus/kg fresh	5000	0.42	Local	Caught by women and men	Sold direct at beach
dry	16000	1.30	Local		
Pen shells fresh/kg	15000	1.25	Local	Collected by women	Sold direct at beach
Ark shells fresh/kg	10000	0.83	Local	Collected by women	Sold direct at beach
Oysters/kg	12000	1.00	Local	Collected by women	Sold direct at beach
Mangrove crab/kg	5000	0.42	Local	Collected by men	
Sweet potato/kg	2000	0.17	Local	Grown in *machambas*	Sold in *machambas*
Cassava leaves/kg	5000	0.40	Local	Grown in *machambas*	Sold in *machambas*
Peanuts/kg	4000	0.33	Mainland		Sold in village market
Coconut (each)	1000	0.08	Local	*Machambas* and plantation	Sold in village market

Onions/kg	25 000	2.08	Mainland	Sold in village market occasionally/Quissanga market
Eggs (each)	1000	0.08	Local	From chickens kept in village and *machambas*	Sold from home
Bread flour/kg	10 000	0.80	Mainland	Imported	Sold in village shop
Maize flour/kg	6000	0.50	Mainland/local	Maize from *machambas*	Sold in village shop
Rice/kg	10 000	0.80	Mainland	Imported	Sold in village shop
Papaya (each)	500–2000	0.04–0.16	Local	Grown in *machambas*	Sold in *machambas*
Mango (each)	100–1000	0.01–0.08	Local	*Machambas* and in plantation	Sold in village market seasonally
Sugar/kg	20 000	1.60	Malawi	Imported	Sold in village shop
Condensed milk	15 000	1.25	Maputo/South Africa	Imported	Sold in village shop
Chicken (each)	15 000–25 000	1.25–2.08	Local	Reared in village or in *machambas*	Sold from home
Capulana (sarong – two pieces)	40 000–80 000	3.33–6.66	Maputo/Tanzania	Imported (cheapest from India)	Sold in village shop
T-shirt	10 000	0.80	Mainland	Imported	Sold in village market occasionally
Trousers	20 000	1.60	Mainland	Imported	Sold in village market occasionally
TB medication (course of treatment)	40 000	3.33	Mainland	Imported	Sold at village clinic

Note: Prices are shown in the local currency of Mozambican meticais and US dollars.

The different economic levels of marine resource use in the Quirimbas
Marine resource collection in the Quirimba Archipelago can be divided into a number of categories depending on the level of investment needed by the fisher:

Little or no personal investment   The main group in this category are the crew of the seine net fishing boats who need to make very little investment to start fishing. Some have their own strong shoes to protect against urchins, others have their own masks but the majority have neither. Young boys go straight into the seine net fishery and strangers to the island are also free to enter this fishery.

The women's invertebrate fishery requires very little personal investment, needing only a bucket or other container. Women collect invertebrates from the intertidal at Santa Maria and Pantopi, and also make trips on boats out to exposed sand banks at low spring tides. Young children who collect small species of mollusc from the upper intertidal, and women who collect these shells for occasional meals can also be included in this category.

Moderate personal investment   Trap fishers need a dugout or plank-constructed canoe with two outriggers (worth between 100 000 and 800 000 MZN, see Table 8.3), 30 to 50 (usually 40) *marema* traps, weights (stones collected locally), 'hippo fat' to waterproof the boats, a pole for punting and a bailer. The total cost for trap fishing equipment ranged from a minimum of 220 000 MZN (US$18) for a second-hand dugout canoe and 30 traps, to over 1 000 000 MZN (US$83) for a large plank-

Table 8.3   The average costs of the major items of gear for seine net and marema *trap fishing*

Item	Cost in Meticais	Cost in US$	Source
Seine net – new	5–10 million	417–833	Imported
Seine net – second-hand	3 million	250	Imported – bought in Pemba
Fishing boat 3–4m	2–3 million	166–250	Made on Quirimba with wood from the mainland
Marema trap	4000 (need 40)	0.33 (13.20 for 40)	Bamboo panels from the mainland
Canoe – dugout	100 000–400 000	8–33	Made on Quirimba
Canoe – timber	400 000–800 000	33–67	Made on Quirimba

constructed canoe and 40 traps. The average initial expenditure was probably around 300000MZN for a standard canoe and 160000MZN for 40 traps, a total of 460000MZN (US$38). *Maremas* also have to be replaced every few months. The average salary for a day's formal work in Quirimba, for example on the coconut plantation or as an ordinary fisher, was 10000MZN (US$.83 – see Table 8.4). Most of this small salary was needed for food and other necessities and saving was not an option for most ordinary people on Quirimba.

Weeks could pass when a fisher could not go fishing because of the weather, because of problems with the net or the boat, or because of illness so any money saved would be used then. Unforeseen expenses included money for medicines for the fisher himself or a member of the family, for example, treatment for an elderly relative with TB might cost 40000MZN (see Table 8.2). Quinine and aspirin for malaria also had to be bought fairly frequently. Most people grew their own staple foods such as maize, sweet potatoes and cassava in their *machambas*, so for most people food was rarely a problem.

Groups of women from the district of Mecufi, south of Pemba, were reported to come to Quirimba every year and fish in the shallows of the reef flat in front of the village using large *capulanas* (the patterned cotton wrap worn by local women and with dozens of other uses). This method of fishing for very small fish is found throughout the region (Comores – Dahalani, 1997; Tanzania – Andersson and Ngazi, 1998). Some village women said that the reason they did not do this type of fishing was that it required a special large *capulana* that they couldn't afford. This type of fishing therefore requires a level of investment difficult for Quirimban women who have few opportunities for earning cash.

Gill nets were used by a few groups of people for fishing on the reef flat on the east coast of Quirimba. These nets are shorter than seine nets and a lot cheaper. Three people are needed to do gill netting so either all contributed to the cost of the net or one person bought it. The gill nets on Quirimba were all owned by ordinary families.

A large initial personal investment The main group in this category were the seine net fishing boat owners. The same person usually owned the net and the boat, which together could be an investment of over 6 million MZN (US$500, see Table 8.3). These boat owners were the entrepreneurs of Quirimba. They had the largest stone-built houses and were amongst the only people to own motorbikes and other major consumer items. Some of the boat owners also had other business concerns apart from trading fish; some had shops in the village, others traded across the border in Tanzania.

Table 8.4 The major forms of employment on Quirimba, average time worked, salaries and location of the work

Employment	Hours Worked	Days Worked per Month	Daily Salary Meticais	Daily Salary US$	Annual Salary Meticais	Annual Salary US$	Location
Trap fisher	4 hours	22–24	15000–40000	1.25–3.33	4–10 million	333–833	Montepuez Bay
Net fisher	5 hours	18–20	10000	0.83	2160000	180	Montepuez Bay
Net fishing captain	5 hours	18–20	14000	1.17	3024000	252	Montepuez Bay
Net fishing boat owner	Various	Various	111000	9.25	23976000	1998	Montepuez Bay
Work in *machamba* (mainly subsistence)	Various – seasonal	Various – could be whole day	Food				*Machambas* (Kumilamba)
Worker in coconut plantation	8 hours	6 days a week	10000	0.83	3138600	262	Plantation
Fish processor	1–2 hours	18–20 (same as net fishers)	4000	0.33	864000	72	Quiwandala
Intertidal invertebrate collector	3 hours	6–8 days around spring tides	20000 +	1.67	1680000	140	Santa Maria, Saja, etc
Making bread and cakes to barter for fish	Various	18–20 (same as net fishers)	Fish				Quiwandala
Trader	Various	Various	Various				Quirimba, Pemba, Tanzania

256

Socioeconomic aspects of the seagrass fishery on Quirimba Island   257

The stories of how fishing boat owners made the money to initially buy the fishing boats varied. Some were local men who had worked their way up from being a fisher crewing a large seine net fishing boat, to trap fishing, to buying their own large boat. A few had family money from land ownership, others had come from the mainland specifically to fish.

The seine net fishing boats were built in Quirimba village with wood imported from the continent. A 3m boat cost from 2 to 2.5 million meticais (around US$200). A 4m boat cost around 3 million meticais. One of the boat owners interviewed had bought his current fishing boat for 1.5 million four years previously. The nets were bought on the mainland, usually in Pemba. A new net complete with floats, weights and ropes reportedly cost 10 million meticais but all the boat owners interviewed had bought their nets second-hand for around 3 million meticais. They expected the nets to last for 10–15 years. Various systems of payment were used on the fishing boats. Some of the boat owners paid their crew 4000MZN per kg of fish caught and said their catch could range from a few kilos to a few hundred kilos. The pay per trip for the crew worked out at 10 000MZN per fisher and 14 000MZN for the captain, and some small share of the catch, usually 1 kg. The owners also had to pay four or so people to clean the fish. They were usually paid one kilo of fish or 4000MZN each day. (See Table 8.4 for salaries for the various fishing jobs on Quirimba and Table 8.5 for the costs and earnings in the seagrass trap and net fisheries.)

Economic gain from fishing
Net fishing The average daily catch per boat was 75kg. If this was sold at 4000MZN per kilo it would be worth 300 000MZN (US$24). Out of this the boat owner paid an average of eight crew 10 000MZN each, the captain 14 000MZN and five people 4000MZN to clean the fish. This would give a total daily expenditure of 114 000MZN, making a total profit of 186 000MZN (US$14.90). Boat owners had responsibility for maintaining boats and nets, buying new ropes and other materials, so some money would go towards this. Ropes in particular needed replacing regularly, perhaps once a year. Around neap tides some boats catch just 25kg of fish, worth 100 000MZN, which is less than the daily expenditure calculated above. Occasionally, when things went wrong and the sail broke or the boat ran aground, the crew would come back with no fish and presumably on these days everyone would go home empty-handed.

Other boat owners organized the fishing in a different way. They took a certain amount of fish every day as their share and the value of the rest was divided between the crew with the captain getting slightly more. In this case it was in the interest of the crew to catch more fish to increase their wage but there did not seem to be an inclination for fishers on boats

Table 8.5 Summary of average values for various aspects of the seagrass trap and net fisheries

	Trap Fishing	Net Fishing
Mean daily catch per boat	7kg	75kg
Total daily catch for fleet	280kg	2250kg
Mean value of daily catch	21 000–35 000MZN US$1.75–2.91	225 000–375 000MZN US$18.75–31.25
Initial expenditure	Boat 300 000MZN US$25 Traps 160 000MZN US$13 Total 460 000MZN US$38	None for individual fishers
Crew's wages	–	8 crew at 10 000MZN per day, total 80 000MZN, US$6.67
Captain's wages	–	14 000MZN, US$1.17
Processors' wages	–	5 processors at 4000MZN/ US$0.33 per day, total 20 000MZN/US$1.67
Total daily expenditure	3000MZN/US$0.25	114 000MZN/US$9.5
Mean daily profit for boat owner	21 000–35 000MZN US$1.75–2.91	111 000–261 000MZN US$9.25–21.75
Total daily value of catch for fleet	630 000–1 400 000MZN US$52.5–116.67	3 375 000–11 250 000MZN US$281.25–947.5
Days fished per year	264	216
Total annual catch	60 tonnes	440 tonnes
Total annual value of catch	180–300 million meticais US$15 000–25 000	1320–2200 million meticais US$110 000–183 333
Total annual catch for both methods	1500–2500 million meticais US$125 000–208 333	

Note: The exchange rate used was US$1 to 12 000MZN.

managed this way to work harder to maximize their earnings. The boats that seemed to work the most efficiently were those with small crews of five or so. They had to work harder than larger crews to haul the nets but they often seemed more efficient at getting their boats out, catching the fish and returning to port. Many of the boats with large crews of 12 or more seemed to be primarily a social occasion that just happened to result in something to eat and some money.

Trap fishing The average daily catch per fisher for trap fishing was 7kg, an average daily earning of 28000MZN, nearly three times the average salary as a labourer or as fishing crew. Trap fishers fished alone and cleaned their own fish and so had no expenditure on a day-to-day basis. If fishers could sell their catch fresh in the village they would get 4000MZN whereas if they sold it all in one go to a fish trader they would only get 3000MZN per kilo (21000MZN for an average catch) but they could organize this in advance and would be guaranteed to sell it all on the day they caught it. Trap fishers usually retained about 1kg of their catch for their own consumption. Very few trap fishers dried their own fish.

Socio-cultural aspects of fish use on Quirimba
Fish was highly valued as food on Quirimba. Although some species were favoured for their taste, particularly lethrinids and siganids, all edible species fetched a similar price. Fish was often used for barter. Small fish (around 10cm) caught by net fishers were bartered for bread rolls or peanut biscuits made by local women, which they brought out to the boats as they came into the landing site at Quiwandala at the end of the day's fishing trip. The women who made these biscuits were usually single or had husbands who were not involved in fishing. There was an inflexible exchange rate of four small fish (often variegated emperors or three-ribbon wrasse) for one biscuit. Fish was commonly shared between family and friends in the complex Quirimban 'kinship' networks (discussed below). Fish were also commonly given as gifts or in exchange for other goods or services.

The small size of the fish eaten was very striking. The capture and consumption of small fishes that would be considered 'trash fish' in many places is a common feature of some tropical artisanal fisheries (Pauly, 1979; Gayanilo and Pauly, 1997). As Munro (1996) states, unless a fish is poisonous, the concept of trash fish does not exist in many poor developing countries. People on Quirimba utilized extremely small fish that would have been discarded elsewhere, for example 5cm-long butterfly fishes, emperors and damsel fishes. Fish was usually added to stews containing coconut milk, chillies and sweet potatoes. The small fish could add flavour and add nutritionally in terms of animal protein, but did not significantly increase the cooking time. Firewood was always in short supply on Quirimba and collecting wood, whether from the mangroves or from scrubland or from around the *machambas*, was hard work done mainly by the women. Small fish could be cooked quickly and required much less fuel than large fish.

There were some unusual uses of fish and fish products. Boxfish (*Ostraciidae*), not eaten in many places because of skin toxins, were well-liked in Quirimba. They were stuffed with rice and cooked directly in the fire, their tough skins protecting the flesh from burning. Other toxic fish such as scorpionfish (*Scorpaenidae*), catfish (*Plotosidae*) and most pufferfish and tobies (*Tetraodontidae*) were discarded but some large pufferfish were retained for sale to Makua people in Pemba and Nampula. Some Makua people eat these fish and know the special methods for preparing them to avoid fatal poisoning.

Fishing traditions
Fishing communities all over the world have long traditions and complex systems of beliefs and superstitions (Ruddle, 1996b), from the South Pacific (Johannes, 1981) to the Isle of Man (pers. obs.; Manx Heritage Foundation, 1991). In southern Kenya, where the habitats and fishing methods are similar to those used on Quirimba and the coastal people have a broadly similar cultural history with strong Arab influences, strong fishing traditions still exist (McClanahan et al., 1997). There were few traditions associated with fishing in Quirimba but it was difficult to ascertain whether this was because there had been a complex system of beliefs that had gradually disappeared or whether there had never been such a belief system. The seine net fishers did not seem to have any traditions or superstitions directly associated with fishing apart from a vague concept of 'luck'. One fisher sometimes wore a seamoth fish (*Pegasidae*) round his neck as a talisman to bring luck. Seeing dolphins during fishing was considered good luck. On the other hand, fishers in some parts of Kenya still hold ritual ceremonies and have sacred sites at sea where offerings are made to spirits to improve fishing. These rituals are gradually being lost, in many instances because Islam is increasing in strength and the younger generation of Muslim fishers see the traditional rituals as against Islam. Although there has been a lot of population movement around the Kenyan coastal area, it probably does not approach the large-scale migration of people around the north of Mozambique as a consequence of the war. McClanahan et al. (1997) suggest that the more recent Islamization of culture in Kenya is linked to the decay of the rich and elaborate cultural traditions of coastal management. This may also be the case to some extent in Quirimba.

There were a number of traditions associated with eating fish or other marine products. When dugongs were caught, and they were fished in the area up until the 1990s, they had to be taken to the mosque for blessing before they were eaten. Dugong meat was given to pregnant women to make their babies 'beautiful'. Other fish species were by tradition not eaten by pregnant women because they were said to harm the baby. (In a place where a large proportion of women lost babies at late stages of pregnancy and where infant mortality was high it was not surprising that there were a lot of superstitions surrounding pregnancy and the health of the unborn baby.)

Although there was no formal system of rules and traditions for fisheries management, conflicts within and between user groups were rare. There were a number of potential sources of conflict – between trap fishers and net fishers using the same fishing sites, between local Quirimban fishers and fishers from Quissanga on the mainland side of the bay, and between Quirimban fishers and Nampula fishers. There were also many opportunities for theft and for 'cheating'. Seine nets, perhaps the most valuable possessions on Quirimba, were always left anchored in the water off the fish landing site overnight and were never stolen during the study period. Traps were rarely stolen or emptied by people other than their owner. In interviews, trap and net fishers were asked if theft was a problem and about conflicts between user groups, but none of those interviewed expressed any concerns. Most net fishers tried to avoid fishing over areas containing traps or lifting nets over traps although some were reported to take the fish out of traps caught in their nets. Trap fishers did not come into conflict over sites fished or empty the traps of other fishers. Dynamite fishing has not been witnessed in the Quirimba Archipelago, but in the past it has been a serious problem in coral areas of Southern Tanzania just across the border (Guard and Masaiganah, 1997).

In southern Kenya, McClanahan et al. (1997) found that where old traditions of respect for sea spirits and customs for behaviour at sea had been lost, theft from set nets and traps and of the nets and traps had increased. This implies that without the specific superstitions in place that were feared by all sectors of the community, people (in that case mainly the younger people to the anger of the elders) would behave without respect for others. It was therefore remarkable how rare conflicts were on Quirimba, where religious and superstitious beliefs seemed to be disjointed and did not form a coherent code of living and where people of a range of different backgrounds, ethnic groups and beliefs were living in the shadow of a 17-year civil war. One of the reasons for the high levels of trust and low incidence of conflict may have been the strong sense of 'kinship'. People had large networks of extended families and friends with whom they shared food and exchanged help and services. Many people married more than once (some ordinary Quirimban fishers had up to three wives, each living in a different household) and had siblings with different sets of parents, linking together disparate parts of the community. This type of complex kinship was also reported from southern Mozambique (Gengenbach, 1998). Parents often do not bring up some of their own children, but entrust them to relatives who may not have children. This complex network of kinship in a community of 3000 meant that anti-social behaviour such as stealing from other people's fishing gear would, first, be likely to be witnessed by a member of the community and, second, might well affect the 'kin' of the thief. Theft was remarkably rare on Quirimba and most thieves came from outside the island; for example large-scale coconut theft in the plantation was almost always done from boats from the mainland.

The main sources of conflict during the study period were land disputes and land-use issues such as the problem of one person's goats grazing on another person's crop. These were taken to the district administrator to resolve. Land disputes are an acknowledged problem in post-war Mozambique (Gengenbach, 1998; West, 1998). All land in Mozambique belongs to the state, and officially can only be acquired on 50-year leases (Massinga and Hatton, 1997). In most cases in rural areas formal leases were not issued and land ownership or the right to farm the land was an informal arrangement. During the war displaced people often took over land left by others who had moved elsewhere because of the conflict. The gradual return of refugees to their homes ('rural resettlement') has led to serious land disputes in many parts of Mozambique (Gengenbach, 1998). The current administration system of Quirimba seemed to be in the process of change. Unlike Tanzania, a few hundred kilometres to the north, local elders and chiefs did not make the important decisions, and the whole system of village elders that was in place before independence does not exist anymore. This was rooted in the decision of the new FRELIMO government at independence in 1975 to abolish the system of chiefs in Mozambique (West, 1998). Traditional local hierarchies of elders were replaced by the system of a Provincial Governor and a District Administrator (or Chef de Post).

The strongest tradition adhered to by virtually all Quirimbans, whether of Mwani or Makua origin, was the belief in the power of witch-doctors and witchcraft. Most people had a little sewn up cotton packet that they wore round their neck or kept in a pocket containing a good luck charm from a witch-doctor. People with illnesses went to a witch-doctor for treatment before they went to the clinic and spells were bought for a variety of reasons, from protecting property against theft to giving people energy. There were well-known witch-doctors in the area, such as Tanzoor on Quisiva Island, and well-known thieves who had powers of witchcraft to help them. However, none of these beliefs were related to the sea and it seems likely that many of them were brought from mainland villages and Makua traditions and superimposed on whatever belief system existed previously in the Quirimbas. It is possible that the Mwani people never had the strong belief system found in some other fishing communities. Many of the communities with the strongest belief systems are isolated island communities such as those in the Pacific (Ruddle, 1996b), where communities had strong cultural and linguistic identities. The Quirimbas have been invaded, colonized and formed an important part of trade routes for Arab and Portuguese traders for over 500 years. This, together with the large-scale destabilization caused by Portuguese colonial rule until 1975 and then the 17 years of civil war that followed directly after do not make for a community with a strong sense of identity or the preservation of tradition. The current isolation of the Quirimba Archipelago is a modern phenomenon.

Total value of the fishery
Total annual catch for net fishing was estimated at 440 tonnes, which would be worth 1320 million meticais or US$110000 (at the minimum price of 3000MZN per kilo), and for the trap fishery 60 tonnes, worth 180 million meticais or US$15000. The total value of the seagrass fin fishery (not including the women's invertebrate fishery) of Montepuez Bay was therefore estimated at 1500 million meticais or US$125000 (1997 exchange rates) per year (see Table 8.5). About half of this, 750 million meticais or US$61560, went directly back into the local economy in the form of wages to fishers and fish processors. The area of seagrass fished by the Quirimba fleet was estimated at around 35km^2. If the total annual value of fish caught was US$125000, then the minimum annual value per square kilometre of seagrass was estimated at US$3570.

Approximately 400 people on Quirimba Island were employed in the seagrass fisheries. The annual individual wages for those involved in the fishery range from 864000MZN (US$72) for fish processors, 2160000MZN (US$180) for ordinary fishing crew, to 3024000MZN (US$252) to 7392000MZN (US$616) for trap fishers. Boat owners could in theory earn from US$1458 to US$4698 annually. Whether they actually did this was unclear – the boat owners themselves were not clear about the kind of annual profit they made, and it is likely that a lot of the 'profit' went back into maintaining boats and gear.

A provincial and recent historical context
In 1986 a beach seine could be purchased in Cabo Delgado for the equivalent of 2731 kilos of fish. In 1994 this had risen to 66915 kilos of fish (Republic of Mozambique State Secretariat of Fisheries, 1994b). In the study described here, the 1997 price of a new seine net was given at between 5 and 10 million meticais. Using the lowest price it was possible to get for fish on Quirimba, 3000MZN, this was the equivalent of between 1667 and 3333 kilos of fish. Using the maximum price for fish, 5000MZN this was the equivalent of 1000 to 2000 kilos of fish, more akin to the 1986 relative prices than those in 1994. The price of fish has not risen much

since 1994 but the prices of nets have evidently fallen as supply networks to the area have improved. In the 1980s the price of fish in Cabo Delgado was kept artificially elevated; the price of fish decreased nine-fold since 1986 from US$3.37 for a kilo of fish in Pemba to just US$0.39 in 1993 (during the 1980s ordinary people in Pemba did not eat local fish, but ate fish imported from around Africa – there was a notorious poisoning incident from freshwater fish – M. Carvalho pers. comm.). Fishers in the area have very low financial incentives and little access to new gear, precluding anything more than subsistence fishing (Republic of Mozambique State Secretariat of Fisheries, 1994a) and also face the storage and transportation problems mentioned above.

National per capita consumption of fish was estimated at 5.1kg per year in 1994 (Republic of Mozambique State Secretariat of Fisheries, 1994b). It is likely that per capita consumption of fish on Quirimba greatly exceeded this national value. Fishers and their families shared about 1kg of fish per fishing day. Each person ate approximately 100g per day on fishing days (16 days per month), giving 1.6kg per month and a minimum of 19.2kg of fish per year. In addition to this dried fish was also often eaten on the non-fishing days.

In 1989 Cabo Delgado had 4539 fishers accounting for 0.7 per cent of the working population, the same proportion of fishers in the population as was seen on a national scale. Quirimba Island had approximately 400 fishers in a population of 3000, so fishers there accounted for at least 10 per cent of the population and a much higher percentage of the working population. Between 1989 and 1994 the number of fishers nationally increased from 52 000 to 80 000 and a similar increase would be expected in Cabo Delgado. This increase has mainly been attributed to the displacement of the population towards the coast because of the civil war (Republic of Mozambique State Secretariat of Fisheries, 1994b; Hatton and Massinga, 1994). With this migration there has been a disproportionate increase in the numbers of people living off subsistence fishing, often the only source of food and employment to displaced people. The pressure on artisanal coastal fisheries has been particularly intense because, although there was a surge of people to the coast in the 1980s and early 1990s, there was no accompanying development of the local fishing industry or related diversification of activities. This situation is described as having led to the local population and the refugees collectively mining as opposed to husbanding the littoral resources (Hatton and Massinga, 1994).

Employment opportunities and salaries
Workers on the coconut plantation or in other daily employment worked six days a week all year, a total of 312 days per year. Their potential

annual salary would be just over 3 million meticais. or US$260, significantly higher than that of the net fishers (see Table 8.4). However, fishing crew had up to ten days off per month to do other work such as on their *machambas* and some were able to supplement their income from net fishing with other types of fishing, for example spear-fishing. Jobs in the coconut plantation were limited, but many people claimed to prefer fishing to plantation work. Fishing may have been chosen despite the less regular income for social reasons. On the fishing boats there was an atmosphere of cooperation and everyone, regardless of age or experience, appeared to be treated equally (with the exception perhaps of the young boys who start off as apprentices and have to do some of the unpleasant jobs). Although the hauling of the fishing nets was extremely hard work, much of the time spent out fishing was spent sailing between fishing sites when fishers talked, sang and seemed to enjoy themselves. The coconut plantation, the only source of formal daily employment on the island offered regular work that did not depend on the weather or the state of a leaky boat. It also offered extra benefits such as the opportunity to buy some foodstuffs cheaper and some free medical treatment. However, the plantation was run by white farmers of European origin and there was some historical resentment of working for white plantation owners, with the memory of ruthless Portuguese plantation and sisal factory owners still fairly recent. The parents of some of the young fishers were employed by the Portuguese in this way before independence.

Fisheries elsewhere in the region, in Tanzania, the Comores and Seychelles for example, have developed as local economies have expanded. Fishers have left traditional fishing methods behind to use outboard engines and more sophisticated gear. The fishers of Quirimba did not use outboard engines on their boats. The fishing boat owners on Quirimba appeared to have been earning enough to potentially be able to buy boat engines in the future, although it wasn't something any of them talked about doing. The main obstacle to this may not have been so much the cost (although this is likely to be extremely high – import tax of 300 per cent is commonly charged on certain imports and the cost of transporting a boat engine from South Africa to the north of Mozambique may also be enormous) but the availability. If no one is importing engines to Pemba, which was the case during the study period, no one will be able to buy them. Much larger sums of money than the local boat owners have access to would be needed for actually importing numbers of engines from South Africa.

Another reason for the lack of engines may have been that even with engines it would not necessarily be efficient to fish on the outer reefs instead of in the seagrass beds. Weather conditions in the area were very changeable and it was often not possible to anchor the project's research boat off the outer reef of Quirimba to do survey dives. It was regularly not possible to get the boat out onto the east coast of the island at all because of strong winds and rough seas. Traditional sailing dhows were almost certainly better suited to fishing in the shallow seagrass beds than boats with outboard engines. There was thus no incentive to 'develop' fishing techniques.

Social structure of the fishing fleet
The crews of seine net fishing boats were a mixture of ages, from young boys of 12 or less to older men in their 50s and 60s. However, the majority of fishers were young, aged between 15 and 25. Many young men came to Quirimba as refugees from the fighting around inland villages during the war. Many came as children and fishing had been their only job, even though it was not an option that was open to their parents in the inland villages they came from. In some more developed fishing communities, for example in the Seychelles (Wakeford et al., 1998) or in parts of the Caribbean (J. Hawkins, pers. comm.), most of the fishers are from older generations whereas the young men try to avoid fishing and attempt to get better salaries for modern, cleaner jobs. In Quirimba there was clearly no lucrative alternative to fishing so it remained the most common job for young men. Some of the Quirimban fishers left fishing to work as traders across the border from Mtwara in Tanzania to Moçimboa on the Mozambican border, or as crew on boats that did this run, but this was often temporary or seasonal work.

Older men were more likely to be trap fishers. Trap fishing was a skilled method of fishing requiring an apprenticeship. It was difficult to learn and required a large investment in terms of learning the skills involved in making and maintaining traps and setting them. If someone bought, assembled and set some traps without guidance from an expert they could quite literally catch nothing. In the first year of study there were signs that trap fishing was in decline. The fishery was dominated by elderly men and very few young men seemed to be going into the trap fishery. However, in 1997 *marema* traps became available locally. Instead of fishers having to make a special journey to the mainland to buy traps, they could buy them in Quirimba village. There was a sudden increase in people buying traps, and even people who did not have a boat to set them from started making traps.

The division of fisheries work between the sexes
Women never worked in the seagrass seine net or trap fisheries as fishers but provided the main labour force in the processing of the fish from the net fishery. The idea of a woman involved in either of these two fisheries was

completely unacceptable to the men and women interviewed. Men cited the weakness of women as a reason, although Quirimban women regularly carried 30-litre water barrels from one end of the island to the other, felled mangroves, carried the wood long distances and did most of the heavy agriculture work. Men also cited women's inability to swim. There was a traditional reluctance to teach girls to swim even though they were likely to spend a lot of time on boats, and I was told by a number of different people that women often drowned in boat accidents. The women interviewed did not have any aspirations to fish in the trap or net fishery but when pressed did suggest that economic factors made it impossible to consider trap fishing. In Tanzania a clear distinction was made between women's territory and men's for marine resource collection (Mtwara – M. Guard pers. comm.; Bagamoyo – Semesi et al., 1998). Women fished in water up to waist deep. In deeper water, where a boat would be needed, men took over. In some Tanzanian coastal communities these designated areas led to conflicts of interest in the community and to members of the same family working against each other over these rights. In Kwale District, Kenya, there was a similarly inflexible division of labour with men fishing and felling mangroves and women collecting invertebrates, growing food and collecting firewood. Men there maintained that this was because women weren't strong enough to do the 'men's' work, whereas women maintained that the reason they didn't do this work was financial (Juma, 1998).

There were a few options for catching fin fish open to women. They could do the *capulana* fishing (mentioned previously), they could catch fish that had been trapped in intertidal pools as part of intertidal gleaning and they could empty their family's *luwando* or fence trap. Women collected three main groups of invertebrates in the intertidal seagrass beds of the Montepuez Bay:

- *Mbari* (*Pinctada nigra*). *Mbari* were collected at low spring tides from the stalks of the large seagrass species *Enhalus acoroides*. They were sold for fresh consumption and also dried for sale on the mainland.
- *Macaza* (*Pinna muricata*). *Macaza* were also collected at low spring tides from intertidal areas of sand and the seagrass *Thalassia hemprichii*. The majority of the animal was discarded and only the small adductor muscle was retained to eat.
- *Ombe* (*Barbatia fusca*). Ark shells were collected on the intertidal flats and in *Thalassia hemprichii* at low tide, the women skilfully detecting their presence beneath the surface of the sand using their toes. Ark shells were also either consumed fresh locally or dried and sold in Pemba.

Some other invertebrates were collected by women and men. *Makome* (*Chicoreus ramosus* – murex shells) were one of the few invertebrate species routinely collected by men. They were collected incidentally by trap and net fishers in the seagrass beds. Abandoned *marema* traps appeared to provide a good aggregation device for murex shells. Some fishers also specialized in *makome* fishing, using canoes and paddling out to deeper seagrass sites to dive for the shells. Shells caught in deeper water by full-time murex fishers were much bigger than those caught incidentally by other fishers. Murex shells were used in a variety of ways. The animal was cooked in the shell for immediate consumption, the opercula were sold by the kilo to Tanzania and the shells were burnt on large pyres with other shells to make lime for painting houses. Despite the large size of many of these shells there was no evidence of their use in the curio trade.

Kome nlume (*Pleuroploca trapezium* – tulip shells) were collected almost entirely incidentally in other fisheries. They were sometimes eaten but their main value was for their opercula, which were also sold by the kilo to Tanzania. A remarkably wide variety of other molluscs were caught by children. These were mainly small molluscs that are found higher than the level of low spring tides and even those found near the top of the beach and in small areas of mangrove. The value of these shells appeared to be solely to supplement the diet of the family and none were sold. Women who are not normally involved in invertebrate collection also collect these species on a casual subsistence basis.

Many of the intertidal invertebrates such as *mbari* (oysters), *macaza* (pen shells) and *ombe* (ark shells) were mainly found at the low tide level and so were only accessible at low spring tides, just three or four days per fortnight. This meant that women who collected these molluscs had a less regular income than net fishers but their catches on these few days were often more valuable than a day's net fishing. Women could also fit a range of activities around the tidal cycle. Some women collected octopus and other organisms associated with the exposed reef flat on the east coast of Quirimba around neap tides, then moved round to the lower littoral species around tides closer to springs. During completely unsuitable tides (high neaps) women worked on their *machambas*. Some women also collected the higher littoral molluscs during these tides, such as *Strombus sp.* (tiger sand conch).

Alternative sources of income
On Quirimba the main alternative source of employment was subsistence agriculture. Most people on the island were involved to some extent in agriculture, tending a *chamba* to provide their family with papaya, sweet potato, cassava and beans. Most agricultural activity took place

in the wet season (November to April) but many people were involved all year round. The food available in the local market was very limited and imported shop-bought food was prohibitively expensive, so most people relied on food grown in their *machambas* for the bulk of their food, with fish or shellfish providing the animal protein. The *machambas* were constantly being expanded into the few remaining areas of forest and scrub by slash and burn clearing methods. On Inhaca Island in southern Mozambique this method has produced land that is high in nutrients for the first year of cultivation but that soon becomes poor in nutrients and becomes impossible to cultivate after 15 years (Serra King, 1995). Erosion is also a problem in land cleared by slash and burn, and this may be something that will impact on the coastal habitats. The cash economy was very limited so there were few shops and it was virtually impossible to buy a meal or cooked food of any kind in the village, so there were few opportunities in service jobs. This situation may be typical of a largely cashless economy for poor rural Mozambicans. In Tanzania many rural areas are much more developed and there are more opportunities for jobs with regular salaries and cash (Andersson and Ngazi, 1998). Street food is common and provides an extra employment opportunity. There was no tourist industry mainly because of the inaccessibility of the island and the lack of basic infrastructure.

The gender roles in the fishery at Quirimba were fairly typical of those found in tropical artisanal fisheries around the world. Women throughout East Africa, South East Asia and the South Pacific glean intertidal flats for invertebrates and the fishery for fish is dominated by men (Palau – Matthews and Oiterong, 1995; South East Asia – Bailey and Pomeroy, 1996; Kenya – Juma, 1998). Although the biology of the women's invertebrate fishery was not included in the present study and can be found in Barnes et al. (1998), the socioeconomics were studied to some extent. Women collected the three main species of mollusc on the intertidal (mainly *Thalassia hemprichii*) and upper subtidal seagrass beds (mainly *Enhalus acoroides*).

The community on Quirimba utilized a wide variety of resources. Although seine net fishing in the seagrass beds was the main single employer, most households used a wide range of sources of subsistence food and additional income, typically a mixture of men's income from fishing by net or trap, food from the *machambas* grown predominantly by women, invertebrates gleaned from the seagrass intertidal by women and some additional invertebrates collected by children. Consequently, although many people were involved in the seagrass-based fish and invertebrate fishery very few people, except maybe some boat owners, were entirely reliant on the seagrass fishery. As Bailey and Pomeroy (1996) found in South East Asia, households and, therefore, communities reliant on a wide range of resources that changed with seasons and with changing weather conditions were much more adapted to cope with long-term changes in the availability of resources. The use of a wide variety of resources by people in coastal communities was also found in Tanzania. On Mafia and Zanzibar (Unguja) islands only 11 per cent of people interviewed relied on a single source of income, with most people combining their main source of income, some sort of fishing, with small-scale agriculture (Andersson and Ngazi, 1998).

On Quirimba most people did combine some sort of marine resource use, whether fishing or invertebrate collection, with subsistence agriculture. However, within the invertebrate fishery a change was taking place. The molluscs (*Pinctada nigra, Pinna muricata* and *Barbatia fusca*) were originally collected by women as a subsistence commodity that was eaten locally. In the community workshops the women who collected them reported that in the last few years the prices they could get for them in Pemba had risen sharply as demand for dried seafood increased. It was therefore much better to sell them on the mainland rather than eat them themselves or sell them on Quirimba. Others in the community associated this change in use of invertebrates with a decline in their numbers and said that it was virtually impossible for an ordinary Quirimban person to afford to buy these preferred species of shellfish. The shellfish produced by the women were therefore the main commercial product of ordinary people in Quirimba, whereas the fish caught by the trap and net fishers, although also forming an important commodity for the boat owners and for dried fish traders, was the main source of subsistence food for ordinary households on Quirimba.

In Tanzanian coastal communities a similar pattern has been observed, with an increase between 1993 and 1998 of the local consumption of fish and vegetables and the complete disappearance of intertidal invertebrates from local diets. There too, prices have increased because of a combination of increased demand due to growing populations and the developing tourism industry and increased scarcity as marine resources are overfished (Andersson and Ngazi, 1998). The situation in Tanzania is perhaps ten or more years ahead of what was recorded in Quirimba in terms of economic development. Cabo Delgado is one of the least economically developed provinces in Mozambique (Hatton and Massinga, 1994). In Inhaca Island, southern Mozambique, the invertebrates collected by the women are eaten by the family and it is the fish that is sold for cash (Wynter, 1990), perhaps reflecting the close proximity of a tourist market for fish and the national capital.

Integrated coastal zone management in northern Mozambique
A comprehensive review of integrated coastal zone management can be found in the proceedings of a national workshop on integrated coastal zone management in Mozambique (Lundin and Lindén, 1997). The only example of integrated coastal zone management in Cabo Delgado province is that of the Mecufi Coastal Zone Management Project (Massinga, 1997).

Along with the other East African and Western Indian Ocean Island States who signed the 'Arusha Resolution on Integrated Coastal Zone Management in Eastern Africa (including Island States)', Mozambique has pledged to develop and implement integrated coastal zone management (ICZM) programmes. Within the main government department involved in coastal zone management, MICOA (Ministry for the Coordination of Environmental Affairs) there is now a Coastal Zone Unit with specific responsibility for coastal management issues. Active coastal zone management has largely been restricted to some large marine protected area projects in the southern sector of the country, for example the Inhaca Island marine park, the Bazaruto Archipelago marine protected area and the Xai-Xai Integrated Coastal Area Management project. However, projects are gradually being developed throughout the country, for example the Mecufi Coastal Zone Management Project just south of Pemba or the Island Management project on Isla de Moçambique, which are currently in progress. City coastal zone management projects are also planned for the city of Nacala in Nampula Province and the city of Beira in Sofala Province.

Having only emerged from the civil war in 1992, Mozambique has the difficult task of reconciling coastal zone management programmes with the need for rapid development of large areas left by the war without even the most basic infrastructure and facilities. It also has to address the problem of a population that, during the civil war, gravitated towards coastal areas. Many refugees have remained and are now settled in their coastal lifestyles. It is estimated that more than 40 per cent of the Mozambican population live in coastal districts (MICOA, 1997). The pressures on coastal resources are manifold: increased populations through migrations, through the natural increase in population that is being experienced throughout the region and also increased exploitation of resources by outside user groups such as foreign fishing fleets and recreational users (Hatton and Massinga, 1994).

Itinerant fishers
Small numbers of itinerant fishers from local mainland villages have fished seasonally in the Quirimbas, but in the five years preceding the

study increasing numbers of fishers from Nampula started coming to Quirimba and other islands in the Archipelago. They inhabited the islands in groups of tens to hundreds of fishers on a seasonal basis, usually staying between three and six months during the dry season. These fishers said that they have been forced to fish in the Quirimbas by the depletion of their own nearshore fish stocks. Nampula has indeed had a much more intensive artisanal fishery sector than Cabo Delgado. In 1994 there were reportedly 180 beach seines in the Province of Cabo Delgado and 10 882 in Nampula. The coast of Nampula is about 50 per cent longer than that of Cabo Delgado but it had a seine netting intensity 600 per cent greater (Republic of Mozambique State Secretariat of Fisheries, 1994b). Fishers working for companies in Nampula fished with an intensity and a commercial intent that was very different from the Quirimbans' subsistence approach. Nampula fishers returned to their home towns with boats full of the dried fish and invertebrates they had collected in the Quirimbas, part of which they kept for their families over the wet season and part they sold commercially in Nampula. Fishers from Nacala (provincial capital of Nampula) were recognized as a major component of the fishing industry in the Macomia district, further north from Quirimba in Ibo district (Wilson et al., 1996).

Up to the end of 1997 no active regulation of fishing activity in the Quirimbas was in place. Far from trying to control the systematic depletion of their marine resources themselves, Quirimban fishers were welcoming the invasions of Nampula fishers despite the gradually emerging problems of wells running dry, traps being caught in the nets of the visiting fishers and other small but significant signs of potential impending conflict. Possible reasons for the Quirimbas welcoming behaviour include a desire to maintain contact with people from a more developed part of Mozambique. One local fisher said he enjoyed the company of the Nampula fishers because they were more educated in Islam than most local people and he found them interesting. The Quirimban fishing community also seemed generally to be a very open, welcoming community (the way they welcomed this study is a good example), maybe an extension of the 'kinship' idea addressed previously.

Highly mobile fishing fleets are not unusual in the East African region and have led to conflicts elsewhere. For instance, in southern Kenya, itinerant fishers from Pemba Island in Tanzania, having reputedly depleted their own fish stocks, fished in traditionally managed reef areas against local management regulations (McClanahan, et al., 1997). In Tanzania, iceboats from Dar es Salaam fish right the way down the coast of Tanzania, often using dynamite and other damaging methods.

One of the problems with this situation of over-exploitation by itinerant fishers in the Quirimbas is that, for the purpose of future management or legislation, it will be very difficult to say who is a Quirimban resident and to whom the marine resources of the Quirimbas actually belong. During the war people were forced to move around a lot to escape the intense fighting in rural areas. Some refugees stayed where they fled to, others have moved back to their home villages. Many people have also continued to move, whether between different opportunities to work or between partners or family members displaced throughout the area. There is also a lot of movement between towns and rural areas. Much trading is done by boat, so as well as the movement of people between the islands and mainland towns and villages there is an additional group of people who are continuously moving up and down the coast between ports from Nampula to Mtwara in southern Tanzania. The Mwani people of the coast are historically fishers and traders and the trading is still an important part of coastal culture.

It seemed to be difficult for people on the islands to think of the surrounding marine resources as theirs. The majority had moved around so much in their lifetime that the idea of having to move on again because of resource depletion did not seem as serious as it perhaps may have seemed to a more settled community. One fairly typical fisher interviewed on Quirimba Island was just 24 years old but had already moved numerous times and had changed his fishing method on each occasion. He had moved from freshwater fishing as a boy in an inland village, to line fishing from a coastal village, octopus fishing on Quisiva, through a variety of other locations and fishing methods to trap fishing on Quirimba. He had family members on four of the other islands and in numerous villages in the mainland. To him the idea of having to leave Quirimba because the 'fish had run out' was not distressing at all. The threat of fish or invertebrates running out was not normally considered, and one interviewee said 'There will be fishers at Quiwandala [the main fish landing site] until the end of the world'.

Threats to sustainable resource use
This study has identified four main threats to sustainable resource use in the area, namely:

Population growth Coastal populations throughout Mozambique are on the increase. During the 1990s the population of the Quirimbas increased through a higher birth rate and immigration of refugees. Quirimba Island had a population of nearly 3000 but no provision for waste management. A growing population will inevitably put increasing pressure on the marine resources. The need for firewood from mangrove areas and for fish to provide protein in local diets will increase. The pressure to exploit resources for sale outside the community to bring in much-needed cash for medicines and other manufactured goods will increase and these communities may already be exceeding sustainable levels of exploitation.

Influx of visiting itinerant fishers from outside the local area   The influx of itinerant fishers was a recent phenomenon and had only really had a major impact on the islands since the beginning of the 1990s (Gessner pers. comm.). It is likely, however, that the numbers of itinerant fishers and the intensity with which they fish will increase as the pressures on fisheries and other resources in their home villages and towns increase and more people seek the reputed wealth of resources in the Quirimbas. The commercial fishing sector is also likely to grow to feed the expanding tourist market.

The difficulties of controlling and managing resource use and development   The Quirimba Archipelago is a remote area and has been neglected for a long time in terms of infrastructure, development and effective law enforcement. There was a low official authoritative presence in the area and no officials in the capacity of marine resource protection and management. Fishing gear regulations were not enforced and most people were not even aware of their existence. The local administration system did seem to be improving, but even local administrators with the best intentions for the protection of resources and the implementation of legislation are unlikely to have the time or resources to do very much about managing resource use on their own.

Because of the lack of official intervention there was great potential for unscrupulous outsiders exploiting the marine resources of the islands to the detriment of the needs of the local people (Massinga and Hatton, 1997). The sea cucumber and shark fin fisheries run in the Archipelago by large companies for export to the Far East are a good example of this. Intermediaries exporting these commodities are likely to have made a lot of money but very little was paid to the local people involved. In both cases the methods of extraction were so intensive that the potential for local people to exploit these resources sustainably in the future may already have been lost. Local fishers reported that sites where they once skin-dived for abundant sea cucumbers now do not have any at all. Sharks were also a rare sight now throughout the Archipelago.

Many foreign business people are interested in exploiting some resource or other that they have heard is abundant in the Quirimbas, from crayfish to the opportunities for tourism. Their attitude seems to be that the resources of the Quirimba Archipelago are there for the taking and it is unlikely that anyone will interfere actively. It is important that the rights of the Quirimban residents to their resources are established. Once they are assured, people of the Quirimbas can begin to take responsibility for the protection or guardianship of those resources before, for example, land is sold to a businessperson as a theme park or sea cucumbers are dried and sent to the Far East as an aphrodisiac.

Indirect threats to the marine ecosystems An important new growth industry in the north of Mozambique was forestry and with two major rivers entering the sea in the vicinity of the Archipelago the potential for the indirect impacts of siltation of reefs, seagrass beds and mangroves was great. As with all other development in the remote north of the country, logging of coastal forests was difficult to monitor and control and it is likely that illegal logging was a serious problem in the area. Large areas of old forest were logged and this was likely to lead to increased erosion leading to silt being washed downriver and into the sea around the Archipelago. Slash and burn agriculture is widespread on the mainland and this will also lead to increases in soil erosion and potential siltation problems. Coral reefs are particularly susceptible to the effects of heavy siltation but seagrasses too are unable to tolerate high levels of sedimentation with low light levels (Bach et al., 1998).

Management implications and problems
The four major threats identified above do not feature strongly in the concerns expressed by the residents of the Archipelago. The majority of the threats have yet to manifest themselves in the form of dramatic decreases in fish or invertebrate catches. The fishers did not complain about the outside fishers using their fishing sites but did complain about the high prices of fishing gear. The main concern of women was that with the growing demand for their dried shellfish products in Pemba and other mainland towns, the favoured species of shellfish were becoming an unaffordable luxury. The majority of people interviewed who were resident permanently in the islands seemed satisfied with the present situation. They had access to agricultural land to grow basic foods such as sweet potatoes, cassava and papayas and they supplemented this diet with fish and shellfish.

On a day-to-day basis, people throughout the Quirimbas suffered through lack of basic health care. Paying for medicines was often difficult in a virtually cashless economy. Minor injuries and illnesses often went untreated and led to more serious conditions and the long-term loss of income. For example, a fisher on Quirimba with a small infected sore did not receive treatment, the injury became more serious and he was unable to fish for a period of months, inflicting hardship on himself and his family. Malaria and tuberculosis are both common potentially life-threatening illnesses, and both can be controlled using medicines that are cheap by Western standards, but even the most basic treatments can be prohibitively expensive. There will be an increasing pressure on the people of the Quirimbas to find ways of earning cash so that they can deal with these emergencies and also so that they have the freedom to buy the consumer goods that are becoming available.

Based on this analysis, four basic needs can be identified that could begin to address the potential problems in the Quirimbas:

Education The general level of education is very low, particularly for women. Boys generally receive a few years of education and learn to read, write and speak Portuguese. Many girls are educated for less than one year and remain illiterate. It is important that all children are educated to a standard where they are literate and able to access educational materials and the media. Children should either be given a firm grounding in Portuguese (whilst it is ensured that local languages and cultures are not suppressed), or educational materials, radio programmes and newspapers should be produced in local languages, particularly Kimwani, which is the main island language and is marginalized at present. It would, of course, be ideal to give children both of these opportunities. If people were given the opportunity to reach a higher level of education the ecological and economic arguments for them to conserve their own resources may be more easily communicated. A higher level of education may also offer people the opportunity to find employment other than fishing and invertebrate collection and would make people more employable in service industries that will accompany the development of tourism. Literate, Portuguese-speaking Quirimban residents could then by employed in the inevitable tourist developments, rather than imported labour from the cities as has happened elsewhere.

Legislation The ever-increasing numbers of itinerant fishers in the Archipelago need to be regulated in some way. On some of the smaller islands in the Archipelago every available space is covered with their makeshift shelters. All are there both to catch fish to sustain their family over the wet season and also to make a profit selling their fish inland. Without some sort of limit or regulation this component of the fishery could lead to its decline. This problem may be one that has to be addressed at its source, in the majority of cases in Nampula, and is therefore in essence a political problem requiring political will to solve. Social or environmental problems that have forced these fishers to look outside their province for additional income on such a huge scale need to be addressed within the province. The use of unacceptably small mesh sizes, in some cases even finer material such as sacking, and other fishing gear infringements also need to be addressed. Active fisheries monitoring in the islands is definitely required.

Research and monitoring More research is needed into the sustainability of the resource use methods currently in use and into potential alternative use of resources. A great positive step would be the establishment of a marine research centre on one of the larger islands. A large GEF (World Bank Global Environment Facility) loan has been allocated to the northern Mozambican coastal area and there are currently plans for a marine station at Pemba to the south of the Archipelago or in Moçimboa da Praia to the north (S. Bandeira pers. comm.). This would give scientists who are based in Maputo a good base from which to conduct research and perhaps most importantly to monitor changes and developments in the area and their impact on the marine ecosystems. At present there is no scientific monitoring. For example, the inaccessibility of the area and lack of marine scientists have meant that the status of the Quirimbas coral reefs during 1998's catastrophic coral bleaching events remains unknown. The establishment of a marine research centre would also play an important role in the education programme. Such a centre could serve as a base for training environmental educators and provide educational facilities for schools and community groups, and an opportunity for local people to actively participate in research and decision-making.

Implementation To implement any gear regulations it would be necessary to provide financial and practical help to fishers to help them purchase more favourable gear types. The education programme must instil in the island residents the idea of the marine resources of the Archipelago as theirs to conserve and use sustainably now so that they will continue to provide food and employment to their children and grandchildren. In this way, enforcement of legislation could come in part from within the community. Only in the role of guardians of their own future resource use will the people of Quirimba be motivated to participate in management. The training of local people as environmental educators to work in schools and with fishers on a long-term basis would be one way of developing environmental awareness in the area and would also overcome the language problem mentioned earlier. An integrated programme of education, research in close cooperation with local people, legislation to protect vulnerable resources and to prevent destructive practices, and a long-term programme of implementation and monitoring are needed.

How would possible changes in fishing patterns affect the economy?
One potential management option would be to move away from seine netting as the main fishing method and increase trap fishing. Each seine fishing boat covers a large area of *Enhalus acoroides* on each trip and the small trap fishing areas required by individual trap fishers would be much smaller than the available area. How would an increase in trap fishing and a decrease in net fishing affect the people involved? From an economic point of view everyone employed in the seine net fishery, with the exception of the fishing boat owners, would have a higher income if they were employed as trap fishers (after the initial expenditure on boats and traps). Net fishers could potentially earn double their salary as trap fishers. They would also catch fewer fish per person so if the same number of men were involved in the fishery, the total fishing intensity would drop (trap fishers catch an average of 6.7kg per trip whereas net fishers catch an average of 9.4kg per person per trip – even taking into account the extra days fished by trap fishers, fewer fish would be caught) so putting less pressure on resources. There would, however, be some problems with this change in fishing strategy. There would be the initial problem of fishers being able to raise the capital to buy a canoe and traps. Trap fishers also had an outlay of about 160 000MZN every few months, for replacing their traps. This worked out as an expense of approximately 3350 MZN per fishing day. If fishers moved from net fishing to trap fishing this would restrict the business of the seine net boat owners. As the net fishing boat owners were the only businesspeople on the island, if they were to lose their livelihood it could have a negative effect on the overall development of the area. Boat owners invested money from fishing in other local businesses such as shops and market stalls. These provided employment for members of the owners' families and other local people, and provided a service. Each boat owner provided a reliable income for anything from five to 12 fishers and also part-time income for the women and children who cleaned fish.

Conclusions
In conclusion, the study reported here showed that fishing was the single most important source of food and employment on Quirimba Island, along with subsistence agriculture and the women's invertebrate fishery, but the majority of people and households relied on a combination of marine resource use and agriculture. Seagrass beds were the main habitat to be exploited on Quirimba and therefore the most valuable habitat to the community. There was a hierarchy of earnings in the seagrass fishery in Quirimba, with boat owners being the highest earners, then trap fishers followed by boat captains and boat crew. Net fishing was the biggest single source of employment on the island and yielded the lowest wages of jobs

for young men. Net fishing may have been preferred to better-paid trap fishing or plantation work because it was sociable, flexible and seemed to be enjoyable. Small fish may have been preferred for social and economic reasons. They could be used easily for barter and were easy to cook, dry and sell. Interestingly, there were remarkably few conflicts between the users of marine resources despite the fact that there were overlapping uses of the resource and many recently arrived fishers. However, an increasing population and a declining resource base will inevitably lead to competition and environmental degradation unless there is investment in the community.

Note

1. There were many people in Quirimba village and in Kumilamba who gave generously of their time to assist in the collection of the information in this chapter. Much of the information was collected with the help of trap fisher Anibal Amade who suggested many useful interviewees, provided translation assistance where necessary and also provided detailed information on the trap fishery. Seine net fishers Mussa and Lamu from Quirimba spent hours explaining the workings of the Quirimba fisheries and introduced me to many other fishers, including reef flat gill netters. Saidi Ndiki, an older gill net fisher, was interviewed on a number of occasions about this fishery. Captains Ibrahimo, Manueli and Jabira from the seine net fishery provided a great deal of useful information. Oscari, Zaidu and Momadi from Santa Maria were key informants on the fisheries. For general information on the economics of life on Quirimba, Awaje Shale and her family were very helpful. The workshop with the invertebrate fishers was organized with Fatima Mussa and other women from OMM (Mozambican Women's Organization) in the village. Mario Carvalho and Alex Corrie of The Frontier-Moçambique project provided assistance with translation and setting up interviews and so on, and the rest of the Frontier staff, visitors and volunteers provided invaluable support at various times. The Quirimbas National Park was designated in 2002.

References

Andersson, J. and Z. Ngazi (1998), 'Coastal communities' production choices. Risk diversification and subsistence behaviour: responses in periods of transition', *Ambio*, **27**(8), 686–93.

Bach, S.B., J. Borum, M.D. Fortes and C.M. Duarte (1998), 'Species composition and plant performance of mixed seagrass beds along a siltation gradient at Cape Bolinao, the Philippines', *Marine Ecology Progress Series*, **174**, 247–56.

Bailey, C. and C. Pomeroy (1996), 'Resource dependency and development options in coastal Southeast Asia', *Society and Natural Resources*, **9**(1), 190–99.

Barnes, D.K.A., A. Corrie, M.W. Whittington, M.A. Carvalho and F.R. Gell (1998), 'Coastal shellfish resource use in the Quirimba Archipelago, Mozambique', *Journal of Shellfish Biology*, **17**(1), 51–8.

Birkeland, C. (1997), 'Disposable income in Asia – a new and powerful external pressure against sustainability of coral reef resources on Pacific Islands', *Reef Encounter*, **22**, 9–13.

Boxer, C.R. (1963), 'The Querimba Islands in 1744', *Studia*, **11**, 343–60.

Chambers, R. (1997), *Whose Reality Counts? Putting the First Last*, London: Intermediate Technology Publications.

Dahalani, Y. (1997), *L'Impact de la pêche au 'djarifa' sur le recrutement des populations des poissons et des crustacés en face des mangroves du littoral côtier de Mayotte (baie de Chiconi)*, Centre d'Océanologie de Marseille and Service des Pêches et de l'Environment Marin de Mayotte.

Dayaratne, P., O. Linden and R. De Silva (1995), 'Puttalam Lagoon and Mundel Lake, Sri Lanka: A study of coastal resources, their utilization, environmental issues and management options', *Ambio*, **24**(7), 391–401.
Gayanilo, F.C. and D. Pauly (1997), *FAO Stock Assessment Tools: Reference Manual*, Rome: ICLARM/FAO.
Gell, F.R. and M.W. Whittington (2002), 'Diversity of fishes in seagrass beds in the Quirimba Archipelago, northern Mozambique', *Marine and Freshwater Research*, **53**(2), 115–21.
Gengenbach, H. (1998), '"I'll bury you in the border!": Women's land struggles in post-war Facazisse (Magude District), Mozambique', *Journal of Southern African Studies*, **24**(1), 7–36.
Guard, M. and M. Masaiganah (1997), 'Dynamite fishing in southern Tanzania: Geographical variation, intensity of use and possible solutions', *Marine Pollution Bulletin*, **34**(10), 758–62.
Hatton, J. and A. Massinga (1994), *The Natural Resources of Mecufi District. Mecufi Coastal Zone Management Project*, Maputo, Mozambique: Ministry of Environment.
Jennings, S. and N.V.C. Polunin (1996), 'Fishing strategies, fishery development and socio-economics in traditionally managed Fijian fishing grounds', *Fisheries Management and Ecology*, **3**, 335–47.
Johannes, R.E. (1981), *Words of the Lagoon*, Berkeley, CA: University of California Press.
Juma, S.A. (1998), 'Men, women and natural resources in Kwale District, Kenya', *Ambio*, **27**(8), 758–9.
Lindén, O. (1993), 'Resolution on integrated coastal zone management in East Africa signed in Arusha, Tanzania', *Ambio*, **22**(6), 408–9.
Lindén, O. and C.G. Lundin (1996), 'The journey from Arusha to the Seychelles: Successes and failures in integrated coastal zone management in Eastern Africa and island states', Proceedings of the Second Policy Conference on Integrated Coastal Zone Management in Eastern Africa and Island States, Seychelles, 23–25 October: World Bank and Sida/SAREC.
Loureiro, N.L. (1998), 'Estudo da ictiofauna coralina e pesqueira do distrito de Mecúfi, Província de Cabo Delgado', Tese de Lienciatura, Universidade Eduardo Mondlane, Maputo, Mozambique.
Lundin, C.G. and O. Lindén (eds) (1997), 'Integrated coastal zone management in Mozambique', Proceedings of the National Workshop, Inhaca Island/Maputo, Mozambique 1996: World Bank and SIDA/SAREC.
Macia, A. and L. Hernroth (1995), 'Maintaining sustainable resources and biodiversity while promoting development – a demanding task for a developing nation', *Ambio*, **24**(7–8), 515–17.
Manx Heritage Foundation (1991), *Manx Sea Fishing*, Douglas, Isle of Man: Manx Heritage Foundation Publications.
Massinga, A. (1997), 'Mecufi coastal zone management project', Proceedings of the National Workshop on Integrated Coastal Zone Management in Mozambique, Inhaca Island and Maputo, Mozambique, May 1996.
Massinga, A. and J. Hatton (1997), 'Status of the coastal zone of Mozambique', Proceedings of the National Workshop on Integrated Coastal Zone Management in Mozambique, Inhaca Island and Maputo, Mozambique, May 1996.
Matthews, E. and E. Oiterong (1995), 'Marine species collected by women in Palau, Micronesia', *Micronesia*, **28**(1), 77–90.
McClanahan, T.R., H. Glaesel, J. Rubens and R. Kiambo (1997), 'The effects of traditional fisheries management on fisheries yields and the coral-reef ecosystems of southern Kenya', *Environmental Conservation*, **24**(2), 105–20.
McManus, J.W. (1993), 'Managing seagrass fisheries in Southeast Asia: An introductory overview', in M.D. Fortes and N. Wirjoatmodjo (eds), *Seagrass Resources in South East Asia*, technical papers from the advanced training course/workshop on Seagrass Resources, Research and Management (SEAGRAM 2), Quezon City, Philippines, 1990.
MICOA (Ministry for the Coordination of Environmental Affairs, Mozambique) (1997),

'Territorial planning of the coastal zone of Mozambique – methodology', Proceedings of the National Workshop on Integrated Coastal Zone Management in Mozambique, Inhaca Island and Maputo, Mozambique, May 1996.
Munro, J.L. (1996), 'The scope of tropical reef fisheries and their management', in N.V.C. Polunin and C.M. Roberts (eds), *Reef Fisheries*, London: Chapman and Hall.
Pauly, D. (1979), 'Theory and management of tropical multispecies stocks: A review with emphasis on the Southeast Asian demersal fisheries', *ICLARM Stud. Rev.* No. 1, 35 pp.
Republic of Mozambique State Secretariat of Fisheries (1994a), *Master Plan*, Maputo, Mozambique.
Republic of Mozambique State Secretariat of Fisheries (1994b), *Sector Report*, Maputo, Mozambique.
Ruddle, K.R. (1996a), 'The geography and human ecology of reef fisheries', in N.V.C. Polunin and C.M. Roberts (eds), *Reef Fisheries*, London: Chapman and Hall.
Ruddle, K.R. (1996b), 'Traditional management of reef fishing', in N.V.C. Polunin and C.M. Roberts (eds), *Reef Fisheries*, London: Chapman and Hall.
Semesi, A.K., Y.D. Mgaya, M.H.S. Muruke, J. Francis, M. Mtolera and G. Msumi (1998), 'Coastal resources utilization and conservation issues in Bagamoyo, Tanzania', *Ambio*, **27**(8), 635–44.
Serra King, H.A. (1995), 'Estudo da dinâmica da matéria organica do solo após corte e queimada, Ilha da Inhaca', Tese de Licenciatura, Universidade Eduardo Mondlane, Maputo, Mozambique.
Sillitoe, P. (1998), 'The development of indigenous knowledge: A new applied anthropology', *Current Anthropology*, **39**(2), 223–40.
Sousa, A.G. (1960), 'As Ilhas Quirimbas', unidentified Portuguese historical journal.
UNDP (2006), *Human Development Report 2006*, http://hdr.undp.org/en/reports/global/hdr2006/, accessed 24 August 2009.
Van der Elst, R., E. Bernadine, J. Narriman, G. Mwatha, P. Santana Afonso and D. Boulle (2005), 'Fish, fishers and fisheries of the Western Indian Ocean: Their diversity and status. A preliminary assessment', *Philosophical Transactions of the Royal Society A*, **363**(1826), 263–84.
Wakeford, R.C., G.P. Kirkwood and C.C. Mees (1998), 'Management of the domestic fishery of the Seychelles', poster at the FSBI Annual International Symposium on Tropical Fish Biology, University of Southampton, UK, July 1998.
West, H.G. (1998), '"This neighbour is not my uncle!": Changing relations of power and authority on the Mueda Plateau', *Journal of Southern African Studies*, **24**(1), 141–61.
White, A.T., L.Z. Hale, Y. Renard and L. Cortesi (eds) (1994), *Collaborative and Community-based Management of Coral Reefs: Lessons from Experience*, West Hartford, CT: Kumarian Press.
Wilson, J.D.K., D.A. Pinto, J.B. Gomes and A. Sumale (1996), 'Comercialização e distribuição de pescado, Distrito de Macomia, Província de Cabo Delgado, Relatório da 1ª e 2ª Fases do Estudo', Instituto de Desenvolvimento da Pesca de Pequena Escala, Maputo, Mozambique.
Wynter, P. (1990), 'Property, women fishers and struggles for women's rights in Mozambique', *SAGE*, **7**(1), 33–7.

9. The link between ecological and social paradigms and the sustainability of environmental management: a case study of semi-arid Tanzania
Claire H. Quinn and David G. Ockwell

Introduction

Over the last two decades, sustainable development has become widely accepted as the goal that environmental management should strive to achieve. The idea of sustainable development has evolved due to the inherent tensions between economic development and the desire to protect the environment. The most commonly used definition comes from the *Brundtland Report*, which defines sustainable development as: 'Development that meets the needs of the present without compromising the ability of future generations to meet their own needs' (WCED, 1987, p. 43).

The argument in favour of sustainable development has often been characterized as a response to the fact that economic development in a capitalist market economy leads to environmental degradation (Carter, 2001; Anand, 2003). This, however, fails to recognize that some approaches to protecting the environment can result in negative economic impacts where such protection prevents previous economic uses of natural resources – a particularly important issue for developing countries where the poorest people often depend most directly on natural resources for their livelihoods. In such situations, the challenge for environmental managers working to achieve sustainable development is how to protect and conserve natural resources at the same time as protecting the livelihoods of poor people. Furthermore, management approaches also need to enable poor people to develop opportunities to improve their economic situation and lift themselves out of poverty in the long term. Recognition of this link between sustainable environmental management and the eradication of poverty is central to the thinking behind goals 1 and 7 of the United Nations Millennium Development Goals, which aim, respectively, to eradicate extreme poverty and ensure environmental sustainability. Within this context, therefore, 'environmental management for sustainable development' needs to protect the environment at the same time as protecting and developing the livelihoods of those people who depend on it.

Using the case study of semi-arid Tanzania, in this chapter we highlight the reciprocal relationship between the environment and society. We demonstrate how the paradigms that define environmental managers' and policy-makers' conceptions of ecological and social problems are integral to defining the policy discourses (see Ockwell and Rydin, Chapter 6, this volume) that shape their choice of management solutions. Furthermore, we argue that if these paradigms fail to recognize the reciprocal link between the environment and society, the resulting management solutions will fail to protect and enhance the livelihoods of poor people. The chapter begins with an analysis of the reciprocal relationship that defines and shapes both the ecology and socioeconomics of semi-arid regions. It then moves on to explore the different paradigms that have dominated conceptions of ecological and social problems in the region and how these paradigms have shaped policy discourses and environmental management. The chapter concludes by highlighting how recognition of the reciprocal relationship between people and the environment within the paradigms that shape policy and management decisions is critical to achieving sustainable development.

The reciprocal relationship between environment and society
Understanding what defines and shapes a particular ecosystem has traditionally been assumed to be a purely ecological question. More recently, however, as well as recognizing the importance of ecological factors in defining semi-arid regions, observers have begun to recognize the defining nature of human activities. This stems from an increased understanding of the reciprocal relationship that exists between the environment and human society. Focusing on East Africa, and Tanzania in particular, to demonstrate this reciprocal relationship we begin by examining the climatic and vegetative characteristics of semi-arid regions. We then show how people and their livelihoods play an important role in shaping the ecology of semi-arid regions and how, in turn, the ecology of the region shapes people's livelihoods.

Climatic and vegetative characteristics of semi-arid regions
Climate, the average weather conditions such as temperature, day length and rainfall (O'Brien, 1993), forms a key attribute of semi-arid regions. More precisely, variability in climate, especially rainfall, since temperature and day length are less variable near the equator, has a huge influence on the nature of semi-arid regions. In sub-Saharan Africa the semi-arid regions are influenced by the Intertropical Convergence Zone (ITCZ), a broad low pressure zone in which the trade winds meet and pick up moisture from the equator and move it north and south towards the tropics

(Griffiths, 1972; Mussa, 1978). In East Africa rainfall is influenced by the movement of trade winds over the Indian Ocean and additionally by the El Niño Southern Oscillation (Lovett, 1993). These major air and ocean current circulations mean that the semi-arid regions in East Africa are subject to highly variable rainfall concentrated in one or two rainy seasons a year separated by relatively long dry seasons (Van Keulen and Seligman, 1992). While parts of East Africa have a single rainy season that stretches from November/December to April/May (Griffiths, 1972; Peberdy, 1972), most of the semi-arid region in Tanzania experiences a short break in the rains so that there are, in fact, two rainy seasons (Kingdon, 1971; Hamersley, 1972). These peaks in rainfall correspond with the passage of the ITCZ as it tracks the sun's movement in the sky caused by the earth's axial tilt (Kingdon, 1971; Griffiths, 1972; O'Brien, 1998). The equatorial location of Tanzania also means that mean monthly temperatures vary by only 3–4°C (Griffiths, 1972; Mussa, 1978). The seasons therefore broadly divide into a hot and dry season from December until March, a cooler rainy season in April and May followed by a long cool dry season with the short rains in November corresponding with a rise in temperature (Hamilton, 1989; Lovett et al., 2002). In addition to large-scale influences on climate, the position of lakes and mountain ranges as well as altitude can create localized weather conditions (Kingdon, 1971; Griffiths, 1972; Maro, 1978, Hamilton, 1982; Lovett, 1993), so that rainfall is highly variable spatially as well as temporally.

Semi-arid regions are subject to both inter-annual and intra-annual variations in rainfall. This means that rainfall fluctuates between years but also during seasons (Mortimore, 1998). Large volumes of rain can fall over quite short periods so that while seasonal and yearly averages may be maintained the patterns of rainfall between seasons and years might be very different. The short rains in November are particularly variable and may not appear at all in some years (Griffiths, 1972). Failure of the rains occurs when the trade winds fail to meet in the right place at the right time (Walter, 1939). Rainfall can also fluctuate by as much as 60 per cent from the average between years (Peberdy, 1972), with both single and multi-year droughts a common occurrence. However, long-term studies of rainfall patterns in sub-Saharan Africa have not shown conclusively a downward trend in the amount of rainfall received in these regions overall, although sub-Saharan Africa has been subject to a period of particularly low rainfall since the 1980s. This period of aridity is not unprecedented but rises in sea temperature caused by anthropogenic climate change could contribute to a situation of long-term aridity (Nicholson, 1993, 2001). It is also possible that an increase in aerosols in the atmosphere from pollution may be causing reduced rainfall by reducing surface solar radiation (Ramanathan et al., 2001; Stanhill and Cohen, 2001; Roderick and Farquhar, 2002). Lower surface solar radiation reduces the energy available for evaporation leading to reduced rainfall and disruption of tropical circulation patterns (Ramanathan et al., 2001).

The climate of semi-arid regions and its variability has a strong influence on the vegetation found in semi-arid regions. Vegetation distribution, diversity and productivity are all influenced in the first order by climate, but also by soils, competition, fire and grazing (O'Brien, 1993; HilleRisLambers et al., 2001). Two large phytogeographical regions are recognized to cover most of the semi-arid region of Tanzania, the Somalia-Maasai region and the Zambezian region as described by White (1983) (see Figure 9.1). The Somalia-Maasai region is the driest vegetation type found predominantly in north and central Tanzania and is comprised of a range of grass-dominated savanna habitats. Savannas vary from completely grass-dominated grass savanna through bushland, thicket and shrubland, to tree savannas and wooded savannas (Jacobs et al., 1999). Trees and shrubs can comprise up to 50 per cent of species composition but the canopy remains open (Trapnell and Langdale-Brown, 1972). Included within the Somalia-Maasai region are the edaphic savanna grasslands found on the Serengeti plains (see Figure 9.1). These grasslands occur on soils derived from volcanic ash, which tend to be alkaline and often have a hard calcareous pan created by the leaching of carbonates through the soil by rainfall (White, 1983). The calcareous pan creates a barrier to root growth, preventing the growth of woody vegetation, and is also impermeable to water so that these areas can be prone to flooding during the rainy season. Savannas are believed to have existed in East Africa for 20 million years and were, in part, created by the development of seasonal rainfall patterns that restricted the distribution of tropical rainforests (Pratt et al., 1966). Today a mixture of climate, soils, natural fires, grazing and anthropogenic influences maintain the open nature of savanna ecosystems.

The Zambezian region consists of miombo woodland dominated by trees from the genus *Brachystegia* (Kingdon, 1971; White, 1983; Lovett et al., 2001). The region forms a crescent in the south of Tanzania around the more arid Somalia-Maasai region (see Figure 9.1). It is less recognized as a semi-arid ecosystem even though it can occur in areas with rainfall as low as 500mm per year. Drier miombo woodland tends to be less diverse while wetter miombo woodland vegetation has to be adapted to periods of flooding and waterlogging followed by long dry seasons with little available water and frequent fires. It has been argued that in some areas miombo forests may be the result of long-term human disturbance (Hamilton, 1982; White, 1983). Certainly, in areas with deeper soils,

286   *A handbook of environmental management*

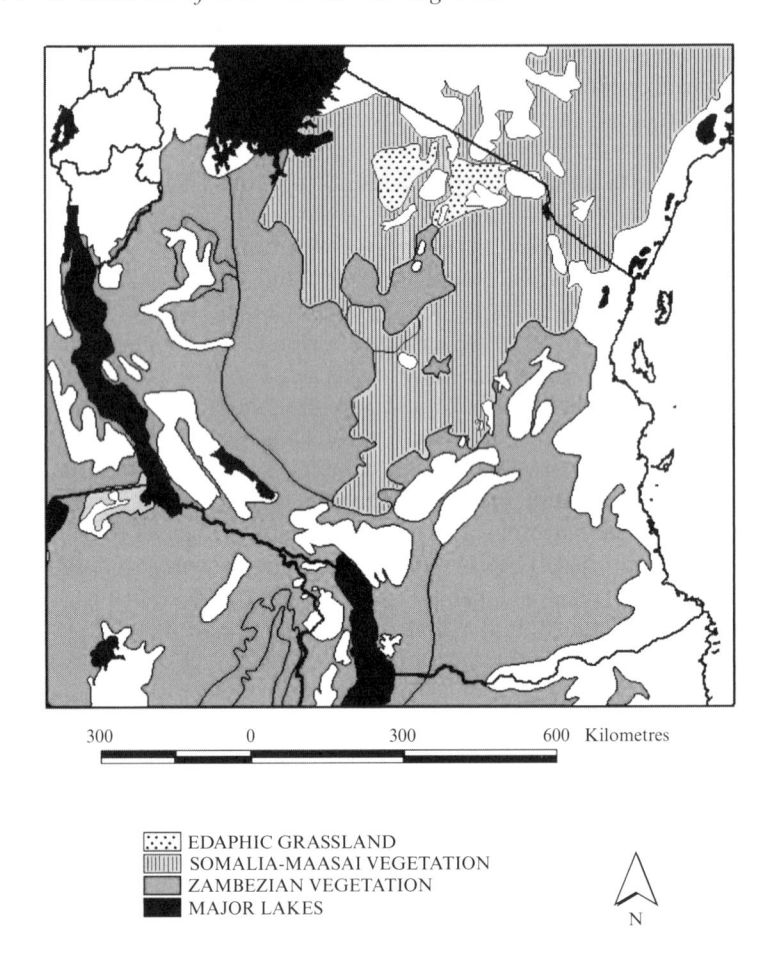

```
EDAPHIC GRASSLAND
SOMALIA-MAASAI VEGETATION
ZAMBEZIAN VEGETATION
MAJOR LAKES
```

Source: White (1983).

Figure 9.1   Major vegetation zones in semi-arid Tanzania

rainfall parameters suggest that tropical dry forests could be supported and so the presence of miombo woodland may be the result of past clearance for cultivation (White, 1983). In Tanzania the Zambezian region is currently less open to occupation and use by indigenous communities because of tsetse fly, which is widespread and responsible for disease transmission to both humans and livestock (Kingdon, 1971; Lovett et al., 2001) but it is still used seasonally and often provides grazing and forage for livestock in times of drought elsewhere.

While the phytogeographical regions are useful for describing vegetation on a larger spatial scale they can overlook the importance of smaller-scale heterogeneity in vegetation distribution. East Africa has been a region of large geological upheaval with volcanoes, block faulting and erosion creating plains, mountains and rift valleys and a variety of soils (Griffiths, 1993) so that vegetation composition can change rapidly over short distances (Van Keulen and Seligman, 1992; Snyman and Fouche, 1993; Armitage, 1996). Water availability is also an important factor for vegetation distribution in semi-arid regions. While rainfall is seasonal and variable in low-lying areas, rainfall in the mountains is more consistent and provides an important source of water (Quinn et al., 2001). As a result, gallery forests along rivers and wetlands can be found in semi-arid regions and provide important variations in vegetation as well as sources of water in the dry season. All these factors of seasonality and geology have created a system in which resources such as grasslands, forests and water vary both spatially and temporally. Primary production tracks rainfall patterns so that not all areas will be productive at the same time over the region.

The reliance on climate as the defining factor for semi-arid regions has posed difficulties in producing an accurate definition. A review of governmental and international organizations working in semi-arid regions has revealed four main definitions of semi-arid regions. The UK Department for International Development (DFID) refers to semi-arid zones in its Renewable Natural Resources Research Strategy as one of six production systems (ODA, 1994). In the DFID definition a semi-arid zone is one with a mean monthly temperature above 18°C, where evapotranspiration exceeds precipitation in one or more seasons and mean annual rainfall is between 400mm and 1200mm. The term 'semi-arid' is also used in the Food and Agriculture Organization (FAO) agroclimatological stratification and defines an area with a growing period of between 75 and 119 days (Bourn and Blench, 1999). This does not correspond directly with the DFID definition as the DFID rainfall parameters cover a growing season of 75 to 180 days (Bourn and Blench, 1999). The FAO agroclimatological definition is also more restrictive than the arid and semi-arid tropics and subtropics definition that the FAO uses in its World Livestock Productions Systems (growing season less than 180 days) (Sere et al., 1995). In addition a fourth definition is used by UNEP (United Nations Environment Programme) in its studies of desertification (Mortimore, 1998). In many empirical studies the semi-arid region has been defined by a minimum rainfall of 400–500mm a year to exclude arid regions, which come with their own set of particular management issues related to extremely low rainfall and low population densities (Pratt and Gwynne, 1977; Mwalyosi, 1995; Mortimore, 1998). Above 400–500mm of rainfall a year rain-fed agriculture, although marginal, is possible and livestock and people are found in much higher densities than in arid regions (Mortimore, 1998). The variation in definition means that areas are included or excluded from the semi-arid zone depending on the definition used and the type of vegetation associated with semi-arid areas.

In Tanzania, a restricted definition based on the FAO agroclimatological stratification, or an upper annual rainfall limit of 900mm, would limit the area considered semi-arid to Somalia-Maasai vegetation, which forms a strip of arid land running from the north to a tip at the junction of the Tanganyika, Nyasa and Eastern Arc rifts (Figure 9.2). The DFID definition of semi-arid regions, on the other hand, where the mean monthly temperature is above 18°C, evapotranspiration exceeds precipitation in one or more seasons and mean annual rainfall is between 400mm and 1200mm, covers a more extensive area of land. This definition covers an area that includes the Zambezian vegetation zone as well as the Somalia-Maasai vegetation zone (see Figure 9.3). For Tanzania the semi-arid region in this context covers approximately 80 per cent of the land area of the country, an area of approximately 704800km^2.

The role of people and livelihoods
In the same way that climate and resources are variable in semi-arid regions, so are the people who live and maintain their livelihoods in them. Mortimore (1998, p. 10) refers to semi-arid regions as the 'heart of the drylands' in Africa. They are extremely important for people, livestock and agriculture. In order to understand semi-arid regions it is not enough to understand the ecology in isolation. It is also necessary to understand the communities who live in these regions. These communities have had, and continue to have, an impact on the ecology of where they live. Equally, the ecological conditions in which these people live have impacted, and continue to impact, on them. It is this that constitutes the reciprocal relationship between people and their environment in semi-arid regions. This relationship has not always been recognized and as a result attempts at sociopolitical and ecological control, first by the colonial governments and later by the independent socialist government, have often led to a decoupling of people and the environment.

Many of the ecological conditions observed today are the result of past people–environment interactions and the subsequent decoupling of the people–environment relationship. For example, savanna ecosystems may be the result of long-term interactions between humankind and the environment while the distribution of tsetse fly may be the result of depopulation and subsequent woodland encroachment in areas once inhabited and cleared for livestock or cultivation (Iliffe, 1979).

East Africa is recognized as important to human history. Archaeological

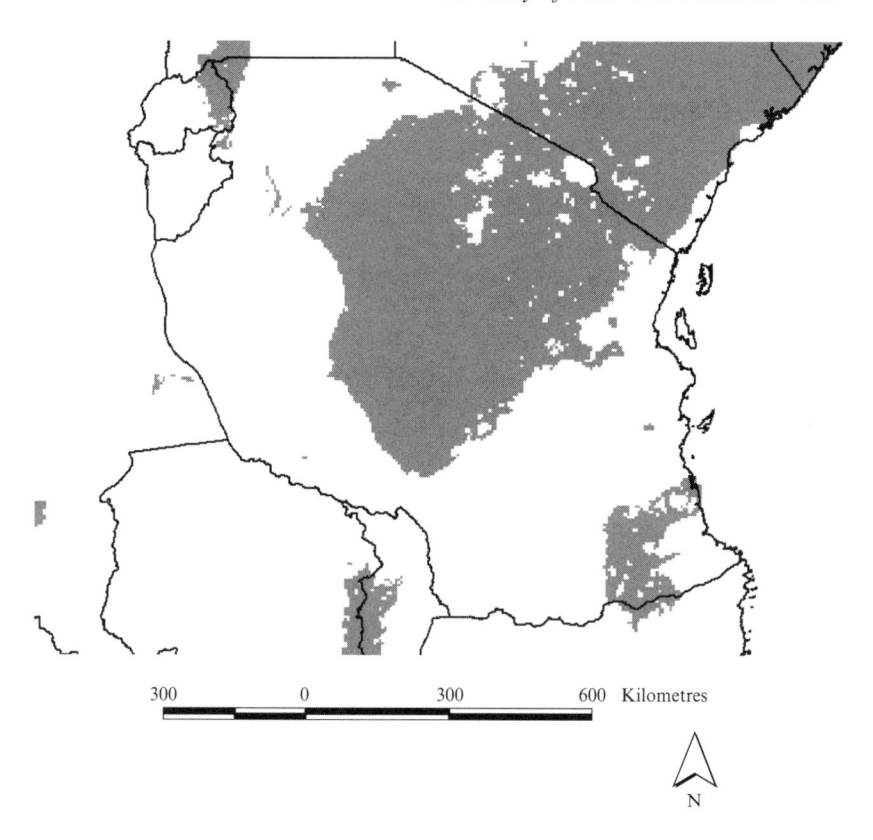

Note: Rainfall upper limit of 900mm per annum, equivalent to the FAO agroclimatological classification of semi-arid regions; growing period 75–119 days. (Seasons are defined on the basis of climatic fluctuations in Tanzania and take into account the long rains and short rains.)

Source: Quinn (2005).

Figure 9.2 The extent of semi-arid regions in Tanzania based on a restricted definition

finds in Kenya and especially Tanzania suggest that modern humans originated in this region (Leakey, 1972). However, it is not a simple task to decipher the origins and history of current ethnic groups in Tanzania. This is because membership of and division into tribes was not and had never been as clear-cut as often believed. Society in East Africa was pluralist (Shorter, 1974), meaning that society was made up of heterogeneous and often fragmented groups of people, reflecting the heterogeneous nature of the environment. However, this did not mean that there were static

290   *A handbook of environmental management*

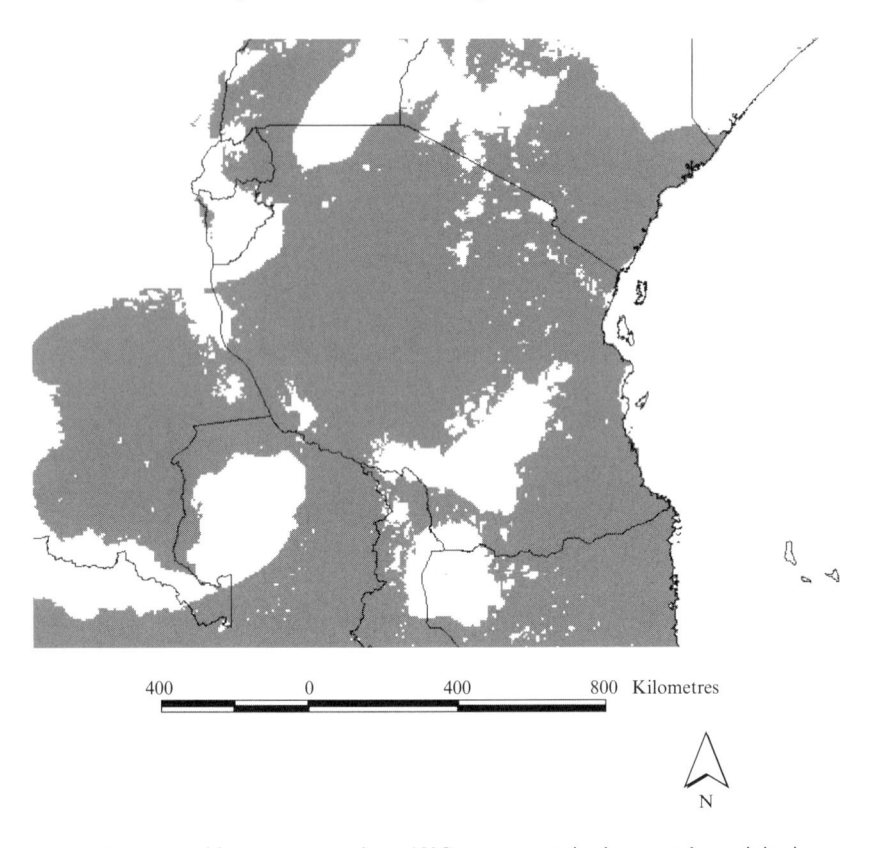

Note: Mean monthly temperature above 18°C; evapotranspiration exceeds precipitation in one or more seasons; mean annual rainfall between 400mm and 1200mm. (Seasons are defined on the basis of climatic fluctuations in Tanzania and take into account the long rains and short rains.)

Source: Quinn (2005).

Figure 9.3   The extent of semi-arid regions in Tanzania based on the DFID definition

and discrete tribes, rather that tribal membership was fluid and the result of adaptation to the landscape (Iliffe, 1979). Tribal membership also did not mean isolation from other groups. Many shared a common heritage and continued to interact through trade, intermarriage and migration (Homewood and Rodgers, 1991).

Kieran (1972), Shorter (1974) and Iliffe (1979) have attempted to discuss in detail the origins and history of people in East Africa. The historical development of societies in Tanzania will be discussed here based on their influence on the people–environment relationship. The broad divisions used are based on linguistic origins. The first inhabitants of the region are believed to be Khoisan in origin and were hunter-gatherers. A low, dispersed and nomadic population using traditional low technology tools such as bows and arrows meant that this livelihood strategy was sustainable in terms of land and resources, even in times of drought (Armitage, 1996). It is unlikely that hunter-gatherers changed the ecology of their surroundings to any great extent because of their low population and primitive technology. Hunter-gatherer livelihoods rely on exploiting wild animal and plant resources already present rather than domestication or cultivation of animals or crops. Once widespread in the region they were displaced or absorbed by other tribes who began to colonize the area from North and West Africa. Only small groups such as the Hadza, who are believed to be the descendants of these early inhabitants, still practise a hunter-gatherer livelihood strategy in Tanzania today (Armitage, 1996).

From about 1000 BC and possibly as early as 2000 BC Cushitic tribes migrated south into Tanzania from a region now in modern day Ethiopia. They were generally cattle herders and occupied the northern central region of Tanzania dominated by Somalia-Maasai vegetation, which was settled because of its suitability for pastoral livelihood strategies. These early pastoralists are still represented by the Iraqw in Tanzania. Possibly one of the largest migrations, and certainly one of the most widespread and influential linguistically in Tanzania, was the arrival of the Bantu-speaking tribes from West Africa. Bantu tribes are thought to have started migrating into the lake regions of Tanzania as early as 500 BC and after AD1000 approximately 90 per cent of the tribes in Tanzania were of Bantu origin. The success of the Bantu tribes is attributed to their use of cultivation, which allowed them to successfully colonize the more fertile highlands before spreading down onto the plains (Shorter, 1974). After the Bantu tribes the fourth major group to migrate into the region were the pastoral Nilotic tribes from Eastern Ethiopia who arrived much later in the 1500s. The Maasai were part of this migration and had settled in Northern Tanzania by the late 1700s.

Both pastoral and agricultural livelihood strategies had a much greater impact on the ecology of Tanzania than the hunter-gatherer strategies of the original inhabitants. Pastoralism was based on transhumance rather than being truly nomadic. This meant that the homestead, comprised of women and very young children, remained in one place for relatively long periods of time. Lactating cows and calves were kept in, and grazed around, the homestead, so often homesteads were widely spaced to allow adequate forage. Young men and boys moved seasonally with the rest of the herd to find adequate grazing throughout the year. Areas were designated and used for dry or wet season grazing, or only used in years of drought. Fire was often used to clear woody vegetation and encourage grass growth in grazing areas, reducing woody biomass and maintaining a more open savanna system. Agricultural groups also used fire to clear areas for growing crops, and brought in new plant species such as banana, sorghum, millet and maize, which they grew in different areas depending on soil quality and rainfall. As a result, settlement patterns depended on soil quality and water availability. Only in the highlands where soils were generally more fertile, rainfall higher and competition for land greater did communities settle for long periods and invest in their fields. In semi-arid regions agricultural communities moved regularly to clear new areas for cultivation in response to declining soil fertility or inadequate rainfall. It is likely that both pastoral and agricultural groups were responsible for clearing woody vegetation in many areas, creating open savanna or secondary miombo woodlands and reducing dry tropical forest cover (White, 1983).

Today some 110 tribes are recognized in Tanzania (Gulliver, 1972). Although they can be grouped according to their linguistic and historical traditions they show divergent patterns in livelihood systems and social organization (Gulliver, 1972). Tribal identification has been fluid and was often related to location; many tribal names can be translated to mean 'highlanders', 'northerners' or 'from the waterless country' (Iliffe, 1979). Therefore, it was possible to change tribal affiliation by moving and changing livelihood strategies. As a result, linguistic origin is not always useful in delineating livelihood strategies. Although Cushitic and Nilotic tribes were generally pastoralists and Bantu, cultivators, in fact many tribes adapted their livelihood strategies according to the ecology and environment in which they lived. The diversity of soils and variability in climate that created ecological heterogeneity in semi-arid regions were also responsible for the heterogeneity found in tribal identity and livelihoods. Both Cushitic and Bantu tribes cultivated crops as well as owning livestock and the relative importance of these strategies varied with location and especially rainfall. Tribal identity was inextricably linked to the local ecology.

Sociopolitical change has had a huge impact on livelihoods and ecology in Tanzania. The influx of pastoral and agricultural groups changed the way that people interacted with the environment and as a result the ecology of semi-arid regions. In the past there was a close link between livelihoods and ecological conditions. More recently, however, a process of decoupling of people and environment has occurred. This process began with the spread of livestock and human diseases into East Africa in the early 1800s. The spread of rinderpest is recognized as being particularly important in its impact on people and environment and the relationship between them (Kjekshus, 1996). Rinderpest was introduced into Northern Africa by Europeans and spread into East Africa, decimating cattle populations. The Maasai were particularly hard hit. It is estimated that approximately 90 per cent of their cattle were killed and two-thirds of the Maasai population died from famine and disease as a result (Iliffe, 1979; Homewood and Rogers 1991; Kjekshus, 1996; Koponen, 1996). The spread of rinderpest was also accompanied by epidemics of smallpox and influenza, among other human diseases, which served to reduce the numbers of people still further. Many areas that had been used for cultivation or livestock were deserted as people died or left in search of food. This depopulation allowed the expansion of miombo woodland and wildlife numbers increased. The distribution of the tsetse fly, which carries sleeping sickness, also expanded as its woodland habitat increased (Coulson, 1976). The spread of tsetse fly made once inhabited areas no longer suitable for settlement.

This process of decoupling continued into the colonial era where sociopolitical change created by war and conflict contributed to depopulation. German colonial rule was characterized by tribal war and resistance. For example, during Maji Maji[1] and its aftermath, the largest and most widespread resistance to the colonial government, it is estimated that some 250 000 people were killed (Iliffe, 1979). As a result of depopulation large areas of Tanzania became uninhabited and were appropriated by the state for plantations or wildlife reserves. Many areas that had been used for livestock grazing or cultivation were now no longer available to local populations. Further reductions in population occurred during the two World Wars when many local people were forced into labour for both the British and German armies, and food appropriated for the armies contributed to widespread famine (Brooke, 1967).

Even during the relatively peaceful British colonial rule following World War II the decoupling of the people–environment relationship continued. People were moved and concentrated into villages in order to facilitate greater socioeconomic control. The process of indirect rule through tribal chiefs advocated by the British also led to social upheaval as power through tribal leadership was contested. The process of land appropriation for wildlife reserves and plantations continued as the British government sought to control the resources of Tanzania as well as its people. All these factors together meant that the relationship between communities and their environment was disrupted. For example, the area now covered by the Selous Game Reserve was once densely populated. Depopulation due to disease and war led to widespread woodland encroachment and expansion of tsetse fly. The population who were left were scattered in small, mobile communities. People were concentrated into villages so that the district government could have greater socioeconomic control and the

Selous Game Reserve was created in order to prevent people from moving back into the tsetse-infested area (Matzke, 1976). The creation of the game reserve also meant that the land was now in the control of the state rather than outside it. Only later did the remit of the reserve change to include conservation aims.

Even with independence the process of social control over rural communities has continued, with its resultant impact on the environment. Ujamaa villages created by the process of villagization were originally conceived as cooperative villages where people would work together to increase agricultural production (Hyden, 1975; McHenry, 1979; Kikula, 1997). These villages would also facilitate the delivery of social and agricultural services to often remote and dispersed communities. However, the independent government, along with the colonial governments who preceded it, failed to recognize the influence of ecology on patterns of settlement. Variable rainfall, soils, wildlife and disease all determined settlement patterns, livelihood choices and population distributions. In particular, rainfall distribution meant that 60 per cent of the population lived in 20 per cent of the land area of Tanzania (Maro and Mlay, 1978). Villagization settled agricultural groups in areas marginal for farming and so contributed to the conflict that already existed between pastoral and agricultural ethnic groups, as these villages were often located on or near to important water sources. Pastoralism was perceived to be a more primitive livelihood strategy and so pastoral communities were encouraged to settle into villages, regardless of the ecological need to move livestock. Although villagization was planned as a way to increase production it was a failure and led to declines in food and cash crop production (Smith, 2001).

To summarize, the ecology of semi-arid regions is characterized by both inter-annual and intra-annual variability in rainfall. Large volumes of rain can fall over short periods, while single and multi-year droughts are common. This variability in climate, along with variability in geology and soils, results in spatial and temporal heterogeneity in vegetation composition, primary production and water availability. Semi-arid regions are therefore characterized by variability. It is now more widely recognized, however, that the ecological conditions presently found in semi-arid regions are, at least in part, the result of long-term human activity in these areas. At the same time, the climatic variability in semi-arid regions is a huge challenge for people in trying to sustain their livelihoods. As a result, in semi-arid Tanzania both pastoral and agricultural livelihoods strategies are found and each has developed flexibility in order to cope with variability. Traditional pastoral livelihoods common to the region allow for mobility of people and livestock so that adequate forage can be found throughout the year, in much the same way as wild ungulate populations migrate seasonally throughout East Africa. Pastoral livelihoods need access to a mosaic of vegetation types in order to provide adequate grazing for their livestock and reduce their vulnerability to variability in climate. In more agricultural livelihoods diversity in location with access to key water resources, crop selection and combinations of both agricultural and pastoral livelihood strategies have been employed as methods for dealing with ecological variability.

This reciprocal relationship between people and the environment has historically been central to the sustainable management of semi-arid Tanzania. Over the last two centuries, however, long-term sociopolitical change has resulted in a decoupling of the people–environment relationship. Depopulation due to disease and war left large areas uninhabited and allowed the spread of woodland and tsetse fly. Depopulation also allowed successive governments to appropriate land for the state through plantations or wildlife reserves, removing land from local control and denying access to key resources. Villagization also encouraged settlement of agricultural groups in traditionally pastoral areas, adding to conflict over important resources such as water. If agricultural groups take over crucial water resources then pastoral communities are denied access and can become more vulnerable, especially during drought years. Semi-arid regions therefore represent a zone of conflict between different livelihood strategies over access to resources, a conflict often exacerbated by changes in land tenure and appropriation by the state that often results in vital resources becoming inaccessible to local communities.

Despite wider recognition in contemporary times of the importance of this reciprocal link between people and their environment, government policy in semi-arid regions is still broadly influenced by past colonial governments and protectionist and preservationist approaches, which perceive traditional livelihood activities as damaging to the environment. In the next section we move on to consider the dominant paradigms that conceptualize thinking around ecological and social problems in semi-arid regions and demonstrate how the nature of these paradigms defines whether or not resulting policy discourses recognize the reciprocal relationship between people and the environment.

Managing semi-arid regions: ecological and social paradigms
The idea of a 'paradigm' refers to the way in which a certain issue is perceived or understood. The paradigms that define policy-makers' and managers' understanding of the ecological and social dynamics of semi-arid regions will also define their perception of appropriate policy or management solutions. In Tanzania, two opposing discourses dominate policy approaches to managing semi-arid regions. One is the 'people

versus environment' discourse where local people are seen as damaging to the environment. This discourse implies a need to manage people and the environment separately. The other is the 'people and environment' discourse that seeks to encourage integrated management of people and resources by recognizing the importance of the reciprocal relationship between people and their environment. These policy discourses are underpinned by two different ecological and social paradigms upon which is defined the conceptualization of management problems.

Ecological paradigms
Until relatively recently the predominant paradigm for understanding the ecology of semi-arid regions, and rangelands in particular, has been equilibrium theory. This paradigm developed from a European and US perspective of rangelands in temperate climates (Clements, 1916). Equilibrium theory is based on the assumption that vegetation composition will progress towards a climax community, depending on climate and soil conditions, through the processes of succession (Begon et al., 1990; Behnke and Scoones, 1993). The climax community is therefore considered to be the most stable plant community that could occur. Disturbance, either natural or anthropogenic, will result in the vegetation being 'pushed back' to an earlier successional stage and once the disturbance ceases or is removed then natural processes will lead back to the climax condition (Briske et al., 2003). The most commonly used example of succession is the clearing of forest for agriculture in temperate regions. Once the land is abandoned then it is assumed that a climax forest community will eventually return (Behnke and Scoones, 1993). These ideas were applied to grazed rangelands in sub-Saharan Africa, as well as elsewhere, where different grass communities were considered to represent different successional stages with certain ecological communities representing climax (Brown and MacLeod, 1996).

An important element of equilibrium theory as applied to grazed rangelands is the concept of carrying capacity. If, as suggested by equilibrium theory, secondary production is linearly related to the successional status of vegetation then it is possible to calculate carrying capacity in terms of wild herbivores or livestock that can be supported on the range. In equilibrium theory the interaction between plants and animals is an important negative feedback mechanism that regulates the ecosystem (Briske et al., 2003). This density dependence relationship means that at ecological carrying capacity herbivore numbers are controlled by primary production so that the number of births equals the numbers of deaths and the vegetation is held back from reaching the climax composition that would occur without such disturbance from grazing. For managers of rangelands the aim would be to maintain the vegetation at some optimal sub-climax composition so as to maximize secondary production, depending on the management objectives. What is important to note is that carrying capacity in this context is dependent on the management objectives and is not an inherent feature of the ecosystem. Different carrying capacities will be desired, for example, for commercial meat production or subsistence pastoralism.

Equilibrium theory as applied to sub-Saharan Africa has been criticized for not adequately describing the reality of rangelands and recently this criticism has led to the development of an alternative paradigm for range ecology: the non-equilibrium theory (Behnke and Scoones, 1993; Warren, 1995; Brown and MacLeod, 1996; Briske et al., 2003). This paradigm suggests that there is no single successional pathway and that rangeland can exist in multiple stable states. Rangelands are then seen as mosaics of plant communities created by a variety of factors rather than evidence of early succession stages caused by disturbance (Sullivan and Rohde, 2002). What is important for non-equilibrium theory is that density-dependent interaction between plants and animals is weak or non-existent and instead abiotic factors, and in particular rainfall, are the dominant driving factors of vegetation production, distribution and composition. This is an important distinction from equilibrium theory because it recognizes that the highly variable climate and rainfall patterns are more important in determining primary production and composition than grazing by herbivores. In non-equilibrium theory fluctuations in rainfall create fluctuations in vegetation productivity, which in turn cause herbivore numbers to fluctuate. Rainfall is such an important driver of the system that vegetation and herbivores do not reach equilibrium and so it is not possible to calculate a carrying capacity.

There is still controversy over whether equilibrium or non-equilibrium paradigms apply to African rangelands. Evidence has been presented that supports both views (Illius and O'Connor, 1999; Sullivan and Rohde, 2002). Illius and O'Connor (1999) in particular present the view that there are still density-dependent interactions between herbivores and vegetation in key resource areas. Therefore carrying capacity is maintained for these areas even when it does not correspond to the whole range. In contrast Sullivan and Rohde (2002) argue that even key resource areas are subject to highly dynamic rainfall patterns so that primary production is more related to rainfall than to grazing. It has been suggested that rather than focusing on the dichotomy the theories should be considered as positions along a continuum where rangeland can exhibit equilibrium or non-equilibrium dynamics, or a combination of both (Briske et al., 2003). Empirical studies are therefore necessary to determine where along this continuum a particular rangeland lies. Briske et al. (2003) argue that in drier regions where the intra-annual variation in rainfall is greatest then rangelands are likely to exhibit non-equilibrium dynamics, however in wetter areas where variation in rainfall is less then equilibrium theory is more appropriate. Following this argument in Tanzania would lead to the Somalia-Maasai region being considered a non-equilibrium system since rainfall is highly variable. In contrast the Zambezian region would be subject to equilibrium dynamics since rainfall averages are higher and more consistent. However, even in the Zambezian region rainfall is highly seasonal and intra-annual variation is still high, although total rainfall is generally greater than in the Somalia-Maasai region. As a result non-equilibrium dynamics cannot be discounted as important for at least some parts of this system.

Social paradigms
The two different ecological paradigms outlined above are central in determining which social paradigm influences policy, because each focuses on a different cause of environmental degradation. In equilibrium theory overstocking and overpopulation are seen as the primary causes of degradation of rangelands and management regimes should seek to control stocking rates and change land tenure so as to maintain rangelands in an optimal condition. Equilibrium theory leads to the pastoral paradigm being the dominant theory that frames understanding of communities. In the pastoral paradigm pastoral communities are seen as unable to adequately manage rangeland resources, leading to degradation (Warren, 1995). Population increases accompanied by inappropriate management are therefore considered to be key problems that require intervention in order to prevent over-use and protect natural resources and wildlife. In Tanzania the predominance of the equilibrium paradigm in understanding rangeland ecology has helped support the view that pastoral ways of life are primitive and economically unproductive (Armitage, 1996). Early successional stages in rangeland composition are given as evidence that pastoral communities overstock the rangelands leading to degradation. Interventions that de-stock and change land tenure are therefore seen as necessary to prevent degradation and ultimately soil erosion and desertification (Ellis and Swift, 1988; Warren, 1995).

In contrast, abiotic factors, particularly rainfall, are attributed the key role in determining range condition in non-equilibrium theory. As a result, managing stocking rates would not necessarily lead to improvements in range condition. In addition, it is now recognized that most development efforts based on equilibrium theory and the pastoral paradigm have ultimately ended in failure (Oba et al., 2000). While it could be argued that failures have been due to resistance of pastoral communities to accept these management changes, there is a mounting body of evidence that suggests that pastoral forms of management do not always lead to degradation and that they are stocking the rangelands at optimum levels for their management objectives and ecological conditions (Homewood and Rodgers, 1987; Behnke and Scoones, 1993; Behnke and Abel, 1996; Oba et al., 2003). This shift in ecological thinking has therefore led to a shift towards a new social paradigm that acknowledges indigenous knowledge as important for resource management. The emphasis is on community participation in development projects and recognition that supporting and adapting existing traditional management may provide a more successful long-term route for aid intervention.

Policy discourses
The choice between the ecological and social paradigms outlined above, together with the long-term sociopolitical changes that have characterized Tanzania's history over the last few hundred years, have been central in defining contemporary policy discourses. Colonial rule in Tanzania has had a profound effect on the subsequent management of resources in these areas and the way in which local communities are perceived. When the British and other European nations began to colonize large parts of East Africa in the late 1800s the received wisdom was that Africa was one of the last great wildernesses untouched by humankind (Anderson and Grove, 1987; Leach and Mearns, 1996; Neumann, 1996). National Parks were created by the British colonial administration in part to protect this wilderness, but they were also created to preserve the aristocratic lifestyle that was diminishing in the United Kingdom as the result of social and political changes in land ownership (Neumann, 1995, 1996). In reality, the first National Parks were less to do with conserving wildlife as it is understood today and more about protecting hunting rights and a particular way of life, and exercising social control through exclusion and settlement of indigenous communities (Anderson and Grove, 1987; Neumann, 1995). These early ideas of separating people and nature became entrenched in the colonial administration and the independent administrations that followed. This 'people versus environment' discourse in policy is inextricably linked with conservation thinking, equilibrium theory and the pastoral paradigm. Today wildlife is perceived as a common heritage that should be conserved for its own sake and for the sake of future generations (Lovett et al., 2001). Maintaining biodiversity through the exclusion of livestock, which under the equilibrium paradigm is believed to out-compete native ungulates for forage, is seen as a necessary part of conservation strategy. Local communities are perceived to be at odds with sustainable management of resources and top-down interventions are advocated. This

discourse focuses on compensating local communities through wildlife conservation programmes (Lovett et al., 2001).

The problem with the initial colonial view of 'wilderness Africa' was that it represented a snapshot of conditions in East Africa at a time of unprecedented crisis in indigenous populations. As discussed earlier in this chapter, East Africa was devastated by the rinderpest epidemic of the 1890s. It rapidly swept through East Africa killing huge numbers of cattle and almost two-thirds of the Maasai population alone died because of famine and disease (Homewood and Rodgers, 1991; Kjekshus, 1996). Oscar Baumann (cited in Iliffe, 1979) described the devastation experienced by the Maasai:

> There were women wasted to skeletons from whose eyes the madness of starvation glared. . .'warriors' scarcely able to crawl on all fours, and apathetic, languishing elders. These people ate anything. Dead donkeys were a feast for them, but they did not disdain bones, hides, and even the horns of the cattle. (Iliffe, 1979, p. 124)

The decimation of once powerful tribes resulted in destabilization of social structures and without cattle people moved out of certain areas in order to find food. When the German administration arrived in the late 1890s, followed by the British after World War II, they found large areas unsettled by people and communities unwilling to trade, intent on raiding other tribes for food. From this they concluded that the area had never been highly populated and that the people were primitive and economically backward (Kjekshus, 1996). In reality many of the ecological conditions found at that time had been created by human settlement. In this situation of crisis large areas were easily alienated from indigenous communities to create national parks, and the subsequent recovery in populations was seen as evidence of overpopulation and the need for intervention.

Re-examination of historical changes in ecological and sociopolitical conditions has created a shift in thinking about the role of people in East Africa (Leach and Mearns, 1996). While in his independence message to TANU (Tanzania African National Union) Nyerere said 'from now on we are fighting not man but nature' (Nyerere, 1969, p. 139), as Iliffe points out 'it [is] more complicated than that' (1979, p. 576). The 'people and environment' policy discourse has grown from the belief that indigenous communities have been subject to inappropriate Western ideals of ecology and management, dominated by the equilibrium paradigm, which has led to land being appropriated for national parks and agriculture and, as a result, the destabilization of traditional forms of management. Non-equilibrium theory has highlighted the dynamic nature of production and resource distribution in semi-arid regions and suggests that current concerns about overstocking and degradation may be misplaced (Homewood and Rodgers, 1987; Stocking, 1996). While it is not suggested that degradation never occurs (Homewood, 1995) more emphasis is placed on rainfall dynamics and the effects of restricting access to rangelands through alienation for national parks and the expansion of agriculture into traditional pasture lands. The emphasis in this discourse is on equity and the rights of indigenous populations (Lovett et al., 2001). This discourse challenges the pastoral paradigm in which local communities are seen as incapable of managing resources and Western people are viewed as knowing best what needs to be done to make better the lives and environments of communities in developing countries (Leach and Mearns, 1996; Stott and Sullivan, 2000). Instead, indigenous knowledge is recognized and local communities are seen as vital for the sustainable management of resources. In this discourse the emphasis is on community participation and community-based initiatives for wildlife and natural resource management.

Nevertheless the 'people versus environment' policy discourse has historically dominated natural resource management in Tanzania. Approximately 25 per cent of Tanzania has been gazetted as national parks, game reserves or forest reserves (IUCN, 1987) with semi-arid regions being particularly affected. This has been accompanied by concerns with overpopulation and overgrazing in areas still used by pastoral and agro-pastoral communities. A prime example is the HADO (Hifadhi Ardhi Dodoma – soil conservation in Dodoma) project in Kondoa, which was set up in 1973 in response to the perceived overstocking of rangelands in the area and resultant soil erosion (Lovett et al., 2001; Quinn et al., 2007). The HADO project instigated complete de-stocking of the rangelands accompanied by improved agriculture and tree planting (Wenner, 1983) and later the introduction of zero-grazing where small livestock are kept and fed on crop residues (Kerario, 1996). Although the HADO project has been considered a success in improving vegetation cover there is the suggestion that soil erosion in the area was the result of deforestation rather than population increase (Mung'ong'o, 1991). Ultimately the HADO project could only be instigated through direct government intervention and control and even then grazing of livestock still occurs (personal observation).

More recent interventions have considered community participation in natural resource management but there is concern that these only pay lip service to local community knowledge (Leach and Mearns, 1996) and criticism can be found of community-based conservation projects elsewhere that have failed to protect wildlife or fully integrate local institutions (Campbell et al., 2001). It seems unlikely that there is the political will to open up national parks to pastoralism and governments still do

Table 9.1 Two conceptual frameworks for resource management in semi-arid regions

Ecological Paradigm	Perceived Causes of Degradation	Social Paradigm	Policy Discourse	Policy Response	Social Relationship to Environment
Equilibrium theory	Human-induced Overstocking, population increase	Pastoral paradigm	People versus environment	Top-down interventions De-stocking, changes in land tenure	Decoupled from the environment Unable to manage resources
Non-equilibrium theory	Natural Abiotic factors	New social paradigm	People and environment	Bottom-up approaches Indigenous knowledge, community participation	Coupled to the environment Able to manage resources sustainably

Source: Quinn (2005).

not trust the ability of local institutions to manage resources sustainably. Those who support the 'people and environment' discourse are often discredited for believing in the idea that there was a peaceful co-existence between people and wildlife before colonialism (Giblin and Maddox, 1996; Kjekshus, 1996). This is not necessarily accepted as the case. The point instead is that interventions so far have destabilized traditional forms of management that were appropriate for the ecological context of semi-arid regions (Oba et al., 2000). New policy responses therefore need to take a more 'bottom-up' approach to integrate indigenous knowledge and community participation into natural resource management projects (see, for example, Ockwell, 2008).

The problem facing Tanzania, and other countries in sub-Saharan Africa, is that there are, in effect, two alternative conceptual frameworks for understanding and managing natural resources in semi-arid regions (Table 9.1). The first conceptual framework is based on equilibrium theory. Equilibrium theory is still widely supported by both researchers and particularly government, resulting in human-induced environmental degradation being high on the agenda and interventions often being designed to reduce the human impact on the environment (Kikula, 1999).

As a result, the pastoral paradigm still persists where local communities are interpreted as primitive and in need of modernization through intervention (Smith, 2001). This approach leads to the 'people versus environment' discourse and so to interventions that tend to be 'top-down' and set people and the environment at odds with each other, continuing the loss of control over the environment experienced by local communities since the late nineteenth century (Maddox, 1996).

In contrast, the conceptual framework based on non-equilibrium theory recognizes that local communities may not be responsible for environmental degradation and that abiotic factors are more important for determining range condition (Sullivan and Rohde, 2002). As a result the new social paradigm recognizes that traditional forms of management can be suitable for sustainably managing natural resources found in semi-arid regions. This then leads to a policy discourse that seeks to recouple people to their environment by taking 'bottom-up' approaches to resource management that make use of indigenous knowledge and involves communities in the development and running of management projects.

The ecological and social paradigms are still highly contested and as a result the policy context is undergoing a period of flux and change. Local communities are now challenging the authority of the government to alienate land for conservation but it seems unlikely that in the present political climate protected areas will be opened up to multiple land uses. There is a high level of mistrust in local communities of policy interventions that they see as not addressing local needs. However, this seems to be matched by the government's lack of trust in local institutions to manage resources. In addition, the desire to protect areas for conservation and tourism has not abated, especially since it provides much-needed foreign capital for central government, if not for local communities. Alienation of land for protected areas has contributed to a concentration of human activities so that increased population densities are creating demands for more land and creating conflicts over key resources such as water. Government policy that favours agricultural livelihoods over pastoral strategies has also contributed to increased resource conflict (Armitage, 1996). Uncertainty and change in the policy context of resource management in semi-arid Tanzania adds to the ecological uncertainties under which local communities must attempt to sustain their livelihoods.

Revisiting poverty and sustainable development
This chapter began by highlighting the centrality of poverty reduction in achieving sustainable development. The reciprocal relationship between ecology and people implies that 'environmental management for sustainable development' needs to protect the environment at the same time as

protecting and developing the livelihoods of those people who depend on it. By focusing on approaches to environmental management in semi-arid Tanzania, this chapter has illustrated how environmental policy discourses are defined by the ecological and social paradigms that underpin them.

These different social and ecological paradigms have fundamental implications in terms of resulting management approaches. In semi-arid Tanzania, and East Africa more generally, there could not be a greater difference between policy responses than between the traditional 'people versus environment' policy discourse, which implies a need for de-stocking and the exclusion of people from the land, and the 'people and environment' policy discourse that implies a need to engage indigenous knowledge through community participation in environmental management.

In the past there has been a tendency for environmental managers to focus solely on ecological theories or paradigms when formulating policy approaches. This chapter has highlighted the fact that these ecological paradigms are often accompanied by underlying social paradigms. If environmental management is to be successful in achieving sustainable development, including poverty reduction, it is therefore essential that environmental managers are properly aware of the ecological and social paradigms that define their thinking and that underpin the policy discourses and accompanying management responses that they support. In the absence of such reflexive, critical self-awareness, environmental managers are exposed to the danger of making decisions based on hidden assumptions that may or may not reflect the myriad of subjective ecological and social realities that shape the ecosystem dynamics that they seek to manage (Ockwell, 2008; Ockwell and Rydin, Chapter 6, this volume). Such unreflexive decision-making is unlikely to be able to rise to the challenge of achieving development that ensures a sustainable future for both the environment and the people whose livelihoods depend on it.

Note

1. Maji is Kiswahili for water. During the conflict water was distributed that was believed to protect the drinkers from German weapons. Unfortunately, the water did not work and many lost their lives as a result.

References

Anand, P. (2003), 'Economic analysis and environmental responses', in A. Blowers and S. Hinchliffe (eds), *Environmental Responses*, Milton Keynes: Open University and Chichester: John Wiley and Sons Ltd.

Anderson, D. and R. Grove (1987), 'The scramble for Eden: Past, present and future in African conservation', in D. Anderson and R. Grove (eds), *Conservation in Africa: People, Policies and Practice*, Cambridge, UK: Cambridge University Press, pp. 1–12.

Armitage, D.R. (1996), 'Environmental management and policy in a dryland ecozone: The Eyasi-Yaeda basin, Tanzania', *Ambio*, **25**(6), 396–402.

Begon, M., J.L. Harper and C.R. Townsend (1990), *Ecology: Individuals, Populations and Communities*, 2nd edition, Oxford, UK: Blackwell Science.

Behnke, R. and N. Abel (1996), 'Revisited: The overstocking controversy in semi-arid Africa', *World Animal Review*, **87**(2), 4–27.

Behnke, R.H. and I. Scoones (1993), 'Rethinking range ecology: Implications for rangeland management in Africa', in R.H. Behnke, I. Scoones and C. Kerven (eds), *Range Ecology at Disequilibrium: New Models of Natural Variability and Pastoral Adaptation in African Savannas*, London: Overseas Development Institute, pp. 1–30.

Bourn, D. and R. Blench (eds) (1999), *Can Livestock and Wildlife Co-exist? An Interdisciplinary Approach*, London: Overseas Development Institute.

Briske, D.D., S.D. Fuhlendorf and F.E. Smeins (2003), 'Vegetation dynamics on rangelands: a critique of the current paradigms', *Journal of Applied Ecology*, **40**(4), 601–14.

Brooke, C. (1967), 'The heritage of famine in Central Tanzania', *Tanzania Notes and Records*, No. 67, 51–9.

Brown, J.R. and N.D. MacLeod (1996), 'Integrating ecology into natural resource management policy', *Environmental Management*, **20**(3), 289–96.

Campbell, B.M., A. Mandondo, N. Nemarundwe, B. Sithole, W. De Jong, M. Luckert and F. Matose (2001), 'Challenges to proponents of Common Property Resource systems: Despairing voices from the social forests of Zimbabwe', *World Development*, **29**(4), 589–600.

Carter, N. (2001), *The Politics of the Environment: Ideas, Activism, Policy*, New York: Cambridge University Press.

Clements, F.E. (1916), *Plant Succession: An Analysis of the Development of Vegetation*, Washington, DC: Carnegie Institute Publishers.

Coulson, A.C. (1976), 'Crop priorities for the lowlands of Tanga region', *Tanzania Notes and Records*, Nos. 81–82, 43–53.

Ellis, J.E. and D.M. Swift (1988), 'Stability of African pastoral ecosystems – alternate paradigms and implications for development', *Journal of Range Management*, **41**(6), 450–59.

Giblin, J. and G. Maddox (1996), 'Introduction', in G. Maddox, J. Giblin and I.N. Kimambo (eds), *Custodians of the Land: Ecology and Culture in the History of Tanzania*, London: James Currey.

Griffiths, C.J. (1993), 'The geological evolution of East Africa', in J.C. Lovett and S.K. Wasser (eds), *Biogeography and Ecology of the Rainforests of Eastern Africa*, Cambridge, UK: Cambridge University Press.

Griffiths, J.F. (1972), 'Climate', in W.T.W. Morgan (ed.), *East Africa: Its Peoples and Resources*, Nairobi: Oxford University Press, pp. 107–18.

Gulliver, P.H. (1972), 'Peoples', in W.T.W. Morgan (ed.), *East Africa: Its Peoples and Resources*, Nairobi: Oxford University Press, pp. 35–40.

Hamersley, A. (1972), 'Agriculture and land tenure in Tanzania', in W.T.W. Morgan (ed.), *East Africa: Its Peoples and Resources*, Nairobi: Oxford University Press, pp. 189–98.

Hamilton, A.C. (1982), *Environmental History of East Africa: A Study of the Quaternary*, London: Academic Press.

Hamilton, A.C. (1989), 'The climate of the East Usambaras', in A.C. Hamilton and R. Bensted-Smith (eds), *Forest Conservation in the East Usambara Mountains, Tanzania*, Cambridge, UK: IUCN, pp. 97–102.

HilleRisLambers, R., M. Rietkerk, F. van den Bosch, H.H.T. Prins and H. de Kroon (2001), 'Vegetation pattern formation in semi-arid grazing systems', *Ecology*, **82**(11), 50–61.

Homewood, K. (1995), 'Development, demarcation and ecological outcomes in Maasailand', *Africa*, **65**(3), 331–50.

Homewood, K. and W.A. Rodgers (1987), 'Pastoralism, conservation and the overgrazing controversy', in D. Anderson and R. Grove (eds), *Conservation in Africa: People, Policies and Practice*, Cambridge, UK: Cambridge University Press, pp. 111–28.

Homewood, K. and W.A. Rodgers (1991), *Maasailand Ecology: Pastoral Development and Wildlife Conservation in Ngorongoro, Tanzania*, Cambridge, UK: Cambridge University Press.

Hyden, G. (1975), 'Development of the cooperative movement in Tanzania', *Tanzania Notes and Records*, No. 76, 51–6.

Iliffe, J. (1979), *A Modern History of Tanganyika*, Cambridge, UK: Cambridge University Press.

Illius, A. W. and T.G. O'Connor (1999), 'On the relevance of non-equilibrium concepts to arid and semi-arid grazing systems', *Ecological Applications*, **9**(3), 798–813.

IUCN (1987), *IUCN Directory of Afrotropical Protected Areas*, Gland, Switzerland and Cambridge, UK: IUCN.

Jacobs, B.F., J.D. Kingston and L.L. Jacobs (1999), 'The origin of grass-dominated ecosystems', *Annals of Missouri Botanical Gardens*, **86**(2), 590–643.

Kerario, E. (1996), 'A note on problems observed during the initial stages of the zero-grazing project in the HADO project area of Kondoa', in C. Christiansson and I.S. Kikula (eds), *Changing Environments, Research on Man–Land Interrelations in Semi-arid Tanzania*, Nairobi: Regional Soil Conservation Unit and Swedish International Development Cooperation Agency, pp. 132–5.

Kieran, J.A. (1972), 'History', in W.T.W. Morgan (ed.), *East Africa: Its Peoples and Resources*, Nairobi: Oxford University Press, pp. 21–34

Kikula, I. (1997), *Policy Implications on Environment: The Case of Villagization in Tanzania*, Uppsala: The Nordic Institute and University of Dar es Salaam.

Kikula, I.S. (1999), 'Lessons from twenty-five years of conservation and seven years of research initiatives in the Kondoa Highlands of central Tanzania', *Ambio*, **28**(5), 444–9.

Kingdon, J. (1971), *East African Mammals: An Atlas of Evolution in Africa*, London: Academic Press.

Kjekshus, H. (1996), *Ecological Control and Economic Development in East African History: The Case of Tanganyika 1850–1950*, London: James Currey.

Koponen, J. (1996), 'Population: A dependent variable', in G. Maddox, J. Giblin and I.N. Kimambo (eds), *Custodians of the Land: Ecology and Culture in the History of Tanzania*, London: James Currey.

Leach, M. and R. Mearns (1996), 'Environmental change and policy', in M. Leach and R. Mearns (eds), *The Lie of the Land: Challenging Received Wisdom on the African Environment*, Oxford, UK: The International African Institute in association with James Currey.

Leakey, L.S.B. (1972), 'Earlier prehistory', in W.T.W. Morgan (ed.), *East Africa: Its Peoples and Resources*, Nairobi: Oxford University Press, pp. 9–14.

Lovett, J.C. (1993), 'Climate history and forest distribution in eastern Africa', in J.C. Lovett and S.K. Wasser (eds), *Biogeography and Ecology of the Rainforests of Eastern Africa*, Cambridge, UK: Cambridge University Press.

Lovett, J.C., C.H. Quinn, H. Kiwasila, S. Stevenson, N. Pallangyo and C. Muganga (2001), *Overview of Common Pool Resource Management in Semi-arid Tanzania*, York: Centre for Ecology, Law & Policy, University of York.

Lovett, J.C., S. Stevenson and H. Kiwasila (2002), *Final Technical Report: Review of Common Pool Resource Management in Semi-arid Tanzania*, London: DFID.

Maddox, G. (1996), 'Environment and population growth', in G. Maddox, J. Giblin and I.N. Kimambo (eds), *Custodians of the Land: Ecology and Culture in the History of Tanzania*, London: James Currey.

Maro, P.S. (1978), 'Land: Our basic resource for socio-economic development', *Tanzania Notes and Records*, No. 83, 31–38.

Maro, P.S. and W.I.F. Mlay (1978), 'People, population distribution and employment', *Tanzania Notes and Records*, No. 83, 1–20.

Matzke, G. (1976), 'The development of the Selous Game Reserve', *Tanzania Notes and Records*, No. 79–80, 37–48.

McHenry, Jr, D.E. (1979), Tanzania's Ujamaa Villages: The implementation of a rural development strategy, Berkeley: Institute of International Studies, University of California.

Mortimore, M. (1998), *Roots in the African Dust: Sustaining the Drylands*, Cambridge, UK: Cambridge University Press.

Mung'ong'o, C.G. (1991), *Environmental Degradation and Underdevelopment in a Rural Setting in Tanzania: The Case of Kondoa District*, Dar es Salaam: Institute for Resource Assessment, University of Dar es Salaam.

Mussa, D.T. (1978), 'Our climate: Light, rainfall and temperature patterns', *Tanzania Notes and Records*, No. 83, 49–60.

Mwalyosi, R.B.B. (1995), 'Agro-pastoralism and biodiversity conservation in East Africa: The case of Maasailand, Tanzania', in P. Halladay and D.A. Gilmour (eds), *Conserving Biodiversity outside Protected Areas: The Role of Traditional Agro-pastoralism*, Gland, Switzerland: IUCN.

Neumann, R. (1995), 'Ways of seeing Africa: Colonial recasting of African society and landscape in Serengeti National Park', *Ecumene*, **2**(2), 149–69.

Neumann, R. (1996), 'Dukes, earls and ersatz edens: Aristocratic nature preservationists in colonial Africa', *Environment and Planning D: Society and Space*, **14**(1), 79–98.

Nicholson, S.E. (1993), 'An overview of African rainfall fluctuations of the last decade', *Journal of Climate*, **6**(7), 1463–6.

Nicholson, S.E. (2001), 'Climatic and environmental change in Africa during the last two centuries', *Climate Research*, **17**(2), 123–44.

Nyerere, J. (1969), 'Independence message to TANU', in *Freedom and Unity: A Selection from Writings and Speeches, 1952–65*, Nairobi: Oxford University Press, pp.138–9.

Oba, G., N.C. Stenseth and W.J. Lusigi (2000), 'New perspectives on sustainable grazing management in arid zones of sub-Saharan Africa', *Bioscience*, **50**(1), 35–51.

Oba, G., R.B. Weladji, W.J. Lusigi and N.C. Stenseth (2003), 'Scale-dependent effects of grazing on rangeland degradation in northern Kenya: A test of equilibrium and non-equilibrium hypotheses', *Land Degradation & Development*, **14**(1), 83–94.

O'Brien, E.M. (1993), 'Climatic gradients in woody plant species richness: Towards an explanation based on an analysis of Southern Africa's woody flora', *Journal of Biogeography*, **20**(2), 181–98.

O'Brien, E.M. (1998), 'Water-energy dynamics, climate and prediction of woody plant species richness: An interim general model', *Journal of Biogeography*, **25**(2), 379–98.

Ockwell, David G. (2008) '"Opening up" policy to reflexive appraisal: A role for Q Methodology? A case study of fire management in Cape York, Australia', *Policy Sciences* **41**(4), 263–92.

ODA (1994), *Renewable Natural Resources Research Strategy*, London: Overseas Development Administration.

Peberdy, J.R. (1972), 'Rangeland', in W.T.W. Morgan (ed.), *East Africa: Its Peoples and Resources*, Nairobi: Oxford University Press, pp.153–76.

Pratt, D.J., P.J. Greenway and M.O. Gwynne (1966), 'A classification of East African rangeland', *Journal of Applied Ecology*, **3**(2), 369–82.

Pratt, J. and H. Gwynne (1977), *Rangeland Management and Ecology in East Africa*, London: Hodder & Stoughton.

Quinn, C.H. (2005), 'Coping with ecological uncertainty in semi-arid Tanzania: Livelihoods, risk and institutions', PhD Thesis, University of York, UK.

Quinn, C.H., K. Forrester, H. Kiwasila, C. Muganga and N. Pallangyo (2001), 'Fieldwork report – village profiles', York: Centre for Ecology, Law & Policy, University of York.

Quinn, C.H., H. Kiwasila, M. Huby and J.C. Lovett (2007), 'Design principles and Common Pool Resource management: An institutional approach to evaluating community management in semi-arid Tanzania', *Journal of Environmental Management*, **84**(1), 100–13.

Ramanathan, V., P.J. Crutzen, J.T. Kiehl and D. Rosenfeld (2001), 'Aerosols, climate and the hydrological cycle', *Science*, **294**(5549), 2119–24.

Roderick, M.L. and G.D. Farquhar (2002), 'The cause of decreased pan evaporation over the past 50 years', *Science*, **298**(5597), 1410–11.

Sere, C., H. Steinfeld and J. Groenewold (1995), *World Livestock Production Systems: Current Status, Issues and Trends*, Rome: Food and Agriculture Organization of the United Nations.

Shorter, A. (1974), *East African Societies*, London: Routledge and Kegan Paul.
Smith, C.D. (2001), *Ecology, Civil Society and the Informal Economy in North West Tanzania*, Aldershot: Ashgate.
Snyman, H.A. and H.J. Fouche (1993), 'Estimating seasonal herbage production of a semi-arid grassland based on veld condition, rainfall and evapotranspiration', *African Journal of Range & Forage Science*, **10**(1), 21–4.
Stanhill, G. and S. Cohen (2001), 'Global dimming: A review of the evidence for a widespread and significant reduction in global radiation with discussion of its probable causes and possible agricultural consequences', *Agricultural and Forest Meteorology*, **107**(4), 255–78.
Stocking, M. (1996), 'Soil erosion: Breaking new ground', in M. Leach and R. Mearns (eds), *The Lie of the Land: Challenging Received Wisdom on the African Environment*, Oxford, UK: The International African Institute in association with James Currey, pp. 140–54.
Stott, P. and S. Sullivan (2000), 'Introduction', in P. Stott and S. Sullivan (eds), *Political Ecology: Science, Myth and Power*, London: Arnold, pp. 1–15.
Sullivan, S. and R. Rohde (2002), 'On non-equilibrium in arid and semi-arid grazing systems', *Journal of Biogeography*, **29**(2), 1595–1618.
Trapnell, C.G. and I. Langdale-Brown (1972), 'Natural vegetation', in W.T.W. Morgan (ed.), *East Africa: Its Peoples and Resources*, Nairobi: Oxford University Press, pp. 127–40.
Van Keulen, H. and N.G. Seligman (1992), 'Moisture, nutrient availability and plant production in the semi-arid region', in T. Alberda, H. Van Keulen, N.G. Seligman and C.T. de Wit (eds), *Food from Dry Lands: An Integrated Approach to Planning of Agricultural Development*, Dordrecht: Kluwer Academic Publishers, pp. 25–81.
Walter, A. (1939), 'A note on the seasonal rains in East Africa and their causation', *Tanganyika Notes and Records*, No. 8, 21–6.
Warren, A. (1995), 'Changing understandings of African pastoralism and the nature of environmental paradigms', *Transactions of the Institute of British Geographers*, **20**(2), 193–203.
WCED (1987), *Our Common Future*, Oxford, UK: Oxford University Press.
Wenner, C.G. (1983), 'Soil conservation in Tanzania: The HADO project in Dodoma Region', a report on a visit in April–May 1983, Dar es Salaam, SIDA.
White, F. (1983), *The Vegetation of Africa: A Descriptive Memoir to Accompany the UNESCO/AETFAT/UNSO Vegetation Map of Africa*, Paris: UNESCO.

10. Exploring game theory as a tool for mapping strategic interactions in common pool resource scenarios
Vanessa Pérez-Cirera

Introduction

The objective of this chapter is to introduce game theory as an analytical tool for understanding and mapping strategic interactions amongst individuals and institutions in the management of common pool resources.

The chapter on the economics of common property resources explores the relation between poverty and property rights in natural resource management and emphasizes the role of transaction costs in the governance structure of common property systems and how these costs shape the outcome of these systems (Adhikari, Chapter 5, this volume). The chapter on the economic valuation of the different forms of land-use in Tanzania emphasizes the importance of identifying and incorporating non-marketed/non-priced values of environmental goods and services and how such valuations can be undertaken so that optimal levels of land-use are identified (Kirby, Chapter 11, this volume). A question that remains open is if and how these optimal solutions can be reached. This section intends to contribute towards this broad question by introducing game theory as a useful analytical tool that helps us understand how decision-making processes are made in the management of common pool resources. The review explains how strategic decision-making can be mapped in a game-theoretic fashion so that variables that are key for arriving at socially optimal solutions can be identified.

The first section of the chapter gives an introduction to game theory as a method for the construction of game-theoretic models, introducing the reader to game-theoretic language and representation forms. The second section will review the most frequently used games for depicting problems encountered in CPR settings and illustrate the use of game theory in analysing binding agreements as institutional solutions to CPR dilemmas. The last section will aim to illustrate how game theory can be applied to understanding decision-making processes and assessing the desirability and viability of policy options with illustrations from semi-arid Tanzania.

Game theory; language and representation forms

Game theory is a mathematical theory used to understand the outcomes of strategic interactions in terms of quantifiable gains and losses from different decisions, to mostly competing players, as if they were playing a game. As a branch of applied mathematics, game theory started to be used as a framework for analysis after the publication of Von Neumann and Morgenstern's (1944) *Theory of Games and Economic Behavior*. As such, the main use of game theory was in economics through the early 1970s, when its use started to spread to other sciences.[1]

There are two major strands within game theory: cooperative and non-cooperative game theory. Cooperative game theory deals with situations in which the players can negotiate before the game is played on what to do in the game. It is assumed that these negotiations result in signing a binding agreement (Binmore, 1994). For this type of game, what is important is not so much the strategies available to the players, but the preference structure of the game itself, since this is what determines which contracts are feasible.

Non-cooperative game theory is based on a different set of principles. Non-cooperative game theory calls for a complete description of the rules of a game so that the strategies available to the players can be studied in detail. The objective is to find a pair of equilibrium strategies to be designated as the solution of the game and this solution might or might not be cooperative (Binmore, 1994). This strand of game theory gives space to agreements. However, these are not conceived as necessarily binding. Agreements can be made after or before a game is played and, depending on how the payoffs and strategies available change, agreements can or cannot be sustained.

At this stage, it is important to point out that game theory is not the same as game models (Snidal, 1985). Game theory sets out an analytical framework for the construction of game models, not for their design. The results of the game will change depending on the way the model is conceived. This means that one could construct a model and give different features, for example, increase the number of players, and the result would be different. Game theory thus can be conceived as a 'metalanguage for [the construction of] game-theoretic models' (Ostrom et al., 1994, p. 24).

As most of the games used for depicting CPR problems emanate from non-cooperative game theory, the next section intends to outline some of the key concepts used in non-cooperative game theory to familiarize the reader with game-theoretic language and forms of representation.

Figure 10.1 Example of an extensive form game

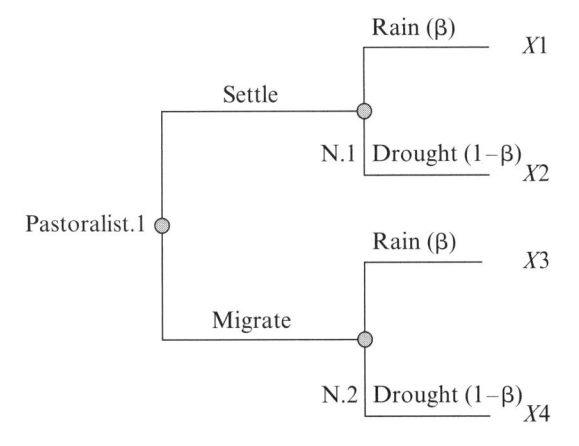

Source: After Bardhan (1993).

Non-cooperative extensive form games
Non-cooperative games can be depicted in two forms: extensive and strategic forms. In the extensive form, the rules of the game are laid out in full detail by drawing a tree (Figure 10.1). To illustrate the way in which extensive games are constructed, consider a simplified decision problem faced by an agro-pastoralist. He or she can decide whether to settle in an area of land, or to migrate. Her or his decision will depend on, among other things, the rainfall patterns she or he expects to encounter in different places. Consider then the following game tree (extensive form game) with two players (player 1 and player 2). Player 1 will represent the agro-pastoralist (P) and player 2 will represent nature (N).

Pastoralist.1 represents the node at the root of the tree from which the first available strategies follow. The strategies are named branch lines in this representation form, and depart from the small circles called nodes. In this game, the two available strategies for the agro-pastoralist are to settle in the current land or to migrate and move to another place. The next nodes of the game N.1 and N.2 represent two different nodes for nature. When two nodes are connected with a dotted line, which is not the case in this example, it is said that both nodes are in the same information set. That is, there is incomplete information about the actions of the last player. Otherwise, as we move forward in the tree from left to right, information about the last player's moves is gained along the way.

In this game, the two available strategies for nature are rain or drought. In reality the level of rainfall could fall at different levels along a continuum ranging from no rain at all to a lot of rain, but to keep this game simple, consider these two extremes. The two nodes for nature (N.1 and N.2) are chance decision nodes as there is uncertainty on what nature will do. In chance decision nodes, all branches deriving from them have a probability distribution equal to 1 (for example, $\beta + 1 - \beta = 1$). The agro-pastoralist can have some information on these probabilities, such as what the rainfall was last year, however, there is still uncertainty on what in fact will happen.

The letters or numbers, in this case X's at the terminal nodes, represent the payoffs[2] of each decision path. These payoffs will depend on the utility functions of the players in the game. In this case only the actions of nature will have an effect on the payoffs for the agro-pastoralist and not vice versa. Therefore, only one payoff is sketched. In the case that two or more players have a payoff contingent on the final scenario, these payoffs will be written at the end of the terminal nodes separated by a comma, the first letter or number being the payoff for player 1, the second letter or number the payoff for player 2 and so forth. In the game considered here, if the agro-pastoralist settles and it rains, she or he will receive $X1$, if he or she settles and there is drought, she will receive $X2$ and so on.

In order to find a solution for this game, we have to compute the expected payoffs at each node, starting from the right and moving to the left. This commonly used method is called solving by backward induction. In this game we thus have to compute the expected payoffs at N.1 and N.2, that is, the expected payoffs from settling and moving respectively. To compute the expected payoffs of a decision under uncertainty, one has to multiply the payoff by the probability of each payoff and add them.[3] For example, the expected payoff from settling in this game would be: $Ep(S) = \beta(X1) + (1 - \beta)(X2)$. The same would have to be done for the other available strategies for player 1. These two payoffs would have to be computed and compared in order to derive a preferred strategy.[4] To identify the preferred strategy, an arrow in the branch that represents the preferred strategy can be drawn. To make this clearer, numerical payoffs to this example are given below.

Consider the hypothetical payoffs for the agro-pastoralist from each of her or his available strategies as those depicted in Figure 10.2. These payoffs would have to include all benefits and costs (for example, monetary costs from moving, opportunity costs, and so on) derived from each of her or his strategies given the actions of nature. That is, the payoffs reflect the net benefits of all contingent scenarios.

Given the payoffs from Figure 10.2, the expected payoff for the agro-pastoralist from settling would be:

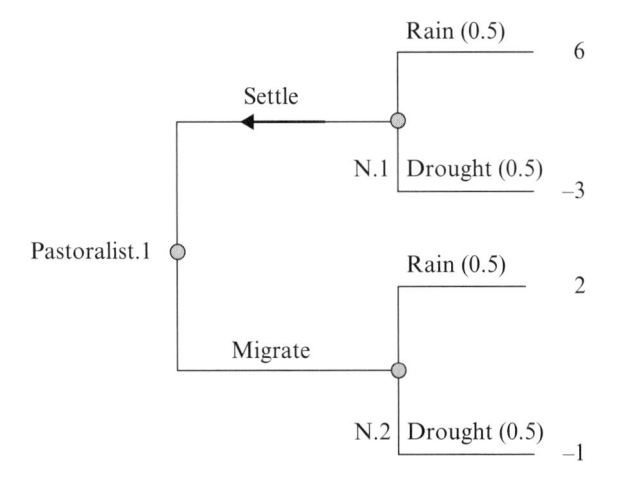

```
                              Rain (0.5)
                         ┌──────────────── 6
                 Settle  │
              ←──────────┤
                         │ N.1 Drought (0.5)
                         └──────────────── −3

Pastoralist.1 ○
                                Rain (0.5)
                         ┌──────────────── 2
                 Migrate │
              ───────────┤
                         │ N.2 Drought (0.5)
                         └──────────────── −1
```

Source: After Bardhan (1993).

Figure 10.2 The 'migration game' with numerical payoffs

$$Ep(S) = 6(0.5) - 3(0.5) = 3 - 1.5 = 1.5$$

Conversely, the expected payoff from migrating would be:

$$Ep(M) = 2(0.5) - 1(0.5) = 1 - 0.5 = 0.5$$

Given these payoffs and probabilities, the best strategy for the agro-pastoralist would be to settle since the expected payoff from settling will be higher than the expected payoff from moving ($Ep(S) > Ep(M)$). An arrow on the 'settle' branch denotes this.

Now, let us consider another approach to this game. What if the agro-pastoralist is not sure about the probabilities of drought or rain? How much would the expected probability have to fall or rise in order for the chosen strategy to still be preferred? To find a solution, we need to compute the same expected payoffs and solve for β:

$$Ep(S) = 6\beta - 3(1 - \beta) = 9\beta - 3$$

And, $$Ep(M) = 2\beta - 1(1 - \beta) = 3\beta - 1$$

For settling to be a preferred strategy, $Ep(S) > Ep(M)$, thus:

$$9\beta - 3 > 3\beta - 1$$

314   *A handbook of environmental management*

Solving for the past inequality, β would have to be higher than 2/6 for settling to be the preferred strategy for the agro-pastoralist. This means that for all probabilities of rain higher than 0.33 it will always be best for the agro-pastoralist to settle.

Clearly this is a simplified view of the available decisions to the agro-pastoralist and of the variables that will shape his or her payoffs. However, once familiar with the way games can be constructed and solved, and the interactions amongst several players, games illustrating real-life situations can be constructed. The next section will introduce the other, usually more often used, form of game in non-cooperative game theory, the 'normal' or strategic form game.

Non-cooperative strategic form games
The other form in which non-cooperative games are depicted is the normal or strategic form (Figure 10.3). This form is useful to depict interactions amongst two or more players in which their payoffs are shaped on what the other player does and for illustrating simultaneous decision-making.

Consider two agro-pastoralists or two groups of agro-pastoralists: agro-pastoralist 1 (player 1) and agro-pastoralist 2 (player 2). In normal/strategic form games, the payoffs for each of the players are depicted within the boxes of the matrix (Figure 10.3). In some normal form representations, the payoffs are written in a parenthesis with the first number being the player 1 payoff and the second number or letter in the parenthesis, the player 2 payoff. In other normal form representations, for example the game in Figure 10.3, the payoffs are depicted at the corners of each

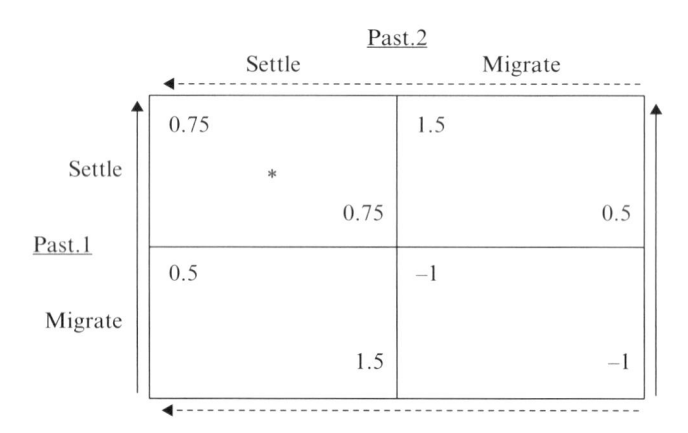

	Past.2 Settle	Past.2 Migrate
Past.1 Settle	0.75 * 0.75	1.5 0.5
Past.1 Migrate	0.5 1.5	−1 −1

Source: After Bardhan (1993).

Figure 10.3   A normal 2 × 2 form of the revisited 'migration game'

Exploring game theory 315

box within the matrix, the payoff to the left top corner of the box being the payoff for player 1 (agro-pastoralist 1) and the one located at the right bottom of the box, being the payoff for player 2 (agro-pastoralist 2). One of the players plays the columns and the other player plays the rows. The names of the players and available strategies are written at the top of the columns and at the side of the rows, respectively.

Drawing on the example presented in the past section, suppose that the payoffs for the agro-pastoralists are not only dependent on what she or he thinks nature will do, but on what other agro-pastoralists do. Suppose then that the payoffs from settling for the agro-pastoralist will be shaped by what other agro-pastoralists do and vice versa. For example, if land is scarce then if both agro-pastoralists settle in the same area the benefits for each agro-pastoralist will be reduced. If both agro-pastoralists decide to settle in the area (settle, settle), the payoff for each agro-pastoralist will no longer be 1.5 as derived in the past example. If both agro-pastoralists decide to stay, each will only receive 0.75. If only one of them decides to migrate, the one who settles will receive 1.5 while the one who migrates will receive 0.5. If, on the other hand both players decide to migrate, each of them will receive -1. That is, both players would carry the costs and risks of migrating and the fact that both will graze in the same area imposes a burden on one another.

In order to derive the contingent or best response strategies for each player given the payoffs for the game and what the other player does, let us start with player 1. What would the best response for agro-pastoralist 1 be if she or he thinks agro-pastoralist 2 will settle (left column)? Given that strategy by player 2, the best response strategy for player 1 is to settle since $0.75 > 0.5$ (the two payoffs at the left top corner of the two boxes in the left column). If, on the other hand, agro-pastoralist 2 migrates (right column), the best response for agro-pastoralist 1 would be to settle since $1.5 > -1$. These best response strategies are illustrated by arrows at the sides of the matrix. The vertical arrows illustrate agro-pastoralist 1's best response strategies for this game. In this game, player 1 (agro-pastoralist 1) has a dominant strategy. A dominant strategy is a strategy that given the rules and payoffs of the game, in any specific node, is preferred by a player no matter what the other player does.

The procedure followed with player 1 now needs to be done with player 2. What would the preferred strategy for player 2 be, given the actions of the other player? If player 1 decides to settle (top row), player 2's best response strategy would be to settle given that $0.75 > 0.5$ (bottom right payoffs from boxes in the top row). If, on the other hand, player 1 decides to migrate (bottom row), the best response strategy for player 2 will still be to settle, since $1.5 > -1$ (bottom right payoffs from boxes in the bottom row). Thus, the dominant strategy for player 2 will also be to settle. The horizontal arrows represent the best response strategies for player 2.

When a game has at least one of a player's arrows pointing to a box and at least an arrow of the other player pointing to the same box, then the pair of strategies is said to be in a Nash equilibrium.[5] A Nash equilibrium is thus defined as any pair of strategies with the property that each player maximizes her or his payoff given the actions of the other player. In this case, the Nash equilibrium for this game is settle/settle.

There are additional names for specific strategies and equilibria. A pure strategy is one that does not involve chance (probabilities). And a pure Nash equilibrium is the equilibrium reached when each player plays a pure strategy. In the game illustrated in Figure 10.3, there is only one Nash equilibrium (S,S) and this is a pure Nash equilibrium, since it does not involve probabilities. Another type of strategy, not present in this game, is mixed strategies. A mixed strategy requires a player to randomize her or his pure strategies in order to keep the opponent guessing. Consequently, a mixed equilibrium is the equilibrium reached when each of the players is playing a mixed strategy.

Two further equilibrium notations are used for specific purposes in game theory. Symmetric equilibrium, in which every player chooses the same strategy and an asymmetric equilibrium in which at least two players choose different strategies. When there is only one equilibrium in the game, this is called the unique Nash equilibrium and when there is more than one equilibrium, it is said that the game has multiple Nash equilibria. In the game presented in Figure 10.3, there is a unique pure symmetric Nash equilibrium. This equilibrium is represented with a star in the middle (note the star in the middle of the box settle/settle), though there are many other identifiable forms for illustrating where the Nash equilibria are.

Having presented a brief review of game-theoretic language and representation forms, next section will review the game-theoretic approach to CPRs.

Game theory, common pool resources (CPRs) and common pool institutions (CPIs)[6]
Game theory has been extensively used as a framework for analysis of CPR problems (see Ostrom et al., 1994; Baland and Platteau, 1996). Before presenting the game-theoretic approach to common pool resources (CPRs), it is important to review the way CPRs have been conceived and the problems arising in the management of these resources.

Common pool resources have been mainly explained in terms of the physical attributes of the goods or resources. CPRs are considered to share two characteristics: (1) the difficulty of excluding individuals from benefiting from the resource and (2) the subtractability of the benefits consumed by one individual from those available to others (Ostrom et al., 1994). These characteristics have been considered to generate two broad problems: that of appropriation and that of provision. The first relates to extracting more than the socially optimal level.[7] The difficulty of excluding others from the use of a CPR creates the problem of effectively limiting use (Ostrom et al., 1994). In the case of renewable natural resources, such as a common property forest land, for example, the concern is that the flow extracted will not exceed its regeneration rate. The second problem refers to the difficulty of investing for the 'adequate' provision of the CPR or in activities related to its improvement or maintenance. Both these problems can be considered to generate negative consequences for other appropriators/users.

Commonly used games for depicting CPR problems
Three main games have been used for depicting CPR problems. These games are: (1) the prisoner's dilemma game[8] (2) the chicken game[9] and (3) the assurance game.[10] The prisoner's dilemma game has been widely used to represent the appropriation problems encountered in CPRs. When limits for resource use cannot be established, all users of the CPR will want to use as much as possible so as to maximize their own profit, resulting in over-appropriation of the resource. In the case of natural resources this over-appropriation can lead to over-exploitation.

The prisoner's dilemma game is probably the most well-known game with two players each having two strategies (Figure 10.4). The two available strategies for each player are to cooperate or to defect. The payoffs considered for this game are such that $c > a$, $d > c$ and $a > d$. Thus, given the payoffs for this game, the best response strategy for each player is to defect. This game has a unique Nash equilibrium that is: defect, defect. Here is where the dilemma lies, since the players could both be better off if both chose to cooperate (C) simultaneously since $a > d$.

In the case of resource provision, the problem arises with free-riders benefiting from the resource (as a public good) provided without having contributed to its provision. For example, time investments for the monitoring and enforcement of communitiy forestry or a commonly managed natural reserve. This problem has been modelled, depending on the provision technology, with the prisoner's dilemma or with the assurance game, the former having no one contributing to the resource provision (Figure 10.4) and the latter having players contributing if, and only if, the others are to contribute (Figure 10.5). The assurance game for which the payoffs are such that $a > c$ and $d > b$ (Figure 10.5) thus depicts situations in which one person's contribution is not enough to gain a collective benefit, but the contribution of both players will result in a joint benefit. That

318 *A handbook of environmental management*

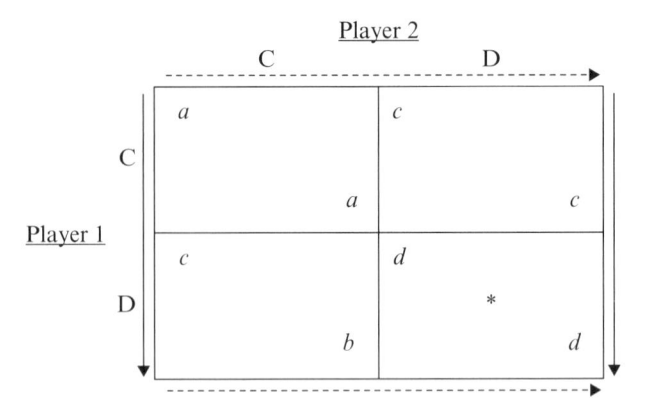

Source: Adapted from Ostrom et al. (1994, p. 53).

Figure 10.4 Adaptation of the prisoner's dilemma game (c > a, d > c and a > d)

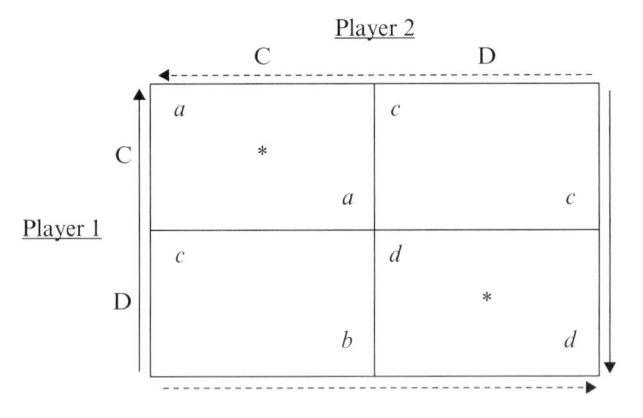

Figure 10.5 Adaptation of the assurance game (a > c, c > d and d > b)

is, both players would be willing to cooperate if, and only if, the other player cooperates. As can be seen in Figure 10.5, this game has multiple equilibria.

Lastly, the chicken game has been used to illustrate assignment problems when there are different resource locations with different richness. Consider, for example, two fishing grounds with different distances to a protected area. This game has multiple equilibria, as does the assurance game. If one of the locations is much better than another, both players will want to use the same location (D,D) and this might not be optimal

Exploring game theory 319

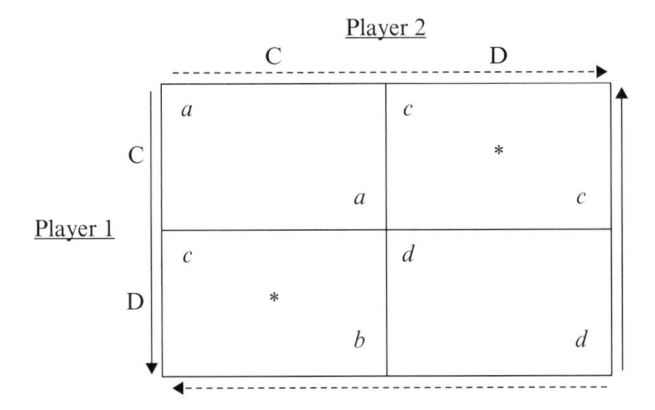

```
                        Player 2
                   C              D
           ┌──────────────┬──────────────┐
           │ a            │ c            │
           │              │       *      │
         C │              │              │
           │              │              │
           │          a   │          c   │
Player 1   ├──────────────┼──────────────┤
           │ c            │ d            │
           │      *       │              │
         D │              │              │
           │              │              │
           │          b   │          d   │
           └──────────────┴──────────────┘
```

Source: Adapted from Ostrom et al. (1994).

Figure 10.6 Adaptation of the chicken game (c > a, c > d and b > d)

for the group as a whole. If both locations are equally good (Figure 10.6), players will be indifferent and will play a mixed strategy leading either to one (C,D) or the other equilibrium (D,C).

Due to the attributes of CPRs, individuals jointly using CPRs are many times stuck in a dilemma. Without previous communication and the lack of institutions (norms and conventions), agents pursuing their own interests will engage in non-rational collective outcomes that result in resource degradation. Institutional choice theorists have focused on identifying viable institutional alternatives by which appropriators of the CPR can choose (1) how much, when, where and with what technology to withdraw units and (2) how much, when and where to invest in the maintenance of the CPR (Ostrom et al., 1994). The next section outlines the game-theoretic logic for such viable institutional solutions.

Institutional solutions to CPR problems
While the use of the prisoner's dilemma game was for many years mistakenly applied to represent all CPR situations (for a critic see Ostrom, 1990; Ostrom et al., 1994), as mentioned in the previous section, it can be very useful for illustrating appropriation and provision problems encountered when managing CPRs.

Consider the prisoner's dilemma game again, now in the extensive form, with the payoffs thought out by Dawes (1973). As we have seen, the prisoner's dilemma structure games have a unique Nash equilibrium which is (D,D). The game in Figure 10.7 depicts this equilibrium in a square with the resulting payoffs of 0,0 respectively.

320   *A handbook of environmental management*

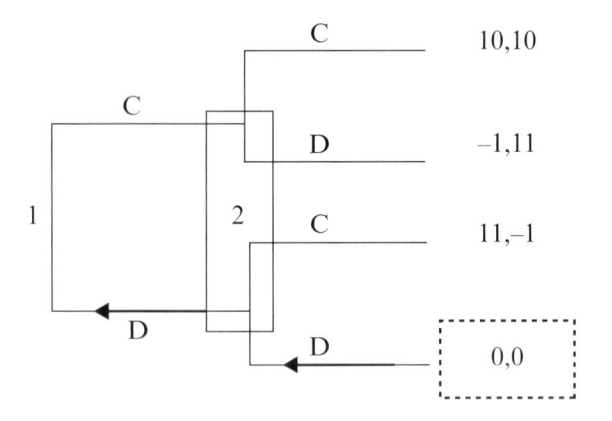

```
                    C ─────── 10,10
           C ┌─────┤
             │     └ D ─────── −1,11
    1 ───┬───┤ 2
         │   │     ┌ C ─────── 11,−1
         │   └─────┤
         ←── D     └ D ─────── ⸤ 0,0 ⸥
```

Source: Dawes (1973).

Figure 10.7 Adaptation of the prisoner's dilemma game in extensive form and numerical payoffs

Now, consider that this game is not finite. That is, there is not only one round to the game, but that the game is played over and over again. This leads to the alternative game set by Ostrom (1990) following the structure of the prisoner's dilemma game but adding two branches for each player at the beginning of the game. These branches are termed: agree (A) or do not agree (~A). That is, players can get together and arrange a contract in order to arrive at a joint payoff (Figure 10.8).

In this game, the issues are: (1) whether the contracts can be binding, that is, there is unfailing enforcement and (2) whether the payoffs are such that the transaction costs of enforcing the contract are lower than the expected gains from the contract. A binding contract is interpreted in non-cooperative game theory as one that is unfailingly enforced by an external actor. The cost of enforcing the agreement is denoted in Figure 10.8 by e.

Given the payoffs and structure of the game, the dominant equilibrium is having both parties agreeing to cooperate. The equilibrium is shown in the rectangle at the bottom of the tree. If these strategies are followed, players will receive the joint benefit from cooperation minus the shared enforcement costs ($10 - e/20$, $10 - e/20$). Note that this equilibrium will hold only if joint enforcement costs (e) are lower than 20.

Encouraged by this and other parallel work, over the past few decades theorists have explored the relationships between rules, institutions, property rights and resource user characteristics for arriving at cooperative outcomes that can enhance collective benefits. The next section will illustrate the use of game theory for identifying key variables and levels of

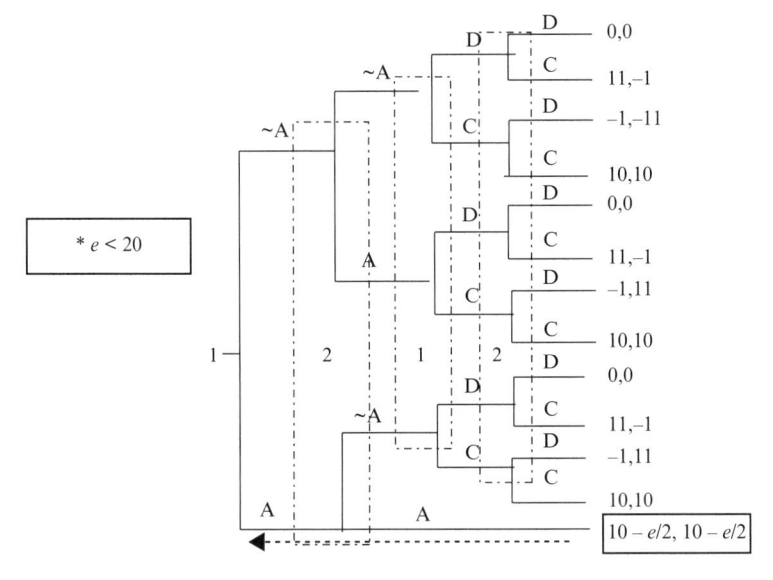

Source: Ostrom (1990, p. 15).

Figure 10.8 Example of a self-financed contract enforcement game

variables that can enhance cooperation and for assessing the viability and desirability of policy options using examples from common pool resource management in semi-arid Tanzania.

Worked examples from semi-arid Tanzania
Much of the Tanzanian land area is still under open access or some form of common property or management arrangement (Quinn and Ockwell, Chapter 9, this volume).

Two different problems will be taken as examples to illustrate the use of game theory for understanding decision-making processes and assessing the viability and desirability of policy options in the management of common pool resources. The first will be the problems encountered by villages that face soil erosion on communally managed land. The second will be the problems faced by communities living next to wildlife conservation reserves and who are prevented from using resources within the reserve.

Erosion prevention/restoration and cooperation in common pool scenarios
Semi-arid ecosystems are characterized by highly variable rainfall patterns, such as short, intense storms, and high evapotranspiration rates. In several areas of semi-arid Tanzania, there are serious problems of land

322   *A handbook of environmental management*

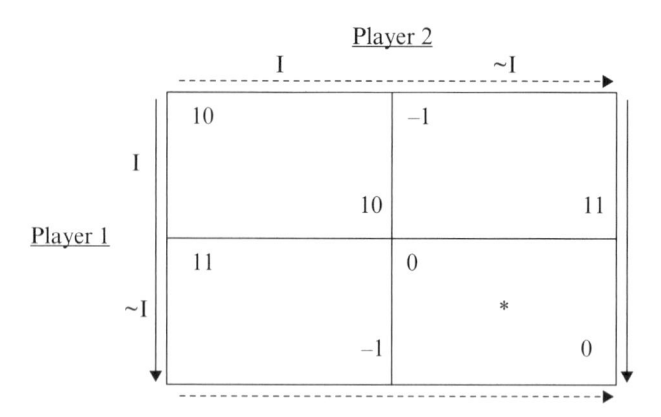

```
                         Player 2
                I                      ~I
         ┌──────────────┬──────────────┐
         │ 10           │ -1           │
       I │              │              │
         │         10   │         11   │
Player 1 ├──────────────┼──────────────┤
         │ 11           │ 0            │
      ~I │              │          *   │
         │         -1   │         0    │
         └──────────────┴──────────────┘
```

Figure 10.9   Adaptation of a prisoner's dilemma structure game for erosion restoration/prevention investments in common pool scenarios

degradation and soil erosion affecting the already low living standards of people living in those areas.

Consider an area with a high degree of land erosion. Erosion can depend on natural factors such as rainfall patterns, wind and topography but can also result from human activities such as deforestation, clearing for cultivation, overgrazing, extensive fuel wood cutting, and so on. Erosion in turn affects the productivity of the natural resource base. The negative impacts of erosion can be ameliorated by investments in the common property land such as afforestation, the construction of diversion ditches, ridge banking, de-stocking, and so on. The implementation of these activities would provide joint benefits for all users; however, problems of provision can arise with free-riders benefiting from the resource without having contributed efforts towards these activities.

Recall that provision problems in CPRs can be modelled, depending on provision technology, with the prisoner's dilemma or with the assurance game, the former having no one contributing to resource provision and the latter having players contributing if, and only if, the others are to contribute. These two games are illustrated in Figures 10.9 and 10.10.

The individual profits from realizing the investments are the individual benefits derived from the investment, minus the costs incurred in the investment, including the opportunity costs of time devoted to the activity. Realization of these investments will depend on having the flow of money or resources required for realizing the investment, but will also depend on what an individual thinks other members of the village will do.

Consider the payoffs and strategies of Figure 10.9. Two different

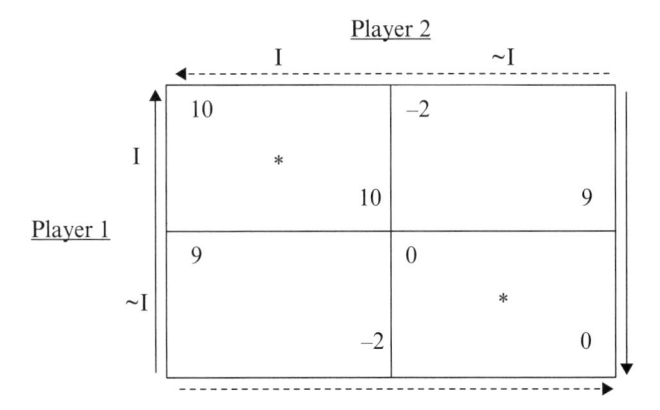

Figure 10.10 Adaptation of an assurance structure game for erosion restoration/prevention investments in common pool scenarios

members of the community are represented as player 1 and player 2 respectively, each having two possible strategies: to invest (I) or not to invest (~I) in restoration/prevention practices. If both players invest in the improvement of the CPR, there will be a joint benefit of 20. However, when maximizing individual payoffs, the resulting equilibrium is no one contributing to the investment (~I, ~I), thereby deriving a joint benefit of zero.

Consider now that the technology is such that it is convenient to contribute only if the other player is to contribute. As seen in the past section, these situations can be modelled by an assurance structure game for which there are two possible equilibria. That is, when a player takes the initiative to invest (I), the best the other player can do is go and help him or her (I), since by doing so her or his payoffs will increase. However, if nobody takes the initiative, there will be no investment (~I). The result, thus, is either investment by both parties (I,I) or non-investment by both parties (~I,~I) as illustrated in Figure 10.10.

As seen in the previous section, there are institutional alternatives by which resource users can organize and arrange for the collective gain to be achieved, transaction costs permitting. The question that remains is the effectiveness of village governance. If communal institutions are weak, should there be a role for third parties, such as the government or NGOs entering into co-management arrangements for the fostering of such agreements? Among the ways in which the fostering of agreements could be effected are providing the initial money for realizing investments, improving information to stakeholders on the benefits that can be derived from agreements and/or reducing the transaction costs of enforcing agreements.

Protected areas, eviction and encroachment, game theory as an analytical tool for assessing the viability of policy options

A central policy in Tanzania has been the creation of protected areas as part of a broader strategy aiming to promote biodiversity conservation and the meeting of local and national development needs. The viability of these policies cannot be isolated from strategic interactions and decisions made by different individuals and institutions at the local level. As will be illustrated in this section, game theory can also be useful in assessing the viability and desirability of policy options in the context of strategic interactions.

Until now, we have considered games made under the assumption that players are symmetric, facing the same decision alternatives and being in the same position, for example, both players being appropriators and facing cooperate or defect strategies. These symmetries, however, can be relaxed to construct games in which the strategies do not need to be the same, nor the players to be assigned to the same positions.

Let us first try to map a decision problem faced by communities who are living adjacent to a reserve and who are completely prevented from the use of resources within the reserve.

Consider first a hypothetical scenario illustrated by Figure 10.11, where M.1 represents the first node for a member or members of the reserve adjacent communities and Ch.1 represents a chance decision node with a probability distribution equal to 1. M.1 has two available strategies: to trespass (t) and use the resources within the area or not to trespass (~t). If the user trespasses and is not detected (~D) he or she will receive a benefit (B) from the additional resources within the reserve (for example, grazing land, water, wood, bushmeat, medical plants, and so on) with a probability of $(1 - \alpha)$. B will be shaped by the availability, quality and dependence of resources inside the reserve relative to those

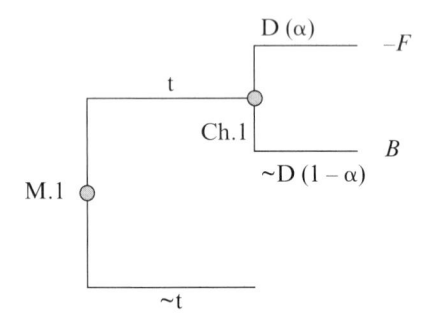

Figure 10.11 The trespassing game

outside the reserve. If, on the other hand, the member is detected (D) she or he will carry the cost of the fine ($-F$) with probability α. If the member does not trespass she or he will remain under status quo (SQ) conditions.

Let us derive the expected probabilities for the unique decision node for the members of the reserve adjacent community:

$$Ep(t) = \alpha(-F) + (1-\alpha)B = -\alpha F + B - \alpha B = -\alpha(F+B) + B$$

$$Ep(\sim t) = 0$$

Thus, for not trespassing to be a preferred strategy either:

$$(1)\ \alpha > B/(F+B)\ \text{or}$$

$$(2)\ F > B(1-\alpha)/\alpha\ \text{or}$$

$$(3)\ B < \alpha F/(1-\alpha)$$

From the previous set of equations, there are four ways of making not trespassing to be a preferred strategy. These are either to:

- increase the probability of getting caught to the level shown in the equation; or
- increase the level of the fine to the level shown in the equation; or
- decrease the level of benefits to the level shown in the equation; or
- a combination.

The use of game theory is not a substitute for assessing side-effects from each of the available strategies. These can be built up in other rounds of the game or subsequent games. Possible side-effects from each of these policy options are, for example, increasing the probability of getting caught would mean having to invest more resources in monitoring activities, increasing the level of the fine might foster corruption and so on. The usefulness of this example rests in illustrating a situation to identify the key variables shaping a decision and their levels.

Let us now consider that complete exclusion is not the only scenario into which reserve adjacent communities can be put. The state, or the managers of the reserve, could have a continuum of possibilities ranging from uncontrolled access to total exclusion (Figure 10.12).

Merging Figure 10.11 and 10.12, we could construct a game with two players: the state (or managers of the reserve) and reserve-adjacent

326 *A handbook of environmental management*

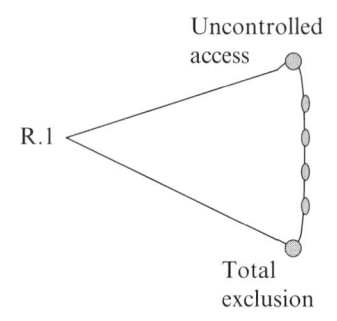

Figure 10.12 Reserves and the continuum of resource use allowances

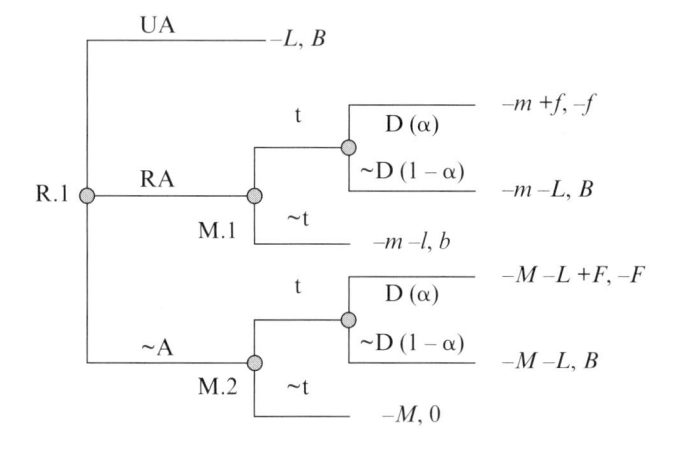

Figure 10.13 The trespassing game with two players

communities (Figure 10.13). R.1 is the first and only decision node for the reserve managers (or the state) in this game and M.1 and M.2 are two different nodes for the members of the resource-adjacent communities. For simplicity consider the three available strategies for the managers of the reserve to be (1) uncontrolled access denoted by UA, (2) regulated access denoted by RA and, (3) total exclusion denoted by ~A (no access). For the adjacent communities, consider the same available strategies at each decision node as those in Figure 10.11.

If the state decides to impose a policy of unregulated access, the reserve will bear a cost from the loss of resources ($-L$) and the adjacent communities will receive a benefit (B) from unregulated access to additional resources within the reserve.[11] If the state decides to regulate access, allowing uses that do not threaten the ecological stability of the reserve, the adjacent communities have two options, to trespass and carry out activities that are not allowed (t) or not to trespass (~t). If they trespass (t), and are detected (D), the reserve will bear the costs of monitoring illegal activities (−m) and will receive the level of the fine (f), whilst the adjacent communities will bear the cost of the fine (−f). If community members are not detected (~D), the reserve will bear the costs of monitoring (−m) anyway and the total losses from activities (L). In this case, the members of the adjacent communities will receive the benefits from the additional resources (B). If the communities do not trespass (~t), the reserve will bear the cost of monitoring (−m) plus a smaller loss (−l) due to allowed activities. The communities in this case will receive a smaller benefit due to the allowed activities (b). In the third strategy, if no access is allowed (~A), the reserve will carry the costs of monitoring for all types of activities (−M).[12] The adjacent communities, if trespassing and detected, will receive the level of the fine (−F), minus the loss from additional resources (−L). If members are not detected (~D), the reserve will bear the no access monitoring costs (−M) minus the loss from resources (−L). Communities in this case will receive the additional benefits (B). If adjacent communities do not trespass, the reserve will bear the monitoring costs of no access (−M) and the adjacent communities will remain under status quo conditions (SQ).

Let us give hypothetical numerical payoffs to solve for this game. Suppose that:

L = 180   B = 150   m = 20   f = 20   l = 20   b = 80
M = 30   F = 40   α = 0.5

Solving by backward induction, in M.1 the best the member can do is not to trespass since her or his expected payoff from not trespassing is higher that that of trespassing:

$$Ep(\sim t) = 80 > Ep(t) = -20(0.5) + 150(0.5) = -10 + 75 = 65$$

If the member were in M.2, the best she or he can do is to trespass since the expected payoff from trespassing is higher than that of not trespassing:

$$Ep(t) = -40(0.5) + 150(0.5) = -20 + 75 = 55 > Ep(\sim t) = 0$$

Given this preferred strategy for the community member, the best the reserve can do is to regulate access (RA) since the expected payoffs from regulating access are higher than those of unregulated access (UA) and higher than those of no access (~A):

328 *A handbook of environmental management*

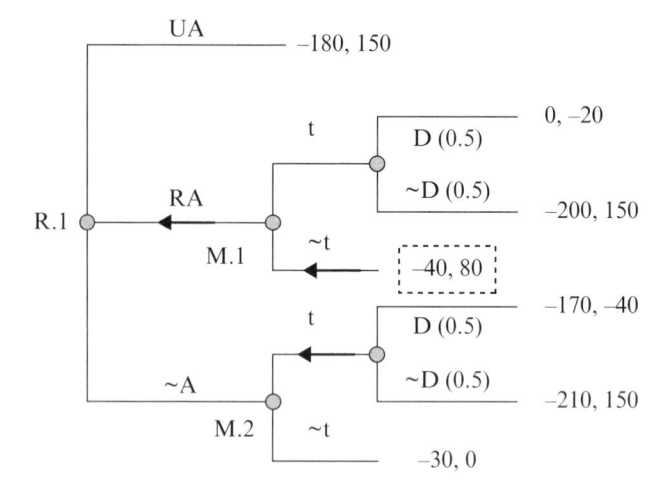

Figure 10.14 The trespassing game with two players and numerical values

$$Ep(RA/\sim t) = -40 > Ep(UA) = -180 > Ep(\sim A/t)$$
$$= -170(0.5) - 210(0.5) = -85 - 105 = -190$$

Note that if the state decided to completely restrict access (~A) the best response strategy for the community would be to trespass (t), the payoffs for these strategies being (−190, 55), yielding a negative joint payoff of −135.

Given the rules, available strategies and values for this game, the result will be to regulate access and not to trespass (RA, ~t) illustrated by a square. Note that from the way in which the payoffs were given, the resulting pair of strategies is also the one that yields the higher social profits, since the joint profits of regulating access and not trespassing (RA,~t) are equal to 40, being the highest joint profit that can be achieved in this game.

Notes

1. Current applications of game theory include biology, engineering, political science, internal relations, computer science and philosophy.
2. In economic jargon, the payoffs represent utilities that comply with the Von Neumann–Morgenstern axiom.
3. If there are subsequent nodes involving probabilities, starting with the payoffs at the terminal nodes, we would move from right to left computing the conditional or independent probabilities, following the rules for independent or conditional probabilities respectively.
4. In some cases knowing that one outcome is preferred over another is enough to derive a preferred strategy.
5. A pair of strategies with these properties is called a Nash equilibrium, after J. Nash (1953) who showed that all games with a finite number of strategies have at least

one equilibrium, provided that mixed strategies are allowed. Mixed strategies will be explained later.
6. This section is adaptively drawn from Ostrom et al. (1994).
7. There are some problems in determining the socially optimal point. In addition to its praxis it will depend on which and how many involved or affected agents are introduced to the equation. In this case, social optimality is considered the result of equating marginal social benefits to marginal social costs of appropriators, taking appropriators to be the individuals sharing the use of the CPR (Ostrom et al., 1994).
8. The 'prisoner's dilemma' was originally framed by Merrill Flood and Melvin Dresher in 1950. Albert W. Tucker formalized the game in an academic presentation in the 1950s with prison sentence payoffs and gave it the 'prisoner's dilemma' name. Imagine that two suspects have been arrested by the police. The police have insufficient evidence for a conviction, and, having separated both prisoners, visit each of them to offer the same deal. If one testifies (defects from the other) for the prosecution against the other and the other remains silent (cooperates with the other), the betrayer goes free and the silent accomplice receives, say, the full ten-year sentence. If both remain silent, both prisoners may be sentenced to only six months in jail for a minor charge. If each betrays the other, they each receive a five-year sentence. Each prisoner must choose to betray the other or to remain silent. Each one is assured that the other would not know about the betrayal before the end of the investigation. How should the prisoners act? (For more details on the origin and formalization of the prisoner's dilemma see Poundstone, 1992).
9. The 'chicken game', also known as the hawk–dove or snowdrift game, was first formalized by John Maynard Smith and George Price in 1973. The principle of the game is that while each player prefers not to yield to the other, the outcome where neither player yields is the worst possible one for both players. The name 'chicken' has its origins in a game in which two drivers drive towards each other on a collision course: one must swerve, or both may die in the crash, but if one driver swerves and the other does not, the one who swerved will be called a 'chicken', meaning a coward.
10. The 'assurance game' is a generic name after Sen (1967), for the game also known as the stag hunt game. The French philosopher Jean Jacques Rousseau, in his 1755 writings on inequality, presented the following situation: two hunters can either jointly hunt a stag (an adult deer and rather large meal) or individually hunt a rabbit (tasty, but substantially less filling). Hunting stags is quite challenging and requires mutual cooperation. If either hunts a stag alone, the chance of success is minimal and thus there is the need of cooperation and trust to achieve a larger joint benefit.
11. Non-priced/non-marketed values could be incorporated to the payoffs of each alternative.
12. It could be argued that M is very similar to m, as the costs of monitoring for additional activities, while monitoring anyway may be very small (decreasing marginal costs). Even if this difference is small, it exists and so, for illustrative purposes, let us denote it by using the letter M.

References

Baland, J.-M. and J.-P. Platteau (1996), *Halting Degradation of Natural Resources. Is There a Role for Rural Communities?*, Rome: FAO and Oxford, UK: Oxford University Press.

Binmore, K. (1994), *Just Playing: Game Theory and the Social Contract*, Cambridge, MA: MIT Press.

Dawes, R.M. (1973), 'The commons dilemma game: an n-person mixed motive game with a dominating strategy for defection', *Oregon Research Institute Research Bulletin*, **13**(2), 1–12.

Maynard Smith, J. and G. Price (1973), 'The logic of animal conflict', *Nature*, **246**, 15–18.

Nash J. (1953), 'Non-cooperative games', *Annals of Mathematics*, **54**(2), 286–95.

Ostrom, E. (1990), *Governing the Commons. The Evolution of Institutions for Collective Action*, New York: Cambridge University Press.

Ostrom, E. and R. Gardner (1993), 'Coping with asymmetries in the commons: self-governing irrigation systems can work', *Journal of Economic Perspectives*, **7**(4), 93–112.
Ostrom E., R. Gardner and J. Walker (1994), *Rules, Games and Common Pool Resources*, Ann Arbor: The University of Michigan Press
Poundstone, W. (1992), *Prisoner's Dilemma, John Von Neumann, Game Theory and the Puzzle of the Bomb*, New York: Anchor Books, Doubleday.
Sen, A. (1967), 'Isolation, assurance and the social rate of discount', *Quarterly Journal of Economics*, **81**, 112–24.
Snidal, D. (1985), 'The Game Theory of International Politics', *World Politics*, **36**(1), 25–57.
Von Neumann, J. and O. Morgenstern (1944), *Theory of Games and Economic Behaviour*, Princeton, NJ: Princeton University Press.
Weissing, F. and E. Ostrom (1991), 'Irrigation institutions and the game irrigators play: rule enforcement without guards', in R. Shelten (ed.) *Game Equilibrium Models II: Methods, Morals and Markets*, Berlin, New York: Springer.

11. Economic valuation of different forms of land-use in semi-arid Tanzania
Deborah Kirby

Introduction
In many ecosystems there is a conflict of interest between land-users who wish to utilize the land for different purposes. In most of these situations, the overriding factor influencing the use of the land is its perceived economic value. A forest or savanna region may produce many different types of goods for human use. Although these goods have a value to humankind, many have traditionally been excluded from estimates of valuation. This has arisen because policy-makers have generally assumed that natural ecosystems such as forests or savannas have no economic value other than their direct use of producing saleable products, and thus non-monetary values from their land-use have been excluded from any economic analysis. Thus, forested areas have only been valued in terms of timber production or land area available for farming (Barreto et al., 1998), and savanna areas for the volume of saleable livestock (for example, Bembridge and Steenkamp, 1976).

A comparison of the relative values of the different forms of land-use, which includes all values of the resource, monetary and non-monetary, can help identify a socially equitable use of land. This chapter considers how valuations of different forms of land-use can be made and emphasizes the importance of establishing linkages between ecological and economic systems to facilitate valuation of biological goods and services in a developing country, Tanzania. In this context, the use of a particular tool in economic analysis, the production function, is described.

The importance of valuation of ecosystems
Many resources provided by ecosystems do not have a direct market value. For example, the indirect values of ecosystem services provided under traditional pastoral systems of land-use are harder to quantify compared with the direct market prices of commercial ranching. Consequently, evaluation of the economic worth of one system of land-use against another will not be equitable if only those products that have a direct market value are compared. One of the principal pressures on natural ecosystems is their conversion into more productive systems in commercial terms (Wilson, 1994). This occurs when decisions over land-use are taken

by individuals or groups of individuals, who are interested in maximizing their private benefits through the production of particular products that have a specific market value to them (for example, Kirby, 2000). However, creating a commercially increased output through a change in land-use may not in reality result in the most productive ecosystem if as a consequence the non-marketed values of the ecosystem are reduced or eliminated. Policy-makers and governments have frequently overlooked the non-market values of ecosystems with the consequence that a change in land-use has occurred in favour of the commercial system of production, thus leading to an overall reduction in the net benefits produced by the ecosystem (Godoy, 1992). This has resulted in policies that have misdirected resource use away from the most economically valuable form of production and towards maximizing financial outputs.

In order to make rational decisions on issues of natural resource management it is therefore important that consideration is given by policy analysts to all aspects of an ecosystem's value. Only if this is undertaken can the total economic worth of the ecosystem be estimated and an effective comparison of its value under different uses made. Valuation of ecosystem productivity can be undertaken by numerous methodologies (see, for example, Dixon and Sherman, 1990; Heywood, 1995). However, this chapter concentrates on one type of methodology that is particularly useful for the valuation of non-marketed and marketed values associated with biological resources that support economic activity. This is the production function approach.

Tanzanian land-use issues
As more detailed descriptions of the land-use problems of Tanzania are given elsewhere (Quinn and Ockwell, Chapter 9, this volume), this chapter gives only a brief overview of the relevant issues. The Tanzanian semi-arid to arid climate has a rainfall ranging from 400 to 1200 mm. The resulting vegetation varies from grasslands to savanna thornbush and wooded savanna in higher rainfall areas. The population density is relatively low and at least 11 per cent of the population are below the poverty line. Much of the land area is still under open access or some form of common property tenure.

The traditional form of land-use in semi-arid Tanzania has been pastoralism, defined as a production system where at least 50 per cent of production (subsistence and marketed) comes from livestock or livestock-related activities. This is predominantly an extensive production system, ranging from nomadism, where no form of cultivation takes place, to semi-transhumance, where part of the family is sedentary and practises cultivation (Niamir-Fuller, 1998).

Economic valuation of land-use in semi-arid Tanzania   333

In recent decades, pastoralists have faced increasing problems of resource degradation as a result of immigration of other populations and changes in land-use. In many parts of the country this has led to apparently unsustainable forms of land-use and the exclusion of many of the indigenous population from their homelands, with a resulting increase in poverty and deprivation. The reasons and consequences of such changes in land-use are discussed below.

The causes and consequences of change in the pastoral lifestyle

- *Settlement policies*. Pastoral settlement leading to agricultural farming is a common government policy in many African countries.
- *Security of land tenure*. Many pastoralists are also settling of their own accord in response to decreasing rangeland areas and also in order to guarantee land tenure.
- *Tourism and wildlife conservation*. Government policies have encouraged the takeover of land for private hunting enterprises. This has resulted in the progressive reduction of rangeland area for grazing and in the loss of wildlife resources important for the pastoral lifestyle.
- *Forestry*. Clearance of forested land has occurred for charcoal and in order to increase grazing area.
- *Land privatization*. A policy of agricultural promotion as the predominant land-use has encouraged the migration of agriculturalists into ecologically fragile areas. With land privatization has come the fencing of areas of rangeland for private ranching. This has resulted in the expropriation of the land for the rich and a system of farming incompatible with the ecological conditions.

In conjunction, these policies have resulted in the deterioration of the dryland ecosystems of Tanzania through habitat loss, wildlife decline, deforestation and overgrazing. Settled agriculture results in active clearance of vegetation to make space for livestock and has thus reduced the tree and plant diversity that has traditionally provided food and medicines for the local population. In addition, lower grazing pressure in distant pastures often results in the invasion of the range in these areas by unpalatable plants.

Settlement of pastoralists has also resulted in the breakdown of traditional systems for managing natural resources, resulting in inefficient management of the habitat and a consequential decrease in livestock productivity and increased impoverishment of this sector of the population.

The predominance of a social and cultural emphasis towards agriculture and against other forms of land-use in Tanzania has enabled changes to occur. Political discrimination against nomadic pastoralism and hunter-gathering has led to the belief by many that this form of lifestyle is backward, primitive and economically unproductive. Pastoralism and hunter-gathering is predominantly seen as an inferior and unproductive use of the land. That much of the land is common land enables those engaged in formal agricultural practices that require privatized land, greater economic status and political control over people involved in more traditional land management practices. This gives the latter groups little or no control over the land-use. In this context, and with the gathering body of evidence that settled agriculture is an ecologically, and thus economically, unsustainable land-use option in semi-arid land, there is a clear need for an accurate assessment of the full values of each form of production. More precise estimation in monetary terms of the value of each system in supporting production will allow a direct and fair comparison of each form of land-use to be made.

Ecosystem values
In the past few decades, attention has been increasingly turning towards recognizing the fact that natural ecosystems such as forest and savanna provide a wide range of products that are highly valuable but may not have an immediately apparent market value (Scoones et al., 1992). It is realized that failure in the past to include these types of ecosystem values in analyses of land-use has led to the undervaluing of ecosystems (Swanson and Barbier, 1991). Scientists and to a certain extent policy-makers are therefore now recognizing that if the total economic value of an ecosystem is to be realized, a monetary value must be assigned to the resources in question, enabling them to be 'counted' in an estimation of the ecosystem's worth.

The lack of a market value, and thus underestimation or complete absence of the resource's value, may be for one of two reasons. First, the resource may not be one that is exchanged in the marketplace. For example, hunting and gathering of wild products for subsistence living will directly use many natural resources but these may be consumed directly rather than sold (Campbell et al., 1997). In the case of livestock, animals may be used for products other than meat that do not have a market value, such as transport or labour (Scoones, 1992). Second, the value of the resource may be in supporting the production of other goods rather than through its own direct utilization.

The remainder of this section describes in detail the differences between different forms of ecosystem products. These products can be separated into three categories, according to their type of use.

Commercial values
Extraction of timber or production of livestock from an ecosystem will directly produce revenue for the land-users. If they are marketed, their value will be equal to the revenue from selling the timber or livestock minus the costs of production and transport to market (Godoy, 1992). If they are not marketed but kept for home consumption they will also have a value although this will be considerably harder to impute, and may not be equal to the value that they would realize if they were sold on the open market (Godoy, 1992).

Non-timber and non-meat values
A savanna or forest ecosystem can potentially provide a great number of non-timber or non-meat goods and services. Studies have shown that in the case of a forest ecosystem, local people living in and around it regularly use a large number of forest products. Many of these products may be marketed and thus have market values, for example, wild fruit and insects, wood for construction purposes, herbs and medicinal plants (Fuentes, 1980; Oommachan and Masih, 1988). However, many goods and services may be extracted and only used for home consumption, in which case they will not have an immediately identifiable market value (Peters et al., 1989). In the case of savanna ecosystems, a large number of products including palms, grasses, reeds, fruits and wild mushrooms have all been shown to be used by the local inhabitants (Campbell et al., 2000a; IIED and HNWCP, 1997). In addition, livestock production may produce a number of benefits besides those of meat, for example, transport, which must be included in an evaluation of the productivity of the system. As with forest products, these products may or may not have a direct market value, depending on the use to which they are put.

Regardless of whether either of these products are marketed or non-marketed, they can be described as direct-use goods. As this name implies, these are goods that are supplied directly from the ecosystem and that may be either domesticated or wild. In this sense, forestry and livestock production as well as products derived from fishing and hunting are included.

External benefits
The third category of goods and services that can be derived from ecosystems are described as indirect-use resources. Many ecosystems will provide invisible benefits to other ecosystems or people, both within the geographical area of the ecosystem, as well as outside it. In the case of both forest and savanna ecosystems, these external benefits can be considered to exist at two separate levels. A number of empirical studies have shown that forest ecosystems provide protection for other ecosystems located downstream from the

forest. In particular, forests have been shown to provide soil protection for downstream agricultural land (Anderson, 1987), protect downstream water courses from increased levels of sedimentation caused by surface run-off and increased erosion and leading to high costs of de-siltation (Panayotou, 1990; Lal, 1997), to impact on water yield and groundwater recharge, and to control watershed hydrology and provide protection against erosion (Rawat and Rawat, 1994; Ataroff and Rada, 2000; Putuhena and Cordery, 2000). For dry savanna ecosystems, the invisible benefits may be as significant as those of forest ecosystems. Hydrological functions of dry savannas have been shown to have a direct impact on the productivity of surrounding areas. The vegetation cover of woodlands is thought to stabilize the local climate and maintain rainfall patterns (Myers, 1995) as well as prevent soil erosion and flooding, and control the recharging of groundwater reservoirs (Huntoon, 1992). These positive external effects, although they may occur outwith the boundary of the ecosystem, are nonetheless specific in their area of impact. As such, there is a limit to the number of individuals upon which they will have an effect.

A second type of externality, however, is also provided by forest and savanna ecosystems, and has a greater range of impact. It is now known that areas of vegetation cover, in particular forested areas, act as import sinks for atmospheric carbon (Harmon, 2001; Pautsch et al., 2001; Uri, 2001). This function is globally important in terms of regulating atmospheric composition and controlling global temperature. Many savanna ecosystems also form the watershed zones of major international rivers, for example, the Zambezi (Kundhlande et al., 2000), which are internationally important for domestic water consumption and power supply. This type of ecosystem service differs from all others in that its use is 'public' in nature, that is, it is used by society as a whole, and individuals do not have a choice over whether or not they wish to use it.

Both of these invisible benefits are termed indirect-use values, since although not consumed directly they are essential for the production of direct-use values. However, their very nature implies that they will have no direct market value and thus are difficult to quantify in a cost–benefit analysis.

Past valuation studies
A number of research studies have focused on attempting to value forest and savanna resources (Hanemann, 1988). In all of these studies, valuation of various aspects of the complete ecosystem has been undertaken. The relative ease of valuing direct-use resources rather than indirect-use resources is immediately obvious and for this reason, although some studies have attempted to value indirect-use resources, the majority of

research has concentrated on the valuation of direct-use resources. Above all, attention has also been placed on valuing tropical forest resources, and although studies have been made across the globe, there has been a disproportionate interest in Latin American forests (Godoy et al., 1993).

Phillips et al. (1994) compared the usefulness to the indigenous population of six floristically distinct tropical forest types in south-east Peru. They attempted to estimate the use values of each forest type through a series of questionnaires to the local people about the relative importance of the different forest plant species. However, the results were produced as a rating of different values of the forest products and no monetary value was placed on the products, thus making quantitative comparisons with other forest values or uses impossible. In a very similar study, Toledo et al. (1995) evaluated the usefulness of plant species in a Mexican tropical forest. Through questioning and working with the local population, they evaluated the number of non-timber products that were used from the forest. Over 3000 products were identified, but as with Phillips et al. (1994), no attempt at placing an economic value on these products was made.

A number of studies have, however, attempted to put a monetary value on the non-timber forest resources. Peters et al. (1989) estimated the economic value of non-timber plant products of a small (1 hectare) area of Peruvian tropical forest. Their calculations were based on empirical fieldwork estimating the amount of standing vegetation in the forest, and on the prices of the products in the local market. To arrive at a valuation figure, they multiplied the standing inventory of certain non-timber forest products by the local market price, thus carrying out a very simple cost–benefit analysis. A value of US$420 ha^{-1} year^{-1} for the products was estimated, however, as was pointed out by Godoy et al. (1993), Peters et al. failed to take into account that the local market prices are a reflection of the supply of the produce. Supplying the market with all the forest resources would result in a decrease in market prices. In addition, the study did not provide an accurate value of the entire forest resources, since only plant products were valued, with faunal products being completely disregarded. Pinedo-Vasquez et al. (1992) estimated the potential value of six fruits and derived a total value of the forest products of US$20 ha^{-1} year^{-1}. Hecht (1992) estimated the opportunity cost (that is, the value of the other products lost from using the forest in this way) of using forests for livestock-rearing in western Amazonia. The final value derived was dependent on the size of the area over which the extraction was carried out, and ranged from US$5–16 ha^{-1} year^{-1}. Nations (1992) considered the value of only three plant products from a Guatemalan forest and derived a value of US$10 ha^{-1} year^{-1} and Anderson and Ioris (1992) estimated a value of US$79 ha^{-1} year^{-1} for only three plant species in tropical Brazil.

From the above-cited studies it can be seen that any comparison of the relative values of forest resources encounters serious difficulties. First, several of the studies do not consider the full economic value of non-timber forest products but only of a few selected species. In particular, none of the studies includes the value of faunal products from the forest. Second, largely differing values were derived for studies that occurred over the same areas (for example, Peters et al., 1989; Pinedo-Vasquez et al., 1992) bringing into question the validity of these studies. In addition, Nations (1992) and Hecht (1992) consider only the gross value of the resources they estimated, making comparisons with studies that estimated net values impossible.

Fewer studies have attempted to value wildlife resources in forest or savanna ecosystems in Africa. Those early studies that attempted to do so concentrated mainly on animal rather than plant resources, for example, Martin (1983), Redford and Robinson (1985) and Barbier et al. (1990). Scoones et al. (1992) reviewed much of the literature on the use and value of wild resources in agricultural systems. Although much literature existed about the use of wild resources, it was shown that almost nothing was known about the value of these products. One of the few studies to undertake the economic assessment of wildlife resources is that of the International Institute for Environment and Development (IIED) and Hadejia-Nguru Wetlands Conservation Programme (HNWCP) (1997). The principal method of data collection was Participatory Rural Analysis (PRA), and the extensive report gives a detailed description of the methods used to obtain the data; which products were used, how they were harvested, the quantities taken and their value. An overall economic value is derived for a number of the wildlife resources based on a simple cost–benefit analysis of value equals quantity sold times price, minus costs of production. However, no assessment of the value of crops and livestock is made in this study.

In addition to wildlife resources, livestock resources are also of high value to the people living in and around many savanna ecosystems. As was discussed earlier in this report, the value of livestock has frequently only been seen in their commercial meat production. This has frequently led to the belief by policy-makers that traditional production systems of livestock are backward and inefficient compared with those of commercial systems such as cattle ranching. In recognition of this shortcoming, research was initiated to attempt to assess the true productive value of traditional forms of livestock production systems (for example, Cossins, 1985). Scoones (1992) recognized that to understand the full value of livestock within agropastoral systems required a detailed study of all of the useful outputs from the livestock in the system. He defined productivity not simply in terms of meat production, which had a marketed value, but also included non-marketed products such as milk and manure. In addition, services provided by the livestock were also valued, such as the provision of labour and transport. Finally, Scoones also included the costs of production of livestock in the study area, and made a comparison between the relative benefits of different types of livestock production. This study was carried out through a simple cost–benefit analysis, that is, the total value of any animal was calculated as equal to the value or price of it and its products minus the costs of production. Data on the production parameters of livestock (for example, births, deaths, milk production, and so on) were obtained though household surveys. A comparison of livestock productivity on different soil types was also made to assess the effect of a change in this ecological parameter. He found that the productivity of communal cattle was higher than that of beef cattle in the same region and that this was at least partly due to the higher stocking rate of the communal cattle.

Campbell et al. (1997) estimated values for both plant and animal wildlife resources in a Zimbabwean savanna. A large number of products that were used by the local population were identified, for example, firewood, wood for construction, fruit, birds and thatch. As with Scoones et al. (1992), the total value of the ecosystem was calculated by multiplying the amount of each good produced by its value. For many of these resources, an economic value was imputed by subtracting the costs of production (for example, labour and transport costs) from their market value. Two different methods were used to estimate how much of a good was produced by each area of land. For products that were regularly used by households, such as firewood and some wild foods, the amount that could be sustainably produced by the forest was taken to be the amount of the product used (an assessment of sustainability was also made). However, for other goods, such as wild fruits, the amount used was judged to be a proportion of the total annual biological production. Thus, in these cases, an ecological analysis was made of the productivity of each plant species each year. Although not explicitly stated, in this latter case a simple production function was therefore set up, where the quantity consumed by households equalled a proportion of the quantity produced by the vegetation. Although the analysis made by Campbell et al. was extremely precise and detailed it did not include a valuation of the livestock and crops produced in the region.

Campbell et al. (2000b) compared the economics of four different cattle production scenarios in communal grazing lands in Zimbabwe. In this study, data on inputs to, and outputs from, livestock production, and data on prices, were obtained through surveys of households. This allowed the researchers to identify relationships between the inputs into production of livestock (for example, feeding, herding labour, dipping) and various outputs (for example, manure, milk, transport). Although not explicitly stated, this approach again followed a production function approach where the outputs from the livestock production were directly related to the inputs. By multiplying the outputs of livestock production by their price, and subtracting the costs of production, the total profitability of each management system was calculated.

In addition to the valuation of direct-use resources, some research has also been directed towards valuing the non-use values of forested and savanna ecosystems. Since in this case, the value of an ecosystem is measured in terms of its contribution to the production of other goods or services, a relationship (or production function) must be identified between the two factors. For example, if the loss of watershed protection leads to an annual decrease in agricultural productivity worth $50, this can be imputed as the value of the watershed for agricultural production. Kundhlande et al. (2000) estimated the value of carbon sequestration and of water supply for a tropical savanna-woodland ecosystem in Zimbabwe. By measuring the standing biomass of vegetation in the region they calculated the volume of carbon that could be sequestered by the land when forested. This value was then compared with an estimate of the decrease in sequestration capacity that would occur if the land-use were changed to agriculture. The difference between the two amounts was then multiplied by the estimated price of carbon to give an overall change in value of carbon sequestration services. In this sense, the value of carbon sequestration was measured as a function of the available biomass. In estimating the value of water in crop production, a crop production function was used that related the amount of crop produced to the amount of rainfall available. The value of water was then measured as the change in crop production with changing levels of rainfall, multiplied by the price of the crop. In addition, although not explicitly mentioned, similar production functions were also used to define the relationships between wild foods and grass production and rainfall. These production functions were derived by a group of biological and agricultural scientists working on the project. In some cases there were documented relationships, whilst in other cases the relationships were based on the best judgements of researchers who were familiar with the local agricultural practices (W.L. Adamowicz, personal communication).

This review has shown that many studies have been carried out that have valued a certain aspect or layer of an ecosystem's resources. It is likely that the relative recentness of this field's development along with the large amount of resources required to carry out such valuation studies have so far precluded evaluation of entire ecosystems. However, although no study has attempted to estimate the total economic value of an ecosystem, comparisons of the values of ecosystems under different forms of land-use have been made. With the progressive increase in data availability and a growing understanding of the requirements for making such studies, it is anticipated that total valuation of different forms of land-use will become progressively easier.

A method for comparing different ecosystem values: the production function approach

One of the key shortcomings of many of the current studies has been that the relative values derived for different forms of land-use have been difficult, if not impossible, to compare because of the different type of valuations that have been made. However, in cases where a comparison may be possible, the valuation of each type of resource is made in terms of the static state of land-use, that is, a forested ecosystem provides a certain value of resources, whilst a deforested agricultural ecosystem provides another value. Thus, the value of the ecosystem is dependent on the area of cover of a particular vegetation type. At first sight, the equitable method of choosing the use of the ecosystem, based on economic value, may be the one that produces the highest value of resources. However, economic theory shows that this is not the case, but that the economic value of the resource may be maximized through a combination of land-uses. Further detailed information on the theory behind profit maximization is given in Common (1996) but a brief outline is given below.

If the value and thus profit of one use of the ecosystem increases with an increasing area of that type of land-use, profit can be expected to be maximized when the area of that cover of land type is maximized. The reverse is also true, in that the profit of another type of land-use will be maximized when its area is maximized. This can be explained graphically as is shown in Figure 11.1. The profit of any enterprise is maximized when the marginal profit (the rate in change of profit with a unit increase in input) is equal to zero.[1] If we take the example of two types of land cover, forested or deforested, the graph in Figure 11.1 shows that individual profits for each enterprise will be maximized at C_3 and C_1 respectively, where marginal profits are zero. However, if an area of forest cover shown at C_2 is chosen, the total value of the area will be greater than at either at C_3 or C_1. This theory is known as a Pareto improvement but will not be discussed in further detail here. In order to identify this level of forest cover, a relationship between the total value of each ecosystem use and levels of vegetation cover must be identified, that is to say that a common denominator must be identified. The technique behind identifying a common denominator driving productivity is known as the production function approach, which has been described in literature on valuation (see Barbier, 1994).

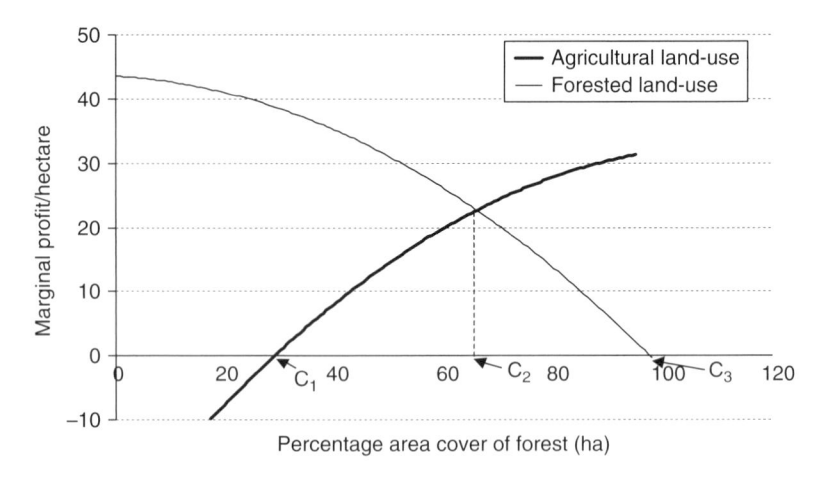

Figure 11.1 Showing the level of marginal profits for different types of vegetation cover

The following section describes the theory of the production function approach, and is followed by a worked example of applying this approach to the valuation of savanna and forest resources.

The production function approach to valuing ecosystems
A production function is a mathematical description of the relationship between the production of a good (for example, sheep, maize), and the amount of inputs (for example, environmental or human-made) required for its production. For example, a production function for clay pots (*P*) might be determined by the amount of clay (*C*), water (*W*) and people (*L*) available to make them, where the way in which these factors interact defines the mathematical form of the function. The development of a production function provides a quantitative description of a relationship between several factors where previously the specific relationship was unknown.

Although initially used in economic analysis to describe the production of goods by firms, the use of this methodology has subsequently been expanded to include measures of environmental factors, and can be used to describe the production of any good where the inputs into its production can be quantified.

Direct-use values
Ecosystems provide many important functions for humankind, which, as has already been described, can be grouped according to their direct-use or indirect-use values.[2] The valuation of an ecosystem is frequently made

according to its tangible outputs (direct-use values), for example, the value of the crop harvested, and where no indirect-use value of the ecosystem exists, this valuation is perfectly acceptable. A production function can then be developed to quantify the components of an ecosystem that contribute towards the production of a good. In doing this, an estimate can be made of the relative values of each of the inputs into production.

If a comparison is being made of the relative values of different forms of ecosystem use, it is imperative that all components of the production function are included in the analysis. For example, the production of livestock and/or game in a particular area (Q), will require inputs of labour (L) and forage (F). However, as was shown by Scoones (1992), the productivity of livestock may also be dependent on factors such as the physical conditions of the environment. Thus, in the case of Scoones' research, livestock productivity was also affected by soil type. If this latter factor is overlooked and excluded from the livestock production function, the result will be an undervaluation of the soil value (S), giving it the incorrect production function:

$$Q = f(L, F)$$

rather than:

$$Q = f(L, F, S).$$

If this production function were defined incorrectly in this way, no value would be attributed to the presence of a specific type of soil cover.

In addition, Scoones (1992) showed that the output of livestock amounted to much more than simply its meat production. Included in the values of livestock in his study were milk production, manure production and the provision of labour. By including these output variables the value of livestock significantly increased. It can therefore be seen that care must be taken to ensure that all inputs into production are incorporated in a production function, and that all the outputs from an activity are also included, if an accurate valuation of the input resources is to be obtained.

Indirect-use values
In agriculture and forestry, which involve the primary production of goods, in addition to providing direct-use goods, in many cases the ecosystem may provide an ecological function that has an indirect value in supporting the production of marketable goods. The ecological function is then a 'factor input' in the production process. As already discussed, the ecological function may have an important input for activities that are far removed, either spatially or temporally, from the production process, and

thus the benefits derived may not be easily ascribed to it. For example, deforestation may result in increased erosion, leading to sediment deposition on downstream farmland, potentially affecting the nutritional content of the deforested soil and crop production on the downstream farmland. Thus, in addition to its value for timber production, the forested land may also have a direct influence on crop and livestock/game production in the immediate area and crop production in a different area. Including forest area as a determinant of livestock/game or crop production may therefore capture some element of the economic contribution of this ecological function. In mathematical terms this can be described as:

$$Q = f(x_i, \ldots, x_n, A)$$

where Q can be considered as the output or marketed good, for example, crop production, $x_i \ldots x_n$ are marketable inputs required to produce that good, for example, labour, seed and so on, and A is the environmental input of interest required for the production of the good, for example, the area of forested land.

In attempting to value these environmental benefits, two underlying relationships must be elucidated. First, the physical effects of changes in a biological resource or ecological function on an economic activity have to be ascertained. That is, an accurate understanding and quantification of the change in the output of a good with a change in land-use patterns must be known. To use our previous example, which other ecosystems are affected by the deforestation of an area and what are these effects? Knowledge of these processes requires detailed ecological understanding of the systems involved. Second, the effect of these environmental changes must be valued by considering the corresponding change in the output of the marketed good, for example, what is the quantitative change in downstream crop productivity and thus farmers' incomes in response to the deforestation of an upstream forest? Again this requires detailed ecological and sociological understanding of the processes in place. In this way, the biological resource is considered as an input into an economic activity, and can be valued in terms of the impact it has on the output of a marketed good.

Limitations to the use of the production function approach: the absence of a monetary value of the output good
When a product, Q, such as crop production, is measurable and a market price is available for it or one can be imputed, determining the values of the environmental inputs (A) into its production is relatively simple.[3] If however, Q can not be measured directly, then a measure of the change in

value of Q with a change in the environmental input, A, cannot be directly obtained. This is often the case in developing countries where the use of the environment is to support a subsistence lifestyle. Since the products do not pass through a market, the benefits accrued do not have a monetary value. In these cases, an accurate comparison in economic terms between different forms of the land-use is difficult to make, given the absence of comparable systems. To overcome this problem, the output of the ecosystem under different uses must be considered in common terms. The standard approach to the valuation of products is in monetary terms and so if goods produced by an ecosystem do not have a monetary value, a proxy must be found for this value. In the case of land-use in Tanzania, land-users include nomads who utilize the natural resources (Q), such as vegetation and game, for their own consumption to sustain a subsistence-level lifestyle, and who do not generally trade goods in the marketplace. Many of these products may have no market value, so to obtain an approximation of their 'worth', an estimate can be made of the market value of an 'equivalent' good.

In general terms, if this solution is unfeasible, the alternative is to find a substitute or complementary resource between S and one of the marketed inputs $x_1 \ldots x_n$. Both this and the previous solution require a detailed knowledge of the ecological interactions between inputs and outputs of goods. It can be realized however, that in a system such as nomadism, this latter solution is implausible since there are no marketed inputs into the production of the goods.

The following section will illustrate, through the use of examples, the employment of the production function approach in estimating resource values for forest and/or savanna ecosystems.

Worked examples

Direct-use values: calculating the direct-use value of agricultural cover
The direct-use value assessed here is the amount of land available for livestock production and for wild fruit production. A hypothetical situation is set up whereby a fixed area of land can either be used to produce livestock (if it is deforested) or wild fruit (if it is forested). A production function is estimated for both products based partially on the area of land available for their production, and the optimal level of land cover of both types of land is identified, in terms of profit maximization.

Step 1: Identifying the production function for livestock The amount of livestock produced is (partially) dependent on the amount of land available for grazing. It will also be dependent on other factors such as the amount of labour used in herding and veterinary costs. This could, for example, be represented as:

$$Q_l = A_s^a L_l^b \qquad (11.1)$$

Where:

Q_l = the amount of livestock produced (output in kg per year);
A_s = the area of land available for livestock production (hectares);
L_l = labour (person hours per year);
a, b = unknown parameter values to be estimated.

In this example, primary data would be required for the input and output variables given in Equation 11.1. These data could be obtained through a questionnaire survey of the local population answering the following type of questions:

Livestock production data:

1. Do you keep livestock?
2. What type of livestock do you rear: sheep, cattle, goats, poultry?
3. What was the calving/lambing, and so on, rate of this livestock in the past year?
4. What was the mortality rate of your livestock in the past year?
5. What are the uses of your livestock other than for meat production, for example, milk, manure?
6. How much of each of these products was produced?
7. How many people were involved in caring for your livestock and for how many hours did each work?

It should be noted that this illustration gives an example of the kinds of questions that should be asked in a questionnaire survey but not of the form the questions should take. To elucidate the correct information, questionnaire surveys must be presented in an objective manner and in a way that can be understood by the recipients of the questionnaire.

A simplified example of the results of such a study is given in Table 11.1 using only two inputs, labour and land area, and three study areas. To make a comparison over time, two different approaches can be taken: either the survey can be longitudinal, that is, carried out over a number of years, or cross-sectional, that is, carried out at the same time but over a number of different study sites. This will then provide variation in the data of the inputs into production.

Table 11.1 Hypothetical results from agricultural questionnaire survey on livestock production

Input into Livestock Production	Amount of Inputs Per Year	Amount of Livestock Produced (kg)
Labour (hours)	200	63
Area (ha)	20	
Labour (hours)	400	89
Area (ha)	20	
Labour (hours)	200	100
Area (ha)	50	

To identify the quantitative relationship between these variables, a multiple regression analysis is carried out using the quantity of livestock produced as the dependent variable and the input data as the independent variables. The results of the multiple regression for this example are given in Equation 11.2.

$$Q_l = L_l^{0.5} A_s^{0.5} \qquad R^2 = 1 \qquad (11.2)$$

The results could, of course, take an alternative form of production such as a linear relationship shown below if the data fitted the multiple regression more accurately in this form.

$$Q_l = 4.5L_l + 2A_s$$

Clear and accessible information on the choice and understanding of different functional forms can be found in Hill et al. (1997).

Step 2: Calculating the value of livestock production Once the production function has been identified, the value of the ecosystem can be calculated by estimating the value of the livestock products. Data are therefore required on the price of livestock products and on the costs of production. Again these data would be collected through a questionnaire survey, either of the householders involved in production or through other individuals who have knowledge of the markets.

Livestock costs data:

1. What was the price of the livestock that was sold?
2. How much of the livestock was retained for home consumption?

348   *A handbook of environmental management*

3. How many hours were spent herding the livestock?
4. What was the cost of this labour?
5. Were there any other costs in involved in rearing the livestock? What were they?

Once again, it should be noted that this illustration gives an example of the kinds of questions that should be asked in a questionnaire survey, but does not give the form that the questions should take.

With this information, a profit function can be written:

$$\Pi_l = P_l Q_l - C_l \qquad (11.3)$$

Where:

Π_l = total profit of livestock($);
P_l = price of livestock ($/kg);
Q_l = quantity of livestock produced (kg);
C_l = costs of production of livestock ($).

Substituting Equation 11.2 into Equation 11.3 gives:

$$\Pi_l = P_l(L_l^{0.5} A_s^{0.5}) - C_l \qquad (11.4)$$

The total profit from livestock production can then be identified for any given levels of land and labour.

Thus, if $P_l = 10$, $L_l = 200$, $A_s = 20$ and $C_l = 500$:

$$\Pi_l = (10 * 200^{0.5} * 20^{0.5}) - 500$$
$$= \$126 \ annum^{-1} \qquad (11.5)$$

Step 3: Identifying the production function for forest goods   In exactly the same manner as described for the production function of livestock, a production function of forest products can also be identified.

The amount of a good, for example, wild fruit, produced by the forest will be partially dependent on the area of forest. It will also be dependent on other factors such as the amount of time spent in collecting it. The methodology required would be the same as that for identifying the livestock production function in that data would be obtained through a questionnaire survey of the local population. Therefore further details are not given here. In Equation 11.6, an example is given of a possible fruit production function:

$$Q_f = 2(A_t - A_s^{0.2}) + 3L_f^{0.7} \qquad (11.6)$$

Economic valuation of land-use in semi-arid Tanzania 349

Where:

Q_f = the amount of fruit collected (kg per year);
A_t = total area of study region (hectares);
L_f = labour for collecting fruit (person hours per year).

Note that the area of land is given as the total area of the study area minus the area of land used for livestock production, that is, the area of forested land.

Step 4: Calculating the value of the fruit production As with the livestock profit function, a fruit profit function can be estimated through identifying the price of the fruit and the costs of its production. This would give a profit function such as that shown for livestock in Equation 11.3. Substituting Equation 11.6 into this will then give the total profit function, as shown in Equation 11.7. An example of the fruit profit function is given in Equation 11.8 for values of $P_f = 5$, $L_f = 200$, $A_t = 50$ and $C_f = 25$.

$$\Pi_f = P_f(2(A_t - A_s)^{0.2} + 3L_f^{0.7}) - C_f \tag{11.7}$$

Where:

Π_f = total profit of fruit production ($);
P_f = price of fruit ($/kg);
C_f = costs of production of fruit ($).

$$\Pi_f = (5 * 2((50 - 20)^{0.2}) + (3 * 200^{0.7}) - 25$$
$$= \$116\, annum^{-1} \tag{11.8}$$

Step 5: Identifying the total value of produce from the region and the optimal level of different vegetation covers To identify the total value of the area, the values of all the forms of production must be summed together. In this worked example, these values are simply the value from livestock production and the value from wild fruit harvesting. Thus, the total value of the region would equal the sum of Equations 11.5 and 11.8, given algebraically in Equation 11.9 and numerically in Equation 11.10.

$$\Pi_{total} = P_f(2(A_t - A_s)^{0.2} + 3L_f^{0.7}) - C_f + P_l(L_l^{0.5}A_s^{0.5}) - C_l \tag{11.9}$$

$$\Pi_{total} = (5*2*(50 - 20)^{0.2}) + (3*200^{0.7}) - 25 + (10*200^{0.5} * 20^{0.5}) - 500$$
$$= \$242\, annum^{-1} \tag{11.10}$$

Following conventional theory on optimization (see, for example, Nicholson, 1992), the optimal level of agricultural land area (A_s) and thus also forested land area, to give the maximum total value of the region (Π_{total}) can be found by taking the partial derivative of the total profit function with respect to land area and setting the value to zero.[4] This is shown algebraically and mathematically in Equations 11.11 and 11.12 respectively.

$$\frac{\partial \Pi_t}{\partial A} = P_f(2* - 0.2*(A_t - A_s^{-0.8}) + 0.5P_l L_l^{0.5} A_s^{-0.5} = 0 \quad (11.11)$$

$$\frac{\partial \Pi_t}{\partial A} = 5(-0.4(50 - A_s^{-0.8}) + 0.5(10*200^{0.5} A_s^{-0.5}) = 0$$

$$A_s = 1.0 \quad (11.12)$$

Thus, for a total study area of 50 hectares, the optimal area of agricultural land is one hectare and hence the optimal area of forested land is 49 hectares.

It is obvious that for reasons of clarity and simplicity, this worked example has been chosen with only a few variables. However, in carrying out the complete research, two critical points must be noted. (1) To identify the total value of the land, all of the products from it must be identified and assessed. Summing the total profit from all of these products would then give the total value of the region. For the sake of brevity, in this worked example only one product from each type of land-use is given, however the method is equally applicable with numerous products from each type of land-use. (2) Acquiring the data outlined here requires a detailed socioeconomic knowledge of the population living within the region. It must be noted that although the questions outlined here for obtaining information on the production of goods give a guideline as to the type of information required, they do not precisely illustrate the methodology for obtaining that information. Much of this information may not be directly obvious but may have to be imputed through other means. For example, to obtain an estimation of the price of wild foods that are not marketed will require more than a simple question. The value may be able to be imputed, for example, by comparison with the value of other goods that are marketed. The reader is referred to the other valuation studies referenced in this report for more detailed descriptions of survey techniques (for example, Scoones, 1992; Campbell et al., 1997; IIED and HNWCP, 1997).

Indirect-use values: calculating the indirect-use value of rainfall
To measure the value of an indirect-use good or resource, economists use changes in the productivity of a direct-use good as an indicator. A productivity change simply examines the physical relationship between the ecosystem and private goods for which values are known or can be imputed. For example, in the case of agricultural production, the economic value of a change in the supply of an environmental resource will be the difference between the profit made from agricultural production before and after the environmental change. It is important to note that this method does not provide the total value of the indirect-use resource, but rather the marginal value. That is, the change in the value of the good produced, for example, a crop, for a change in the amount of input of the indirect-use resource. The following example shows how this evaluation can be carried out.

By a very similar method as described for estimating direct-use values of ecosystems, indirect-use values can also be measured. The indirect-use value assessed here is the amount of water present in the ecosystem, measured as annual rainfall. It is known that the presence of woodland and forest vegetation in a region maintains rainfall patterns and the local climate (for example, Myers, 1995). Thus it can be assumed that changing the area of vegetation cover in a region will have an impact on the rainfall patterns in that area.

Step 1: Identifying the production function The amount of a crop produced is (partially) dependent on the amount of water available through rainfall. It will also be dependent on other factors such as the amount of labour and fertilizer used.

This could, for example, be represented as:

$$Q = A^a W^b L^c F^d S^e \qquad (11.13)$$

Where:

Q = the amount of crop produced (output in kg per year);
A = the area of land used (hectares);
W = average annual rainfall (millimetres of rainfall per year);
L = labour (person hours per year);
F = fertilizer (kg applied per year);
S = amount of seed sown (kg per year);
a, b, c, d, e = unknown parameter values to be estimated.

Estimates have already been made of the quantitative relationship between crop production and water for certain African dry tropical

352   *A handbook of environmental management*

ecosystems (Coe et al., 1976; Kundhlande, 2000). However, if the analysis is to be accurate, it is essential that this relationship is estimated as precisely as possible. Data may therefore be required to estimate this relationship for a particular ecosystem or study area.

In this example, primary data would therefore be required of the input and output variables given in Equation 11.13. With the exception of the data on rainfall, these data could be obtained through a questionnaire survey of the local population answering the following type of questions:

Crop production data:

1. Do you own or rent land?
2. What is the total size of your land holding?
3. What do you produce on your land?
4. What was the total amount you produced of each crop last year?
5. What is the total amount of seed you sowed last year for each crop?
6. How many people were involved in caring for that crop and for how many hours did each work?
7. Did you apply fertilizer to any of your land? If so, what type of fertilizer, to which crops, and how much?
8. Did you irrigate your crops with groundwater? If so, how much water was used?

It should be noted that again this illustration gives an example of the kinds of questions that should be asked in a questionnaire survey, but not of the form the questions should take.

In addition to this data, the level of rainfall must be known for the study area. To make a comparison over time, two different approaches can be taken: either the survey can be longitudinal, that is, carried out over a number of years, or cross-sectional, that is, carried out at the same time but over a number of different study sites. This will then provide variation in the data of rainfall and the other inputs into production. A simplified example of the results of such a study is given in Table 11.2 using only three inputs, rainfall, labour and land area, and three study areas.

To identify the quantitative relationship between these variables, a multiple regression analysis can be carried out using the quantity of crop produced as the dependent variable and the input data as the independent variables. The results of the multiple regression for this example are given in Equation 11.14.

$$Q = W^{0.7} L^{0.2} A^{0.1} \qquad R^2 = 1 \qquad (11.14)$$

Table 11.2 Hypothetical results from agricultural questionnaire survey

Input into Crop Production	Amount of Input Used Per Year	Amount of Crop Output (kg)
Rainfall (mm)	1000	898
Labour (hours)	1500	
Area (ha)	20	
Rainfall (mm)	500	503
Labour (hours)	800	
Area (ha)	20	
Rainfall (mm)	100	467
Labour (hours)	1000	
Area (ha)	50	

Step 2: Calculating the value of the ecosystem in terms of water production Once the production function has been identified, the value of the ecosystem can be calculated by estimating the value of the agricultural products. Data are therefore required on the price of agricultural products, and on the costs of production. Again these data would be collected through a questionnaire survey, either of the householders involved in production or through other individuals who have knowledge of the markets.

Crop costs data:

1. What was the price of the crops that were sold?
2. How much of the crops were retained for home consumption?
3. What was the price of labour and other inputs into production, for example, fertilizer?

With this information, a profit function can be written:

$$\Pi = P_C Q - C \qquad (11.15)$$

Where:

Π = total profit ($);
P_C = price of crop ($/kg);
Q = quantity of crop produced (kg);
C = costs of production of crop ($).

Substituting Equation 11.14 into Equation 11.15 gives:

$$\Pi = P_c(W^{0.7}L^{0.2}A^{0.1}) - C \tag{11.16}$$

The change in the value of crop production for changing levels of rainfall can be identified for any given levels of rainfall, labour, area and costs. It is important to note that these other variables must be held constant and can be chosen to be at any level.

Thus, if $P = 10$, $W = 1000$, $L = 800$, $A = 50$ and $C = 5000$:

$$\Pi = (10 * 1000^{0.7} * 800^{0.2} * 50^{0.1}) - 5000$$
$$= \$2087 \; annum^{-1}$$

But reducing the rainfall level by half to $W = 500$ mm per annum gives:

$$\Pi = (10 * 500^{0.7} * 800^{0.2} * 50^{0.1}) - 5000$$
$$= \$ - 636 \; annum^{-1}$$

Thus, the value of rainfall changing from 1000 mm per annum to 500 mm per annum would be estimated as $-2723 ($-636 - $2087).

This example does not quantify a relationship between the amount of forest cover in a region and the changes in crop productivity but simply quantifies the relationship between rainfall levels and crop production. Including the valuation of the area of forest cover in the region would require a second production defining the relationship between forest cover and average annual rainfall. If this relationship is known, it can be simply incorporated into the original production function. For example, if:

$$W = F^{0.5} \tag{11.17}$$

Where:

W = average annual rainfall (millimetres of rainfall per year);
F = area of region with forest cover (hectares).

Substituting Equation 11.17 into Equation 11.16 will give Equation 11.18. Exactly the same calculations can then be carried out as before but with changing forest cover values rather than rainfall levels to provide a value of the change in crop production for a change in forest levels in the region.

$$\Pi = P_c[(F^{0.5})^{0.7}L^{0.2}A^{0.1}] - C \tag{11.18}$$

The above example demonstrates the technique of using a production function approach to value indirect-use resources of an ecosystem. In this example, the value of water on crop production is used; however, crop production is only one of a number of factors that will be affected by the level of rainfall in a region. To make a complete assessment of rainfall value or forest cover value in the region, all factors affected by water availability should be assessed in the same manner. These could include livestock production, wild food production, fuel wood production and human health (Kundhlande et al., 2000).

Conclusions

Any research project that aims to undertake work in evaluating different land-use systems would be required to consider several different areas of research. The research would require the integration of ecological, sociological and economic skills in order to gain a full understanding of the problem. To assess the total economic productivity of the ecosystems of Tanzania under different forms of land-use, for example, nomadic systems, sedentary agricultural systems, ranching and commercial hunting/tourism, each system of land-use must be considered in turn, and an estimate made of the total economic value of that system. As described in previous sections of this report, total economic value comprises the direct-use and indirect-use values of the ecosystem. Therefore an analysis must incorporate an estimate of both types of value.

To assess the primary products utilized by individuals from the ecosystem, that is, the direct-use value of the system, studies will be required to estimate the off-take from the environment of natural products, both vegetation and animal, as well as the number of people sustained by this system (population density). This can be achieved through, for example, sociological studies that directly measure the quantity and type of natural resources used. The productivity of agricultural goods can also be measured in this manner.

To estimate the value of products that are not marketed, a market value must be assigned to them. This can be through comparison with values of similar resources that have a market value or, alternatively, an estimate may be made of the compensation that individuals would have to receive in order to provide themselves with similar products to the ones provided by their ecosystem. For systems of land-use where the products are marketed a direct estimate can be made of their economic value.

To estimate the indirect-use values of the ecosystem, an understanding of the role of ecological support functions of an ecosystem must be gained through, for example, ecological, geographical or hydrological studies. With this knowledge, a production function approach, as outlined earlier, can then be used to translate these ecological relationships into economic values.

Achieving a complete economic valuation of an ecosystem requires a substantial amount of research. However, even if complete valuation is not initially achievable, an estimation of some of the indirect values of ecosystems may be sufficient to allow a comparison of different land-uses. Optimization techniques can then be used to estimate the socially optimal level of various natural resources required to maximize social welfare.

Notes

1. For a detailed explanation of the mathematics of profit maximization, see Dowling (1992).
2. Non-use values have also been identified but are not considered in this case. Estimation of their values is extremely difficult and not deemed relevant here. See Heywood (1995) for further information.
3. The mathematics of the problem-solving are not given here but are relatively simple to carry out.
4. Further details of how to carry out partial differentiation can be found in Dowling (1992).

References

Anderson, A.B. and E.M. Ioris (1992), 'The logic of extraction: resource management and income generation by extractive producers in the Amazonian estuary', in K.H. Redford and C. Padoch (eds), *Conservation of Neotropical Forests: Working from Traditional Resource Use*, New York: Colombia University Press, pp. 175–99.

Anderson, D. (1987), *Economic Aspects of Deforestation: A Case Study in Africa*, Washington, DC: The Johns Hopkins University Press.

Ataroff, V. and F. Rada (2000), 'Deforestation impact on water dynamics in a Venezuelan Andean cloud forest', *Ambio*, **29**(7), 440–44.

Barbier, E.B. (1994), 'Valuing environmental functions: tropical wetlands', *Land Economics*, **70**, 155–73.

Barbier, E., J. Burgess, T. Swanson and D. Pearce (1990), *Elephants, Economics and Ivory*, London: Earthscan Publications.

Barreto, P., P. Amaral, E. Vidal and C. Uhl (1998), 'Costs and benefits of forest management for timber production in eastern Amazonia', *Forest Ecology and Management*, **108**(1/2), 9–26.

Bembridge, T. and J. Steenkamp (1976), 'An agroeconomic investigation of beef production in the Matabeleland Midlands provinces of Rhodesia', *Rhodesian Agriculture*, **73**(2), 100–112.

Campbell, B.M., M. Luckert and I. Scoones (1997), 'Local-level valuation of Savanna resources: a case study from Zimbabwe', *Economic Botany*, **51**(1), 59–77.

Campbell, B.M., R. Costanza and M. van den Belt (2000a), 'Special section: Land-use options in dry tropical woodland ecosystems in Zimbabwe: Introduction, overview and synthesis', *Ecological Economics*, **33**(3), 341–51.

Campbell, B.M., D. Dore, M. Luckert, B. Mukamuri and J. Gambiza (2000b), 'Economic comparisons of livestock production in communal grazing lands in Zimbabwe', *Ecological Economics*, **33**(3), 413–38.

Coe, M.J., D.M. Cumming and J. Phillipson (1976), 'Biomass and production of large African herbivores in relation to rainfall and primary production' [cited in Campbell et al., 2000a], *Ecology*, **22**(4), 341–54.

Common, M.S. (1996), *Environmental and Resource Economics: An Introduction*, London: Longman.

Cossins, N. (1985), 'The productivity and potential of pastoral systems', *ILCA Bulletin*, No. 21, 10–15.

Dixon, J.A. and P.B. Sherman (1990), *Economics of Protected Areas. A New Look at Benefits and Costs*, London: Earthscan Publications Ltd.
Dowling, E.T. (1992), *Introduction to Mathematical Economics*, New York: McGraw-Hill.
Fuentes, E. (1980), 'Los Yanomami y las plantas silvestres' (The Yanomami and wild plants), *Antropologica*, **54**(3), 3–138.
Godoy, R. (1992), 'Some organizing principles in the valuation of tropical forests', *Forest Ecology and Management*, **50**(1/2), 171–80.
Godoy, R., R. Lubowski and A. Markandya (1993), 'A method for the economic valuation of non-timber tropical forest products', *Economic Botany*, **47**(3), 220–33.
Hanemann, W. (1988), 'Economics and the preservation of biodiversity', in E.O. Wilson and F.M. Peters (eds), *Biodiversity*, Washington, DC: National Academy Press, pp. 193–9.
Harmon, M.E. (2001), 'Carbon sequestration in forests – addressing the scale question', *Journal of Forestry*, **99**(4), 24–9.
Hecht, S. (1992), 'Valuing land-use in Amazonia: colonist agriculture, cattle and petty extraction in comparative perspective', in K.H. Redford and C. Padoch (eds), *Conservation of Neotropical Forests: Working from Traditional Resource Use*, New York: Colombia University Press, pp. 379–99.
Heywood, V.H. (1995), *Global Biodiversity Assessment*, New York: Cambridge University Press.
Hill, C., W. Griffiths and G. Judge (1997), *Undergraduate Econometrics*, New York: John Wiley and Sons.
Huntoon, P.W. (1992), 'Hydrogeologic characteristics and deforestation of the stone forest karst aquifers of South China', *Ground Water*, **30**(2), 167–76.
IIED and HNWCP (1997), 'Local-level assessment of the economic importance of wild resources in the Hadejia-Nguru wetlands, Nigeria', *Sustainable Agricultural Programme Research Series*, **3**(3).
Kirby, D.K. (2000), 'An ecological economic approach to upland heather moorland management', DPhil thesis, York: University of York.
Kundhlande, G., W.L. Adamowicz and I. Mapaure (2000), 'Valuing ecological services in a savanna ecosystem: a case study from Zimbabwe', *Ecological Economics*, **33**(3), 401–12.
Lal, R. (1997), 'Deforestation effects on soil degradation and rehabilitation in western Nigeria. IV. Hydrology and water quality', *Land Degradation and Development*, **8**(2), 95–126.
Martin, G. (1983), 'Bushmeat in Nigeria as a natural resource with environmental implications', *Environmental Conservation*, **10**(2), 125–32.
Myers, N. (1995), 'The world's forests – need for a policy appraisal', *Science*, **268**(5212), 823–4.
Nations, J.D. (1992), 'Xateros, chicleros and pimenteros: harvesting renewable tropical forest resources in the Guatemalan Petén', in K.H. Redford and C. Padoch (eds), *Conservation of Neotropical Forests: Working from Traditional Resource Use*, New York: Colombia University Press, pp. 208–19.
Niamir-Fuller, M. (1998), 'The resilience of pastoral herding in Sahelian Africa', in F. Berkes, C. Folke and J. Colding (eds), *Linking Social and Ecological Systems*, Cambridge, UK and New York: Cambridge University Press.
Nicholson, W. (1992), *Microeconomic Theory*, Orlando: Dryden Press.
Oommachan, M. and S.K. Masih (1988), 'Multifarious uses of plants by the forest tribals of Madhya Pradesh: wild edible plants', *Journal of Tropical Forestry*, **4**(2), 163–9.
Panayotou, T. (1990), 'The economics of environmental degradation. Problems, causes and responses', Development and Discussion Paper No. 335, Harvard Institute for International Development, Cambridge, MA.
Pautsch, G.R., L.A. Kurkalova, B.A. Babcock and C.L. Kling (2001), 'The efficiency of sequestering carbon in agricultural soils', *Contemporary Economic Policy*, **19**(2), 123–34.
Peters, C.M., A.H. Gentry and R.O. Mendelsohn (1989), 'Valuation of an Amazonian rainforest', *Nature*, **339**(6227), 655–6.

Phillips, O., A.H. Gentry, C. Reynel, P. Wilkin and C. Galvezdurand (1994), 'Quantitative ethnobotany and Amazonian conservation', *Conservation Biology*, **8**(1), 225–48.

Pinedo-Vasquez, M., D. Zarin and P. Jipp (1992), 'Economic returns from forest conversion in the Peruvian Amazon', *Ecological Economics*, **6**(2), 163–73.

Putuhena, W.M. and I. Cordery (2000), 'Some hydrological effects of changing forest cover from eucalypts to *Pinus radiata*', *Agricultural and Forest Meteorology*, **100**(1), 59–72.

Rawat, J.S. and M.S. Rawat (1994), 'Accelerated erosion and denudation in the Nana-Kosi watershed, Central Himalaya, India. 1. Sediment load', *Mountain Research and Development*, **14**(1), 25–38.

Redford, K.H. and J. Robinson (1985), 'Hunting by indigenous peoples and conservation of their game species', *Cultural Survival*, **8**(4), 41–4.

Scoones, I. (1992), 'The economic value of livestock in the communal areas of southern Zimbabwe', *Agricultural Systems*, **39**(4), 339–59.

Scoones, I., M. Melnyk and J.N. Pretty (1992), *The Hidden Harvest: Wild Foods and Agricultural Systems*, Sustainable Agricultural Programme, London: International Institute for Environment and Development.

Swanson, T. and E. Barbier (1991), *Economics for the Wilds. Wildlife, Wildlands, Diversity and Development*, London, Earthscan Publications Ltd.

Toledo, V.M., A.I. Batis, R. Becerra, E. Martinez and C.H. Ramos (1995), 'The useful forest – quantitative ethnobotany of the indigenous groups of the humid tropics of Mexico', *Interciencia*, **20**(4), 177–87.

Uri, N.D. (2001), 'The potential impact of conservation practices in US agriculture on global climate change', *Journal of Sustainable Agriculture*, **18**(1), 109–31.

Wilson, E.O. (1994), *The Diversity of Life*, London: Penguin.

12. Economic growth and the environment
Dalia El-Demellawy

Introduction

This chapter critically reviews empirical literature on the relationship between economic growth and the environment. It starts with outlining the debate around the relationship and then presents basic models used in the empirical studies. Finally, the various empirical studies are discussed.

Study of the relationship between economic growth and the environment goes back to the 1960s. It started in developed countries when considerable concern arose regarding the impact that economic development was having on their environment. This growing concern led to the 1972 United Nations World Conference on the Human Environment in Stockholm. This conference highlighted the conflict of interest between developed and developing countries regarding the relationship between economic growth and the environment. The developing countries were more concerned with development at the expense of the environment. In other words, slowing economic development to protect the environment was not appreciated. Even in developed countries there was conflict of interest among different groups in society, with workers ranking improvements of their standard of living associated with technological advancements above environmental concerns (Beckerman, 1992). Despite conflict of interest, concerns about the natural environment and its link to problems of economic development were explicitly stated. It was argued that exhaustion of the environmental resource base, in terms of minerals and food production, would provide limits for future economic growth. However, it was not thought that economic collapse was inevitable. On the contrary, it was concluded that the world economic system could be sustainable into the future provided that radical changes in the way it was being run were adopted (Meadows et al., 1972).

Awareness of the relationship between the environment and economic growth gained prominence in 1987 for a number of reasons. First, the widespread environmental and economic problems in the world; particularly high levels of poverty and famines in various countries. Second, the urgent and complex problem of the national and global scale of environmental pollution; especially that associated with climate change due to increasing emissions of greenhouse gases (GHGs) causing global warming. This, in turn, was considered to lead to serious unfavourable effects on (1) agriculture, with output falling as the interior of most continents became drier, (2) sea levels rising and flooding coastal communities, (3) other effects, such as greater need for the use of air-conditioning, loss of forests and possible increase in the frequency of storms. Third, the serious local environmental problems in developing countries; these problems included supply of safe drinking water, access to decent sanitation, deterioration of air quality and urban degradation, which had damaging effects on health and human welfare (Brundtland, 1987 and IBRD, 1992).

It was recognized that alleviation of poverty would not be accomplished by ceasing economic development or by redistribution of wealth from rich to poor countries, but by reviving economic growth in developing countries. Also, due to the interdependence of the world economy, the prospects of the developed countries would depend on the economic growth of the developing ones. Finally, economic growth had to be based on policies that would sustain the resource base of the biosphere and such growth would alleviate the problems of poverty and underdevelopment (Brundtland, 1987).

The economic-growth–environment relationship is one of the prominent debates of the international community at the present time. At one extreme, some social and physical scientists argue that, despite the rise in income, higher economic growth exhausts natural resources, causes higher accumulation of waste and increases pollution (Meadows et al., 1972). This, in turn, results in environmental degradation and thereby, decline in a human welfare. At the other extreme, some social and economic environmentalists argue that higher income increases the demand for better environmental quality (Beckerman, 1992). In other words, economic growth and environmental improvements follow the same path. In between these two views, some economists and environmentalists argue that the relationship between economic growth and the environment is not fixed over time. They argue that environmental quality deteriorates with economic growth until it reaches a certain threshold, after which improvements in environmental quality take place. In other words, the relationship follows an inverted U-shaped path, known as the Environmental Kuznets Curve – EKC (Kuznets, 1955; Grossman and Krueger, 1991; Shafik and Bandyopadhyay, 1992; Panayotou, 1993; Selden and Song, 1994).

Determining the relationship between economic growth and the environment is crucial as it has serious policy implications. A positive relationship between economic growth and environmental degradation requires strict environmental policies in order to limit economic growth to a level within the carrying capacity of the environmental base (Arrow et al., 1995). On the other hand, a negative relationship requires policies that accelerate economic growth, rather than restrictive environmental policies, in order to achieve environmental improvements (Beckerman, 1992). However, if the relationship is positive at lower levels of income and then turns negative at higher levels of income, then policies that accelerate economic development, although improving environmental quality in the long run, will have serious damaging effects on the environment in the short and medium run. Therefore, caution is necessary in introducing the right policies at the different stages of economic development without sacrificing environmental improvement at any level. Economic development policies should be set and viewed in relation to their effects on the environment. Both economic and environmental policies should complement each other. In fact, good environmental protection policies will help and sustain economic development. However, promoting economic growth without taking into account the impact on the environment can bring economic development to a halt (IBRD, 1992).

In that respect, the enactment and enforcement of international legislation by the international community is vital in controlling and minimizing the problems resulting from economic development and its damaging effect on the environment for a number of reasons. First, they tend to moderate the growth rates of pollution associated with economic growth by encouraging firms and/or consumers to use less polluting technologies. This is essential because individuals and firms are interested in maximizing profits, and therefore have few incentives in curbing pollution, such as air emissions in urban centres, dumping wastes in public waters or the severe overuse of land (Congleton, 1992). Moreover, the gains from a better environment, such as improved human health, higher economic productivity and enhanced amenities, are sometimes difficult to measure compounded with the lengthy time lag between taking the proper actions and realizing the effects of environmental improvements (Goodstein, 1999). Furthermore, individuals and firms can make serious trade-offs in their decision-making due to the lack of consideration of the earth's limited resource capacity and pollution to the environment, which, in turn, will impede sustainable development. This has an intergenerational effect as it results in compromising the average standard of living of future generations in comparison with the present one (Beckerman, 1992). This is also known as the intergenerational displacement of environmental costs, where costs are transferred across a great distance or to a remote future (Roca, 2003).

Second, the costs of environmental protection are high and are often unaffordable by poorer countries. Accordingly, the international community sets policies whereby rich countries bear a reasonable share of the costs of improving the environment in poor countries. This is due to the fact that the benefits from a cleaner environment, such as the protection of tropical forests and biodiversity, will accrue not only to the poor countries but to the rich ones as well. Also, some of the problems facing the world, such as global warming and ozone depletion, stem from high consumption in the richer countries. Moreover, the international community allows and facilitates the access to cleaner technologies and lessons learned from developed countries by less-developed ones (IBRD, 1992).

Third, the nature of environmental protection, being a public good, makes it difficult to control due to transboundary and free-rider effects. A public good is defined as 'non-rival and non-excludable in use' (Common, 1995, p. 129). The transboundary effect results when environmental degradation crosses national boundaries. This can take a number of forms. The first is where neighbouring countries share a common resource and the action of one country affects the others. Examples that fall in this category are acid rain and the management of regional seas and international rivers. The second is when one country's action affects the world's global environmental resources, such as the atmosphere, and thereby has an effect on all of the other countries. Examples that fall in this category are ozone layer depletion and global warming. The third form is when resources belonging to one country have high values for the international community. Examples that fall in this category are tropical rainforests, individual species and special ecological habitats. The free-rider effect, where countries have the incentives to take advantage of the efforts of others to provide the public good without paying their share or contributing to this step, also necessitates the existence of institutions and policies (IBRD, 1992; Goodstein, 1999; Stavins, 2000).

Fourth, the international community negotiates international environmental standards and sets them based on common principles and rules of collaboration among the various countries. It gives proper weight to the interests of all countries, including the poor and politically weak ones. In addition, it monitors, enforces and abides by multilateral agreements on environmental matters (Congleton, 1992; IBRD, 1992).

The relationship between economic development and environmental pollution is complex as it involves a lot of different interrelated factors. Among these factors are: the size of the economy, the sectoral structure, the age of technology, the public demand for environmental quality and the level and quality of environmental protection expenditure (Goodstein, 1999). However, the relationship mainly centres around five questions. First, does the EKC relationship between income and environmental degradation exist, and if yes, what is the income turning point? Second, what are the other factors affecting this relationship, such as population growth/density, international trade, policy, geography and income distribution and what roles do they play? Third, how reliable are the results derived from cross-sectional country or panel data in forecasting the environmental path for an individual country? Fourth, what is the irreversible damage to the ecological thresholds as a result of economic growth in terms of the carrying capacity of the biosphere, ecosystem resilience and sustainability? Fifth, what is the role played by institutions and policies in (1) explaining the shape of the environment–income-growth relationship, (2) reducing environmental damage and (3) ensuring sustainability of outcomes in the future (Panayotou, 2000)?

Empirical models on economic growth and the environment
Most of the empirical models used in the studies on the economic-growth–environment relationship consisted of reduced-form equations relating environmental indicator(s) as the dependent variable(s) to a measure of income as the independent variable. They basically captured the net impact of income on the environment, without understanding the various influences underlying this relationship. Some studies also controlled for other variables, such as population density/growth, international trade, income distribution, institutional policies and geography. The environmental indicators were either in the form of pollutant emissions (such as sulphur dioxide (SO_2)), particulates, nitrogen oxides (NO_x), carbon dioxide (CO_2), carbon monoxide (CO) or ambient concentrations of various pollutants as recorded by monitoring stations. The measure of income that was commonly used was income per capita, however, some studies used income data converted into purchasing power parity (PPP) or income at market exchange rates. The shape of the relationship between income and environmental degradation was econometrically estimated using cross-sectional or panel data and the functional equation used was usually quadratic, log-quadratic or cubic form. The reduced-form approach turned the income–environment relationship into a 'black box' as it did not consider the underlying determinants of environmental quality and which, in turn, limited its use in policy formulation. This led to studies using decomposition analysis in order to disentangle other effects, such as scale or level of production, composition and structure, abatement technology, trade, preferences, the role of institutions and environmental policies to understand the controlling effects behind the relationship between income and environmental degradation (Panayotou, 2000).

The types of models used in these empirical studies differed. As segregated by Panayotou (2000), the simplest model was the one examining the relationship between per capita income and an environmental indicator(s). In some cases, these models included a time trend and were usually in linear, quadratic, log-linear or log-quadratic forms (Shafik and Bandyopadhyay, 1992; Hettige et al., 1992; Rothman, 1998; Kahn, 1998):

Linear $E_{it} = \beta_0 + \beta_1 y_{it} + \varepsilon_{it}$;
Quadratic $E_{it} = \beta_0 + \beta_1 y_{it} + \beta_2 (y_{it})^2 + \varepsilon_{it}$;
Log-linear $E_{it} = \beta_0 + \beta_1 \ln(y_{it}) + \varepsilon_{it}$;
Log-quadratic $E_{it} = \beta_0 + \beta_1 \ln(y_{it}) + \beta_2 (\ln y_{it})^2 + \varepsilon_{it}$;

where:
E = environmental indicator;
y = per capita income;
β = parameter to be estimated;
t = time trend;
ε = error term.

Other models included population density as another independent variable, commonly in a log-quadratic form equation (Panayotou, 1993; Selden and Song, 1994; Cropper and Griffiths, 1994; Roberts and Grimes, 1997; Vincent, 1997; Carson et al., 1997):

Log-quadratic $E_{it} = \beta_0 + \beta_1 \ln(y_{it}) + \beta_2 \ln(P_{it}) + \beta_3 (\ln y_{it})^2 + \beta_4 (\ln P_{it})^2 + \varepsilon_{it}$;

where:
E = environmental indicator;
y = per capita income;
P = population density;
β = parameter to be estimated;
t = time trend;
ε = error term.

A third specification of models was used that included geographical characteristics as independent variables, in addition to per capita income and population density, usually using a quadratic form equation (Grossman and Krueger, 1991, 1995):

Quadratic $E_{it} = \beta_0 + \beta_1 (y_{it}) + \beta_2 (P_{it}) + \beta_3 (G_{it}) + \beta_4 (y_{it})^2 + \beta_5 (P_{it})^2 + \beta_6 (G_{it})^2 + \varepsilon_{it}$;

where:
E = environmental indicator;
y = per capita income;
P = population density;
G = geographic characteristic;
β = parameter to be estimated;
t = time trend;
ε = error term.

A fourth type of model included trade variables and per capita income as independent variables, commonly using a quadratic form equation. Examples of trade variables are export or import manufacturing ratio, commerce intensity and prices for goods, such as wood or steel (Cropper and Griffiths, 1994; Cole et al., 1997; Suri and Chapman, 1998; Kaufmann et al., 1998):

Quadratic $\quad E_{it} = \beta_0 + \beta_1 y_{it} + \beta_2 (y_{it})^2 + \beta_3 T_{it} + \varepsilon_{it};$

where: $\quad E$ = environmental indicator;
y = per capita income;
T = trade variable;
β = parameter to be estimated;
t = time trend;
ε = error term.

A fifth type of model used institution variables in addition to per capita income, mostly using a linear form equation. Examples of institution variables were political rights and civil freedom and macro-policy variables, such as proportion of indebtedness to GDP and black market premium on the exchange rate (Torras and Boyce, 1998; Bhattarai and Hammig, 2001):

Linear $\quad E_{it} = \beta_0 + \beta_1 y_{it} + \beta_2 I_{it} + \beta_3 M_{it} + \varepsilon_{it};$

where: $\quad E$ = environmental indicator;
y = per capita income;
I = institutions related variable;
M = macro-policy related variable;
β = parameter to be estimated;
t = time trend;
ε = error term.

The last and more comprehensive type of models included policy variables in addition to population density, growth and per capita income as independent variables, applying a cubic form equation (Panayotou, 1997):

Cubic $\quad E_{it} = \beta_0 + \beta_1 (y_{it}) + \beta_2 (y_{it})^2 + \beta_3 (y_{it})^3 + \beta_4 (P_{it})$
$\quad\quad\quad + \beta_5 (P_{it})^2 + \beta_6 (P_{it})^3 + \beta_7 (g_{it}) + \beta_8 (g_{it})(y_{it})$
$\quad\quad\quad + \beta_9 (p_{it}) + \beta_{10} (p_{it})(y_{it}) + \varepsilon_{it};$

where: E = environmental indicator;
y = per capita income;
P = population density;
g = economic growth rate;
p = policy variable;
β = parameter to be estimated;
t = time trend;
ε = error term.

Empirical studies on economic growth and the environment
The shape of the relationship between economic growth and the environment centres around four main explanations. It does not, however, imply that there is a causal relationship between economic growth and environmental quality (Perrings and Ansuategi, 2000). The first explanation is the sectoral composition of the economy. The theory is that as the economy moves along the production structure from agriculture to industry, local air and water quality is bound to deteriorate due to the nature of industrialization. Industry in the low-income countries is mainly based on highly polluting industries, such as chemicals, textiles, food processing, and so on. As income rises, the economy moves towards expanding the services sector, and thereby, environmental quality improves (Grossman and Krueger, 1991; Panayotou, 1993, 1997; De Bruyn, 1997).

The second explanation is the technology shifts that accompany economic growth. The interpretation is that as the economy becomes richer, companies tend to use cleaner technology, which, in turn, improves environmental quality (Seldon and Song, 1995; De Bruyn, 1997; Andreoni and Levinson, 1998).

The third explanation is the income elasticity of demand for environmental quality. It is interpreted that, since environmental quality is a luxury good, as per capita income rises, the demand or preferences for environmental amenity increase, and therefore new institutions and regulations are created to protect the environment. However, given the nature of environmental quality being a non-marketable good, changes in its demand can be captured indirectly through changes in technology, consumption, policy and regulations as well as their direct impact on the environment (Antle and Heidebrink, 1995; McConnell, 1997).

The fourth explanation is the environmental constraints or the carrying capacity of the biosphere. It is argued that economic growth is constrained by the carrying capacity of the environment. In other words, there is a limit to pollution or depletion resulting from growth that can be absorbed by the environment, beyond which economic growth will come to a halt, especially if the effects of growth on the environment are irreversible. This, in turn, stimulates policy responses to improve the environmental quality. The argument here is that, at low levels of income, the impact of economic growth is within the carrying capacity of the biosphere. As income rises, the environmental constraints tighten, and therefore induce regulation and policy responses to reduce environmental degradation (Arrow et al., 1995).

Basic studies of Environmental Kuznets Curve
The relationship between economic development and environmental quality is a recent controversial debate, which started at the beginning of the 1990s. It centres around the hypothesis that there is an empirical relationship between income per capita and environmental quality in terms of pollution or resource depletion. There have been several observations by a number of economists and ecologists that, as income increases, environmental degradation increases up to a point, after which environmental quality improves. In other words, the relationship has an 'inverted-U' shape. This relationship is referred to as the 'Environmental Kuznets Curve'.

The name of the relationship is derived from the similarity of the curve to the relationship between income inequality and income growth in the income distribution theory by Simon Kuznets in 1955, and which is termed the 'Kuznets Curve'. Given the obvious analogy, the bell-shaped relationship between per capita income and pollution is named the Environmental Kuznets Curve (EKC, hereafter).

Empirical literature on the EKC goes back to a paper prepared by Grossman and Krueger (1991) for a conference on the US–Mexico Free Trade Agreement. In this paper, which was later published in 1993, Grossman and Krueger investigated the implications of economic growth as a result of the North American Free Trade Agreement (NAFTA) on environmental degradation as there were claims that the agreement might increase environmental degradation. They studied the effect of reduction in trade barriers on the environment in Mexico and assessed the relative magnitudes of change in the level of pollution resulting from further trade liberalization of three effects: expanding the scale of economic activity, altering the composition of economic activity and changing the production techniques. They carried out their study using a cross-country panel of data on urban air pollution in two or three locations in each group of cities in 42 countries from 1977 to 1988. The three pollutants were suspended particulate matter (SPM), SO_2 and dark matter (smoke). The data were taken from the Global Environmental Monitoring System (GEMS) published by the World Health Organization (WHO). The income data used were real per capita GDP measured in 1985 US PPP from Summers and Heston (1991) (Penn World Tables). They found that concentrations for sulphur dioxide and smoke increased with higher per capita GDP at low levels of national income and decreased with higher levels of income, with a turning point between $4000 and $5000, measured in 1985 US dollars of GDP. For the mass of suspended particles, in a given volume of air, they found that the relationship was monotonically decreasing. They also found that all three pollutants increased at an income level of $10 000 to $15 000.

In another study, Grossman and Krueger (1995) examined the relationship between per capita income and four types of environmental indicators: urban air pollution (SO_2, smoke and heavy particles), the state of the oxygen regime in river basins (dissolved oxygen, biological oxygen demand (BOD), chemical oxygen demand (COD) and concentration of nitrates), faecal contamination of river basins (faecal coliform and total coliform) and contamination of river basins by heavy metals (concentration of lead, cadmium, arsenic, mercury and nickel). They used panel data across many countries for the 14 environmental indicators obtained from the GEMS. They found evidence that, for most indicators, environmental quality initially deteriorated with economic growth, followed by a subsequent phase of improvement. They also found that the turning points for the different pollutants varied but, in most cases, were below an income per capita level of $8000, measured in 1985 US dollars of GDP. They also found that for seven indicators there was a statistically significant positive relationship between environmental quality and per capita income GDP of $10 000. Only in the case of total coliform did they find a significant adverse relationship at this level of income. They also highlighted that there were a few points to be taken into account with regard to their findings. First, there was no reason to believe that improvement in environmental quality as a result of economic development was an automatic process. Second, the reason for the downward slope and inverted U-shaped curve could be due to countries ceasing to produce pollution-intensive goods as they imported these goods from other countries. Third, the relationship observed in previous research reflected the technological, political and economic conditions existing at that time, from which poor countries could learn and avoid the mistakes of earlier growth experiences.

Shafik and Bandyopadhyay (1992) explored the relationship between economic growth and environmental quality in a background paper for the *World Development Report*, which was later published by Shafik in 1994. They conducted a systematic analysis on environmental quality data from 149 countries for the period 1960 to 1990. They used ten environmental indicators: lack of clean water, lack of urban sanitation, ambient levels of SPM, ambient sulphur oxides (SO_2), change in forest area and annual rate of deforestation between 1961 to 1986, dissolved oxygen in rivers, faecal coliforms in rivers, municipal waste per capita and carbon emissions per capita. The income data used were GDP per capita measured in 1985 US PPP. In order to overcome the comparability problem of data across countries arising from differences in definitions and inaccurate measurement sites, they applied a simple modelling technique, on a consistent basis, to a large number of environmental quality indicators and countries. They highlighted that there were four determinants affecting environmental quality in any given country. The determinants were (1) climate and location, which was referred to as endowment, (2) per capita income, which reflected the production structure, urbanization and consumption patterns of private goods, (3) exogenous factors, which were available to all countries, but change over time, such as technology, and (4) policies, which reflected the social decisions affecting the provision of environmental public goods. Shafik and Bandyopadhyay in their study focused on the relationship between per capita income and environmental quality, while taking into account the other determinants. The study showed that some environmental indicators improved with higher income per capita, such as water and sanitation, others worsened then improved, such as particulates and sulphur oxides, and others deteriorated steadily, such as dissolved oxygen in rivers, municipal solid wastes and carbon emissions. For those environmental indicators that followed a bell-shaped curve with rising income, the turning points varied. For instance, the turning point for particulates was around $3280 per capita, while that of sulphur oxides was around $3670 per capita.

Shafik (1994) concluded that individuals and countries tended to address environmental problems at different stages of economic development depending on the relative costs and benefits of the environmental problem. For instance, water and sanitation were the first to be addressed, given the relatively low costs and high private and social benefits involved. Local air pollution came next as it imposed external but local costs and was relatively costly to abate, with high benefits. They were also addressed when countries reached a middle level of income where they became more severe, as middle-income economies were often energy-intensive and industrialized. Environmental problems that could be externalized, and therefore, borne by poor or other countries, came last, such as solid wastes and carbon emissions, as there were few incentives to bear the associated significant costs to abate. Moreover, the level of technology had a positive impact on certain environmental quality indicators. If the costs of degradation were localized, such as water sanitation and air pollution, there would be high demand for improvements, and therefore technology was crucial. However, where costs could be externalized and there was uncertain knowledge of their detrimental effects, such as carbon emissions, there was little demand for technological improvements. Furthermore, he highlighted that environmental improvements were not automatic. They required effective environmental policies and investments to be put in place to reduce environmental degradation.

Panayotou (1993) examined the relationship between nominal per capita of GDP and four environmental indicators: SO_2, NO_x, SPM and deforestation using cross-sectional data. The income data used were GDP per capita in 1985 nominal US dollars. There were 54 countries in the pollution sample and 68 in the deforestation sample and the three pollutants were measured in terms of emissions per capita on a national basis. He found evidence of EKC for the four indicators with turning points per capita of around $823 for deforestation, $3000 for SO_2, $5500 for NO_x and $4500 for SPM, measured in 1985 US dollars of nominal GDP. Panayotou concluded that:

> at low levels of development both the quantity and intensity of environmental degradation are limited to the impacts of subsistence economic activity on the resource base and to limited quantities of biodegradable wastes. As economic development accelerates with the intensification of agriculture and other resource extraction and the take off of industrialization, the rates of resource depletion begin to exceed the rates of resource regeneration, and waste generation increases in quantity and toxicity. At higher levels of development, structural change towards information-intensive industries and services, coupled with increased environmental awareness, enforcement of environmental regulations, better technology and higher environmental expenditures, result in levelling off and gradual decline of environmental degradation. (Panayotou, 1993, p.1)

In another study on deforestation, Cropper and Griffiths (1994) examined the relationship between per capita income and the rate of deforestation in three regions: Africa, Latin America and Asia. They used pooled cross-sectional and time series data for each region for the period 1961 to 1988. Their sample included 64 developing countries. They also examined the effects of population pressure on deforestation, which they captured by including the rural population density and the rate of population growth as independent variables in their equation. They reported two findings. First, an EKC relationship for both Africa and Latin America with turning points of $4760 and $5420, respectively. These turning points were higher than those found by Panayotou, possibly because of the use of panel data. Second, the rural population density for Africa shifted the relationship between income and the rate of deforestation upwards. However, none of the variables included for Asia were statistically significant. They concluded that reducing the rate of population growth was not necessarily the best method for reducing the rate of deforestation in developing countries. They also said that deforestation was the problem of market failure, which resulted from the lack of defining and enforcing property rights, where people had no right in land ownership, and therefore no incentive to make efficient decisions in the use of land.

Antle and Heidebrink (1995) examined the relationship between income and two classes of environmental amenities: parklands and forests in low- and high-income countries. The environmental indicators used were total area of parks and protected areas (PARKS), deforestation (DEFOR), afforestation (AFFOR) and total forest area (FOR). They gathered their income data and area values from the *World Development Report 1987*, PARKS data from the *1989/90 Environmental Data Report* by the United Nations Environment Programme (UNEP) and FOR and DEFOR data from *World Resources 1990/91*. The income data used were per capita GNP in 1985 nominal US dollars. They found a U-shaped relationship and the income elasticity of demand for environmental quality was near zero for countries with income per capita of around $1200–$2000, while being positive and generally exceeding unity for countries with higher per capita income levels.

However, in an empirical study done by Koop and Tole (1999) in which they examined the relationship between deforestation and per capita GDP, using less restrictive model specifications, they did not find an EKC relationship between the two variables. They used data for 76 tropical developing countries for various years during the period 1961–92. The deforestation indicator used was the loss in forest cover derived from the *FAO Production Yearbook*. The per capita GDP data were taken from the Penn World Table. Also they included two demographic measures; the population density and population change, which were derived from the *FAO Production Yearbook* and the Penn World Table. The authors concluded that the extreme assumptions used in previous studies that assumed uniformity across the countries led to the existence of the EKC pattern between deforestation and per capita income. Upon adopting random coefficient specifications, as they did in their study, the inverted U-shaped relationship did not exist. They added that their approach was more realistic given the wide variety of physical and social characteristics across the countries.

With respect to urban environmental quality, a large number of studies concentrated on air pollution. Selden and Song (1994) carried out an empirical study using a cross-national panel of data on aggregate emissions to examine the relationship between pollution and economic development. They used aggregate emissions instead of urban air quality data as they believed that the latter was valid if the objective was to understand the factors underlying the pollution faced by urban dwellers. On the other hand, they were trying to understand the environmental impact beyond urban areas, and therefore the analysis using aggregate emissions was perceived to provide greater insight. They concentrated on four air emissions: SPM, SO_2, NO_x and CO across 30 countries and across time. Out of the 30 countries, 22 were categorized as high income, six were middle income and two were low income. The emissions data constructed from fuel consumption figures were obtained from the World Resources Institute (1991) and were measured in terms of kilograms per capita on a national basis. The income data used were per capita GDP in 1985 US dollars. Their findings showed a U-shaped curve of per capita emissions for all four air pollutants and per capita GDP. However, they found substantially higher turning points for the similar pollutants used by Grossman and Kreuger. The estimated turning points were $8709 for SO_2, $11 217 for NO_x, $10 289 for SPM and $5963 for CO. They considered this to be reasonable given their use of aggregate emissions data as opposed to urban atmospheric concentrations. On the other hand, their results were widely different from those found by Panayotou, who also used aggregate emissions data. As the turning points were high enough above the per capita income of most countries, they concluded that global emissions of these pollutants would continue to increase in the foreseeable future.

Cole et al. (1997) examined the relationship between per capita income and a wide range of environmental indicators, which had indirect impact on the environment using cross-country panel data sets. These indicators were carbon dioxide, CFCs and halons, methane, nitrogen dioxide, sulphur dioxide, suspended particulate matter, carbon monoxide, nitrates, energy consumption, municipal waste and traffic volume. They arrived at three main conclusions. First, the EKC existed only for local air pollutants, while indicators with more global or indirect impact either increased monotonically with income or had turning points at very high levels of per capita income with large standard errors, unless they had been subjected to a 'multilateral policy initiative', which implied that the process was not automatic. Second, concentrations of local air pollutants in urban areas peaked at a lower per capita income level than total emissions, which implied that it was easier to improve the quality of urban air than to reduce national emissions. Third, transport-generated local air pollutants peaked at a higher per capita income level than total emissions per capita, which implied the severity and need to tackle the problem in the transport sector.

On the other hand, in a number of studies it has been found that CO_2 follows an EKC pattern. In a study by Holtz-Eakin and Selden (1995) in which they examined the relationship of CO_2 emissions and GDP, they found that carbon emissions increased monotonically, with a turning point of $35 428 GDP per capita. They used a panel data of 130 countries for the years 1951 to 1986. They also used the data to forecast global CO_2 emissions. They obtained the CO_2 emissions data from the Oak Ridge National Laboratory (ORNL), while the income and population data were gathered from the Penn Mark V World Tables in Summers and Heston (1991). The income data were per capita GDP in 1985 nominal US dollars. Using the results of their study, they arrived at four conclusions. First, the marginal propensity to emit (MPE) carbon dioxide diminished as GDP per capita increased. Second, global emissions would continue to increase at 1.8 per cent per annum. Third, the reason for the continued growth of CO_2 was the rapid population and economic growth in middle- and lower-income countries, which had the highest MPE. Fourth, economic development did not significantly alter the future of annual or cumulative flow of CO_2 emissions.

Also Roberts and Grimes (1997) examined the relationship between per capita GDP and national CO_2 emissions for the period 1961 to 1991. They also tracked the changes in the amount of carbon dioxide emitted (in kilograms) per unit of their GDP, known as the National Carbon Intensity (NCI), for low-, middle- and high-income countries, using data from the World Bank and the Carbon Dioxide Information Analysis Center (CDIAC). The GDP figures were adjusted for inflation using the World Bank's GDP deflator and the 1987 exchange rate. They found that CO_2 intensity had steadily worsened for low- and middle-income countries, but to a lower extent in the latter, however, it improved for high-income countries. Their results were also statistically significant in the early 1970s and then since 1982. They concluded that the EKC for carbon emissions was not due to countries passing through stages of development but rather due to a relatively small number of wealthy countries that became more efficient after 1970, while the rest of the world worsened.

Schmalensee et al. (1998) conducted a study to project the emissions of carbon dioxide from the combustion of fossil fuels to 2050. They used national-level panel data for 47 countries for the period from 1950 to 1990, which focused on around 80 per cent of CO_2 emissions caused by human activity. This was considered the most important source of potential future warming. The GDP data were in purchasing power parity of 1985 US dollars and were taken from the Penn World Table. They found evidence of a U-shaped relationship between per capita carbon emissions and per capita income. Their model also captured two typical patterns. First, for developing countries, there was continuous, or even accelerating growth of per capita CO_2 emissions as per capita income increased. Second, for highly developed countries, the relationship between per capita CO_2

emissions and per capita income followed an inverted U-shaped pattern. For instance, the turning point for the United States was $10 000 GDP per capita. In other words, per capita CO_2 emissions increased, then flattened and then reduced at higher levels of economic development.

Similar results to those of Schmalensee et al. were found in a study by Panayotou et al. (1999). They studied the relationship between per capita income and CO_2 emissions using panel national data with 3869 observations for 127 countries, with populations over 1 million, for the period 1960 to 1992. The sample accounted for approximately 95 per cent of the world's population and 90 per cent of the global CO_2 emissions from fossil fuels. The GDP data were obtained from the Penn World Table (1992) and were in 1985 US dollars purchasing power parity. The CO_2 data were taken from Marland et al. (1999). They found evidence of an inverted U-shaped relationship between per capita income and per capita CO_2 emissions, with a turning point of $12 000. They also found that the highest additions to CO_2 emissions were for income levels in the range of $1500 to $4000.

In a study by Ravallion et al. (1997), they found that CO_2 emissions and per capita GDP followed an inverted U-shaped pattern. They studied the relationship between per capita GDP and carbon emissions for 42 countries using the data from the Oak Ridge National Laboratory. The data on fuel use were obtained from the United Nations statistical division, where the Marland and Rotty (1984) method was applied in order to convert fuel consumption and cement manufacturing into carbon emissions. The per capita GDP data used were in PPP-adjusted values based on 1985 US dollars. They found that at low average income levels, carbon emissions tended to increase, but then the relationship flattened out at middle- to high-income levels and reversed at high levels of income. The authors also found that at higher income inequality, there was a sizeable positive impact on the aggregate income elasticity of carbon emissions, which indicated that both average levels of income and inequality had effects on carbon emissions.

Also Galeotti and Lanza (1999) attempted to shed further light on the issue of the greenhouse gas CO_2, using data covering 108 countries around the world over the period 1971 to 1995, using new functional forms developed by the International Energy Agency (IEA). In 1995, the 108 countries accounted for 88 per cent of the CO_2 emissions generated by fuel combustion. Their sample consisted of 2700 annual observations, of which 700 observations were from 28 OECD countries and 2000 observations were from 80 non-OECD countries. In addition to analysing the 108 countries in the sample, they also analysed the sub-samples of OECD and non-OECD countries in order to account for the different stages of economic development, technological position and other structural differences. Their findings showed a reasonable EKC relationship between the country GDP and CO_2 emissions. Under the Gamma functional form, the turning points were $13260 for all countries, $17868 for non-OECD countries and $15582 for OECD countries, while under the Weibull functional form, they were $13648, $17079 and $15709, respectively. They concluded that when using non-nested tests and applying new functional forms, such as the Gamma and Weibull, to describe the reduced-form relationship between GDP and CO_2 emissions, the emergence of a bell-shaped EKC was possible with reasonable turning points.

In a recent study, Cavlovic et al. (2000) conducted a statistical meta-analysis to synthesis the results of 25 existing EKC studies, with over 120 observations, and calculated new income turning points (ITPs) for 11 different pollutants to determine whether credible conclusions from previous studies could be made. These pollutant categories were toxic emissions, urban air quality, deforestation, heavy particulates, urban quality, water quality/pollution, heavy metals, SO_2, combustion by-products, hazardous waste and CO_2. They found the new ITPs ranged from $3020 for toxic emissions to $199345 for CO_2. They concluded that (1) ITP results were not necessarily representative across nations as their ITPs were higher than those of previous studies for developed countries, (2) ITPs for emissions and ambient concentrations of a particular pollutant were not comparable, (3) inclusion of trade effects variable, rather than income alone, would tend to increase the ITPs and (4) methodological choices would significantly affect the magnitude of ITPs.

In a study by Hettige et al. (1997) on water pollution, they examined the relationship between economic development and industrial water pollution. They used panel data collected at the factory level from national and regional environmental protection agencies (EPAs) in 13 countries: Brazil, China, India, Korea, Finland, Indonesia, Mexico, the Netherlands, the Philippines, Sri Lanka, Taiwan, Thailand and the United States of America. They basically tested the effect of income growth on three determinants of pollution: the share of industry in total output, the industry composition/share of polluting sectors and end-of-pipe pollution abatement. In order to test the impact of regulation on the demand for labour, they did a complementary study where they added a measure of regulatory strictness to cross-country labour intensity. They found evidence of EKC for the manufacturing share only. The sectoral composition improved to middle-income level and then stabilized, while end-of-pipe pollution intensity strongly declined with growth of income. They attributed these results to (1) the application of stricter regulations as income increased and (2) the complementarity of pollution and labour in production. The results of their study also showed that income elasticities for both the pollution and labour intensity were almost negative. Accordingly, they concluded that the sector's pollution/labour ratio was constant across all countries throughout the development process.

In an earlier study by Grossman et al. (1994), they tried to study the relationship between the composition and level of local economic activity, in terms of current income, average lagged income and sectoral composition of income and concentrations of six criteria of air pollutants across countries in the United States using panel data from the Aerometric Information Retrieval System maintained by the US Environmental Protection Agency (EPA). The air pollutants examined were total suspended particulates (TSPs), SO_2, NO_x, CO, ozone (O_3) and lead (Pb). The income data used were per capita GDP in 1985 nominal US dollars. They tried to separate scale, composition and technique effects in examining the relationship between income and pollution. They concluded that composition of output had a significant effect on local air quality, however, changes in the composition did not account for the observed relationship between output levels and pollution in the case of the United States. With respect to separating scale and technique effects, their results were inconclusive.

Policy and political institutions
As mentioned above, enactment of international legislation and enforcement of environmental policies and regulations are vital in controlling and minimizing the problems resulting from economic development and its damaging effect on the environment. Despite this, very few authors have included a policy variable in their studies of the economic-growth–environment relationship. Congleton (1992) was among the first to explore the effect of political institutions on the enactment of environmental policies and regulations. Using cross-sectional data from 118 countries, he tested the hypothesis that political institutions, rather than resource endowments or market structure, determine enactment of environmental policies and regulations to control pollution. His model depicted the level of preferred environmental standard as a function of the policy-maker's share of national income, time horizon, market institutions in the country and its resource base.

In his analysis, Congleton made several proxies and assumptions due to the unavailability of data. With respect to the data on domestic environmental policies and regulations he assumed that domestic politics determine whether or not a country would sign a particular covenant or protocol, so he used the data on signatories of two international agreements. The first was the 1985 global convention of the United Nations in Vienna whereby countries were committed to enact domestic legislation to reduce emissions of the ozone-depleting substances, mainly CFCs. The second was the 1987 Montreal Protocol that obliged developed countries to reduce their CFC emissions to half 1986 levels by June 1999 and limits the future consumption of CFCs by developing countries. He used the type of regime as a proxy to the personal characteristics of decision- or policy-makers. Gastil's classification of countries (1987) had been used to construct the type of regime, market structure, shares of national income and time horizon of policy-makers. The country's area was used as a proxy for its resource endowments and population was proxied for labour force. Both of these data were taken from the *World Fact Book* (1988). Finally, the data on real GNP in 1987 US dollars were gathered from the World Petroleum Institute (1990/91). Congleton arrived at three main conclusions. First, authoritarian regimes faced a higher price for pollution abatement than democratic governments. Second, authoritarian regimes had uncertain career paths and shorter terms of office, and thereby would tend to adopt less stringent environmental policies and regulations than democratic ones. Third, they were less willing to sign international agreements to protect their environments. The author concluded that political institutions, rather than economic endowments, technology or market institutions, would determine the level of domestic and international environmental policies and regulations in the country.

Also Panayotou (1997) presented a study that focused on the policy scope and provided a more analytical and structural approach to the income–environment relationship. The roles of economic growth and population density were also explored. He examined the relationship between income growth rate and policy variables against the level of sulphur dioxide using panel data for 30 developed and developing countries for the period 1982 to 1994. The data for SO_2 were gathered from the GEMS, while those for GDP were obtained from the Penn World Table (1995) and were measured in terms of 1985 constant US dollars as purchasing power parity (adjusted GDP figures) was not available beyond 1992. He used proxies for the quality of institutions to represent environmental policies. He used five indicators to represent the quality of institutions obtained from Knack and Keefer (1995). These indicators were enforcement of contracts, quality of bureaucracy, the rule of law, corruption in government and the risk of expropriation. Panayotou found an inverted U-shaped relationship with a per capita turning point just under $5000. He also found that the emission elasticity with respect to the improvements in the quality of institutions was much higher than those of economic growth and population density: a 50 per cent improvement in the efficacy of institutions and policies at income levels between $10 000 and $20 000 could result in 50 per cent reduction in the concentrations of SO_2. On the other hand, at lower income levels, the same improvement did not result in a similar reduction of SO_2. Panayotou came to three main conclusions. First, environmental quality was influenced by the level of income rather than its growth rate. Second, in the case of SO_2, governmental policies and social institutions played an important role in significantly reducing environmental degradation at low income levels and speeding up the improvements at high income levels, and thus mitigating the environmental externalities of economic growth. Hence, improvements in environmental quality were not automatic but depended on policies and institutions. Third, the structural changes of supply and demand for goods and services throughout the development process played an important role in determining the when and how of environmental improvements. Therefore, markets as well as policies determined the 'environmental price' of economic growth. In conclusion, Panayotou noted that improvements in the quality of policies and institutions had higher payoffs for pollution abatement at higher levels of income as it tended to improve monitoring activities. Therefore, efforts should be focused on improving the quality of policies and institutions rather than slowing down economic or population growth.

De Bruyn (1997) investigated the origins of change in emissions using decomposition analysis in order to empirically explain the mechanisms underlying the U-shaped relationship between SO_2 emissions and income in developed economies during the 1980s. In other words, he attempted to determine the factors that shape the patterns of emissions over time to explain why a U-shaped relationship occurred between economic growth and pollution. The emissions data were obtained from OECD, Environmental Data Compendium (1995), Klaassen (1995) and World Resources (1994–95), while the income data were obtained from the World Bank, World Table (1995) for 1993 in thousands of US dollars based on market exchange rates. He highlighted that the main three arguments explaining the EKC hypothesis according to the literature were valid, but scarce. First, as income increased, the positive income elasticities of demand for environmental quality and a more open political system resulted in effective environmental policies. De Bruyn explained that the argument was valid but was likely to occur in democracies, which are scarce with contradictory policies. Second, the composition changes of production and consumption were associated with rising incomes, which assumed that economic development passed through transition stages with respect to the structure of production. In other words, a shift took place in countries from agriculture to industry to service-oriented economies as income increased and that, in turn, resulted in the pattern of the inverted U-shaped curve with the highest point occurring in the industrial stage. Third, if changes in the production structure were not accompanied by changes in the consumption structure, then the EKC relationship resulted from dumping of pollutants from developed countries into the economies of developing countries.

De Bruyn's findings showed that structural changes failed to give evidence of being the determinant of the reduction in SO_2 emissions of developed economies during the 1980s. However, by assuming that environmental policy targets were representative of environmental policy efforts, he found that richer countries had more ambitious environmental policies. Therefore, he concluded that the scale effect of economic growth resulted in environmental degradation unless corrected by appropriate environmental policies. Hence, environmental policies, fostered by international agreements, were the most important determinant for the reduction of SO_2 emissions at higher income levels. He also added that income was only a minor determinant of environmental policy and that the current state of the environment was an important determinant in reducing SO_2 emissions.

Hilton and Levinson (1998) examined the relationship between automotive lead emissions and national income for 48 countries, with 1990 populations exceeding 10 million, over a period of 20 years (1972–92) using data from Octel's Worldwide Gasoline Survey. The survey reported the average lead content of gasoline for over 150 countries, on a biannual basis. The total gasoline consumption data were compiled from various OECD publications, while income and population data were obtained from the Penn World Table documented in Summers and Heston (1991). The income data used were real per capita GDP. They broke down total emissions into 'pollution intensity', which was measured as the lead per gallon of gasoline, and 'pollution activity', being the total gasoline consumption.

The authors arrived at three main conclusions. First, a U-shaped curve relationship existed between lead emissions and income. Second, the peak of the curve was dependent on the functional form estimated and the time period considered. Third, lead pollution was the product of pollution intensity and polluting activity and the declining portion of the curve depended on reducing the gasoline lead content and not gasoline use. In other words, the existence of regulations to reduce pollution intensity, as a policy response, was a requirement for the improvement in environmental quality that accompanied growth in income.

Hilton and Levinson examined the relationship between income and two separate factors of pollution: pollution activity and pollution intensity. They explained that the observed inverse-U relationship might be due to two reasons. The first reason they attributed to what was called the 'composition effect'. This was referred to as the pattern of economic development, which was considered to go through transition stages, starting with agriculture, which was not pollution-intensive, to manufacturing with its high pollution, especially in the early stages, to the less polluting service industries. One of the reasons for this pattern could be that wealthier countries dumped their polluting processes in poorer countries. This, in turn, meant that it was not possible for all nations to experience improvements in environmental quality. The second reason they gave was that the EKC might be based on two separate relationships. First was the 'scale effect', where higher economic activity generated more pollution and wealthy countries, with more polluting activity, generated more pollution. This was referred to as pollution activity. Second was the 'technique effect', where citizens in wealthy countries demanded more environmental quality in the form of regulations to protect the environment and to reduce the amount of pollution per unit of activity. This was referred to as pollution intensity. Therefore, the shape of the pollution–income relationship was the product of pollution activity and pollution intensity. So their conclusion was that polluting activity increased with income, while polluting intensity decreased with income and the product of the two followed an inverse U-shape.

Midlarsky (1998) examined the relationship between democracy and the environment. He used six environmental indicators as the dependent variables: CO_2 emissions, deforestation, soil erosion by water, land area protection, freshwater availability and soil erosion by chemicals. He found that democratic countries had higher CO_2 emissions, deforestation and soil erosion. However, they tended to protect a higher percentage of their land area. For the last two environmental indicators he found no significant relationship and concluded that there was no uniform relationship between democracy and the environment.

Tuan (1999) found similar results for CO_2 emissions to those found by Panayotou for SO_2. He presented a paper on the relationship between income and environmental quality, mainly CO_2 emissions. He used panel data for six countries at different stages of economic development. The countries examined were Vietnam, Thailand, Korea, France, Japan and the United States. In addition to the income variable, he included in his model other social variables, such as population, economic growth and institutional capacity in order to examine the underlying impacts on the EKC trajectory. The environmental data for the three developed countries were obtained from the OECD Environmental Data Compendium for the years 1993, 1995 and 1997. The energy data for the three Asian countries were obtained from the ENERDATA database. The per capita GDP data were in 1987 constant US dollars and were obtained from the database of ENERDATA as well as the data on economic growth rates and demographic density. For the political or institutional capacity variable, the author used four indicators from the Business Environmental

Risk Intelligence (BERI), a private international investment risk service. The indicators were bureaucratic delays, enforceability of contracts, nationalization potential and infrastructure quality.

Tuan tested four hypotheses. The first was the existence of an inverted U-shaped relationship between per capita income and CO_2 emissions. The second was the impact of the rate of economic growth on the EKC trajectory. The third was the impact of population density on EKC. He hypothesized that the higher the population density, the lower the turning point of the EKC, and hence, a better environmental quality. The fourth hypothesis was the role of institutions and policies in determining the EKC path. Again, he hypothesized that effective policies and institutions would lower the EKC turning point, and therefore would improve the environment along with economic growth. Tuan found an EKC relationship between per capita income and CO_2 emissions with a turning point of approximately $18 000. With respect to the impact of economic growth and population rate on CO_2 emissions, the author found that their impact varied according to the level of income and the country's stage of development. The author's main conclusion was that the quality of policies and institutions played an important role in smoothing out the EKC, however, the payoffs were not the same for developing and developed countries. He noted that a good institutional capacity and sound environmental policies could, to a large extent, contribute to a better environment.

Similar results were also found in a study by Bhattarai and Hammig (2001) for deforestation. They studied the relationship between income and deforestation in 21 countries in Latin America for the period 1972 to 1995. They used data from various sources, such as the FAO, WRI and the UNEP. They also used the Penn World Table from Summers and Heston (1991) for the national income data, exchange rates and trade, the Freedom House Tables and Knack and Keefer (1995) for the index measures for sociopolitical institutions and the *World Development Report* (1998) for other variables. The per capita GDP data were adjusted 1998 US dollars PPP. Their findings confirmed the existence of an EKC in Latin America with turning point around $6800. They concluded that (1) macroeconomic factors, such as indebtedness, inflation and exchange rate policies would shift the intercept of the EKC and (2) the present level of deforestation in Latin America would be reduced by strengthening the sociopolitical institutions in the region.

A number of studies examined the relationship between economic growth and the environment across different political systems. Torras and Boyce (1998) investigated the causal linkages between changes in income and pollution levels. They hypothesized that more equitable distribution of power was an essential link for the improvement of some air and water pollutants as a result of higher per capita income. The air and water pollutants examined were sulphur dioxide, smoke, heavy particles, dissolved oxygen, faecal coliform, access to safe water and access to sanitation. For the first five pollutants, they used data from the GEMS for the period 1977 to 1991. The air pollution data contained observations from 19 to 42 countries and that for water pollution contained observations from 58 countries. The percentages of population with safe water and sanitation were obtained from the United Nations Development Programme (1994) and were national-level variables with no time series dimension. The per capita income used were measured in real PPP adjusted for 1985 US dollars. The income data for per capita GDP were taken from Summers and Heston (1991), while the national income levels data were taken from the UNDP (1994).

The authors predicted that the higher the power inequality, the higher the levels of pollution, so they included other variables as proxies for power inequality, such as Gini ratio of income inequality, literacy rates and political rights and civil liberties. Four of the seven environmental indicators showed EKC patterns when excluding the power inequality variable. These indicators were sulphur dioxide, smoke, access to safe water and access to sanitation with turning points of $3890, $4350, $11 255 and $10 957, respectively. Heavy particles monotonically decreased with rising income while dissolved oxygen and faecal coliform monotonically increased. When including the power equality variable, only three pollutants showed EKC patterns, which were sulphur dioxide, dissolved oxygen and access to safe water with turning points of $3360, $19 865 and $6900, respectively. Access to sanitation increased monotonically, while the relationships for smoke, heavy particles and faecal coliform were statistically insignificant. They arrived at two main conclusions. First, equitable distribution of power, in the form of equitable income distribution, wider literacy and greater political liberties and civil rights, would lead to better environmental quality, especially in low-income nations. Second, improvements in environmental quality would not automatically accompany continued growth in per capita income.

Also, Deacon (1999) argued that the form of government in a country played a crucial role in the way economic growth would affect the environment and that omitting it as a determinant of environmental quality could lead to biased results and interpretation. He presented a study to examine the effect of the government regime on the provision of environmental public goods and environmental policies using cross-sectional analysis of 118 countries. Deacon classified governments into groups with similar political attributes using compiled data sets from two sources. The first was the Cross-national Time-series Data Archive, which was first compiled by Arthur Banks and listed attributes of the government systems in almost all countries. The second was the Polity III database maintained by Keith Jaggers and T.R. Gurr (1995) that had information on the political systems of all countries with populations over 500 000. The data sets in these two sources went back to the nineteenth century and up to the late 1990s.

He examined two environmental public goods; the rural population with access to sanitation facilities and safe drinking water, using data from the World Health Organization reported in the World Resources Institute database. Deacon also examined two other public goods; roads and public education. The data for roads were taken from the International Road Federation and those for public education were taken from Banks. The environmental policy measure he used was the lead content of gasoline as reported by Octel Corporation. He showed that there was an inverted U-shaped relationship between pollution and income, however, the shape of the curve was determined by the form of government. In other words, income elasticities of provision of public goods were affected by the type of governance. He concluded that income levels and government regimes were highly correlated. He showed that incomes were higher in countries with democratic governments than autocratic ones and environmental quality tended to be lower in the latter. His explanations were that (1) property rights were less defined and the rule of law was absent in less democratic countries, which in turn suppressed investment in environmental quality and other public goods, and (2) the governments were generally corrupted, and hence drained any economic surplus that could be produced by the private sector.

Barrett and Graddy (2000) made an attempt to determine the relationship between a selected number of air and water pollutants and per capita income. The air pollutants were sulphur dioxide, smoke and heavy particles, while the water pollutants were dissolved oxygen, biological oxygen demand, chemical oxygen demand, nitrates, faecal coliforms, total coliforms, lead, cadmium, arsenic, mercury and nickel. They relied on the data used by Grossman and Krueger, which were gathered by the Global Environmental Monitoring System and they added freedom variables that were constructed from the civil and political indices developed by Freedom House. The civil freedoms index reflected constraints imposed on the level of freedom of the press and individuals, while the political freedoms index reflected the level of freedom of choice of government, existence of opposition, and so on. The income data were taken from Summers and Heston (1991) and were adjusted for differences in PPP.

Barrett and Graddy's hypothesis was that improvement in air and water pollutants was due to improvement in the political system in a country, as they believed that politics played an intermediate role between income and pollution. Their findings showed that as income per capita increased, certain measures of pollution deteriorated then later improved. They also found that for some measures of pollution, environmental quality improved according to the level of civil and political freedoms allowed in the countries, while for other measures environmental quality did not seem to depend on freedoms. For air pollutants, they found three main observations. First, income and freedom were positively correlated. In other words, the high-freedom countries were rich, while the low- or medium-freedom countries were poorer. Second, with the exception of concentrations of sulphur dioxide, pollution levels and freedom were negatively correlated. Third, as civil and political freedoms were highly correlated, the correlations between freedoms and income and freedoms and pollution were the same for both civil and political freedoms. On the other hand, the relationship between freedom and water pollution levels was not always negative. Some of the water pollutants increased with higher freedom, such as biological and chemical oxygen demand, nitrates and total coliform. Faecal coliform, on the other hand, reduced with increased freedom. With respect to heavy metals, freedoms were jointly significant in the case of arsenic and cadmium, where the former reduced with higher freedom, and therefore was consistent with their hypothesis of an induced policy response, while the latter increased. The combined effect of freedom and income was very significant in the case of lead, with a negative relationship, which also might be the result of an induced policy response. However, the relationships for nickel and mercury were statistically insignificant. They concluded that effective environmental protection required both economic and political reforms. Nevertheless, they could not explain why freedoms affected some measures of environmental quality and not others.

The role of international trade
A number of empirical studies attempted to examine the role of international trade and direct investment in the income-growth–environment relationship. They hypothesized that the shift of pollution dumping from richer to poorer countries either through international trade or direct investment was the reason behind the reduction of pollution in the richer countries. Among the authors who examined this hypothesis were Hettige et al. (1992). They examined the relationship between the level of economic development and toxic intensity of industrial production in 80 countries for the period 1960 to 1988. They used industrial data gathered by the United Nations to calculate the shares of total manufactured output for 37 sectors as defined in the International Standard Industrial Classification

(ISIC). The income data were GDP per capita in 1985 nominal US dollars. They found an inverted U-shaped relationship for pollution intensity per unit of GDP, however, they did not find the same pattern for toxic intensity per unit of industrial output. On the contrary, they found that manufacturing output increased steadily with income. They concluded that this was the result of development in these countries, and, therefore, a shift in the production structure from industry towards services, which were lower in pollution. They also analysed the relationship of toxic intensity and trade policy in less developed countries (LDCs) and they found that the closer the country to international trade, the more rapid the growth of toxic intensity of manufacturing. They concluded that, for the majority of countries, there would be an increasing long-term trend of industrial emissions, with respect to both GDP and manufacturing output. This would also be higher in the poorer countries mainly due to the tighter environmental regulations in the industrialized nations since 1970.

Grossman and Krueger (1991) also included a trade variable in their study as they studied the effect of reduction in trade barriers on the environment in Mexico and assessed the relative magnitudes of change in the level of pollution resulting from further trade liberalization. They found significant evidence that trade helped to reduce the level of pollution for only one of the three pollution variables, which was urban concentrations of SO_2. Shafik and Bandyopadhyay (1992) tested the hypothesis that trade induced the use of cleaner technology, and thereby would reduce pollution. However, they found weak evidence between the two variables.

Rock (1996) studied the relationship between trade policy and the environment in rich and poor countries using cross-country data over the period 1973 to 1985. He used the pollution data on toxic chemical intensity of GDP, which were previously used by Hettige et al. (1992) in their study. However, he included two other variables: the manufacturing share of GDP and the energy intensity of GDP. The first variable was included to separate the effect of the broad composition of output on pollution intensity from that of trade policy, while the latter was included to separate the effect of energy price policy from trade policy. The trade orientation data were taken from the World Bank (1987) where countries were grouped into four categories depending on their trade policies. The four groups were strongly outward-oriented, moderately outward-oriented, moderately inward-oriented and strongly inward-oriented. He found evidence of higher pollution intensity in countries with more open trade policies and that the most inward-oriented developing countries had the lowest toxic chemical pollution intensity of GDP.

The relationship between per capita income and consumption of primary commercial energy was examined by Suri and Chapman (1998) using pooled cross-country and time series data in order to quantify the impact of the actual movement of goods between countries. They used observations from 33 countries over the period 1971 to 1990. The energy data, expressed in oil equivalents, were obtained from the IEA, while the income and trade data were obtained from the Penn World Table and the income data were expressed in 1985 US dollars PPP. The authors presented two types of models. The first model implicitly captured the impacts of both structural change and international trade, while the second model explicitly analysed the effect of international trade on commercial energy use. The variable representing structural change was the share of total manufacturing in GDP (that is, manufacturing for domestic market and production for exports). International trade variables were represented by import- and export-manufacturing ratios. In both models, the EKC pattern was depicted, with a turning point of $55 000 in the first model and $224 000 in the second one. Accordingly, they concluded that, all things being equal, international trade would tend to increase the pollutant emissions related to commercial energy consumption. The authors also found that energy use increased in both industrialized and industrializing countries, however, it was substantially higher in the latter. Moreover, as opposed to industrializing countries, whose imports were largely intermediate and capital goods, which were essential for building an industrial base, industrialized nations benefited from importing manufactured goods from industrializing countries. Therefore, their relative increase in energy requirements to GDP was substantially lower. The authors concluded that the imports of manufactured goods, in addition to the structural change from manufacturing to non-energy-intensive service sector, were the result of this decline.

Some authors concentrated on trade patterns and consumption, rather than production, in examining the income–environment relationship. Rothman (1998) studied the relationship between per capita GDP and consumption activities using data on eight categories of consumer goods from the United Nations International Comparison Programme (1994). Among the commodities used were food, beverages and tobacco, clothing and footwear, gross rent, fuel and power, medical care and services, transport and communications, recreation, entertainment and education. The unit for each commodity was the quantity that could be bought for $1 at average international prices. The data used for the per capita GDP were in 1985 US dollars PPP. The only commodity they found showing evidence of EKC was food, beverages and tobacco with a turning point of $12 889. Other commodities, in terms of shares, also changed with income, such as clothing and footwear, gross rent, fuel and power and medical care and services, with turning points of $35 263, $23 278 and $47 171, respectively.

However, this was due to changes in relative share rather than a decline in consumption.

The extent of the so-called pollution haven hypothesis (PHH) in explaining the EKC relationship was assessed by Cole (2004). The PHH was defined as the migration or displacement of pollution-intensive industries from developed to developing countries due to more stringent environmental regulations in the former countries. The author estimated the EKC model for ten air and water pollutants for a sample of OECD countries over the period 1980 to 1997. The air pollutants were CO_2, NO_x, SO_2, CO, SPM and volatile organic compounds (VOC), while the water pollutants were the level of nitrates, the level of phosphorous, BOD and the levels of dissolved oxygen. To capture the PHH effects, the author included the share of pollution-intensive exports and imports to and from non-OECD countries in total exports and imports. He also included the share of manufacturing in GNP to capture the effect of structural change within the economy on pollution. To reflect trade openness/intensity, the author used the ratio of the sum of imports and exports to GNP. Cole found that, with the exception of VOC and CO, all the other eight pollutants exhibited an EKC relationship with per capita income, with turning points ranging between $1973 and $35 140. He also found a positive relationship between the manufacturing share of GNP and environmental quality for eight pollutants, of which seven were statistically significant. This indicated that structural change within the economy was partially responsible for pollution reduction at higher income levels. With respect to PHH, Cole found mixed results. There was a negative relationship between the share of pollution-intensive imports to total imports and environmental quality for seven pollutants (five air pollutants and two water pollutants), of which five were statistically significant. On the other hand, there was a positive relationship between the share of pollution-intensive exports to total exports and environmental quality for five pollutants (one air pollutant and four water pollutants), of which four were statistically significant. Nevertheless, upon omitting the pollution haven variables, the EKC turning points were lower for eight out of the ten pollutants, suggesting that these effects contributed to the reduction of emissions at higher levels of income. With respect to trade openness, the author found a negative relationship between trade openness and environmental quality for eight out of the ten pollutants, of which six were statistically significant.

The author concluded that the reduction in emissions at higher levels of income was due to higher demand for environmental regulation and investment in abatement technologies, trade openness, structural change within the economy and increased imports of pollution-intensive industries. Nevertheless, the author could not ascertain that the developing countries would be able to follow an income–pollution path similar to that of the developed countries as the demand of the developed countries for pollution-intensive products was being met by the developing countries. Therefore, the developing countries did not have anyone to pass on the production of such products to.

The role of population density
In the economic-growth–environment literature, there are a number of empirical studies that included population density variables as it is perceived that population density affects the relationship between income and the environment. As mentioned earlier, Cropper and Griffiths (1994) examined the effects of population pressure on deforestation by including the rural population density and the rate of population growth as independent variables in their equation. They found an EKC relationship for both Africa and Latin America with turning points of $4760 and $5420, respectively. They found that the higher the rural population density, the higher the deforestation rate. An increase in rural population density by 100 persons per 1000 hectares implied an increase of the deforestation rate by 0.33 per cent in Africa. However, none of the variables included for Asia were statistically significant. The authors also found that the rural population density for Africa shifted the relationship between income and the rate of deforestation upwards, which implied the existence of a large trade-off between per capita income and rural population density. For instance, a country with a population density of 0.1 persons per hectare, being the average population density for the sample of African countries, had a peak deforestation rate of 1.26 per cent per year at a per capita income level of $4760. For a country with a population density of 0.7 persons per hectare, it would need a per capita income level of $11 650 per year to achieve the same deforestation rate. Nevertheless, the authors concluded that deforestation was the problem of market failure, which resulted from the lack of defining and enforcing property rights. Therefore, reducing the rate of population growth was not necessarily the best method for reducing the rate of deforestation in developing countries.

Also as mentioned earlier, Panayotou (1997) explored the role of population density using two approaches: the reduced-form analysis and the income decomposition analysis with its three constituents, scale, composition and abatement. Under the reduced-form approach, Panayotou found an inverted U-shaped relationship with a per capita turning point of just under $5000. However, he found that the higher the population density, the higher the level of ambient SO_2 at every income level. With respect to the decomposition approach, the author found that, controlling for all other factors, low levels of population density of under 50 persons per square kilometre were associated with high levels of ambient SO_2 emissions, being 70 kg/km^3. As population density increased, SO_2 levels dropped, reaching their lowest level of 45 kg/km^3 at a population density level of 170 persons per square kilometre and then started to increase again. The author attributed the higher SO_2 levels at lower population density to the lack of pressure in countries with spare population to control emissions. However, as population density increased, more pressure was exerted to control emissions as more people were exposed to pollution. This continued up to the level where the household use of coal and non-commercial fuels exceeded the pressure for pollution abatement.

Similarly, as mentioned before, Tuan (1999) tested the impact of population density on the EKC trajectory. He hypothesized that the higher the population density, the lower the turning point of the EKC, and, thus, a better environmental quality. He found an EKC relationship between per capita income and CO_2 emissions with a turning point of approximately $18 000. However, with respect to the impact of economic growth and population rate on CO_2 emissions, Tuan found that their impact varied according to the level of income and the country's stage of development. At income levels below $2000 per capita, a highly dense population had a negative impact on the environment. On the other hand, at income levels of $2000 per capita or more, a highly dense population had a positive effect on the environment. For instance, at an income level of $1000 per capita, an increase in the population density of 67 per cent resulted in a 12 per cent increase in CO_2 emissions. However, at an income level of $10 000 per capita, a similar increase in the population density resulted in a 22 per cent reduction in CO_2 emissions. The author concluded that at higher income levels, the higher the population density, the more pressure was being exerted to control emissions as more people were being exposed to the pollution.

Vincent (1997) presented an analysis for a single country, Malaysia, in an attempt to examine the pollution–income relationship. Unlike previous studies that conducted an analysis on cross-sectional or panel data for a sample of developed and developing countries, Vincent used a panel data set for the 13 Malaysian states and he analysed the relationship between income and a number of air and water pollutants over time (1970s to 1990s). He also measured the impact of population density on the various pollutants. The pollutants were TSPs, BOD, COD, ammoniacal nitrogen, pH and suspended solids. The data used were ambient levels measured by monitoring stations.

He chose Malaysia for several reasons. First, the country had a rich data set with nearly two decades of readings on ambient air and water pollution. Second, the country's economy had been one of the fastest growing in the world since the 1970s, and therefore its income data contained ample variation over time, even within the states as growth had not been equally rapid across all states. Third, income data overlapped between the states. Fourth, the borders of most states were determined by ridges that separated the major water basins and, to a certain extent, air basins. Fifth, the states shared the same pollution policies, which varied across pollutants.

According to Vincent's study, there was no evidence of EKCs for any of the six income–pollution relationships in Malaysia. This result was inconsistent with previous research conducted using cross-country relationships. With respect to population density, Vincent found that the net impact of population density on TSP concentrations was positive and the interaction term between population density and time was statistically significant but negative. He attributed the positive impact to the various household activities, which were important sources of TSP concentrations, such as heating, cooking, rubbish disposal and transportation. The negative interaction term indicated a reduction in the per capita TSP concentrations due to the pressure exerted through the enforcement of effective anti-pollution regulations. For the water pollutants, and while holding income constant, Vincent found that higher population densities were associated with worse water quality for BOD and ammoniacal nitrogen, and better water quality for suspended solids. He attributed this to the higher sewage discharge as a result of higher population, in the case of BOD and ammoniacal nitrogen, and the movement of people from rural to urban areas in the case of suspended solids. The interaction term between population density and time was also significant. It was negative in the case of BOD and ammoniacal nitrogen as a result of the introduction of regulations to reduce BOD discharge and the percentage of population without access to sanitary facilities. The interaction term was positive for suspended solids indicating agriculture and settlement movements to steeper and erosion-prone areas. In conclusion, the author said that his findings did not mean that EKC did not exist anywhere, but it meant that policy-makers should not assume that economic growth was an automatic solution for the air and water pollution problems.

Kaufmann et al. (1998) attempted to explore the effects of income and spatial intensity of economic activity on the atmospheric concentration of SO_2. They used a panel of international data for 23 countries with observations between 1974 and 1989. Out of the 23 countries, 13 were developed nations, seven were developing nations and three were centrally planned economies. The data on SO_2 were obtained from the *United Nations Statistical Yearbook* (1993) and GDP per capita and population data were obtained from Summers and Heston (1993). The income data were expressed in 1985 nominal US dollars. They used two variables to proxy the spatial intensity of economic activity, one for cities and the other one for the entire nation. The variable for spatial intensity of economic activities in cities was the product of a city's population density and the per capita GDP of the nation in which the city was located, while that for the nation was the GDP divided by the nation's area. The result of their study indicated a U-shaped relationship between both per capita GDP and atmospheric concentration of SO_2. The concentration of SO_2 tended to reduce as per capita income rose from $3000 to $12 500, however, it increased beyond $12 500. On the other hand, there was an inverted U-shaped relationship between spatial intensity of economic activity and atmospheric concentration of SO_2 with turning points of $6.7 million per square mile for the nation variable and $153 million per square mile for the city variable. They suggested that the impetus for policies and technologies to reduce SO_2 atmospheric concentration was spatial intensity of economic activity rather than income. This implied that atmospheric concentration of SO_2 would decline faster than indicated in previous studies, however, it would depend on the rate of income growth relative to population growth. They suggested that the inverted U-shape found in previous studies might be due to the omission of the variables representing changes in the mix and spatial intensity of economic activity, and therefore the pattern was a proxy for changes in the mix of economic activity associated with changes in per capita GDP.

Household preferences and income elasticity of demand
Income elasticity of demand for environmental quality changes as income rises. One of the explanations given for the downward sloping section of the EKC is that as per capita income increases, the demand or preference for environmental amenity also rises. A number of studies attempted to study the role of preferences in the economic-growth-environment relationship.

In a study, McConnell (1997) tried to understand the role of preferences, and in particular income elasticity of demand for environmental quality in the EKC relationship, by decomposing the reduced-form effect of income changes on pollution. He used panel data of approximately 50 countries from the 1970s to 1990s. The types of environmental indicators were for ambient urban air and water pollutants, such as sulphur dioxide, suspended particulates, nitrogen oxides, carbon monoxide, dark matter, dissolved oxygen, faecal coliform and emissions, in the case of toxic substances and carbon dioxide. He concluded that:

> preferences that were consistent with a high income elasticity of demand were neither necessary nor sufficient for pollution to grow with income. Simple

models showed that preferences that were consistent with a high income elasticity of demand for environmental quality were attenuated by high and rising abatement costs or a high impact of pollution on production. Hence, it would be feasible for pollution to decline with a zero income elasticity of demand for environmental quality, or to increase with a high income elasticity of demand for environmental quality.

In another study, Kahn (1998) examined the effect of per capita income on hydrocarbon emissions of vehicles. He used the 1993 Random Roadside Test created by the California Department of Consumer Affairs Bureau of Automotive Repairs, whereby vehicles at random are selected to have an emissions test. He used the 1990 Census of Population and Housing, which included the zip code median household income as a proxy for household income. The hydrocarbon emissions were measured in parts per million (ppm) and household income was measured in thousands of US dollars. The author tried to study the total annual contribution of a household to local pollution in relation to its income. He presented evidence of an inverted U-shaped relationship between income and emissions, with a turning point between $25 000 and $35 000. He also found that average emissions for vehicles owned by households with income levels below $35 000 were twice as high as those for households with income levels over $45 000, showing a higher demand for better vehicle quality at higher levels of income.

Pfaff et al. (2001) examined the linkage between income and household choices and its impact on the environment. The model they used emphasized two main features. First, environmental degradation was a by-product of the household activities. Second, households could recognize their activities, and therefore would substitute for more expensive, less environmentally degrading commodities, with higher income. They highlighted that this would occur with the assumption that environmental quality was a normal good and given the elimination of natural constraints of desirability and feasibility of substitution. They found a non-monotonic relationship between income and environmental quality, generating an EKC at the household level as substitution constraints were eliminated.

In another paper, Pfaff et al. (2002) provided an explanation for the EKC using their household-choice framework. They showed that at low income levels, the marginal rate of substitution between household consumption and the environment made abatement undesirable for the household. As income increased, both consumption and abatement expenditure rose. They concluded that both household consumption preferences and abatement technologies were necessary and sufficient conditions to generate the EKC path. They also applied their model to a set of externalities and multiple agents who voted for environmental spending and taxation.

They found evidence for the EKC. As the chosen tax rate increased with income, abatement occurred after a certain range, and therefore environmental quality deteriorated at first and then improved.

Also Chaudhuri and Pfaff (2002), using the above model, studied the linkages between household fuel-choice preference and income on the environment, specifically indoor air pollution, in Pakistan. The individual and house-level data were obtained from the Pakistan Integrated Household Survey (1991), which covered the energy consumption of 4800 households. To derive their sample, they used a multi-state stratified sampling procedure from the Federal Bureau of Statistics based on the 1981 census. The authors found significant evidence of an inverted U-shaped relationship between monthly household income and monthly household emissions, in terms of the quantity of the monthly fuel consumption due to the transition in the fuel-choice behaviour of the household. As per capita income increased, the household moved from using traditional fuel and invested in cleaner modern fuel. However, for all modern fuel consumption, their quantities increased with higher incomes, forming a concave curve.

In a recent study by Khanna and Plassmann (2004), they examined the impact of demand for environmental quality on the relationship between income and pollution. They argued that the ability of the consumers to separate themselves from the source of pollution played a decisive role in the consumer's decision to limit their exposure to pollution as income increased. Also they argued that the turning point of the EKC would be lower for pollutants where spatial separation was possible. Unlike Shafik and Bandhopadhyay (1992) and Suri and Chapman (1998) who used multi-country panel data that combined the effects of other factors, such as structural changes, technology and consumer preferences for environmental quality, Khanna and Plassmann used cross-sectional census-tract-level data for the United States to isolate the effects of changes in income from those in other factors. The census-tract-level data were the smallest geographical unit for which detailed socioeconomic data were available. The authors tested their hypothesis on five air pollutants: SO_2, particulate matter (PM), CO, ground level O_3 and NO_x. They used ambient concentrations data from the EPA AIRS database. They found an EKC relationship for SO_2 and PM, with turning points of $8653 and $11412, respectively. For the other three pollutants, the income–pollution relationship continued to increase.

The authors concluded that the EKC hypothesis was an equilibrium relationship between income and pollution based on the interaction between consumer preferences and technology. For pollutants where spatial separation was relatively straightforward, the opportunity cost of pollution abatement was relatively low, and therefore the turning point of the EKC or the equilibrium income elasticity of pollution, as referred to by the authors, changed its sign from positive to negative at lower income levels. However, for pollutants, where spatial separation was more difficult and costly, the equilibrium income elasticity of pollution remained positive.

The role of technology
Technology shifts is one of the explanations conjectured for the economic-growth–environment relationship. Hence, some of the empirical results were attributed to the use of cleaner technology. Komen et al. (1997) examined the increase in income and its role in promoting new technologies to improve environmental quality. They gathered data from 19 OECD countries for the period 1980 to 1994. The countries were Canada, United States, Australia, Austria, Belgium, Denmark, Finland, France, Germany, Greece, Ireland, Italy, Norway, Spain, Sweden, United Kingdom, the Netherlands, Portugal and Japan. They carried out an empirical analysis focusing on the relationship between real budget per capita allocated for public research and development (PRD) and real gross per capita GDP. The PRD, for each country, was measured as nominal expenditures converted to 1991 US dollars at prevailing purchasing power parities. Similarly, GDP for each country was measured in thousands of US dollars at prevailing purchasing power parities. The results of their study showed a positive income elasticity of public research and development funding for environmental protection. The income elasticity was approximately equal to unity, which implied that emissions of some pollutants might decline after an income-level threshold was reached. The authors concluded that their results should be interpreted with caution due to (1) the small size of funding allocated to public research and development relative to the size of overall spending on environmental protection, (2) the limited ability of the country to substitute between public and private research and development spending and alternative policy instruments and (3) the possibility that funding for public research and development might act as a form of industrial subsidy in some countries. The study made two main contributions. First, unlike previous research that focused on the overall linkage between per capita income and environmental degradation, this study focused on research and technology development, which were the underlying individual components. Second, it helped to explain why the EKC had a negative slope for some pollutants in relatively industrialized countries.

Carson et al. (1997) did not study the role of technology per se in the income–environment relationship, however, they attributed their findings to the use of the latest-vintage technology. Carson et al. examined the existence of an EKC relationship between income and seven types of air emissions across the 50 US states. In order to overcome the problem of comparability and quality of available environmental data, they moved away from developing countries and used data from the 50 US states. They also used state-level emissions for the seven major air pollutants (greenhouse gases, air toxics, carbon monoxide, nitrogen oxides, sulphur dioxide, volatile organic carbon and particulate matter). They estimated the relationship using a panel data set consisting of the different states over time. In addition, access to technology and air pollution control regulations were likely to be the same across the United States, or at least the differences would be less than that across different countries. The income data used were expressed in thousands of 1982 US dollars. They arrived at two main findings. First, all seven pollutants decreased with increase in income even when industrial composition and population density/urbanization were controlled for. Second, high-income states had low per capita emissions while emissions in low-income states were highly variable. They attributed this finding to two main factors. First, richer countries produced goods using less polluting technology per unit of output, either in the form of designing technology for this purpose or using the latest technology, which was usually efficient in terms of energy consumption. Second, consumers demanded more environmental quality as they became wealthier, such as shifting their consumption to less polluting products, moving to cleaner areas or demanding that their government strictly regulate the output of pollution.

De Bruyn et al. (1998) examined the relationship between economic growth rate and environmental quality for the Netherlands, West Germany, the United Kingdom and the United States for various time intervals between 1960 and 1993. They used emissions data for CO_2, NO_x and SO_2. The CO_2 emissions data were taken from the Oak Ridge National Laboratory (1994), while the SO_2 and NO_x data were taken from the US Environmental Protection Agency (*National Air Quality and Emission Trends Report*, 1984) and OECD (*Environmental Data Report*, 1993). They found a positive and significant effect of economic growth on the growth of emissions in all of the cases, with the exception of SO_2 in the Netherlands. The analysis suggested that in only half the cases was the rise in income the reason for the reduction of emissions. They also found a positive correlation between time patterns of emissions and economic growth. The authors concluded that there was no evidence in their country analysis to confirm that economic growth would improve environmental quality. However, the relationship between income and emissions that could be reduced by using cleaner technology would follow an inverted U-shaped pattern when using panel data. The decline in emissions over time would be due to technological and structural changes.

Decomposition and structural models of the income–environment relationship

A number of studies attempted to understand the underlying determinants of the relationship between economic growth and the environment. In doing so they used more structural and analytical models in their studies instead of the reduced-form models, in order to identify the structural forces that would influence the relationship. Panayotou (1997), in examining the relationship between income and the level of SO_2 using panel data for 30 countries, identified three forces/effects influencing environmental quality: (1) the scale effect, (2) the composition effect and (3) the abatement effect. He found that expansion of the scale of an economy increased the level of SO_2 concentrations monotonically, but at a diminishing rate. It was particularly strong at income levels up to $3 million per square kilometre. The composition effect, represented by the industry share of GDP, also monotonically increased the level of SO_2 concentrations as its share of GDP increased from 20 per cent to 43 per cent at income levels up to $8000, then declined to 37 per cent when income levels reached $17 000 and then started to rise again. The abatement effect, which is the 'pure' income effect after being stripped out of the scale and composition effect, had a negative relationship with SO_2 concentrations up to income levels of about $13 000.

Similarly, in a study, Islam et al. (1999), identified the same three forces in examining the relationship between GDP per unit of area and ambient level of SPM using data from the GEMS. The data contained 901 observations from 23 countries, including 56 cities, for the period 1977 to 1988. They also used data from the World Bank Tables. The per capita GDP data were in 1985 US dollars purchasing power parity. The authors found that the level or scale effect increased monotonically, showing a positive relationship between ambient level of SPM and the level of GDP per unit of area. The composition effect showed an inverted U-shape with a hump-shaped relationship between the level of SPM and the share of industry in GDP. However, the peak was reached at a very high industry share level at which the SPM level increased almost steadily. The abatement effect generally declined.

Moomaw and Unruh (1997) tried to provide insight into the processes that generated the changes in environmental quality and national economic growth. They examined the structural transition changes in per capita CO_2 emissions and per capita GDP over 16 countries for the period 1950 to 1992 and compared the structural transition models to EKC models. They used the data from the Oak Ridge National Laboratory (1995) for the CO_2 emissions and the Penn World Table for the income data in real per capita GDP, measured in 1985 US dollars. The 16 countries were characterized by having sustainable income growth with a stable or decreasing level of CO_2 emissions per capita over time. From their findings they made three conclusions. First, the transition initiated in the countries did not correlate to changes in income levels but to historic exogenous events, such as oil price shocks in the 1970s, and policies. Second, the positive CO_2 emissions elasticities (also called marginal propensity to emit) in these countries happened as a sudden, discontinued transition, instead of a gradual change, and therefore, the decrease in CO_2 emissions did not correlate to specific income levels, but to specific points in time. Third, the relationship had an N-shape, which was the result of data aggregation instead of income dependence. The turning points were $12 813 and $18 333, which implied a very narrow range for CO_2 to decline as emissions increased again once the second turning point was passed.

They concluded that neither the U- nor the N-shaped relationship between CO_2 emissions and income provided a reliable indication that at low income levels, environmental quality would deteriorate and then would improve above a certain income threshold or transition value, and therefore the model was inappropriate for forecasting future emissions behaviour. They also added that CO_2 emissions would continue to rise as economic growth was pursued in countries. Moomaw and Unruh provided two main explanations for their conclusions. First, individuals would not sacrifice consumption for investment in environmental quality and that was why pollutant emissions increased at low income levels. This was because the environment was assumed to be a luxury good, and therefore individuals would only be willing to trade consumption for improvements in environmental quality at high levels of income. Second, CO_2 emissions created global and not local disutility to the public, who would only demand controls on the level of pollutants that would create disutility at the local level.

Also Moomaw and Unruh (1998) evaluated whether income was the determining variable for the reduction of CO_2 emissions by applying techniques of non-linear dynamical analysis. This technique was used in order to account for temporal patterns and discontinuous changes that might have taken place. The study was conducted on 16 OECD countries that showed an EKC pattern in previous studies using data from Penn World Table (1994). The countries were Austria, Belgium, Canada, Denmark, Finland, France, West Germany, Iceland, Italy, Japan, Luxembourg, the Netherlands, Sweden, Switzerland, United Kingdom and United States. The per capita income used was measured in 1985 US dollars PPP. The authors arrived at two conclusions. First, income was not the determining variable for reducing CO_2. They added that emissions were expected to follow a regular, incremental path until they were subjected to a shock that would lead to a new trajectory. In the case of CO_2, the oil price shock was the impetus that changed its trajectory. Second, a national capacity existed for a rapid and persistent change under the proper stimuli. In the case of CO_2, the change in its trajectory happened within a single year and continued at a particular emission level despite the continued economic growth and the decline in oil prices.

Stern and Common (2001) criticized the EKC models by saying that they lacked adequate specification due to the omissions of a number of variables that were correlated with GDP. Their criticism was based on the results of statistical tests from their study on the relationship between sulphur emissions and per capita GDP. They applied fixed and random effects models to a subset of panel data from 73 countries over the period 1960 to 1990. They used data from the ASL and Associates (1997) database, consisting of data on sulphur emissions for most of the countries of the world over the period 1850 to 1990. An EKC relationship was found in both models, however, the turning points were much lower when using data for only the OECD countries (fixed effects: $9239, random effects: $9181), than when estimating EKC for the whole sample (fixed effects: $101 166, random effects: $54 199). Similarly the estimated turning points were extremely high for non-OECD countries (fixed effects: $908 178, random effects: $343 689), implying a monotonic EKC relationship between per capita income and sulphur emissions in the case of non-OECD countries and the whole sample. They said that their findings suggested that the inclusion of trade variables would be important as OECD countries could outsource the production of pollution-intensive products to the rest of the world. However, inclusion of global macroeconomic trends and shocks, such as the oil crisis of the 1970s, seemed unimportant.

In a study by Halkos (2003) using the same sample data from ASL and Associates (1997) as Stern and Common (2001), he empirically tested the relationship between environmental damage from sulphur emissions and per capita GDP. He applied two econometric methods: random coefficients and the Arellano–Bond Generalized Method of Moments (A–B GMM). The per capita GDP data were derived from the Penn World Table in Summers and Heston (1991), together with the population data. The sample of countries chosen by Halkos in his study represented 81 per cent of the world's population. He found an EKC pattern relationship in the case of an A–B GMM model with turning points well within the sample for all cases (global, OECD and non-OECD), ranging from $2805 to $6230. However, unlike the results found by Stern and Common (2001), the turning point for only the OECD countries was higher than those for the global and non-OECD countries. On the other hand, no support for the EKC hypothesis was found using the random coefficients model. The author concluded that the econometric techniques adopted in the empirical studies were crucial to the results and the level of the turning points of the EKC.

Ecological thresholds and sustainability
There have been a number of criticisms of the Environmental Kuznets Curve as it suggests that countries can overcome their environmental problems for certain pollutants simply by economic growth without paying attention to the environment and ecological thresholds. The criticisms were started by Arrow et al. (1995), who attributed the existence of the relationship between income and a selected set of pollutants to the fact that people in poor countries could not afford to emphasize environmental amenities over material well-being. As their standard of living improved and reached a sufficiently high level, their demand for environmental amenities increased. This led the governments to create environmental legislation and institutions to protect the environment. They also said that while the inverted U-shaped curve indicated that while improvements of some environmental indicators could be associated with economic growth, it did not mean that economic growth was sufficient to induce environmental improvement. Moreover, the effects of economic growth on the earth's resource base should not be ignored as it could not support indefinite economic growth. In addition, irreversible degradation in this base could put economic development at risk.

Furthermore, Arrow et al. highlighted that in interpreting the inverted U-shaped curves, a number of factors should be taken into account. First, research showed that the inverted U-shaped relationship between income and environmental quality was only valid for pollutants involving local and short-term costs (for example, sulphur, particulates and faecal coliforms), not for those involving long-term and more dispersed costs (such as CO_2). The latter type of pollutants showed in most research that they had an increasing function with income. Second, the relationship was valid for emissions of pollutants and not stocks of resources of the earth's base, such as soil, forests and other ecosystems. Third, the inverted-U curve relationship did not explain the effects of reductions in emissions on the wider system, in terms of increasing other emissions in the same country or transferring pollutants to other countries. Fourth, in most cases the reduction in emissions as income increased was due to local institutional reforms, such as environmental legislation and governmental policies, which often ignored the international and intergenerational consequences on other countries or future generations.

A number of environmentalists and ecological economists commented on the article by Arrow et al., including Ayres (1995). He rejected the proposition that economic growth improved the environment due to reasons other than the ones mentioned by Arrow et al. He referred in his argument to two well-known relationships: (1) the close correlation between economic growth and the increase in energy consumption and (2) most of the environmental problems were directly traceable to the unsustainable use of fossil fuels and/or other materials, such as toxic heavy metals and chlorinated chemicals, being a potential waste according to the basic law of conservation.

Stern et al. (1996) critically examined the concept of the EKC and they concluded that there were three major generic problems to the estimation and testing of EKCs. First was the assumption of unidirectional relationship from economic growth to environmental quality and no feedback from the environment to production. Second was the assumption that international trade associated with development did not affect environmental quality, while standard trade theory implied that under free trade, developing countries would specialize in producing labour-intensive goods and use their natural resources, while developed countries would specialize towards human capital and manufacture of capital-intensive goods. This specialization and the existence of environmental regulation would reduce environmental degradation in developed countries while increasing it in developing ones. In other words, developed countries would migrate polluting activities to poorer countries by increasing substitution towards importing manufactured goods. They also added that historical experiences of some economies could not be extrapolated to the whole global economy because once poor countries adopted similar levels of environmental policies, they would be faced with the problem of abating these activities as there would be no unregulated countries to migrate pollution to. Therefore, the whole world could not achieve a similar transformation. Third, the data on environmental problems were patchy in coverage and poor in quality, and therefore were insufficient to provide conclusions or to project future trends. They suggested that in order to project future trends using econometric analysis, they had to take the form of structural models rather than the generally reduced-form equations used to examine the EKC hypothesis. Such models would inform choices for policy-makers and would require policy adjustments in order to sustain environmental development of the global economy.

Farber (1995) also criticized the notion that economic growth induced improvements in environmental quality as naive as it was based on observations of only a limited number of environmental variables, which were local pollutants, and ignored global pollutants and their impact on the environment. Moreover, most findings resulted from empirical research and observations using cross-sectional country data. This was very limited in terms of the long-term, time-series relations across countries. In other words, any country could not be expected to follow the cross-country data points as its income increased. Furthermore, all countries with low income levels could not be expected to grow as (1) their environmental capital resource could be so depleted that economic growth would not be feasible and (2) the empirical relation did not measure the extent to which high-income countries substitute environmental degradation in poorer countries in the process of economic growth. Farber concluded that environmental and economic policies could not be set based on this relationship.

The answer to the U-shaped relationship between income and environmental quality was attributed by Page (1995) to the idea of harmony. This concept occurred in a number of studies, such as population studies, where there was an empirical claim of demographic transition at a low rate of economic development, then the rate of population growth would slowly increase up to a certain level beyond which further increase in economic growth and development would lead to reduction in the population growth rate to environmentally sustainable levels. Another example was the Laffer curve, which suggested that with very low tax rates there would be little government revenue, with increasing tax rates the government revenue would increase, but with further increasing tax rates the government revenue would decrease. A third example was the scarcity and growth argument by Barnett and Morse (1961), which stated that as depletion in natural resources continued, people would anticipate future scarcities, therefore prices would rise, and thereby incentives would be created for developing new technologies and substitutes, and hence the resource base would be renewed. In other words, natural competition among the individual players in the market would take care of the larger environment. A fourth example is Coase's policy (1990), which stated that if bargaining costs were kept low enough, the results would be close to efficient and good for the society as a whole.

The idea of harmony in economics goes back to Adam Smith's invisible hand (1776 [1994]). Nowadays, this idea is also strongly present in the form of structural adjustment and trade liberalization and other reforms, which aim at lower taxes, less regulation, freer markets and free trade. In the last decade, this resulted in shifting economic policies in International Financial Institutions (IFIs), such as the World Bank, International Monetary Fund, and so on, towards developing countries. It also resulted in large policy shifts in the world as a whole.

Concern regarding the structural adjustment programmes and economic reforms that took place in poor countries to accelerate economic growth was expressed by Munasinghe (1999). He argued that economic reforms could be at the expense of violating ecological thresholds, especially during the crossing-over stage of the EKC. The author said that economic reform was good for both the economy and the environment only if the reform introduced complementary environmental measures as part of the reform package in order to address the imperfections associated with it to mitigate environmental harm. Addressing such imperfections, which include policy distortions, market failures and institution constraints, would allow the reform programme to go forward without adverse impacts on the environment. Finally, the author recommended (1) caution in introducing policy reforms with great consideration to the timing and sequencing of the policies introduced and (2) addressing specific distortions in the economy rather than a wide economic reform programme with fewer environmental gains.

Panayotou (2000) argued that the EKC relationship between economic growth and the environment was not optimal or inevitable. He gave three reasons. First, the environment did not improve because of higher income levels, but as a result of environmental legislation and institutions due to the higher demand for better environmental quality. It was the policy intervention that determined the turning of the shape and height of the EKC. Second, the damage that could accumulate in a country until the environmental improvements were realized could exceed by far the present value of economic growth and better environmental quality. Again environmental policy was necessary to mitigate the environmental damage at early stages of development. Third, the height of the EKC, which reflected the environmental damage or what the author called 'environmental price' as a result of economic growth, was determined by the effectiveness of policies and efficiency of markets. The higher the externalities, undefined property rights and harmful subsidies, the higher the environmental price as a result of economic growth. In other words, market and policy failures resulted in unnecessary environmental degradation and economic inefficiency. Panayotou argued that better management of the resource base, especially in the presence of ecological thresholds that could be irreversible, would result in an optimal economic growth and environmental improvement.

The process of dematerialization, which was defined as the 'unlinking of income and the use of nature', was examined by Canas et al. (2003). Dematerialization was related to the concept of strong sustainability: the maintenance of natural capital. It could mean an absolute reduction in the use of material (strong dematerialization) or just a reduction in the material intensity of income (weak dematerialization). They tested the EKC hypothesis using the direct material input (DMI) per capita as the dependent variable and per capita GDP as the independent variable for

16 industrialized countries, using panel data over the period 1960 to 1998. Their results showed robust support for weak dematerialization in industrialized countries, whereby the aggregate material intensity, measured in mass units, reduced as per capita income levels increased. The authors suggested that this relationship could be due to (1) the higher material use at lower levels of income to respond to infrastructural needs and (2) the structural change within the economy towards an increase in the service sector as income increased.

Dinda (2004) reviewed a number of studies on the EKC hypothesis and ended up by saying that the subject was open-ended and the existence of the EKC was inconclusive. He also highlighted several areas that required further research. First, the choice of economic models was important to properly reflect the physical and ecological aspects of the economy. Second, unfolding the underlying factors behind the EKC was a priority for any research. Third, moving towards structural rather than reduced-form models and decomposition analysis might be needed. Fourth, using of time series rather than panel data was essential to identify the development of pollution associated with economic growth in individual countries. Finally, determining the policy measures and regulations necessary to allow sustainable development.

Conclusion

This chapter reviewed the extensive research conducted in examining the relationship between economic growth and the environment. Earlier studies on the income–environment relationship used the reduced-form single-equation approach to show the relationship between economic growth and environmental degradation, without understanding the dynamics of this relationship. In order to test whether income was the determinant of the behaviour of the environmental indicator(s), recent studies moved towards decomposition analysis and more structural and analytical models to identify the real factors or influences behind the income–environment relationship. Most of the studies used cross-sectional country or panel data instead of time series data across countries due to the difficulty and limited availability of the latter. Furthermore, although some research explored the role of policies and institutions, a limited number of empirical studies examined the influence of international environmental legislation on the relationship between economic growth and the environment.

References

Andreoni, J. and A. Levinson (1998), 'The Simple Analytics of the Environmental Kuznets Curve', National Bureau of Economic Research, Inc, NBER Working Paper No. 6739.

Antle, J.M. and G. Heidebrink (1995), 'Environment and Development: Theory and International Evidence', *Economic Development and Cultural Change*, **43**(3), 603–25.
Arrow, K., B. Bolin, R. Costanza, P. Dasgupta, C. Folke, C.S. Holling, B.-O. Jansson, S. Levin, K.-G. Maler, C. Perrings and D. Pimentel (1995), 'Economic Growth, Carrying Capacity, and the Environment', *Ecological Economics*, **15**(2), 89–95.
Ayres, R.U. (1995), 'Economic Growth: Politically Necessary but not Environmentally Friendly', *Ecological Economics*, **15**(2), 97–9.
Barnett, H. and C. Morse (1961), *Scarcity and Growth: The Economics of Natural Resource Availability*, Baltimore, MD: Johns Hopkins Press.
Barrett, S. and K. Graddy (2000), 'Freedom, Growth, and the Environment', *Environment and Development Economics*, **5**(1), 433–56.
Beckerman, W. (1992), 'Economic Growth and the Environment: Whose Growth? Whose Environment?', *World Development*, **20**(4), 481–96.
Bhattarai, M. and M. Hammig (2001), 'Institutions and the Environmental Kuznets Curve for Deforestation: A Cross-country Analysis for Latin America, Africa and Asia', *World Development*, **29**(6), 995–1010.
Brundtland, G. (1987), *Our Common Future: The World Commission on Environment and Development*, Oxford, UK: Oxford University Press.
Canas, A., P. Ferrao and P. Conceicao (2003), 'A New Environmental Kuznets Curve? Relationship between Direct Material Input and Income per Capita: Evidence from Industrialized Countries', *Ecological Economics*, **46**(2), 217–29.
Carson, R.T., Y. Jeon and D.R. McCubbin (1997), 'The Relationship between Air Pollution Emissions and Income: US Data', *Environment and Development Economics*, **2**(4), 433–50.
Cavlovic, T.A., K.H. Baker, R.P. Berrens and K. Gawande (2000), 'A Meta-analysis of Environmental Kuznets Curve Studies', *Agricultural and Resource Economics Review*, **29**(1), 32–42.
Chaudhuri, S. and A. Pfaff (2002), 'Economic Growth and the Environment: What Can We Learn from Household Data', School of International and Public Affairs, Columbia University, Paper No. 0102-51.
Coase, R.H. (1990), *The Firm, the Market, and the Law*, Chicago: University of Chicago Press.
Cole, M.A. (2004), 'Trade, the Pollution Haven Hypothesis and the Environmental Kuznets Curve: Examining the Linkages', *Ecological Economics*, **48**(1), 71–81.
Cole, M.A., A.J. Rayner and J.M. Bates (1997), 'The Environmental Kuznets Curve: An Empirical Analysis', *Environment and Development Economics*, **2**(4), 401–16.
Common, M. (1995), *Sustainability and Policy: Limits to Economics*, Cambridge, UK: Cambridge University Press.
Congleton, R.D. (1992), 'Political Institutions and Pollution Control', *Review of Economics and Statistics*, **74**(3), 412–21.
Cropper, M. and C. Griffiths (1994), 'The Interaction of Population Growth and Environmental Quality', *Population Economics*, **84**(2), 250–54.
De Bruyn, S.M. (1997), 'Explaining the Environmental Kuznets Curve: Structural Change and International Agreements in Reducing Sulphur Emissions', *Environment and Development Economics*, **27**(2), 485–503.
De Bruyn, S.M., J.C.J.M. Van Den Bergh and J.B. Opschoor (1998), 'Economic Growth and Emissions: Reconsidering the Empirical Basis of Environmental Kuznets Curves', *Ecological Economics*, **25**(2), 161–75.
Deacon, R. (1999), 'The Political Economy of Environmental Development Relationships', Santa Barbara, Preliminary Framework Working Paper, University of California.
Dinda, S. (2004), 'Environmental Kuznets Curve Hypothesis: A Survey', *Ecological Economics*, 49(4), 431–55.
Farber, S. (1995), 'Economic Resilience and Economic Policy', *Ecological Economics*, **15**(2), 105–7.

Galeotti, M. and A. Lanza (1999), 'Desperately Seeking (Environmental) Kuznets', Discussion Paper, International Energy Agency.

Gastil, R.D. (1987), *Freedom in the World: Political Rights and Civil Liberties 1986–1987*, New York: Greenwood Press.

Goodstein, E. (1999), *Economics and the Environment*, Englewood Cliffs: Prentice Hall.

Grossman, G.M. and A.B. Krueger (1991), 'Environmental Impacts of a North American Free Trade Agreement', Discussion Papers in Economics, Woodrow Wilson School of Public and International Affairs, Princeton, NJ.

Grossman, G.M. and A.B. Krueger (1995), 'Economic Growth and the Environment', *Quarterly Journal of Economics*, **110**(2), 353–77.

Grossman, G.M., J.A. Laity and A.B. Krueger (1994), 'Determinants of Air Pollution in US Countries', Discussion Papers in Economics, Woodrow Wilson School of Public and International Affairs, Princeton, NJ.

Halkos, G.E. (2003), 'Environmental Kuznets Curve for Sulfur: Evidence Using GMM Estimation and Random Coefficient Panel Data Models', *Environment and Development Economics*, **8**(4), 581–601.

Hettige, H., R.E.B. Lucas and D. Wheeler (1992), 'The Toxic Intensity of Industrial Production: Global Patterns, Trends, and Trade Policy', *American Economic Review*, **82**(2), 478–81.

Hettige, H., M. Mani and D. Wheeler (1997), 'Industrial Pollution in Economic Development: Kuznets Revisited', Policy Research Working Paper No. WPSIF76, World Bank.

Hilton, H.F.G. and A. Levinson (1998), 'Factoring the Environmental Kuznets Curve: Evidence from Automotive Lead Emissions', *Journal of Environmental Economics and Management*, **35**(2), 126–41.

Holtz-Eakin, D. and T.M. Selden (1995), 'Stoking the Fires? CO_2 Emissions and Economic Growth', *Journal of Public Economics*, **57**(1), 85–101.

IBRD (1992), *World Development Report, 'Development and the Environment'*, New York: Oxford University Press.

Islam, N., J. Vincent and T. Panayotou (1999), 'Unveiling the Income–Environment Relationship: And Exploration into the Determinants of Environmental Quality', Working Paper, Department of Economics and Harvard Institute for International Development.

Jaggers, K. and T.R. Gurr (1995), 'Tracking Democracy's Third Wave with the Polity III Data', *Journal of Peace Research*, **32**, 469–82.

Kahn, M.E. (1998), 'A Household Level Environmental Kuznets Curve', *Economics Letters*, **59**(2), 269–73.

Kaufman, R., B. Davidsdotter and D. Garnham (1998), 'The Determinants of Atmospheric SO_2 Concentration: Reconsidering the Environmental Kuznets Curve', *Ecological Economics*, **25**(2), 209–20.

Khanna, N. and F. Plassmann (2004), 'The Demand for Environmental Quality and the Environmental Kuznets Curve Hypothesis', *Ecological Economics*, **51**(3/4), 225–36.

Klaassen, G. (1995), 'Trading Sulfur Emission Reduction Commitments in Europe: A Theoretical and Empirical Analysis', International Institute for Applied System Analysis (IIASA), Laxenburg, Austria.

Knack, S. and P. Keefer (1995), 'Institutions and Economic Performance: Cross Country Tests Using Alternative Institutional Measures', *Economic and Politics*, **7**(3), 207–27.

Komen, M.H.C., S. Gerking and H. Folmer (1997), 'Income and Environmental R&D: Empirical Evidence from OECD Countries', *Environment and Development Economics*, **2**(4), 505–15.

Koop, G. and L. Tole (1999), 'Is there an Environmental Kuznets Curve for Deforestation?', *Journal of Development Economics*, **58**(1), 231–44.

Kuznets, S. (1955), 'Economic Growth and Income Inequality', *American Economic Review*, **45**(1), 1–28.

Marland, G. and R.M. Rotty (1984), 'Carbon Dioxide Emissions from Fossil Fuels: A Procedure for Estimation and Results for 1950–1982', *Tellus*, **36**(B), 232–61.

Marland, G. R.J. Andres, T.A. Boden, C. Johnson and A. Brenkert (1999), *Global Regional*

and National CO_2 Estimates from Fossil Fuel Burning, Cement Production, and Gas-flaring: 1751–1996, Oak Ridge, TN: Carbon Dioxide Information Analysis Center.
McConnell, K.E. (1997), 'Income and the Demand for Environmental Quality', *Environment and Development Economics*, **2**(1), 383–99.
Meadows, D.H., D.L. Meadows, J. Randers and W.W. Behrends (1972), *The Limits to Growth*, London: Earth Island Limited.
Midlarsky, M.I. (1998), 'Democracy and the Environment: An Empirical Assessment', *Journal of Peace Research,* Special Issue on Environmental Conflict, **35**(3), 341–61.
Moomaw, W.R. and G.C. Unruh (1997), 'Are Environmental Kuznets Curves Misleading Us? The Case of CO_2 Emissions', *Environment and Development Economics*, **2**(4), 451–63.
Moomaw, W.R. and G.C. Unruh (1998), 'An Alternative Analysis of Apparent EKC-type Transactions', *Ecological Economics*, **25**(2), 221–9.
Munasinghe, M. (1999), 'Is Environmental Degradation an Inevitable Consequence of Economic Growth: Tunneling through the Environmental Kuznets Curve', *Ecological Economics*, **29**(1), 89–109.
Page, T. (1995), 'Harmony and Pathology', *Ecological Economics*, **15**(2), 141–4.
Panayotou, T. (1993), 'Empirical Tests and Policy Analysis of Environmental Degradation at Different Stages of Economic Development', World Employment Programme Research Working Paper No. WEP2-22/WP 238, Geneva: International Labour Force.
Panayotou, T. (1997), 'Demystifying the Environmental Kuznets Curve: Turning a Black Box into a Policy Tool', *Environment and Development Economics*, **2**(4), 465–84.
Panayotou, T. (2000), 'Economic Growth and the Environment', CID Working Paper No. 56, Center for International Development, Cambridge, MA: Harvard University.
Panayotou, T., J. Sachs and Peterson (1999), 'Developing Countries and the Control of Climate Change: A Theoretical Perspective and Policy Implications', CAER II Discussion Paper No. 45, August.
Perrings, C.A. and A. Ansuategi (2000), 'Sustainability, Growth and Development', *Journal of Economic Studies*, **27**(1/2), 19–55.
Pfaff, A.S.P., S. Chaudhuri and H.L.M. Nye (2001), 'Why Might One Expect Environmental Kuznets Curves? Examining the Desirability and Feasibility of Substitution', Columbia University, 9 July, Paper No. 0102-45.
Pfaff, A.S.P., S. Chaudhuri and H.L.M. Nye (2002), 'Endowments, Preferences, Abatement and Voting: Microfoundations of Environmental Kuznets Curves', Columbia University, Paper No. 0102-46.
Ravallion, M., M. Heil and J. Jalan (1997), 'A Less Poor World, but a Hotter One? Carbon Emissions, Economic Growth and Income Inequality', Working Paper No. 13, Washington, DC: World Bank.
Roberts, J.T. and P.E. Grimes (1997), 'Carbon Intensity and Economic Development 1962–91: A Brief Exploration of the Environmental Kuznets Curve', *World Development*, **25**(2), 191–8.
Roca, J. (2003), 'Do Individual Preferences Explain the Environmental Kuznets Curve?', *Ecological Economics*, **45**(1), 3–10.
Rock, M. (1996), 'Pollution Intensity of GDP and Trade Policy: Can the World Bank Be Wrong?', *World Development*, **24**(3), 471–9.
Rothman, D.S. (1998), 'Environmental Kuznets Curves – Real Progress or Passing the Buck? A Case for Consumption-based Approaches', *Ecological Economics*, **25**(2), 177–94.
Schmalensee, R., T.M. Stoker and R.A. Judson (1998), 'World Carbon Dioxide Emissions: 1950–2050', *The Review of Economics and Statistics*, **80**(1), 15–27.
Selden, T.M. and D. Song (1994), 'Environmental Quality and Development: Is There a Kuznets Curve for Air Pollution Emissions?', *Journal of Environmental Economics and Management*, **27**(2), 147–62.
Selden, T.M. and D. Song (1995), 'Neoclassical Growth, the J Curve for Abatement, and the Inverted U Curve for Pollution', *Journal of Environmental Economics and Management*, **27**(2), 162–8.

Shafik, N. (1994), 'Economic Development and Environmental Quality: An Econometric Analysis', *Oxford Economic Papers*, **46**(5), 757–73.

Shafik, N. and S. Bandyopadhyay (1992), 'Economic Growth and Environmental Quality: Time-series and Cross-country Evidence', World Bank Working Papers No. WPS 904, Washington, DC: World Bank.

Smith, A. (1776 [1994]), *The Wealth of Nations*, New York: Random House.

Stavins, R.N. (2000), 'Economic Analysis of Global Climate Change Policy: A Primer', John F. Kennedy School of Government, Harvard University and Resources for the Future – Working Paper No. 00-003.

Stern, D.I. and M.S. Common (2001), 'Is There an Environmental Kuznets Curve for Sulphur?', *Journal of Environmental Economics and Management*, **41**(2), 162–78.

Stern, D.I., M.S. Common and E.B. Barbier (1996), 'Economic Growth and Environmental Degradation: A Critique of the Environmental Kuznets Curve', *World Bank Development*, **24**(7), 1151–60.

Summers, R. and A. Heston (1991), 'The Penn World Table (Mark 5): An Expanded Set of International Comparisons, 1950–1988', *The Quarterly Journal of Economics*, **106**(2), 327–68.

Suri, V. and D. Chapman (1998), 'Economic Growth, Trade and Energy: Implications for the Environmental Kuznets Curve', *Ecological Economics*, **25**(2), 195–208.

Torras, M. and J. Boyce (1998), 'Income, Inequality, and Pollution; A Reassessment of the Environmental Kuznets Curve', *Ecological Economics*, **25**(2), 147–60.

Tuan, N.A. (1999), 'Evidences of Environmental Kuznets Curve from CO_2 Emissions in Six-country Analysis', Working Paper, Institute d'Economie et de Polique de L'Energie (IEPE) BP 47, 38040 Grenoble Cedex 09.

Vincent, J.R. (1997), 'Testing for Environmental Kuznets Curves within a Developing Country', *Environment and Development Economics*, **2**(4), 417–31.

World Bank (1987), *World Bank Development Report*, New York: Oxford University Press.

13. Biodiesel as the potential alternative vehicle fuel: European policy and global environmental concern
Mahesh Poudyal and Jon C. Lovett

Introduction

The widely accepted principles of sustainable development are that the present generation should be able to meet its own needs without compromising the needs of future generations. Essentially this implies that we can continue to have economic growth, but the means by which we achieve this should not do so much damage that our children look back in anguish and question our actions. In the last two centuries economic growth has been powered by burning fossil fuels with the consequent release of carbon dioxide. In recent years this release has been increasing rapidly: between 1961 and 2002 humanity's carbon footprint grew more than 700 per cent (Kitzes et al., 2007). In the last few decades there has been widespread concern that observed increases in atmospheric carbon dioxide and global temperatures are causally linked. Confidence in the general scientific consensus has reached the point where policy-makers are willing to take firm action, for example the 10th Session of Working Group I of the Intergovernmental Panel on Climate Change (IPCC) in Paris (February 2007) concluded that:

> Most of the observed increase in globally averaged temperatures since the mid-20th century is *very likely* due to the observed increase in anthropogenic greenhouse gas concentrations. . . . Discernible human influences now extend to other aspects of climate, including ocean warming, continental-average temperatures, temperature extremes and wind patterns. (Intergovernmental Panel on Climate Change 2007, emphasis original)

Policy instruments for tackling greenhouse gas (GHG) induced warming have been in place for some time. The United Nations Framework Convention on Climate Change (UNFCCC) entered into force on 21 March 1994 and led to the 1997 Kyoto Protocol. Ratification of the Kyoto Protocol by the European Union (EU) in 2002 meant that all the Member States came under obligation to cut their GHG emissions. The Protocol entered into effect on 16 February 2005 following ratification by Russia in November 2004. Although it is still not ratified by the United

States, a major carbon emitter, and big emitters like China and India are not required to reduce their emissions at present, the policy and practices within the EU – the largest economic bloc in the world – provide a much-needed credibility to the Kyoto Protocol and impetus to development of technological innovations that will allow present generations to have their energy supply without leaving a drastically altered planet for the future.

Among various measures to cut GHG emissions within the EU, the use of biofuels derived from agricultural or forestry products is considered a viable alternative to fossil fuels, especially for the transport sector. As such, the EU Directive 2003/30/EC set targets for biofuels to be used in EU transport at 2 per cent by the end of 2005 and 5.75 per cent by the end of 2010. This directive with other complementary directives, resolutions and legislation on renewable fuels could play a major role in the EU's attempts to reduce GHG emissions from the transport sector. This drive has been further strengthened by the recent announcement about the minimum efficiency requirements set for vehicle manufacturers that limit the emissions from the vehicles to be manufactured in the future.[1]

Substitution of biofuels for fossil fuels also helps to fulfil other policy objectives. First, the world's main oil reserves are located in geopolitically sensitive areas. Climate change notwithstanding, this is perhaps the main incentive for switching from oil to a more costly fuel supply. The economic rise of countries such as China and India mean that there are major players in the global quest for natural resources that are not necessarily inside the traditional European sphere of political influence. Environmental considerations are thus in line with the strategic needs of national security. Second, agricultural subsidies have historically been a major European expenditure and source of considerable debate within the Union. Support for growth of biofuel feedstocks could provide a bridge to resolve differences between the Member States. Third, rapid expansion of the European Union has led to economic disparities between the new and old members. The political need for harmonization requires investment from the rich and markets for the poor. For example, Bulgaria and Romania, who joined the Union in January 2007, have abundant agricultural land and could potentially become major biofuel suppliers. Fourth, biofuels open a huge new market for developing countries that have spare land and cheap labour. Biofuel production could help these countries meet the Millennium Development Goal of poverty alleviation through economic growth based on primary production.

These potential multiple benefits of biofuel use make them attractive to policy-makers, but there have been criticisms too. For example, although some calculations demonstrate a positive energy balance for biofuel use (Hill et al., 2006), other work suggests that the GHG footprint of biofuel production may not be as attractive as hoped because biofuel production requires considerable energy and fertilizer inputs and so does not meet the primary policy goal of reduced GHG emissions (Dias de Oliveira et al. 2005; Crutzen et al., 2007). Indeed, it may be that avoiding deforestation for agriculture and restoring forests is a better option (Righelato and Spracklen, 2007). Second, a switch to intensive biofuel production on the scale needed to supply energy markets will result in a transformation of species-rich habitats such as tropical rainforests in biodiversity hotspots (Koh, 2007) and so perhaps have a greater negative effect on biodiversity than that of global warming. Third, replacing food production with biofuel crops could cause market distortions and increase food prices (Doornbosch and Steenblik, 2007). However, the situation here is complicated by global changes in eating habitats, for example, increased intensive livestock-rearing and meat consumption in China have been driving the rise in soya bean production (a biofuel feedstock) and export in countries such as Brazil (Naylor et al., 2005).

The outline of this chapter is as follows. The next section looks at the EU renewable fuel policy broadly, and at its biofuels policy specifically. The biofuel targets set by the EU and the policies in place to meet such targets are analysed. Furthermore, we discuss the importance of biodiesel in helping to meet the EU's biofuels targets, and the issues related to production of biodiesel within the EU and in other parts of the world in the context of the EU's biofuels policy. We then assess the costs and benefits of biodiesel production and use, mainly in the transport market to which it is geared. In addition, we critically review the results from a number of studies that have looked into the emissions from vehicles using biodiesel. Then we ask whether biofuels in general, and biodiesel in particular, have the potential to provide the double benefit of being a secure and cheap source of energy and at the same time being environmentally friendly – as is often argued by its promoters. We assess the potential of biodiesel to achieve this based on our review of the costs and benefits of biodiesel, as well as on our analysis of the promotion of biofuels in the EU market and its policies on biofuels. We conclude the chapter by summarizing the main issues surrounding biodiesel at present, and looking into the future of this particular type of renewable fuel, in the context of EU policy, climate change issues and global energy demand.

EU renewable fuels policy and biodiesel
The European Union has been promoting production and use of renewable energy within its Member States, especially during the last decade. In the past few years, a number of Council resolutions and directives have been passed in its drive to promote renewable energy production and use

in the Member States (for example, The Council of the European Union, 1998; The European Parliament and the Council of the European Union, 2001, 2003). The Council Resolution of 1998 on renewable sources of energy states:

> there is need to promote a sustained and substantially increased use of renewable sources of energy throughout the Community in the light of the valuable contribution renewables can make to environmental protection and the implementation of the commitments under the Kyoto Protocol, to security of supply and the preservation of finite energy resources, and to economic and social development generally, including in relation to employment and the strengthening of the economic structure of the outermost, isolated and island regions. (The Council of the European Union, 1998, p. 1)

The 1998 Council Resolution was influenced by the Commission's 'White' and 'Green' Papers on renewable sources of energy, and endorsed an indicative target of 12 per cent renewable energy use for the Community as a whole by 2010 as a useful guidance for all Member States pursuing policies towards increasing renewable energy production and use (The Council of the European Union, 1998, p. 1). This resolution also encouraged the Member States to 'choose the most appropriate means of promoting use of renewables', suggesting a number of instruments, such as subsidies, preferential tariffs and purchase obligations to name a few.

The 1998 resolution was followed by a number of other directives promoting use of renewable energy sources, including one on biofuels or other renewable fuels for transport – the Council Directive 2003/30/EC (The European Parliament and the Council of the European Union, 2003). Among various measures to cut GHG emissions within the EU, the use of biofuels has been considered a potential alternative to reducing emissions, especially from transport, as biofuels are seen as the most viable alternative to replace or complement fossil fuel. As such, the European Directive 2003/30/EC not only aims to promote the use of biofuels in the transport sector, but also sets out the target for biofuels to be used in transport in the EU Member States. According to the first article of the directive, it

> aims at promoting the use of biofuels or other renewable fuels to replace diesel or petrol for transport purposes in each Member State, with a view to contributing to objectives such as meeting climate change commitments, environmentally friendly security of supply and promoting renewable energy sources. (Article 1, Directive 2003/30/EC)

Thus, the directive is especially concerned with the promotion of biofuels as the viable alternative to the diesel or petrol used in transport in the EU, and considers this as one of the means to meet its climate change commitments as set out in the Kyoto Protocol. The directive lists ten products it classifies as biofuels, including biodiesel and bioethanol – the biofuels that are already available and used in significant proportions around the world. In Article 3, this directive also sets minimum targets for all the Member States, and encourages them to set their own 'indicative targets'.[2]

Furthermore, the directive provides flexibility as to how the biofuels for transport are made available in the market, allowing for the supply of pure biofuels (biodiesel and bioethanol) or blended with fossil fuels. Where the biofuels are supplied as blends, the directive requires specific labelling of such fuels only when the quantity of biofuels in such blends exceed 5 per cent (Directive 2003/30/EC, *OJ L* 123, p. 45). Finally, the directive sets out the guidelines for reporting the annual progress in meeting the set targets by the Member States, before 1 July each year. The directive stated that the Commission would draw up an evaluation report on the progress made in the first phase by 31 December 2006, adding that:

> If this [evaluative] report concludes that the indicative targets are not likely to be achieved for reasons that are unjustified and/or do not relate to new scientific evidence, these proposals shall address national targets, including possible mandatory targets, in the appropriate form. (Directive 2003/30/EC, *OJ L* 123, p. 46)

So, although the targets set by the directive were non-binding and more as guidelines, if the Member States were not doing enough to promote the use of biofuels in transport during the first phase as set out by the directive, without justifiable reasons, the directive clearly stated that the Commission could make the targets mandatory in the second phase. Indeed, despite a relatively low minimum target set by the directive for the first phase, it is now clear that the 2 per cent target for the minimum proportion of biofuels and other renewable fuels that should have been placed on the markets of the EC Member States by the end of 2005 has not been met by the Member States except for Germany and Sweden (European Commission, 2006, 2007b). According to the 2006 progress report, the share of biofuels in the transport market for the 21 Member States where biofuels are in use was only 1 per cent by the end of 2005. Although this 1 per cent market share is 'a good rate of progress – a doubling in two years' according to the 2006 progress report, it is only half the target set by Directive 2003/30/EC, and less than the 1.4 per cent share that would have been achieved if all the Member States had met their annual indicative targets (European Commission, 2007b, p. 6). Furthermore, both the progress among the Member States and the rate of adoption of biofuels seemed very uneven – with only Germany (3.8 per cent) and Sweden (2.2

per cent) exceeding the 2 per cent target. In terms of adoption of biofuels, biodiesel was significantly ahead of ethanol – with biodiesel achieving a 1.6 per cent share of the diesel market, whereas ethanol achieved only a 0.4 per cent share of the petrol market (European Commission, 2007b, p. 6).

The main policy instrument to facilitate meeting this target was the EU Directive 2003/96/EC (The Council of the European Union, 2003) on energy products taxation, which allowed Member States to reduce, or apply total or partial exemption in the level of taxation to fuels from renewable sources. Moreover, schemes like Energy Crop Payments – a premium payment of €45/ha for growing energy crops under the 2003 Common Agricultural Policy (CAP) reform – and allowing energy crops to be grown on set-aside land, were thought to encourage farmers in producing energy crops to meet the demand for biofuel feedstocks. In addition, increasing prices of fossil fuel and insecurity of supply, particularly in the past two to three years, either due to natural disasters like Hurricane Katrina or due to conflicts in the Middle East and elsewhere (such as the supply disputes between Russia and Ukraine or Russia and Georgia), should have provided further incentives, in theory, to invest in, produce and promote biofuels in the Member States. Even with these biofuel-favourable policies, and geopolitical conditions, the share of biofuels on the transport fuel market has been negligible in most of the Member States.

The next milestone in the EU Directive 2003/30/EC on promotion of the use of biofuels or other renewable fuels for transport is the target of 5.75 per cent – the proportion of biofuels and other renewable fuels that should be placed on the transport fuel market by the end of the year 2010. As in the first phase, most of the Member States have set their own annual indicative targets for the second phase in the promotion of biofuels in the transport fuel market. The indicative targets set by the 19 Member States, if achieved, will increase the share of biofuels in the transport market to 5.45 per cent by 2010, slightly less than the 5.75 per cent target. However, the 2006 progress report concludes, judging by the progress made by a majority of the Member States during the first phase of biofuels promotion in their transport fuel markets, that 'the biofuels directive's target [of 5.75 per cent] for 2010 is not likely to be achieved' (European Commission, 2007b, p. 6).

Although a number of the EU Member States have put in place 'biofuel obligations',[3] the 2006 biofuels progress report calls for legislative measures to support the promotion of biofuels in transport in the EU, arguing that such a legislative measure will 'send a signal of the [European] Union's determination to reduce its dependence on oil use in transport' (European Commission, 2007b, p. 14). The report further argues that the legislative framework in favour of biofuels will give support to national, regional and local authorities working towards the objective of reducing dependence on oil use in transport; give confidence to companies, investors and scientists who are working on more efficient ways to do this; and give pause to those who believe that European consumers will always remain hostage to oil prices, whatever the price. (European Commission, 2007b, p. 7)

Positing that 'a signal in the form of legally binding targets is stronger than a purely voluntary commitment', the biofuels progress report calls for the EU to set minimum targets for the future share of biofuels in transport, and suggests a minimum target of 10 per cent biofuels in the transport market by 2020 (European Commission, 2007b, p. 8). Moreover, the report presses for 'efficiency in biofuel policy' so as to build investor confidence, reduce administrative burden and encourage production of biofuels such that it helps meet the directive's objectives (ibid.).

As mentioned above, biodiesel is the major biofuel adopted within the EU Member States with a significantly higher share of total transport market fuels than bioethanol. Of the two Member States exceeding the 2 per cent target by the end of 2005, Germany's biofuel market is mainly biodiesel, whereas that of Sweden is mainly bioethanol (European Commission, 2007b). However, in aggregate the total proportion of biofuels in the EU transport fuel market is very skewed towards biodiesel. The main reason for higher use of biodiesel is the availability of the biodiesel feedstocks, such as oilseed rape, within the EU. Furthermore, biodiesel requires very little adjustment in the engine as well as the supply infrastructure, which is a great incentive for the investors. In fact Rudolph Diesel's first diesel engine was designed to run on vegetable oil (Demirbas, 2003). On the other hand, bioethanol is produced mainly from corn and sugarcane at this early stage of technological progress, and has to be imported from countries like Brazil. Although the next generation (the so-called 'second generation') of biofuel production technologies are currently being developed and tested, such as the production of ethanol from straw and wood chips, it might be a number of years before their production is commercially viable and they are supplied to the market (Herrera, 2006; Schubert, 2006). Thus, in the meantime, EU Member States have to either import bioethanol, as Sweden is doing from Brazil, or rely largely on biodiesel produced within the EU to meet their biofuels targets.

Production of biodiesel within the EU Member States should be able to supply enough fuel to meet the 2 per cent target set for 2005, however, it will be difficult to meet the 5.75 per cent target set for 2010 from domestic production alone. Apart from increasing the production of the biofuels within the EU, meeting this target will be further complicated by the fact that all of the EU member countries will not be able to grow all of the feedstocks required to produce the biofuels to meet this target unless food-crops are replaced by biofuel crops.[4] For example, the United Kingdom could produce enough feedstocks to meet the 2 per cent target by cultivating oilseed crops in less that 1 million hectares of arable farmland, which could be available by using part of the set-asides, part of grasslands under five years old (which are considered arable land) and without replacing a significant proportion of the food-crops. However, to produce enough feedstocks to meet the target of 5.75 per cent, the United Kingdom will require around 2 million hectares of arable farmland, which in a country with total arable land of just 5.8 million hectares[5] (of which about 1.2 million hectares is grassland under five years old), is not feasible without significantly altering the production of food-crops and the agricultural landscape. The area of the set-asides is only about 0.5 million hectares, so using part of this land will not contribute a significant proportion of the total land required to produce the feedstocks. Thus, the only alternative would be to import the biofuels and/or biofuel feedstocks from countries outside the EU – mainly from countries like Indonesia and Malaysia (for biodiesel) and Brazil (for bioethanol).

The EU Strategy for Biofuels (European Commission, 2006), and the Biomass Action Plan (European Commission, 2005) call for 'supporting developing countries' to develop internal and export markets for biofuels, making sure minimum sustainability standards are met in their cultivation. The potential environmental and socioeconomic impacts from the cultivation of biofuel feedstocks and the production (and consumption) of biofuels are similar in the developing countries to in the EU. However, increasing demand for biofuels from the EU and other developed countries is expected to not only provide opportunities for developing countries to benefit from their production and export, but also to generate negative environmental and socioeconomic consequences, especially where there are few laws in place to ensure sustainable production of biofuels. Although the biofuels progress report refutes the claim that Europe's biodiesel consumption has caused tropical deforestation in palm-oil-producing countries such as Malaysia and Indonesia, and suggests that the global palm oil demand is mainly driven by the food market (European Commission, 2007b), the increasing use and demand for such oils, not only for foods but for biofuels in the West as well as in fast-growing economies like China and India, is likely to be responsible for tropical deforestation both in the Brazilian Amazon and in Indonesia (Casson, 2003; Monbiot, 2004, 2005; Mortished, 2006b). Hence, even if the EU restricts its import of biofuels and biofuel feedstocks to those produced meeting its sustainability criteria, other large-scale importers such as China may not apply similar restrictions. Unless the countries producing biofuels have in place strict environmental regulation with regard to biofuels production, there will be an increasing possibility of negative environmental consequences in those countries as the demand for biofuels increases globally in the future.

The biofuels progress report, however, sees more benefits than costs for both the EU and its trading partners from growth in the biofuels industry. In addition to the much discussed benefits from the reduction of greenhouse gas emissions, it not only sees short- and long-term security of energy supply in the EU through the increasing use of biofuels, it also posits that the demand for biofuel imports from the EU can help improve trade relations and provide opportunities for developing countries to produce and export biofuels at competitive prices (European Commission, 2007b, p.10). In the likely absence of second generation biofuels in the highly competitive fuel market for another decade or so, Europe has to rely heavily on countries such as Malaysia, Indonesia (for biodiesel) and Brazil (for bioethanol) to meet its growing biofuel needs, at least in the foreseeable future. Taking a similar line to the EU Strategy for Biofuels (European Commission, 2006) and the Biomass Action Plan (European Commission, 2005), the biofuels progress report calls for non-discriminatory access to the biofuels market for both domestic production and imports, as long as they meet sustainability criteria in the production of biofuels (European Commission, 2007b). Furthermore, the report argues that the target of 10 per cent biofuels by 2010 could be met with limited use of the second generation biofuels through (1) further development of rapeseed cultivation in the EU and its neighbours to the east; (2) proper incentives to biofuels producers, both in the EU and other countries, for environmentally friendly production of biofuels and (3) 'implementation of the balanced approach to international trade in biofuels, so that both exporting countries and domestic producers can invest with confidence in the opportunities created by the growing European market' (European Commission 2007b, p.13). However, the report argues that the likely failure to meet the directive's biofuels target for 2010 (as a whole in the EU) cannot be described as 'justified', and hence calls for the revision of the directive to make it more effective. It calls for the biofuels directive to be revised to (1) reiterate the EU's determination in reducing its dependence in oil for transport; (2) set minimum targets for the share of biofuels and (3) discourage unsustainable production of biofuels in favour of those produced in an environmentally friendly fashion (European Commission, 2007b). Nonetheless, it recognizes that revision of the biofuel directive will not work by itself, and that the changes will require 'sustained effort on the part of industry, agriculture and Member States as well as the EU' to make them work (European Commission, 2007b, p.13).

Finally, the biofuels progress report is also in line with a broader policy document from the EU – the Renewable Energy Road Map (European Commission, 2007a). Like the biofuels progress report, the Renewable Energy Road Map also calls for revision to existing voluntary targets for use of biofuels in the EU, proposing instead a 'mandatory (legally binding) target of 20% for renewable energy's share of energy consumption in the EU by 2020', along with the proposal for necessary legislative frameworks to make sure such targets are met (p. 3). Like most other documents from the EU relating to use of renewable energy, the Renewable Energy Road Map is inherently optimistic: 'The challenge [that is, meeting renewable energy targets] is huge, but the proposed target can be achieved with determined and concerted efforts at all levels of government assuming the energy industry plays its full part in the undertaking' (European Commission, 2007a, p. 3).

This document also sees biofuels as the 'only available large-scale substitute' for transport fuels, recognizing barriers in trade as one of the key factors that could hinder the EU Member States in meeting their biofuels target, especially if they are dependent on biofuels imports from outside the EU (European Commission, 2007a, p. 7). Furthermore, the document is keen to stress the need to take an incentive-based approach, such as certification schemes with preferential trade agreements for biofuels, when it comes to importing biofuels from developing countries, in order to 'avoid rain forest destruction' and other forms of losses in biodiversity.

Biodiesel: benefits and costs

Some of the main arguments in favour of the use of biofuels in general, and biodiesel in particular, is their potential to reduce greenhouse gas emissions, their potential to supply a secure energy source, and their perceived carbon neutrality and provision of a positive net energy balance. Although the potential of biodiesel to reduce GHG emissions is well established and much less contested, the latter two issues are subject to a lot of debate. In order to assess these claims, it is essential to look into the actual benefits and costs of biodiesel, from the early stages of production to end use. In this section, we try to summarize the arguments surrounding the benefits and costs of biodiesel based on a number of earlier studies. We assess both the economic and environmental costs and benefits, and try to outline whether biodiesel is or could be a beneficial alternative fuel at present and in the future.

A number of life-cycle analyses have looked into the direct environmental impacts of biodiesel, arising mainly from vehicle emissions. A general conclusion from these studies is that biodiesel fuel provides a 'net positive energy balance'[6] and reduced CO_2 emissions (Poitrat, 1999;

Mortimer et al., 2003; Puppan, 2002; Booth et al., 2005; Bozbas, 2005). Individual studies differ in many respects however, making it difficult to generalize all aspects of environmental impacts of biodiesel. Using life-cycle assessment of RME (rapeseed methyl ester) and conventional diesel, Franke and Reinhardt (1998) present a comparative overview of the environmental impacts of both these diesel fuels. Of the six environmental impacts considered, three are shown to favour RME and the remaining three to favour fossil fuel diesel. Although RME could have a potentially lower greenhouse effect (CO_2 equivalents), low eco and human toxicity (NO_x) and lower resource demand (finite energy), it is shown to have a significant ozone depletion potential due to nitrous oxide (N_2O) emissions. Furthermore, potential eutrophication and acidification from the use of fertilizers, pesticides and herbicides in rapeseed production adds to the potential negative environmental impacts of biodiesel.

Franke and Reinhardt (1998) is one of the few studies that can be termed 'complete' life-cycle assessment in that they cover almost all aspects of the life cycle of the products in question for both RME and conventional diesel. As such, the results from the study also seem very balanced with regard to which fuel is more ecologically sound. Nevertheless, the authors conclude – 'under certain assumptions, RME has or can have an "overall" ecological advantage against diesel oil' (Franke and Reinhardt, 1998, p. 1032).

The carbon balance and ecological footprint of ethanol as a fuel were analysed by Dias de Oliveira et al. (2005) for the examples of Brazil and the United States. Controversially they conclude that using ethanol as a substitute for petroleum is not environmentally sustainable when emissions from agricultural inputs (in the case of sugarcane) and conversion (in the case of maize) are taken into account. In contrast, a study on ethanol from maize and biodiesel from soybeans showed that ethanol yields 25 per cent more energy than that invested in its production and biodiesel yields 93 per cent more (Hill et al., 2006). The situation is further complicated by the use of nitrogen fertilizers in biofuel production because N_2O release from fertilizer application significantly contributes towards GHG emissions (Crutzen et al., 2007), though if nitrogen demand is reduced by using second generation biofuel feedstocks such as grasses and woody coppice, this problem is alleviated.

One of the major environmental impacts from the burning of fossil fuel diesel comes from the emission of sulphur. Biodiesel, on the contrary, produces virtually zero sulphur emissions, making it much cleaner compared with fossil fuel diesel (Puppan, 2002; Bozbas, 2005). With biodiesel, there is also a significant reduction in other emissions, such as carbon monoxide, hydrocarbons and soot (Puppan, 2002; Mortimer et al., 2003; Nwafor,

2004; Schmidt, 2004). Although NO_x emissions from vehicles using biodiesel are found to be generally higher than from those using conventional diesel, studies have shown that this can be reduced by a slight modification of the diesel engine (Schmidt, 2004; Booth et al., 2005). A major drawback in using biodiesel compared with conventional diesel seems to be the higher emission of N_2O, both from production processes (that is, from feedstock production) and vehicle emissions, which has stratospheric ozone-depleting potential (Franke and Reinhardt, 1998; Poitrat, 1999). Although Poitrat (1999) contests this drawback, stating that the overall greenhouse effect of biodiesel is four or five times less than that of conventional diesel, Franke and Reinhardt (1998, p. 1037) are more cautious and state that this drawback actually favours conventional diesel over RME given the 'high to very high ecological importance' of this ozone depletion potential.

Many different studies have come up with different figures as to the level of emissions and emission reductions for various gases and particulate matters when using biodiesel. While Ryan et al. (2006) state that for biodiesel the CO_2 equivalent emissions savings could range from 36 per cent to 83 per cent compared with conventional diesel, Mortimer et al. (2003) conclude that savings in total CO_2 emissions of 72 per cent to 86 per cent are possible by using biodiesel derived by conventional and modified production respectively compared with ultra low sulphur diesel. Similarly, net energy balance (measured as NER – Net Energy Ratio) for biodiesel use has been reported to be from between 1.59 and 2.08 (Turley et al., 2002) through to between 1.9 and 2.7 (Poitrat, 1999).

Feedstocks, comprising mainly oilseed rape and other competitor oils like soybean, sunflower and palm oil, are the main input in the production of biodiesel fuel. The production of these feedstocks could have two major environmental impacts, namely an impact on land use and an impact on biodiversity. The general consensus in terms of the potential impacts of biodiesel feedstock production on the environment is that there would be little or no negative impact if the feedstock were grown on existing agricultural land (Anderson et al., 2004). However, Puppan (2002) warns of dangers to the environment from large monoculture oilseed farms, as well as land and water pollution from excess use of fertilizers and pesticides in commercial oilseed plantations. In terms of the impact of biodiesel feedstock production on biodiversity, Anderson et al. (2004) argue that production of oilseed rape could have a positive impact on biodiversity, especially for farmland bird species that feed on seeds and invertebrate (that is insect) species within the crop. However, they warn against using large areas of set-aside land for oilseed production, as that could have negative impacts on the birds inhabiting such land (Anderson et al., 2004; Anderson and Ferguson, 2006).

Franke and Reinhardt (1998) posit that the production of biodiesel feedstocks could be quite an intensive land use, often with greater transportation efforts compared with fossil fuels. Furthermore, commercial production of these feedstocks could pollute water from leaching of fertilizers, pesticides and herbicides – adversely impacting local and regional biodiversity. Turley et al. (2002, 2003) state that biodiesel feedstocks grown in set-asides or on other uncultivated lands could increase traffic impacts in rural areas, whereas those grown on already cultivated lands by replacing current crops will have little or no such impact. In addition to increased traffic impacts, the use or possible use of the set-asides to grow biodiesel feedstocks could have negative environmental impacts according to a number of studies (for example, Turley et al., 2002, 2003; Anderson et al., 2004; Anderson and Ferguson, 2006).

In terms of the impacts of biodiesel on human health, emissions from biodiesel are reported to pose lower direct or indirect risks compared with fossil fuel diesel (Poitrat, 1999). However, reports are conflicting on the impacts of biodiesel feedstock production on human health. Major health concerns from the production of biodiesel feedstocks, such as oilseed rape, are the allergies related to these crops – complaints and concerns about which seem to come mainly from the British public rather than from those in other oilseed rape growing countries like Canada, France and Germany (Hemmer, 1998; Turley et al., 2002). As most of these complaints (from the British public) seem to be from anecdotal rather than scientifically proven ill-health impacts, Hemmer (1998, pp. 1327–8) questions if there is 'prejudice because the expansion of this crop is subsidised by the EU' or if 'people simply dislike [oilseed rape's] intense smell and flashy yellow flowers', concluding that there is 'little evidence to incriminate a versatile crop of economic importance as a cause of ill health'.

Most studies comparing emissions from biofuel-run vehicles with those of fossil-fuel-run vehicles conclude that biofuel could be significantly less harmful, implying significant environmental benefits. However, the carbon neutrality and provision of net energy balance by biodiesel is a contested issue. Although combustion of biodiesel itself is carbon neutral in that it releases no more carbon than that sequestered during growth, inputs in production, transportation and marketing of biodiesel usually increase net carbon emissions due to the use of fossil fuels in these operations. Thus, strictly speaking, biodiesel cannot be 100 per cent carbon neutral unless biofuels themselves are used for the energy component in all parts of the manufacturing process. It is, however, possible to reduce the net carbon emissions from biodiesel further by reducing the use of inputs, such as nitrogen fertilizers, in the production of the feedstocks, the production of which emits the highest amount of CO_2 among the agricultural inputs

(Powlson et al., 2005). Furthermore, reduction in the amount of transportation necessary to move biodiesel feedstocks to oil refineries, by building refineries as close to the feedstock production sites as possible, could help reduce CO_2 emissions further (Turley et al., 2002, 2003).

An equally contested issue is that of the energy balance of biodiesel. We have reported studies earlier in this section that have shown the net energy balance of biodiesel varying from around 1.6 to 2.7 (Poitrat, 1999; Turley et al., 2002). However, in a controversial paper, Pimentel and Patzek (2005) report that the net energy balance of biodiesel is on the negative side, meaning that production of biodiesel requires more energy inputs than the energy output from the biodiesel produced. Presenting their analysis on the energy requirement for the production of biodiesel from soybean and sunflower in the United States, they show that biodiesel production using soybean required 27 per cent more fossil energy, and that using sunflower required 118 per cent more fossil energy than the energy output from the biodiesel produced. Pimentel and Patzek (2005) argued that a number of earlier studies showing net positive energy balance from biofuels were incomplete as they had omitted some of the energy inputs in the production system. However, Wesseler (2007), commenting on Pimentel and Patzek (2005), argues that the latter's study was flawed as it ignored opportunity costs, especially those associated with the alternative use of the land used to produce feedstocks or alternative use of the feedstocks themselves. Revising Pimentel and Patzek's analysis by including the opportunity costs associated, Wesseler (2007) shows a positive energy balance for the biodiesel produced from both soybean and sunflower.

Moving away from the environmental costs and benefits of biodiesel and focusing solely on the economics of biodiesel production, we find that in the majority of cases, the cost of a litre of biodiesel is significantly higher than the cost of a litre of fossil fuel diesel (Barnwal and Sharma, 2005; Demirbas and Balat, 2006). Demirbas and Balat (2006) report that, at the time of writing, pure biodiesel is 120–175 per cent more expensive than fossil fuel diesel, though of course fossil fuel prices can change rapidly in response to demand and geopolitical events. Most of the biodiesel currently produced comes from soybean, sunflower, palm or rapeseed oil. All of these oils are widely used in the food production market, resulting in a higher price. Thus, the use of these oils in biodiesel production increases the costs of production, thereby increasing the market price of biodiesel. Even in countries like India, the cost of biodiesel is significantly higher than the cost of diesel, mainly because of the higher cost of vegetable oils used in biodiesel production (Barnwal and Sharma, 2005). Hence, at present, biodiesel cannot compete with diesel in the market in terms of price without government subsidies. Studies suggest that the cost of biodiesel could be reduced by using low cost oils, such as used frying oils from restaurants and non-edible oils that are usually cheaper than edible oils (Azam et al., 2005; Barnwal and Sharma, 2005; Demirbas and Balat, 2006).

Although the higher environmental benefits from the use of biodiesel could justify a higher cost (price), for the majority of the customers it is the price they pay that matters. Thus, unless the price of biodiesel is competitive with that of fossil fuel diesel, it is unlikely that the average consumer will look into using biodiesel instead of diesel. Environmentally aware 'green consumers' may be willing to pay a higher price for biodiesel, but their use will be negligible in comparison with the use of diesel by the majority of other consumers. Provision of government subsidies could lower the price of biodiesel and increase its adoption in the short run, but subsidies are not a long-term solution. Thus, the only way to increase adoption of biodiesel by an average consumer in the long run would be to make it competitive with fossil fuel diesel in terms of price.

What potential for biodiesel to provide double benefits?
During the past few years biofuels have grown out of being a niche market fuel into being a mainstream market fuel – more so in countries such as Brazil, which developed a biofuels programme in the 1970s because of increased oil prices combined with a low sugar price (Oliveira, 2002). However, the adoption of biofuels has generally not been very encouraging, especially in developed countries. Of the EU states that were required to supply their transport fuel market with 2 per cent of biofuels by the end of 2005, only two countries (Germany and Sweden) met the target. Biodiesel has been the more popular biofuel in the EU, taking over 80 per cent of the biofuels market. However, even with a number of incentives, such as fuel tax subsidies, agricultural subsidies and so on, the adoption of biodiesel has been poor in EU countries to say the least. In a context where biodiesel has been touted as a reliable source to supply growing fuel demands, at the same time as being environmentally friendly, it is essential to ask if this optimism about biodiesel providing double benefits – that is, being a cheaper and more secure source of fuel supply, at the same time as providing environmental benefits – is justifiable. In this section we explore this issue from various angles – the demand and supply of biodiesel, the policies driving its production and use or potential use, and whether this optimism about environmental benefits is actually well founded.

It is fair to say that the demand for biodiesel is growing steadily over the years. This has come about due to a number of reasons: (1) due to a great insecurity surrounding the production and supply of fossil fuels from all major oil-producing regions; (2) due to growing environmental awareness of governments as fuel consumers in terms of GHG emissions and global warming and (3) due to policies driving the use of renewable fuels in general and biodiesel in particular in the EU and in other regions. Security of supply has been a major cause of concern for fossil fuel diesel for decades now. Most of the largest oil-producing regions can, in one way or another, be classified as conflict zones. The Middle East is an obvious example, but a recent dispute between Russia and Ukraine and Russia and Georgia unexpectedly threatened supply of oil to Western Europe. In this context, alternative fuel sources, especially those that can ensure the security of supply in the long run, have become very important. A number of renewable energy sources, such as biomass, wind and solar power and biofuels are seen as potential alternatives and secure sources of energy.

For the transport sector biodiesel has the potential to become a secure source of renewable energy in the EU as it can be 'homegrown'. One of the most common feedstocks for biodiesel production, rapeseed, can be easily grown within the EU Member States, which increases the security of supply. However, the EU has a limited area of arable land that can be used to produce rapeseed or other feedstocks for biodiesel, as the production of biodiesel feedstocks will be competing with food production, which for obvious reasons will always have a higher priority. Thus, although most of the biodiesel used in the EU at present is produced within the EU, it is clear that the higher proportions of biodiesel required to meet future targets cannot be supplied internally. At some point, the EU has to rely on imported biodiesel just as it relies on imported fossil fuel diesel now. Thus, guaranteeing the security of supply of biodiesel could become as much of an issue as the security of supply for fossil fuel diesel at present.

The tropical countries, such as Brazil, Malaysia, Indonesia and a number of African countries could be the potential source of feedstocks for the future EU market. A major concern, however, seems to be the potential deforestation and land use changes due to increased demand for arable lands to grow biofuel feedstocks like palm oil, oilseed rape, soybean and sugarcane in these tropical countries. The public perception about biodiesel being the 'green' alternative fuel is influenced more by reports and opinion pieces in the newspapers than by articles in scientific journals. And, judging by the number of news reports and opinion pieces on how biodiesel could bring an environmental disaster rather than being a greener alternative to fossil fuels (for example, Monbiot, 2004, 2005; Mortished, 2006a, 2006b), it is likely that it will take a lot more effort to make people switch from fossil fuel diesel to biodiesel, even those with a concern for the environment. The European Commission (2005) seems to share this concern as it fears that public support for biofuels will dwindle if worries about possible deforestation and destruction of natural habitats are not addressed properly. Moreover, if biofuels lose their reputation as the 'cleaner, greener fuel' compared with fossil fuels, meeting and sustaining any target level will be much more difficult.

There are also moral and ethical dimensions to the issue when it comes to sourcing biodiesel feedstocks from countries in the tropics that have low food security. The livelihood impacts of switching from cereal production for domestic use to biodiesel feedstocks production for export, especially in the countries that are not self-sufficient in food grains, could be devastating. Economists call for the most efficient use of the resource or mode of production, so according to them, it makes perfect sense to produce biodiesel feedstocks as cash crops in these developing countries and import cheaper grain from those who produce in surplus. However, this is disputed by those concerned with sustainability, and also on moral and ethical grounds. Local livelihood effects of changes in crop yields and prices can be devasting, the examples of farmers committing suicide in the Indian state of Kerala due to cash crop failures have become all too common in recent years to justify the switch from cereal to cash crops in such countries.[7]

The review of cost and benefits of biofuel production above shows that just relying on financial cost–benefit analysis is not sufficient to assess the true costs and benefits from the biodiesel. Although the cost of biodiesel is significantly higher than that of fossil fuel diesel at present, it is offset by significantly greater environmental benefits from biodiesel because of lower GHG emissions and carbon neutrality. However, from both an economic and policy perspective, the costs of switching to a more expensive fuel source are incurred in the present, but the environmental benefits of the reduced GHG from biodiesel may not accrue until 40 or 50 years from now. This makes analyses of costs and benefits difficult and complex because of different time horizons involved. It is also not surprising that in the early days of adoption, biodiesel, like every new technology, is likely to incur high short-run costs and low adoption rates, usually in favour of long-term benefits. Furthermore, despite growing environmental awareness among consumers, lack of competitive prices of biodiesel compared with fossil fuel diesel is slowing the adoption rate. It is essential to have a higher adoption rate of biodiesel if it is to make a significant impact on GHG emissions reduction.

In light of the growing production and use of biodiesel in the EU and biofuels in general in the EU, as well as elsewhere in the world, it is not an over-exaggeration to say that biofuels have a real prospect for providing double benefits, especially when issues such as sustainable production are resolved. There are both positives and negatives from the use of biofuels on a small scale and at a large scale. On a small scale, production costs of biofuels are usually higher, thereby making them more expensive compared with fossil fuels, which enjoy an economy of scale. However, the short- and long-term benefits of biofuel use even at a small scale could be significant with regard to savings in carbon emissions and those of other greenhouse gases. Moreover, competition for land use to grow feedstocks for biofuels on a small scale is minimal, which means it is less likely that the problems related to land use and land degradation will be significant at this level of biofuels production and use.

In large-scale production and use of biofuels the savings on carbon emissions will certainly be significantly higher, especially when production does not involve intensive use of fossil fuels. Moreover, greater production and use of the biofuels is likely to reduce their production costs, and hence their market price, encouraging even more use of such fuels as substitutes for fossil fuel. However, there is a growing concern that this large-scale production of biofuels would encourage extensive land conversion to produce feedstocks, such as soybean, corn, rapeseed and sugarcane, especially in the tropics. If this is the case then benefits of reduced greenhouse emission from such fuels are easily cancelled out by tropical deforestation and biodiversity loss resulting from land conversion to provide feedstocks for biofuel production. In this case, biofuels, instead of being an environmentally friendly alternative to the fossil fuels, could become an environmentally harmful source of energy. This means they might not be able to provide double benefits as they are currently promoted as being able to do, at present and in the future.

Conclusion

Growing demands for fuel, uncertainty and conflicts surrounding fossil fuels, and increasing concern for the environment from the use of fossil fuels have inevitably led to the search for alternatives to fossil fuels, especially renewable sources of fuels. Biofuels have been considered one of the best alternatives to fossil fuels for the transport sector. The use of biofuels in transport has been growing worldwide with bioethanol and biodiesel leading the way. Although the price of biofuels is generally higher than fossil fuels at this stage of biofuel production and use, incentives provided by governments, such as subsidies, have helped promote the growing use of biofuels. Favourable policies are in place, especially in the EU, to promote use of biofuels as transport fuels. Despite the Council directive on the use of biofuels in transport, which sets the target for the Member States to ensure a minimum supply of biofuels in their transport fuel market up to 2010, and favourable policies to go with the directive, the target set by the directive has not been met and according to the Council's own progress report, the target set for 2010 is not likely to be met either.

This shows that legally non-binding targets, such as the biofuels targets, will be difficult to meet despite favourable policies and economic incentives – the reason why the Council's biofuels progress report calls for a legislative framework to make such targets legally binding (European Commission, 2007b). Furthermore, the EU is proposing legislation to force vehicle manufacturers to make technological improvement in vehicles to reduce emissions.[8]

There is clear evidence of environmental benefits from using biofuels generally, and biodiesel particularly, when reductions in greenhouse gas emissions by vehicles running on biodiesel are considered. However, the benefits are not so clear-cut when the environmental impacts of biofuels production are accounted for, especially production of biofuel feedstocks such as palm, soybean, rapeseed, sugarcane and corn – mainly in the tropics. In its policy documents, such as the Biomass Action Plan, and the Strategy for Biofuels, the EU sets out policies for promoting biofuels production in developing countries, setting guidelines for 'sustainable production' and supply of such fuels, and appropriate incentives, such as guaranteed investments, for such 'good practices'. However, it is not guaranteed that producer countries will follow the EU guidelines, when they are faced with growing demands for fuels from the rapidly industrializing countries such as China and India. Indeed, lack of stringent environmental policies and/or lack of enforcement of environmental policies in place has been blamed for the growing loss of tropical rainforest in Malaysia and Indonesia to make way for palm oil plantations. Recent years have seen a massive growth in such plantations mainly to meet the demands for oil, so far mainly for the food industry, from China. However, countries like China will have to rely more and more on alternative fuel sources in the future to meet their insatiable demand for energy, which would mean more pressure on the production of biofuels among other sources of energy. This will no doubt increase pressure on land use, especially in the tropics where most of the biofuels feedstocks come from.

Certification schemes, which will guarantee that the biofuels produced in any country, especially in the developing countries, are produced in an environmentally sustainable way, are seen as a possible solution to concern about potential tropical deforestation and habitat destruction during the production of such fuels (European Commission, 2005). Restrictions on the imports of biofuels and biofuel feedstocks to the EU based on such certification schemes could encourage producers to produce biofuels in an environmentally sustainable way; however, implementing such a scheme is not without problems. International trade agreements and WTO rules require that such certification schemes be non-discriminatory between domestically produced biofuels and imports (European Commission, 2005), however, it is obvious that the developing countries, where most of the biofuels are or will be produced, do not always have the same environmental sustainability standards in production as the EU. Making sure that biofuel producers in developing countries comply with the same environmental sustainability standards as EU producers will increase the costs of production of biofuels in those developing countries. These increased costs could lead to a bifurcation in the biofuels market – one market for the export of standard-compliant biofuels to the EU, and another for standard-non-compliant biofuels to the countries, mostly fast-growing and needy developing countries, with less stringent environmental regulations.

Countries like China, with an ever-growing fuel demand, are already investing heavily in palm oil industries in Indonesia, without due consideration to the potential tropical deforestation caused by the expansion of such industries (Perlez, 2005). And the companies supplying palm oil, either for food products or as processed biodiesel to China, which has less stringent environmental regulations, are unlikely to practice environmentally sustainable production schemes at a greater cost when they are not required to do so. In fact this could lead to a shortage in biofuels supply to regions like the EU, with more stringent environmental regulations, from countries like Indonesia in the future, when they could have easy access to large markets like that of China. Thus, the challenge for the EU in meeting biofuels targets such as 5.75 per cent of all transport fuels by 2010 is twofold. First, meeting as much demand for biofuels as possible from the domestic supply without negatively affecting the agriculture and food production sector, also making sure biofuels are produced in an environmentally sustainable manner. Second, putting effective certification schemes in place that guarantee the environmentally sustainable production of the biofuels and making sure they are followed when importing biofuels from outside the EU, mainly from the developing countries.

Many of the arguments regarding biofuels providing double benefits are not well founded if one considers the total environmental impacts of biofuels, from production to their combustion. It is too early to conclude that the biofuels could be the panacea for the issues surrounding greenhouse gas emissions and global warming, or even pollution from vehicles. However, sustainable production of biofuels could help supply a cleaner fuel to the transport market to a certain extent and help in greenhouse gas emissions reduction in the process, thereby providing double benefits. The prospect is there for biofuels to make a difference, but it is up to all concerned – the state, the market and the consumers – on how to make it work.

Notes

1. http://news.bbc.co.uk/1/hi/world/europe/6334327.stm, accessed 31 August 2009.
2. 1. (a) Member States should ensure that a minimum proportion of biofuels and other renewable fuels is placed on their markets, and, to that effect, shall set national indicative targets.
    (b) (i) A reference value for these targets shall be 2%, calculated on the basis of energy content, of all petrol and diesel for transport purposes placed on their markets by 31 December 2005.
        (ii) A reference value for these targets shall be 5.75%, calculated on the basis of energy content, of all petrol and diesel for transport purposes placed on their markets by 31 December 2010.
    (Article 3, Directive 2003/30/EC, *OJ L* 123, pp. 44–5)

3. A measure that requires the oil suppliers to put a certain percentage of biofuels in the fuel they supply to the market.
4. It is estimated that 17 million hectares of arable land will be needed to produce biofuels to meet this target completely from the domestic production. Given the total arable land in the EU of 97 million hectares, meeting the biofuels target set for 2010 entirely from the EU domestic production is thought to be 'technologically feasible in principle'. However, due to WTO regulations and other trade agreements, the EU cannot block the import of (potentially cheaper) biofuels and biofuel feedstocks (mainly from developing countries). 'Therefore, the scenario of 100% domestic production is a theoretical one and would not be possible in practice' (European Commission, 2005).
5. http://statistics.defra.gov.uk/esg/quick/agri.asp, accessed 31 August 2009.
6. Net energy balance is the difference between energy in the biofuels and the energy equivalence of fossil fuel inputs used in producing, harvesting and processing the fuel. Net energy balance is usually calculated as a ratio – called Net Energy Ratio (NER) – of biofuel energy to the fossil fuel energy used in production of the biofuel. NER greater than 1 shows a 'positive' net energy balance, which indicates that the biofuel generates more energy than that used to produce it.
7. http://news.bbc.co.uk/1/hi/world/south_asia/4988018.stm, accessed 31 August 2009.
8. http://news.bbc.co.uk/1/hi/world/europe/6334327.stm, accessed 31 August 2009.

References

Anderson, G.Q.A. and M.J. Ferguson (2006), 'Energy from biomass in the UK: sources, processes and biodiversity implications', *Ibis*, **148**(1), 180–83.

Anderson, G.Q.A., L.R. Haskins and S.H. Nelson (2004), 'The effects of bioenergy crops on farmland birds in the UK: a review of current knowledge and future predictions', in K. Parris and T. Poincet (eds), *Biomass and Agriculture: Sustainability, Markets and Policies*, Paris: OECD.

Azam, M.M., A. Waris and N.M. Nahar (2005), 'Prospects and potential of fatty acid methyl esters of some non-traditional seed oils for use as biodiesel in India', *Biomass and Bioenergy*, **29**(4), 293–302.

Barnwal, B.K. and M.P. Sharma (2005), 'Prospects of biodiesel production from vegetable oils in India', *Renewable and Sustainable Energy Reviews*, **9**(4), 363–78.

Booth, E., J. Booth, P. Cook, B. Ferguson and K. Walker (2005), *Economic Evaluation of Biodiesel Production from Oilseed Rape Grown in North and East Scotland*, report prepared by SAC Consultancy Division.

Bozbas, K. (2005), 'Biodiesel as an alternative motor fuel: production and policies in the European Union', *Renewable and Sustainable Energy Reviews*, **12**(2), 542–62.

Casson, A. (2003), *Oil Palm, Soybeans & Critical Habitat Loss: A Review Prepared for the WWF Forest Conservation Initiative*, Zurich, Switzerland: WWF Forest Conservation Initiative.

Crutzen, P.J., A.R. Mosier, K.A. Smith and W. Winiwarter (2007), 'N_2O release from

agro-biofuel production negates global warming reduction by replacing fossil fuels', *Atmospheric Chemistry and Physics*, **7**(4), 11191–205.
Demirbas, A. (2003), 'Biodiesel fuels from vegetable oils via catalytic and non-catalytic supercritical alcohol transesterifications and other methods: a survey', *Energy Conversion and Management*, **44**(13), 2093–109.
Demirbas, M.F. and M. Balat (2006), 'Recent advances on the production and utilization trends of bio-fuels: a global perspective', *Energy Conversion and Management*, **47**(15/6), 2371–81.
Dias de Oliveira, M.E., B.E. Vaughan and E.J. Rykiel (2005), 'Ethanol as fuel: energy, carbon dioxide balances, and ecological footprint', *BioScience*, **55**(7), 593–602.
Doornbosch, R. and R. Steenblik (2007), *Biofuels: Is the Cure Worse than the Disease?* Paris: Organisation for Economic Co-operation and Development.
European Commission (2005), *Communication from the Commission – Biomass Action Plan* {SEC(2005) 1573}, Brussels: European Commission.
European Commission (2006), *Communication from the Commission – An EU Strategy for Biofuels* {SEC(2006) 142}, Brussels: European Commission.
European Commission (2007a), *Communication from the Commission to the Council and the European Parliament – Renewable Energy Road Map, Renewable energies in the 21st century: building a more sustainable future* {COM(2006) 845 final} {Sec(2006) 1719} {Sec(2006) 1720} {Sec(2007) 12}, Brussels: Commission of the European Communities.
European Commission (2007b), *Communication from the Commission to the Council and the European Parliament – Biofuels Progress Report: Report on the progress made in the use of biofuels and other renewable fuels in the Member States of the European Union* {COM(2006) 845 final} {SEC(2006) 1721} {SEC(2007) 12}, Brussels: Commission of the European Communities.
Franke, B. and G. Reinhardt (1998), 'Environmental impacts of biodiesel use', paper read at BioEnergy '98: Expanding BioEnergy Partnerships, 4–8 October 1998, at Madison, Wisconsin.
Hemmer, W. (1998), 'The health effects of oilseed rape: myth or reality?', *British Medical Journal*, **316**(7141), 1327–8.
Herrera, S. (2006), 'Bonkers about biofuels', *Nature Biotechnology*, **24**(7), 755–60.
Hill, J., E. Nelson, D. Tilman, S. Polasky and D. Tiffany (2006), 'Environmental, economic, and energetic costs and benefits of biodiesel and ethanol biofuels', *Proceedings of the National Academy of Sciences*, **103**(30), 11206–10.
Intergovernmental Panel on Climate Change (2007), 'Climate Change 2007: The Physical Science Basis – Summary for Policymakers', Contribution of Working Group I to the Fourth Assessment Report of the Intergovernmental Panel on Climate Change (Draft), Geneva, Switzerland: IPCC Secretariat. Available online at http://www.ipcc.ch/, accessed 31 August 2009.
Kitzes, J., M. Wackernagel, J. Loh, A. Peller, S. Goldfinger, D. Cheng and K. Tea (2007), 'Shrink and share: humanity's present and future ecological footprint', *Philosophical Transactions of the Royal Society B*, **363**(1491), 467–75.
Koh, L.P. (2007), 'Potential habitat and biodiversity losses from intensified biodiesel feedstock production', *Conservation Biology*, **21**(5), 1373–5.
Monbiot, G. (2004), 'The adoption of biofuels would be a humanitarian and environmental disaster for the planet', *The Guardian*, 22 November 2004.
Monbiot, G. (2005), 'The most destructive crop on earth is no solution to the energy crisis', *The Guardian*, 6 December 2005.
Mortimer, N.D., P. Cormack, M.A. Elsayed and R.E. Horne (2003), *Evaluation of the Comparative Energy, Global Warming and Socio-economic Costs and Benefits of Biodiesel*, report for DEFRA/, Sheffield Hallam University.
Mortished, C. (2006a), 'Food prices would soar in biofuels switch, says Unilever', *The Times*, 7 August 2006.
Mortished, C. (2006b), 'Clean and green, but is biofuel a winner?', *The Times*, 7 August 2006.

Naylor, R., H. Steinfeld, W. Falcon, J. Galloway, V. Smil, E. Bradford, J. Alder and H. Mooney (2005), 'Losing the links between livestock and land', *Science*, **310**(5754), 1621–2.

Nwafor, O.M.I. (2004), 'Emission characteristics of diesel engine operating on rapeseed methyl ester', *Renewable Energy*, **29**(1), 119–29.

Oliveira, J.A.P. (2002), 'The policymaking process for creating competitive assets for the use of biomass energy: the Brazilian alcohol programme', *Renewable and Sustainable Energy Reviews*, **6**(1/2), 129–40.

Perlez, J. (2005), 'The end of Borneo's tropical forests?', *International Herald Tribune*, 28 April 2006.

Pimentel, D. and T.W. Patzek (2005), 'Ethanol production using corn, switchgrass, and wood; biodiesel production using soybean and sunflower', *Natural Resources Research*, **14**(1), 65–76.

Poitrat, E. (1999), 'The potential of liquid biofuels in France', *Renewable Energy*, **16**(1–4), 1084–9.

Powlson, D.S., A.B. Riche and I. Shield (2005), 'Biofuels and other approaches for decreasing fossil fuel emissions from agriculture', *Annals of Applied Biology*, **146**(2), 193–201.

Puppan, D. (2002), 'Environmental evaluation of biofuels', *Periodica Polytechnica Ser. Soc. Man. Sci.*, **10**(1), 95–116.

Righelato, R. and D.V. Spracklen (2007) 'Carbon mitigation by biofuels or by saving and restoring forests?', *Science*, **317**(5840), 902.

Ryan, L., F. Convery and S. Ferreira (2006), 'Stimulating the use of biofuels in the European Union: implications for climate change policy', *Energy Policy*, **34**(17), 3184–4.

Schmidt, L. (2004), 'Biodiesel vehicle fuel: GHG reductions, air emissions, supply and economic overview', Discussion Paper No. C3-015.

Schubert, C. (2006), 'Can biofuels finally take center stage?', *Nature Biotechnology*, **24**(7), 777–84.

The Council of the European Union (1998), 'Council Resolution of 8 June 1998 on renewable sources of energy', *Official Journal of the European Communities*, **C198**, 1–3.

The Council of the European Union (2003), 'Council Directive 2003/96/EC of 27 October 2003 restructuring the Community framework for the taxation of energy products and electricity' (text with EEA relevance), *OJ L* 283, 51–70.

The European Parliament and the Council of the European Union (2001), 'Directive 2001/77/EC of the European Parliament and of the Council of 27 September 2001 on the promotion of electricity produced from renewable energy sources in the internal electricity market', *OJ L* 283, 33–40.

The European Parliament and the Council of the European Union (2003), 'Directive 2003/30/EC of the European Parliament and of the Council of 8 May 2003 on the promotion of the use of biofuels or other renewable fuels for transport', *OJ L* 123, 42–6.

Turley, D., G. Ceddia, M. Bullard and D. Martin (2003), *Liquid Biofuels – Industry Support, Cost of Carbon Savings and Agricultural Implications*, report prepared by CSL, ADAS, Ecofys for DEFRA, Organic Farming and Industrial Crops Division.

Turley, D.B., N.D. Boatman, G. Ceddia, D. Barker and G. Watola (2002), *Liquid Biofuels – Prospects and Potential Impacts on UK Agriculture, the Farmed Environment, Landscape and Rural Economy*, report prepared by Central Science Laboratory for DEFRA, Organics, Forestry and Industrial Crops Division.

Wesseler, J. (2007), 'Opportunities (costs) matter: a comment on Pimentel and Patzek "Ethanol production using corn, switchgrass, and wood; biodiesel production using soybean and sunflower"', *Energy Policy*, **35**(2), 1414–6.

Index

Aarhus Convention 214
abatement costs 392
abatement effect 388, 396
Abbot, J. 37, 64
abdomen 222
Abel, N. 299
ability to pay 151
Aborigines' policy stance on burning 176–7, 178, 180, 181, 182
   discourses adopted in support of 182–3, 185–8, 192
absolute forest cover 13
acanthurids 248
accountability 178
acid rain 199, 215, 362
Adamowicz, W.L. 340
Adams, W.M. 31, 33
Addy, N.D. 93
Adhikari, B. 139, 143, 309
afforestation 371, 410
Africa
   biodiversity conservation priorities in 14, 20
   biodiversity in managed landscapes in 84–5, 86, 87, 94
   biofuel feedstocks sourced from 423
   economic growth and deforestation in 370, 388
   entitlement to products of the commons in 161
   fuel wood shortage in 2, 4, 171–2
   Integrated Conservation and Development Projects (ICDPs) in 30–70
   savanna valuation studies in 338–40
   see also under names of individual African countries, e.g. Kenya; Tanzania
African whitespotted rabbitfish (*Siganus sutor*) 247
agency and motivation 176
   in Cape York seminar transcript 187, 188, 189, 190, 191

Agenda 21 (Rio Declaration on Environment and Development) 218, 229
Aggarwal, R.M. 140, 144, 148–9
agricultural cover, direct use value of 345–8
agricultural land coverage, percentage of 75
agricultural products, prices of 252–3
   see also food prices
agricultural subsidies 7, 97, 105, 106, 409, 413, 422, 425
agri-environmental programmes 148
Agrostis capillaris 94
Agrostis castellana 85
Agrostis-Festuca 87, 94
Ahlén, I. 88
air pollution *see under names of individual air pollutants*, e.g. carbon dioxide (CO_2) emissions; sulphur dioxide (SO_2) emissions
   see also greenhouse gas emissions
air toxics 395
Albania 221
Albertine Rift, Central Africa 42
Alchian, A.A. 120, 135
Alés, Fernandez R. 84
Algeria 221
Alliance for Zero Extinction 16
Alonso, J.A. 83
Alpert, P. 31
alpine forests 93–4
Altieri, M. 103
altruism 99
Amazonian rain forests 14, 337
ammoniacal nitrogen 389, 390
amphibians 17, 20
Anand, P. 282
anarchy, cooperation under 200, 201–2, 219
Anderson, A.B. 337
Anderson, D. 31, 299, 336
Anderson, G.Q.A. 419, 420

431

Andersson, J. 255, 269, 270
Andes 14
Andreoni, J. 366
Andresen, S. 206
Ansuategi, A. 366
Antarctic Minerals Treaty 215
Antle, J.M. 366, 371
aquatic systems 21
Araucaria cunninghamii 93
Araújo, M.B. 22
Arcese, P. 31
Aredo, D. 120
Arellano–Bond Generalized Method of Moments (A–B GMM) 398
argumentation theory 175
ark shells (*Barbatia fusca*) 248, 252, 267, 268, 270
Arkansas National Wildlife Refuge 102
Armitage, D.R. 287, 291, 298, 303
Arrow, K.J. 98, 135, 360, 367, 399
arsenic 368, 384
arthropods 17
Arusha Resolution on Integrated Coastal Zone Management in Eastern Africa (including Island States) 271
Ashley, C. 43
Asia 241, 370
  see also central Asia; South Asia; South East Asia
ASL and Associates database 298
aspen 92, 93
asset specificity 136, 138, 141, 159
Assetto, V.J. 202
assurance game 317–18, 322, 323
  see also stag hunt game
asymmetric equilibrium 316
asymmetric information 136, 138, 141, 143
Ataroff, V. 336
Athens 227
Attwell, C.A.M. 30
Australia
  biodiversity conservation priorities in 14
  economic growth and pollution in 394
  ecosystem productivity in managed landscapes in 93
  fire management in Cape York in 4–5, 176–94
  functional diversity in 79
  structural transition model (STM) used in 91
Australian Labor Party 215
Austria 394, 397
Ausubel, J.H. 213–14
automotive emissions 379, 383, 392, 409, 417–18, 419, 420, 426, 427
average cost 129
Axelrod, R. 125
Ayres, R.U. 399–400
Azam, M.M. 422

Bach, S.B. 275
backward induction 312, 327
Baden, J.A. 143
Bailey, C. 269
Baillie, J.E.M. 8, 17, 20, 21, 22
Bakker, K. 173
Baland, J. 121, 128, 150–51, 154, 155, 157, 161, 316
Balat, M. 421, 422
bald eagle 102
Baldock, D. 82
Balmford, A. 8, 12, 13, 14, 22
balsam fir (*Abies balsamea*) 93
Bamenda 58
Bamenda Highlands, Cameroon 56–64, 68, 69
Bamenda Highlands Forest Project 69, 70
banana cultivation 52–3, 55
banded wattle-eye (*Platysteira laticincta*) 56
Bandyopadhyay, S. 360, 363, 368–9, 385, 393
Banks, Arthur 383
Bannerman's turaco (*Tauraco bannermani*) 56, 63
Bantu tribes 291, 292
Barbier, E.B. 76, 97, 99, 102, 105, 107, 334, 338, 341
Barcelona Convention 217, 219, 220–26, 229–30, 231, 232, 233
  Dumping Protocol 221–3, 229
  Hazardous Wastes Protocol 223, 226
  LBS (Land-Based Sources) Protocol 222, 224, 229, 233

Offshore Protocol 223, 225
Prevention and Emergency Protocol 222, 223–4
SPA (Specially Protected Areas) and Biodiversity Protocol 222, 225
Bardhan, P. 119, 150, 152, 153, 155, 157, 158, 161, 311, 313, 314
bargaining costs 135, 138, 146, 148, 149, 401
Barnes, D.K.A. 245, 269
Barnett, H. 401
Barnwal, B.K. 421, 422
Barreto, P. 331
Barrett, C.B. 31
Barrett, S. 214, 383–4
barter 259, 279
Barzel, Y. 121
Bateman, I. 101
Baumann, O. 300
Bazaruto Archipelago marine protected area 271
beavers 78
Beckerman, W. 359, 360, 361
beech (*Fagus silvatica*) 89–90
Begon, M. 296
Behnke, R.H. 296, 297, 299
Beira 271
Belgium 394, 397
Bell, F.W. 102
Bembridge, J. 331
Benham, A. 140
Benson, J.F. 100, 105
bequest value 99
Berkes, F. 126, 132, 143
Bernauer, T. 206
Betula 88
Bevir, M. 172
Bhatia, A. 158
Bhatia, Z. 48, 52
Bhattarai, M. 365, 381
Biermann, F. 203
Binmore, K. 310
biodiversity conservation in managed landscapes *see* managed landscapes, biodiversity conservation in
biodiversity conservation priorities, global *see* global biodiversity conservation priorities
biodiversity conservation treaties 199

biodiversity hotspots 9, 10, 16, 18, 410
biofuels
 benefits and costs of 7, 409–10, 416, 417–22, 424, 426
 certification schemes for 426–7
 EU renewable fuels policy and 7, 409, 410–17, 425–7
 potential to provide double benefits 422–5, 427
 price of 421–2, 424–5
 products classified as 412
biogeographic spatial units 13–14
biological diversity, definition of 76
biological oxygen demand (BOD) 368, 384, 387, 389, 390
Biomass Action Plan 415, 416, 426
bird species
 biofuel feedstock production and 419
 conservation of 9, 10, 12, 14, 17, 18, 20, 22–3, 35, 56, 62, 63, 83, 87, 88, 92, 102, 104
bird watching 98
BirdLife International 23, 59, 60, 69
Birkeland, C. 241
Birner, R. 136, 137, 140, 141, 142, 143, 144, 145, 146, 147
Bishop, R.C. 102, 107, 126, 131, 132
black list 222, 223
Black Sea 221
blackcap (*Sylvia atricapilla*) 78
blacktip mojarra fish (*Gerres oyena*) 247
Blench, R. 287
Blomley, T. 31
Blomquist, W. 150
Blue Plan 217, 230, 231
Blue Plan Regional Activity Centre (BP/RAC) 228, 230
Bokdam, J. 84
Booth, E. 418, 419
Borchert, M.I. 87
Borrini-Feyerabend, G. 30
Bosnia and Herzegovina 221
Bosphorus 221
Botswana 97
Boulding, K.R. 176
bounded rationality 135
Bourn, D. 287
Bowes, M. 100

Bowker, J. 102
Bowman, D.M.J.S. 176–7
Boxer, B. 231
Boxer, C.R. 243
boxfish (*Ostraciidae*) 259–60
Boyce, J. 365, 381–2
Boyle, K.J. 102
Bozbas, K. 418
BP/RAC (Blue Plan Regional Activity Centre) 228, 230
Brachystegia trees 285
Bradshaw, R. 89, 93
branch lines 311
Brandon, K. 31, 32
Brazil
  biofuels sourced from 410, 414, 415, 416, 418, 422, 423
  economic growth and pollution in 375
  economic valuation of forests in 337
Brazilian Atlantic forest 14
Briggs, J.C. 21
Briske, D.D. 296, 297, 298
broadleaved species 88, 89
Brokaw, N. 92
Bromley, D.W. 124, 126, 127, 128, 131, 132, 139, 158, 160
Brooke, C. 293
Brooks, T.M. 11, 13, 14, 15, 20, 21, 23
Brookshire, D. 101
Brown, G.M. 75, 99, 102
Brown, J.R. 296, 297
Brown, M. 31
Brown Weiss, E. 212
Brummit, N. 12, 14, 17, 21, 22
Brundtland, G. 360
Brundtland Report 282
Bruner, A. 30
Bryant, D. 9, 12, 13, 16
Buchan, D. 30
Buckley, P. 48, 52
Bulgaria 409
bureaucracy, quality of 377, 381
Burgess, J. 170
Burgess, N.D. 17, 35, 48, 49, 53, 56
burning 60, 64, 92, 95, 269, 275
  management in Cape York, Australia 4–5, 176–94
Business Environmental Risk Intelligence (BERI) 380–81

Buszko, J. 14
Buttel, F.H. 173

C-PLAN 41
Cabo Delgado province 242, 243, 244, 250, 251, 263, 264, 270, 271, 272
Cactoblastis cactorum 79
cadmium 368, 384
CAFNEC (Cairns and Far North Environment Centre) 180, 181, 191
Cairngorm mountains, Scotland 93
Caldecott, J. 31, 40
California 14
California Department of Consumer Affairs Bureau of Automotive Repairs 392
Callaway, R.M. 87
Callmander, M.W. 20
CAMCOF (CAmeroon Mountains COnservation Foundation) 62
camels 85
Cameroon 35
  Integrated Conservation and Development Project in Bamenda Highlands of 56–64, 68, 69
Cameroon Ministry of Forestry and Wildlife (MINFOF) 58, 60–61, 63
Cameroon Ministry of the Environment and Forestry 69
Campbell, B.M. 301, 334, 335, 339–40, 350
Campbell, R.W. 87
CAMPFIRE (Communal Areas Management Programme for Indigenous Resources) 87
Canada 14, 394, 397
Canas, A. 402–3
Canney, S. 17, 21, 22
Cape York (Australia), fire management in 4–5, 176–94
Cape York Peninsula Development Association (CYPDA) 190
Cape York Peninsula Sustainable Fire Management Programme 177–8
capitalism 202, 219–20, 282
Caprinus 88
capulana (cotton wrap) fishing 255, 267
carbon balance 417, 418, 420–21, 424

carbon dating 188
carbon dioxide (CO_2) emissions
  biofuels and 417–18, 419, 420–21
  economic growth and 363, 372–5, 380–81, 387, 389, 391, 395, 396–8, 399, 408
  see also greenhouse gas emissions
Carbon Dioxide Information Analysis Center (CDIAC) 373
carbon emissions 369, 370
  see also carbon dioxide (CO_2) emissions; carbon monoxide (CO) emissions; greenhouse gas emissions
carbon monoxide (CO) emissions 363, 372, 376, 387, 391, 393, 395, 418
carbon payments 43
carbon sequestration 42, 100, 336, 340
Cardillo, M. 16
CARE development agency 48, 49, 50, 55, 67
care-intensity 137–8, 142, 143
Caribbean 21
Carney, D. 32
Carrascal, L.M. 83, 92
carrying capacity 296, 360, 363, 366–7
Carson, R.T. 101, 364, 394–5
Carter, N.T. 215, 216, 282
Carvalho, M. 264
Casey, D. 92
Casson, A. 415
Catchment Forest Project 54
catfish (*Plotosidae*) 260
cattle 86–7, 88, 89, 93, 178, 184, 246, 291, 293, 338, 339–40
Caucasus 14
Cavlovic, T.A. 375
Ceballos, G. 17
Census of Population and Housing 392
central Asia 14
central Europe 93–4
centres of plant diversity 9, 10, 18
certification schemes 426–7
CFC emissions 214, 372, 376–7
chamba (plot of land) 246, 268
Chambers, R. 31, 242
chance decision nodes 312
Chapman, D. 365, 385–6, 393
Chaudhuri, S. 393
Chavas, J.-P. 103

chemical oxygen demand (COD) 368, 384, 389
Cheung, S.N. 121, 135, 148
chicken game 201, 317, 318–19
children, fishing by 249, 268, 269
Chile 14
China 14, 375, 409, 410, 415, 426, 427
Chong, D. 170
Cincotta, R.P. 13, 22
Ciriacy-Wantrup, S.V. 107, 126, 128, 131, 132
Clapp, J. 202–3
Clarke, G.P. 48
Clawson, M. 100
Crayfish (*Panulirus spp.*) 249
Cleaner Production Regional Activity Centre (CP/RAC) 228
Clements, F.E. 296
climate change 13, 21, 176, 202, 359–60, 408
  see also global warming
climax community 296
Clout, M.N. 92
clover 86–7
Club of Rome 175
Coase, R. 96, 119, 120, 122, 135, 213, 401
Coase theorem 120–21, 213
coconut plantations 246, 252, 255, 256, 262, 264–5
Coe, M.J. 352
Cohen, S. 285
Cole, E. 88
Cole, M.A. 365, 372, 387–8
Coleman, J.S. 162
collective action approach to institutional analysis 119–20
co-management 323
  transaction costs of 142–6, 149–50
command and control regulations 107–8
commercial vehicles 335
Common, M. 95, 97, 107, 341, 362, 398
Common Agricultural Policy (CAP) 97, 106, 413
common goods 127
common pool resources (CPRs)
  economic valuation of different forms of land use 331–56

game theory as a tool for mapping strategic interactions in 120, 121, 125, 160, 316–28
  see also game theory
 role of social institutions in management of 4, 119–62
  see also institutions and the management of environmental resources
common property regimes
 efficiency of 124, 125, 139
 group heterogeneity and participation in 150–58, 161
 open access versus 124–7, 131
 theoretical aspects of 130–34, 160
 transactions costs of 139, 142, 145
  empirical studies of 146–50, 160–61
communication spaces 170
community-based monitoring 63, 67, 150
community-based natural resource management
 distinction between ICDP and 31–2
 see also common property regimes
community forestry 146–8, 149, 155, 156, 157–8
complete information condition 97–8
complete set of markets condition 97
compliance in international climate regimes 213–14, 230
composition effect 388, 396
concession purchase models 68
Congleton, R.D. 361, 362, 376–7
Congo 14
conifers 88
Connolly, J. 86
consensus, creation of 170–71
conservation goals
 definition of 35
 development activities consistent with 39
conservation targets
 definition of 35, 37
 direct payments to communities in return for 68
 formulation of 38–41, 50–53, 60, 62–3
consultants 38, 148
consumer surplus 98, 100
consumption, relationship between income and 386–7
contingent valuation method (CVM) 100–102
contracts 4, 119, 138, 320, 377, 381
Convention on Biodiversity 20, 23, 76, 107
conversion 190
Conway, A. 86
cooperative game theory 310
Cordery, I. 336
core conservation areas 32, 37, 54, 56–7, 60
'core values' of conservation 33
corruption in government 377, 383
Cossins, N. 338
cost–benefit analysis 337, 338, 339, 424
Costanza, R. 94
Cotterill, F.P.D. 30
Coulson, A.C. 293
Council of the European Union
 Council Directive 2003/30/EC 409, 411–12, 416, 425
 Council Directive 2003/96/EC 413
 Council Resolution 1998 on renewable sources of energy 411
counterfactual analysis 209–11, 234–5
Cousins, S.H. 78
Cowling, R.M. 23, 90
Cox, S.J.B. 132
CP/RAC (Cleaner Production Regional Activity Centre) 228
Craig, S. 101
cranes 83
crisis ecoregions 9, 10, 18
Critical Ecosystem Partnership Fund 16
Croatia 221, 228
Crocker, T.D. 148
crop diversity, measuring value of 103
crop production function 340, 344
 rainfall as input into 351–5
Cropper, M. 364, 365, 370–71, 388
Cross-national Time-series Data Archive 382–3
Crowley, G.M. 179
Crozier, R.H. 77
Crutzen, P.J. 410, 418
Cumbria, UK 89
Cumming, D.H.M. 86, 87

Cushitic tribes 291, 292
Cyprus 221

Dactylis glomerata 85
Dahalani, Y. 255
Dahlman, C.J. 135–6
Daily, G.C. 75, 76, 78, 102
Dalton, R. 16
Dambach, C.A. 92
Danielsen, F. 43, 67
Danish International Development Agency (DANIDA) 47, 48, 51, 52, 69
Dansk Ornitologisk Forening (DOF) 48, 49, 69
Dar es Salaam 48, 272
Darcy, M. 168
Darwall, W. 21
Dasgupta, P.S. 128, 158, 161
Dauvergne, P. 202
Davies, J. 146–8
Dawes, R.M. 319, 320
Dayaratne, P. 242
Dayton-Johnson, J. 150, 152, 155–6, 158
De Bruyn, S.M. 366, 378–9, 395
De Miguel, J.M. 84
Deacon, R. 382–3
Debussche, M. 78
deciduous tree species 88
decision costs 135, 141–4 *passim*, 149
decision support software 41
decomposition analysis 363, 378, 388–9, 396–9
deer 92
 see also giant Irish deer; red deer (*Cervus elaphus*); roe deer (*Capreolus capreolus*)
deforestation 49, 64, 97, 296, 333, 341, 410
 biofuel crop production and 7, 415, 423–4, 425, 426, 427
 economic growth and 368–9, 370–71, 380, 381, 388
 open access regimes and 129
 soil erosion caused by 301, 322, 344
DeGraaf, R.M. 92
dehesas 82–3, 85
deliberation 170
DeLuca, K. 169

demand curves 100
demand for labour 375–6
dematerialization 402–3
Demirbas, A. 414
Demirbas, M.F. 421, 422
democracy 378, 380, 383
Demsetz, H. 121, 122, 123, 124, 131, 135
Denmark 93, 394, 397
desertification 287, 298
Deudney, D. 215
development assistance funds 32, 44, 67
Di Falco, S. 103
Diamond, P.A. 101
Dias de Oliveira, M.E. 410, 418
Díaz, M. 82
Dinda, S. 403
Dinerstein, E. 9, 12, 13, 16, 21, 22
direct land purchase 68
direct material input (DMI) 402–3
direct payments to communities 68–9
direct use value 99–100, 335
 measurement of 101–2, 336–40, 342–3, 345–50, 355
discourse, definition of 168
discourse analysis 4, 168–94
 arguments for a discursive approach to policy analysis 168–72
 and fire management in Cape York, Australia 4–5, 176–94
  applying Dryzek's analytic framework 185–92
  applying Hajer's analytic framework 181–5
  conclusions drawn from 192–4
 theoretical perspectives on 172–6
  Dryzek 174–6
  Hajer 172–4
discourse coalitions 173, 181–5 *passim*, 186, 189, 190, 192
discursive affinities 173, 184, 190, 191, 192, 193
discursive closure 173
discursive contamination 173
dissolved oxygen 368, 369, 382, 387, 391
distribution school view of property rights 123
Dixon, J.A. 332

DNR (Queensland Department of Natural Resources, Mines and Energy) 178, 179, 180
Dobson, A.P. 17
Dobzhansky, T. 77
Doggart, N. 48
domestic politics 214–15
dominant strategy 125, 315, 316
Doornbosch, R. 410
Douglas fir (*Pseudotsuga menziesii*) 104
Dove, M.R. 132
Dowling, E.T. 356
Downing, J.A. 80
Drake, L. 148
Drennan, L.G. 135, 136–7
Dresher, M. 329
dried fish 250, 251, 252, 259, 264, 270, 272
drought 80, 83, 90, 294
  decision problem dependent on pattern of 311–14
Dryzek, J.S. 168, 171, 173, 174–6, 185, 186, 187, 190, 192, 193
Dubois, O. 32
dugongs 260
Dulvy, N.K. 21
dumping of pollution 221–3, 229, 379, 380, 384, 387–8, 400
dynamite fishing 261, 272

ecological paradigms 296–8
ecological thresholds 91, 94, 363, 399–403
Economic Commission for Europe (ECE) 227
economic efficiency
  conditions for 95, 96, 97–8
  distinction between equity and 95
  institutions and 119, 121, 122, 124, 125, 129–30, 139, 143–5, 152, 153, 154
    international agreements 207, 213
economic growth
  discourse of 169–70, 175
  environmental effects of 6, 359–403, 408
  in Asia 241
    basic studies of environmental Kuznets curve 367–76
    debate concerning 359–63
    decomposition analysis and 363, 378, 388–9, 396–9
    empirical models on 363–6
    empirical studies on 366–7
    income elasticity of demand for environmental quality 366, 371, 378, 391–4
    role of ecological thresholds 363, 399–403
    role of household preferences 391–4
    role of international trade 365, 367–8, 375, 384–8, 398, 400
    role of policies and institutions 202–3, 363, 365–6, 369, 376–84, 399, 402
    role of population density 364, 370, 371, 377, 381, 388–91
    role of technology shifts 366, 369–70, 385, 394–5
economic incentives 97, 104–6
  see also agricultural subsidies; taxes
economic rents 122, 135, 136, 159
economic school view of property rights 122–3
ecoregions 13–14
ecosystem disturbance and biodiversity 87–91
  moderate ecosystem disturbance and biodiversity 91–5
ecosystem services 22, 100
  valuation of 76, 102–4, 331–2, 335–6, 340, 343–4, 351–5
ecotourism 39, 43
Edelman, M. 168
education, access to 276, 383
education and awareness programmes 57, 58, 61, 149, 277
Edwards, K.J. 83
effort-intensity 137
Eggertsson, T. 122, 135, 158
Egypt 221
Eken, G. 23
El Niño Southern Oscillation 284
Elliott, L. 175
Ellis, G.M. 76, 102
Ellis, J.E. 298
Elton, C. 79
emission reductions 210

endangered species 102, 136, 137, 179, 185
endemic bird areas 9, 10, 12, 14, 18
endowment 369, 377
ENERDATA database 380
energy consumption 372, 385–6, 400, 410
  see also net energy balance
Energy Crop Payments 413
enforcement costs 136, 138, 142, 145, 146, 149, 150
England 88
environment and society
  ecological and social paradigms and 5–6, 295–304
  reciprocal relationship between 283–95, 303–4
environmental effectiveness 206, 208, 233, 234, 235
Environmental Impact Assessment (EIA) 221, 223, 225
environmental Kuznets curve 6, 360, 362
  basic studies of 367–76
  decomposition analysis and 378, 388–9, 396–9
  ecological thresholds and 399–403
  household preferences and 391–4
  international trade and 367–8, 375, 384–8, 398, 400
  policies and political institutions and 369, 376–84, 399, 402
  population density and 370, 371, 377, 381, 388–91
  technology shifts and 369–70, 385, 394–5
environmental price 402
Environmental Protection Act (Queensland, 1994) 180
environmental protection agencies (EPAs) 375
  see also Queensland Environmental Protection Agency (EPA); US Environmental Protection Agency
Environmentally Sensitive Area (ESA) scheme 105–6
EPA (Queensland Environmental Protection Agency) 178, 179–80
EPA AIRS database 393

epistemic communities 5, 203, 205, 220, 231–2, 233, 235
equal-area grids 13
equilibrium theory, ecological
  paradigms based on 5, 296–8, 302–3
  policy discourse associated with 6, 299–300, 301, 303, 304
equity
  development projects and 65
  distinction between economic efficiency and 95
  institutions and 119, 137, 145–6, 149, 152, 153, 157–8, 161
  non-equilibrium theory and 301
  see also income and wealth inequality
Er, K.B.H. 13
Erdelen, M. 92
Erwin, T.L. 21
Espelta, J.S. 87
ethanol 412, 413, 414, 415, 416, 418, 425
Ethiopia 291
EU Directive 2003/30/EC 409, 411–12, 416, 425
EU Directive 2003/96/EC 413
EU Strategy for Biofuels 415, 416, 426
European Commission 411, 412, 413–14, 415, 416, 417, 423–4, 426, 427, 428
European Convention on Long-range Transboundary Air Pollution (LRTAP) 211
European Habitats Directive 83
European Parliament 411
European Union
  Common Agricultural Policy (CAP) 97, 106, 413
  funding for ICDPs 48, 69
  participation in agri-environmental programmes in 148
  participation in Mediterranean Action Plan 217, 221, 231
  policy on biofuels 7, 409, 410–17, 425–7
  potential for double benefits from biofuel use 422–5, 427
evening grosbeak (*Hesperiphona vespertina*) 104

evergreen oak forests 82
evolutionary processes 21–2, 94
exclusion of non-members 127, 131, 132, 133
exclusivity in consumption 126–7
　see also non-exclusivity in consumption
existence value 99
exit options 153
expropriation, risk of 377
externalities 96, 120, 121, 125, 128, 129, 133, 134, 137, 153, 335–6

Fa, J.E. 22
faecal coliform 368, 369, 382, 384, 391, 399
Falconer, K. 139, 140
Faliński, J.B. 88
FAO (Food and Agriculture Organization) 75
Faraco, A.M. 84
Farber, S. 400–401
Farquhar, G.D. 285
faunal products, valuation of 337, 338
Fay, P.A. 89
Feeny, D. 131, 132, 134
fence traps 248, 251, 267
fences 91, 139, 155, 333
Fenoaltea, S. 137
Ferguson, M.J. 419, 420
Ferraro, P.J. 33, 68
Ferrier, S. 20
fertilizers 418, 419, 420
Festuca rubra 94
financial costs of conservation 105, 106
Finland 375, 394, 397
Fire Wardens 179
fires 83, 84, 87, 91, 92, 94–5, 285, 292
　management in Cape York, Australia 4–5, 176–94
　see also burning
fish prices 250, 251–3, 257, 259, 263–4, 270
Fisher, A.C. 76, 102
Fisher, K.C. 98
Fisher, R.J. 31
fisheries 125, 128, 129, 149–50
　see also seagrass fishery on Quirimba Island, socioeconomic aspects of

fishing boat owners 250, 255–8, 263, 265–6, 269, 278
fishing regulations 225, 244, 274, 277
fishing traditions 260–63
fixed effects model 398
Fjeldså, J. 22
flagship species 77
Flannery, T. 176
Flintan, F. 31
Flood, M. 329
Florida mangroves 94
floristic diversity 82, 92
Flournoy, A.C. 171
focal biological elements, definition of 35, 37
Folke, C. 132
fon (traditional chief) 60, 61
Fonseca, G.A.B. da 16, 17
Food and Agriculture Organization (FAO) 218, 227, 287, 288, 371, 381
food prices 410
Forboseh, P. 63
Ford, H.A. 88
FORECE model 93
forest boundaries, demarcation of 57, 59
forest conservation
　Bamenda Highlands, Cameroon 56–64, 68, 69
　development activities consistent with 39
　Uluguru Mountains, Tanzania 45–56, 67, 68–9
　value of 42, 45, 55, 61, 64–5
Forest Management Institutions (FMIs) 59, 61, 62, 63
Forest Officers 54, 60–61
Forest Reserves 45, 46, 47, 48, 52, 53, 54, 56, 60, 62, 68, 301
forest tent caterpillars 93
Forestry Commission 101
forests, economic valuation of 102, 331, 334–8, 340, 341, 344, 348–9
FORGRA simulation model 88–9
fossil fuels 373–4, 400, 408, 409, 411, 412, 418–19, 420, 421–2, 423–5
Foucault, M. 172, 173
Fox, L.R. 89

France
  economic growth and pollution in 380, 394, 397
  role in Mediterranean Action Plan (MAP) 219, 221, 228, 230, 234
Franke, B. 418, 419, 420
Franks, P. 31, 32, 35
free-riding 125, 130, 132, 154, 317, 322, 362
freedom, effect on income–pollution relationship 384
Freedom House Tables 381
FRELIMO government 262
French, D.D. 93
frequency of transactions 136, 137, 159
freshwater biodiversity 21
frontier forests 9, 10, 13, 18
Frontier-Moçambique Quirimba Archipelago Marine Research Programme 242, 243
fruit production function 348–9
fuel wood
  collection by locals 52, 53, 54, 259, 273–4, 339
  harvesting and production of 61
  shortage in Africa 2, 4, 171–2
Fuentes, E. 335
Fuller, R.J. 88
functional diversity 77–9
functional redundancy 78
fur trade 123

Galeotti, M. 374–5
Galicia, Spain 89
game models 310
game theory
  CPR problems analyzed using 120, 121, 316–28
    commonly-used games for depicting CPR problems 125, 160, 317–19
    institutional solutions to CPR problems 319–21, 323
    worked examples from semi-arid Tanzania 321–8
  and international environmental agreements 200, 201–2, 210
  language and representation forms 310–16
Gardner, R. 152

Garrod, G.D. 101
Garton, E.O. 104
gasoline lead content 379, 383
Gastil, R.D. 377
Gayanilo, F.C. 259
Gaze, P.D. 92
GEF (Global Environment Facility) 16–17, 48, 49, 69, 277
Gell, F.R. 242
gene pool 77
general equilibrium theory 95
genetic diversity 76–7
Gengenbach, H. 261, 262
Genista scorpius 84
GeoNetWeaver 41
Georgia 423
Germany
  biofuel usage in 412–13, 414, 422
  economic growth and pollution in 394, 395, 397
  position in international negotiations 215
Gessner, J. 243, 244, 274
Ghimire, K.B. 30
giant Irish deer 89
Gibbs, J.N. 124, 132
Giblin, J. 302
Gill, A.M. 177, 178
gill nets 247, 248, 255
Ginsberg, J. 10
Gleichman, M. 84
Global 200 ecoregions 9, 10, 18
global biodiversity conservation priorities 3, 8–23
  challenges facing global prioritization 17–22
  from global to local priorities 22–3
  global prioritization in context 10–16
    measures of irreplaceability 11–12, 17
    measures of vulnerability 12–13, 17
    spatial patterns 14–16, 18–19
    spatial units 13–14
  impact of global prioritization 16–17
  nine major institutional templates of 8–10
Global Conservation Fund 16
Global Environment Facility 16–17, 48, 49, 69, 277

Global Environment Monitoring System (GEMS) 367, 368, 377, 396
Global Plan of Action (GPA) 229
global warming 6–7, 173, 359–60, 362, 373, 408, 410, 423, 427
  see also climate change
globalization 202–3, 234, 241
goats 84, 243
Godoy, R. 332, 335, 337
golden shouldered parrot (*Psephotus chrysopterygius*) 179
González Bernáldez, F. 83
Goodstein, E. 361, 362
Gordon, H.S. 124–5
gorillas 35, 42
Government of Tanzania (GOT) 50, 69
government regime, effect on income–pollution relationship 382–3
government scientists' policy stance on burning 178–80, 181, 182
  discourses adopted in support of 182, 183, 185, 186, 187, 188–9, 192
Graddy, K. 383–4
Graham, R.T. 89
Grainger, A. 97
grazing 59, 60, 61, 132, 156, 262, 286, 293, 295, 333
  environmental effects of 83, 84, 86–7, 88–91, 92–4, 97, 285, 296–9, 300–301, 333
Greece 221, 394
Green Party 215
green radicalism discourse 175, 185, 186
greenhouse gas emissions 6, 359, 395
  biofuels and 409–10, 411, 416, 417–18, 419, 423, 424, 425, 426, 427
  Kyoto protocol and 7, 214, 408–9, 411–12
  see also carbon dioxide (CO_2) emissions; CFC emissions
grey list 222, 223
Grice, T.C. 178
Griffin, J.R. 87
Griffin, R.C. 136
Griffiths, C. 364, 365, 370–71, 388

Griffiths, C.J. 287
Griffiths, J.F. 284
Grima, A.P.L. 143
Grimes, P.E. 364, 373
Grossman, G.M. 360, 364, 366, 367–8, 376, 383, 385
ground parrot 189
group heterogeneity, effect on collective action 150–58, 161
Grove, R. 30, 31, 299
Guard, M. 261, 267
Guatemala 337
Guggenheim, S. 152, 157
Gulliver, P.H. 292
Gurr, T.R. 383
Gwynne, H. 287

Haas, P.M. 200, 201, 202, 205, 206–7, 215, 216, 217–18, 219–20, 231–2, 234, 235
Habermas, J. 171
habit forming cooperation 151
Hackel, J.D. 30, 42
Hadejia-Nguru Wetlands Conservation Programme (HNWCP) 335, 338, 350
HADO (Hifadhi Ardhi Dodoma – soil conservation in Dodoma) project 301
Hadza tribes 291
Haigh, N. 107
Haiku poetry 187, 190
Hajer, M.A. 171, 172–4, 181, 182, 184, 192, 193
Halkos, G.E. 398–9
Halpern, B.S. 16
Hamersley, A. 284
Hamilton, A.C. 284, 285
Hammig, M. 365, 381
Hanemann, W.M. 101, 336
Hanley, N. 96, 101, 105, 106, 107
Hanna, S. 138, 139, 158
Hannah, L. 31, 37
Hansen, L.A. 48
Hardin, G. 1, 68
Harmon, M.E. 336
harmony 401
Harper, N. 16
Hartley, D. 47, 52
Hartnett, D.C. 89

Hastings, A. 168, 172, 194
Hatton, J. 244, 262, 264, 270, 271, 274
Hausman, J.A. 101
Hawaii 14
Hawkins, J. 266
hazard reduction burning 178, 179
hazardous wastes 223, 226
Heal, G.M. 128
Healey, P. 170, 171, 174
health care, access to 275–6
health risks from biodiesel 420
Heath, M.F. 83, 88
heather moorland 97, 106
Hecht, S. 337, 338
Hechter, M. 142
hegemonic stability theory 201, 234
Heidebrink, G. 366, 371
Hein, D. 92
Heller, M. 194
Helm, C. 209, 210
Hemmer, W. 420
Henry, W. 102
Hernandez, M.P.G. 89
Hernroth, L. 244
Herrera, J. 87
Herrera, S. 414
Hester, A.J. 89
Heston, A. 368, 373, 379, 381, 382, 383, 390, 398
Hettige, H. 363, 375–6, 384–5
Heywood, V.H. 332, 356
high-biodiversity wilderness areas 9, 10, 16, 18
Hill, C. 347
Hill, I. 158
Hill, J. 409, 418
Hill, R. 176, 177
HilleRisLambers, R. 285
Hillier, J. 170
Hilton, H.F.G. 379–80
Himalayas 14
historical materialism 202–3, 219–20, 234
HNWCP (Hadejia-Nguru Wetlands Conservation Programme) 335, 338, 350
Hobbes, Thomas
Leviathan 200
Hobohm, C. 14
Hoekstra, J.M. 9, 12, 13, 16

Hoffman, M.T. 90, 91
Holling, C.S. 78, 79, 80, 87, 91, 93, 94
holm oak *dehesas* 83, 85
Holmes, R.T. 87
Holtz-Eakin, D. 372–3
Homer-Dixon, T. 215
Homewood, K. 290, 293, 299, 300, 301
honey 57, 59, 61
Horrill, A.D. 90
horses 88
Hough, J.L. 31
household preferences, role in income–pollution relationship 391–4
Hovi, J. 211
Hughes, R. 31
Hughes, T.P. 21
Hulme, D. 31, 33, 42
Hulme, P.D. 87, 94
Human Development Index 244
Humphries, C.J. 12
Hunter, M.L., Jr. 8
hunting 42, 45, 52, 54–5, 61, 88, 98, 176, 225, 291, 299, 333, 334
Huntoon, P.W. 336
Hutchinson, A. 8
Hutchinson, G.E. 78
hybrid private sector governance, transaction costs of 143–6
Hyden, G. 294
hydrocarbon emissions 392, 418
Hymas, O. 52, 53

IAEA (International Atomic Energy Authority) 227
Ibo Island 242, 243
IBRD (International Bank for Reconstruction and Development) 360, 361, 362
ICBP (International Council for Bird Protection) 9
ICDPs *see* Integrated Conservation and Development Projects (ICDPs)
Iceland 397
identity, link between language and 169, 191
IDRISI 41
IIED (International Institute for Environment and Development) 31, 55, 335, 338, 350

Ijim Ridge 56
Iliffe, J. 288, 290, 292, 293, 300
Illius, A.W. 297
implementation constraints 67
implementation costs 138, 141–5 *passim*
imputed values for non-marketed goods 345, 350, 355
income effect 121
income elasticity of demand for environmental quality 366, 371, 378, 391–4
income and wealth inequality
  and economic growth 367
  and participation in collective action 151–8, 161
  and pollution 374, 382
independent evolutionary history (IEH) 77
independent scientists' policy stance on burning 180, 181, 182
  discourses adopted in support of 182, 184–5, 186, 187, 191–2
India
  biodiversity conservation priorities in 14
  biofuel cost in 421
  biofuel demand from 415, 426
  economic growth and pollution in 375
  entitlement to products of the commons in 158, 161
  exempt from reducing carbon emissions 409
  transaction costs associated with group-owned wells in 148–9
Indian Ocean 243, 263, 284
indirect use value 100, 331–2, 335–6
  measurement of 102–4, 340, 343–4, 351–5
Indonesia 14, 375, 415, 416, 423, 426, 427
industrialism 175, 186
INFO/RAC (Information and Communication Regional Activity Centre, previously ERS/RAC) 229
information asymmetry 136, 138, 141, 143
information costs 135, 139, 146, 148, 149
information set 311
infrastructure quality 381
  see also roads
Inhaca Island 269, 270, 271
Innes, J.L. 13
insects 17–20, 87, 91, 93, 104
institutional choice theory 319
institutional effectiveness 206, 208, 233, 234, 235
institutional failure 132, 159
institutions and the management of environmental resources 4, 119–62
  definitions of institutions 120, 204
  effect of policies and institutions on income–pollution relationship 363, 365–6, 369, 376–84, 399, 402
  group heterogeneity and emergence of local management institutions 150–58, 161
  international agreements *see* international environmental regime effectiveness
  new institutional economics (NIE) and 119–21, 159
  open access versus common property regimes 124–7, 131
  property rights transformation and resource management 121–4
  theoretical aspects of common property regimes 130–34, 160
  theoretical aspects of open access regimes 127–30, 159–60
  transactions costs and natural resource management 134–50, 159, 160–61
institutions and the structuring of discourses 173–4, 176
integrated coastal zone management (ICZM) programmes 271
Integrated Conservation and Development Projects (ICDPs) 3, 30–70
  advantages of 32
  Bamenda Highlands of Cameroon 56–64, 68, 69
  creation of 30–32
  critiques of 32–4
  improving the model 34–45
  issues arising from case studies 64–8

Uluguru Mountains of Tanzania 45–56, 67, 68–9
interest groups 59, 61, 125
intergenerational displacement of environmental costs 361
Intergovernmental Panel on Climate Change (IPCC) 408
International Atomic Energy Authority (IAEA) 227
International Bank for Reconstruction and Development (IBRD) 360, 361, 362
International Council for Bird Protection (ICBP) 9
International Energy Agency (IEA) 374, 386
international environmental regime effectiveness 5, 198–237
　defining and measuring 205–12, 234–5
　　new approach to 234–7
　domestic politics and 214–15
　institutional economics and 213
　Mediterranean Action Plan (MAP) 5, 216–33, 235–7
　regime theory and 199–205, 219–20, 234
　security considerations and 215–16
　transparency, openness and participation and 214
　verification of compliance and 213–14, 229–30
International Institute for Environment and Development (IIED) 31, 55, 335, 338, 350
International Maritime Organization 224, 228
International Monetary Fund (IMF) 401
international political economy approach 202–3, 234
International Road Federation 383
International Standard Industrial Classification (ISIC) 384–5
international trade, effect on income–pollution relationship 365, 367–8, 375, 384–8, 398, 400
International Union for Conservation of Nature and Natural Resources (IUCN) 9, 11, 13, 16, 77, 301

International Whaling Commission 209
Intertropical Convergence Zone (ITCZ) 283–4
invertebrate fishery 245, 248, 252, 254, 267–8, 269, 270, 272, 278
invertebrates 17, 20
Ioris, E.M. 337
IPCC (Intergovernmental Panel on Climate Change) 408
Iraqw tribes 291
Ireland 83, 89, 394
irreplaceability
　measures of 11–12, 17
　relative to vulnerability 10–11, 14–16, 18–19
irreversible loss 137, 142, 208, 363, 399
irrigation systems 153, 155–6
Isenmann, P. 78
Isla de Moçambique 271
Islam, N. 396
Islam 243, 245, 260, 272
Isle of Man 260
Israel 93, 221
Italy 103, 221, 229, 394, 397
itinerant fishers 271–3, 274, 276–7
IUCN (International Union for Conservation of Nature and Natural Resources) 9, 11, 13, 16, 77, 301
IUCN Species Survival Commission 20
ivory trade 243

Jack, M. 32
jacks (*Carangidae*) 248, 251
Jackson, T. 107
Jacobs, B.F. 285
Jacobs, K. 172
Jacobson, K. 212
Jaggers, Keith 383
James, A. 8, 17
James, F.C. 92
Jane, G.T. 89
Japan 161, 380, 394, 397
Jeanrenaud, S. 31
Jeftic, L. 231
Jeltsch, F. 89
Jennings, S. 242
Jensen, O.B. 173
Jepson, P. 14, 17, 21, 22

Joffre, R. 85
Johannes, R.E. 260
Johns, A.D. 88
Johnson, L.A. 102
Johnson, O.E.G. 121
joint forest management (JFM) 158
Jones, R. 176
Jorritsma, I.T.M. 88
Juma, S.A. 267, 269
Just, R.E. 103

K-strategist species 91–2
Kaare, S. 47, 52
Kahn, M.E. 363, 392
Kalahari 89
Kanbur, R. 153
Kant, S. 141, 151, 157
Kareiva, P. 21, 22
Kaufmann, R. 365, 390–91
Keefer, P. 377, 381
Keeley, J. 173
Kelly, C.K. 17
Kenya 43, 260, 261, 267, 272, 289
Keohane, R.O. 150, 200, 204
Kerala 424
Kerario, E. 301
keystone species 78, 94
Khanna, N. 393–4
Kherallah, M. 120
Khoisan tribes 291
Kienast, F. 89, 93
Kieran, J.A. 290
Kikula, I.S. 294, 302
Kilum-Ijim forest 56–64 *passim*, 68, 69
Kilum-Ijim Forest Project 63, 64, 69, 70
Kimwani dialect 243, 245, 247, 248, 249, 276
Kingdon, J. 284, 285, 286
Kingery, J.L. 89
kinship networks 261–2
Kirby, D. 309, 332
Kirsten, J. 120
Kiss, A. 33, 68
Kitzes, J. 408
Kjekshus, H. 293, 300, 302
Klaassen, G. 378
Klein, P.G. 135
Klieman, D.G. 43
Knack, S. 150, 377, 381

Knetsch, J. 100
Knoll, A.H. 21
Knox, A. 137
Koh, L.P. 410
Köhler-Rollefson, I. 85
Komen, M.H.C. 394
Kondoa 301
Koop, G. 371
Koponen, J. 293
Korea 375, 380
Koziell, I. 32
Kramer, R. 30
Krasner, S.D. 203–5, 234
Kremen, C. 43
Kress, W.J. 12
Krueger, A.B. 360, 364, 366, 367–8, 383, 385
Kruess, A. 91
Krupnick, G.A. 12
Krutilla, J.V. 99, 100
Kullenberg, G. 216
Kumilamba 242, 245–6
Kumm, K.I. 148
Kundhlande, G. 336, 340, 352, 355
Kunich, J.C. 17
Küper, W. 17, 20
Kuperan, K. 149–50
Kütting, G. 200, 201, 203, 206, 208, 217, 233, 234, 235
Kuznets, S. 360, 367
Kuznets curve 367
   see also environmental Kuznets curve
kwifon (ruling council) 60, 61
Kyoto Protocol 7, 214, 408–9, 411–12

Laffer curve 401
Lal, R. 336
Lamoreux, J.F. 12, 14
land abandonment 75–6, 83–4
land disputes 262, 295
land tenure 13, 36, 295, 298, 333, 380
landscape-level ICDP design 34–6, 49–50, 58–9
Langdale-Brown, I. 285
language use *see* discourse analysis
Lanza, A. 374–5
large-scale mono-cultures 7, 419, 425
Larson, P.S. 31, 33
last of the wild 9, 10, 18

Late-glacial Interstadial (11000–12000 BP) 89
Latin America
  economic growth and deforestation in 370, 381, 388
  forest values in 337–8
Leach, M. 2, 30, 33, 43, 171, 299, 300, 301
lead (Pb) pollution 368, 376, 379, 383, 384
lead states 215
Leader-Williams, N. 42
Leakey, L.S.B. 289
Lebanon 221
Lees, L. 193
Leffler, K. 148
Levinson, A. 366, 379–80
Levy, M.A. 204
Libecap, G. 120, 122, 123–4, 136, 158, 161
Libya 221
Liebman, M. 103
Lin, J.Y. 119
Lindberg, K. 75
Lindén, O. 241, 271
Linder, P. 92
line fishing 248, 251, 273
Liu, J. 13
livestock movements, restrictions on 90–91
livestock production function 339–40, 343, 344, 345–8
Local Authority Forest Reserves 54
Logan, V.S. 14
Logical Framework Approach 39–40, 44, 52, 60, 65–6
Lolium perenne 85
Londo, G. 83
Long, A. 8, 14
Long, J. 103
Loreau, M. 22
Loureiro, N.L. 251
Lovejoy, T.E. 79
Lovett, J.C. 22, 48, 139, 143, 180, 284, 285, 286, 299, 300, 301
Lowore, J. 32
Lughadha, E.N. 12, 14, 17, 21, 22
Luguru tribe 48, 49, 52
Lukumbuzya, K. 42
Lundin, C.G. 241, 271

Luxembourg 397
Lyamuya, V.E. 50
Lynch, J.F. 88, 92

Maasai tribes 291, 293, 300
MacArthur, J.W. 92
MacArthur, R.H. 79, 91, 92
Mace, G.M. 8, 12, 17, 21, 22
machairs 83
machambas (allotments) 245–6, 252, 253, 255, 256, 259, 265, 268, 269
Machiavelli, Niccolò
  The Prince 199–200
Macia, A. 244
MacKinnon, K. 31
MacLeod, N.D. 296, 297
Madagascar 14, 21, 22
Maddox, G. 302, 303
Mafia Island 270
Main, A.F. 79
Maisels, F. 63
maize 418
Maji Maji 293
Majone, G. 170
Makonde people 243, 244
Makua people 243, 244, 260, 262
Malawi 43
Malaysia 14, 389–90, 415, 416, 423, 426
Maler, K.G. 158
Mali 156–7
malleefowl 189
Malta 221, 224, 228
mammal herbivore species 86
mammal species 17, 20, 63
managed landscapes, biodiversity conservation in 3–4, 75–108
  ecosystem disturbance and biodiversity 87–91
  moderate ecosystem disturbance and biodiversity 91–5
  genetic and functional diversity 76–9
  importance of biodiversity in managed landscapes 79–82
  landscape management and biodiversity 82–4
  landscape management and ecosystem properties 84–7
  markets and 95–108
management agreements 105, 106
mangrove crabs (*Scylla serrata*) 249, 252

mangrove whelk (*Terebralia palustris*) 249
Manx Heritage Foundation 260
manufacturing
  share in the economy 366, 375, 378, 379–80, 385, 386, 387, 388, 396
  toxic intensity of 384–5
MAP *see* Mediterranean Action Plan (MAP)
Maputo 244, 277
Maquis, R.J. 87
Marañón, M. 82, 85, 86
marema trap fishing 246, 247, 251, 254–5, 258, 259, 261, 263, 266–7, 268, 269, 270, 273, 278
marginal benefits 129, 130
marginal cost 129, 130
marginal propensity to emit (MPE) 373, 397
marginal rate of substitution 392
marginal value 351
Margolius, R. 31, 40, 43, 56
Margules, C.R. 10, 11, 12, 16
marine biodiversity 21, 222, 225, 228
marine parks 222, 225, 271
marine pollution
  definition of 221, 236–7
  measuring 212
  oil pollution 215, 218, 222, 224
  prevention in Mediterranean Sea *see* Mediterranean Action Plan (MAP)
market failure 96–8, 371, 402
  correcting with command and control regulations 107–8
  correcting with economic instruments 104–6
markets and biodiversity 95–108
  indirect use values and ecosystem function 102–4
  market failure 96–8
    correcting with command and control regulations 107–8
    correcting with economic instruments 104–6
  measuring diversity values 100–101
  policy failure 97
  value of species and habitats 101–2
  valuing biodiversity 98–100
Marland, G. 374

Marlow, D. 178, 180
Maro, P.S. 284, 294
Marrs, R.H. 90
Martin, G. 338
Marvier, M. 21, 22
Masaiganah, M. 261
Masih, S.K. 335
Mason, M. 170
Massinga, A. 244, 262, 264, 270, 271, 274
Massoud, M.A. 231
Mate, I.D. 83
Matthews, E. 269
Mattson, W.J. 93
Matzke, G. 294
May, R.M. 79
Mbaya, S. 42, 68
McAllister, D.E. 21
McClanahan, T.R. 260, 261, 272
McClean, C.J. 13
McConnell, K.E. 366, 391–2
McConnell, M.L. 231
McDonald, M.J. 171
McHenry, D.E., Jr. 294
McKean, M. 152, 158, 161
McManus, J.W. 242
McNaughton, S.J. 86, 90, 103
McNeely, J.A. 97, 105
MCSD (Mediterranean Commission on Sustainable Development) 227, 229
McShane, T.O. 31, 43, 67
Meadows, D.H. 175, 359, 360
Mearns, R. 2, 171, 299, 300, 301
meat production 42
Mecufi Coastal Zone Management Project 271
Mecufi district 251, 255
MED POL programme 217, 224, 227–30, 231
medicines 255, 275–6, 333, 335
Mediterranean Action Plan (MAP) 5, 216–33
  Barcelona Convention and 217, 219, 220–26, 229–30, 231, 232, 233
  contracting parties 217, 221
  effectiveness of 231–3, 235–7
  international environmental cooperation and the creation of 218–20

objectives of 217, 232
origins, negotiation and formation 217–18
structure and components 220–31
Mediterranean Commission on Sustainable Development (MCSD) 227, 229
Mediterranean region
biodiversity conservation priorities in 14
biodiversity in managed landscapes in 82–3, 84, 85, 93
keystone species in 78
see also Mediterranean Action Plan
Mediterranean Trust Fund 231
megadiversity countries 9, 10, 13, 18
Meinzen-Dick, R. 137
Menard, C. 119, 141
mercury 368, 384
Meshack, C.K. 139
Mesoamerica 14
meta-analysis 375
metaphors 176
in Cape York seminar transcript 187, 188
methane emissions 372
Metrick, A. 77
Mexico 155–6, 337, 367–8, 375, 385
MICOA (Ministry for the Coordination of Environmental Affairs, Mozambique) 271
microbes 17
Middle East 423
Midgley, G.F. 21
Midlarsky, M.I. 380
Miles, E.L. 209
Mill, G.A. 101
Millennium Development Goals 7, 282, 409
Millennium Ecosystem Assessment 8
Miller, H.G. 104
Miller, K.R. 75
Miller, T.E. 92
MINFOF (Cameroon Ministry of Forestry and Wildlife) 58, 60–61, 63
mining 125, 128
miombo woodland 285–6, 293
Mitchell, F.J.G. 88, 89, 93
Mitchell, R.B. 206, 211–12, 234, 235

Mitchell, R.C. 101
Mittermeier, R.A. 9, 11, 12, 13, 14, 16, 20, 21
mixed equilibrium 316
mixed strategy 316, 319
Mlay, W.I.F. 294
Moçimba da Praia 277
Molinas, J.R. 142, 155, 162
Molinia caerulea 94
Monaco 221
Monbiot, G. 415, 423
monitoring costs 136, 138, 142, 143, 144, 145, 150, 212, 327
Montgomery, C.A. 102
Montreal Protocol (1987) 107, 209, 214, 377
Moomaw, W.R. 396–8
moose 88
Moran, D. 13
Morgan, R.K. 87
Morgenstern, O. 310
Morgenthau, H. 215
Morocco 221
Morris, R.F. 87
Morrow, P.A. 89
Morse, C. 401
Mortimer, N.D. 418, 419
Mortimore, M. 284, 287, 288
Mortished, C. 415, 423
Mott, J.J. 176
Mount Cameroon 62, 63
Mount Oku 56
mountain beech (*Nothofagus solandri*) 89
Mozambique
civil war in 244, 262, 264
integrated coastal zone management in 271
see also seagrass fishery on Quirimba Island, socioeconomic aspects of
Mozambique Ministry for the Coordination of Environmental Affairs (MICOA) 271
multiple Nash equilibria 316, 318
multispecies game systems 87
multispecies herbivore systems 85–7
Munasinghe, M. 158, 401–2
Mung'ong'o, C.G. 301
municipal waste 369, 372

Munro, J.L. 259
murex shells (*Chicoreus ramosus*) 249, 268
Murphree, M. 42
Mussa, D.T. 284
Mwalyosi, R.B.B. 287
Mwani people 243, 262, 273
mycorrhiza 103–4
Myers, N. 8, 9, 12, 16, 21, 107, 336, 351
Myerson, G. 193

Nabil, M.K. 119
Nacala 271, 272
Nampula province 250, 260, 261, 271, 272, 273, 276
Nardus stricta 94
Nash, J. 328–9
Nash equilibrium 210, 316, 317, 318, 319
National Carbon Intensity (NCI) 373
National Parks 16, 33, 34, 60, 68, 100, 134, 180, 299, 300, 301
national security 200, 215–16, 234, 409
nationalization 124, 127, 381
Nations, J.D. 337, 338
natural relationships, assumptions relating to 176
  in Cape York seminar transcript 187, 188–9, 190, 191
Nature Forest Reserves 54, 68
NatureServe DSS 41
Naveh, Z. 83
Naylor, R. 410
Nee, S. 17
negotiating costs *see* bargaining costs
neoliberal institutionalism 203–5, 234
neorealism 200–202, 219, 234
Nepal 127, 129, 149, 156
net energy balance 417, 419, 420, 421
Netherlands 88–90, 375, 394, 395, 397
Neumann, R.P. 30, 31, 299
New Guinea 14
New Forest, England 88
new institutional economics (NIE) 119–21, 159
New Zealand 89, 93
Newmark, W.D. 31
Ngazi, Z. 255, 269, 270
Niamir-Fuller, M. 332

niche relations 92
Nicholson, S.E. 284
Nicholson, W. 350
nickel 368, 384
Nilotic tribes 291, 292
nitrates 372, 384, 387
nitrogen mineralization 85
nitrogen oxides (NO_x) emissions 363, 370, 372, 376, 387, 391, 393, 395, 418, 419
nodes 311, 312
Nolan, T. 86
non-convexities in production functions 154–5
non-cooperative extensive form games 311–14
non-cooperative game theory 125, 310
  key concepts used in 311–16
non-cooperative strategic form games 314–16
non-equilibrium systems 90, 94–5
non-equilibrium theory, ecological and social paradigms based on 5–6, 297–9, 303
  policy discourse associated with 6, 300–301, 303, 304
non-exclusivity in consumption 127, 128, 131, 133, 134, 160, 316–17, 362
  see also exclusivity in consumption
non-governmental organizations (NGOs) 31, 32, 40, 48, 58, 63, 65, 171, 214, 228, 323
non-linear dynamical analysis 397
non-rivalry in consumption 362
  see also rivalry in consumption
non-use values 99, 100
Norris, K. 16
North, D.C. 122, 131, 135, 159, 213
North American Free Trade Agreement (NAFTA) 367
North–South divide 202, 219–20, 233
northern spotted owl 102
Norway 394
Nossif, R. 174
Novotny, V. 17
Nugent, J.B. 119
nurse effects 87
Nwafor, O.M.I. 418–19
Nyerere, J. 300

O'Brien, E.M. 283, 284, 285
O'Connor, C. 13
O'Connor, T.G. 297
O'Neill, R.V. 80
Oak Ridge National Laboratory (ORNL) 373, 374, 395, 396
Oakerson, R.J. 127
Oates, J.F. 30, 33, 43
Oba, G. 298, 299, 302
Ockwell, D.G. 2, 169, 180, 194, 283, 302, 304, 321, 332
Octel's Worldwide Gasoline Survey 379
octopus (*Octopus vulgaris*) 248, 252, 268, 273
ODA (Overseas Development Administration) 287
Odling-Smee, L. 22
OECD Environmental Data Compendium 378, 380, 395
oil pollution 215, 218, 222, 224
oil price shocks 397, 398
oil reserves 128, 129
oil supplies, security of 423
oilseed rape 414, 415, 416, 418, 419, 420, 421, 423, 425, 426
Oiterong, E. 269
Oliveira, J.A.P. 422
Olson, D.M. 9, 12, 13, 14, 16, 21
Olson, M. 125, 154
Olsson, O. 120, 159
OMM (*Organizaçao da Mulher Moçambicana* – Mozambican Women's Organization) 245
ontology of a discourse 176
Oommachan, M. 335
Opdam, P.F.M. 88
open access regimes
 theoretical aspects of 127–30, 159–60
 versus common property regimes 124–7, 131
openness 214
opportunistic behaviour 138
opportunity costs 105, 106, 139, 140, 143, 149, 312, 337, 393, 421
option value 98
Opuntia 79
Orkney 97, 106
Orme, C.D.L. 12, 17

Oslo-Potsdam solution 201, 202, 209–11
Ostrom, E. 124, 125, 132, 150, 151, 152, 156, 158, 170, 310, 316, 317, 318, 319, 320, 321, 329
outboard engines 265–6
Ovadia, O. 14
Overseas Development Administration (ODA) 287
ownership by capture 128–9, 134
oysters (*Pinctada nigra*) 248, 252, 267, 268, 270
ozone depletion 199, 362, 376, 418, 419
ozone pollution 376, 393

paddocks 90
Page, T. 401
Pakistan 103, 393
Pakistan Integrated Household Survey 393
palm oil 415, 419, 421, 423, 426, 427
Panayotou, T. 336, 360, 363, 364, 365, 366, 370, 374, 377–8, 388–9, 396, 402
PAP/RAC (Priority Actions Programme Regional Activity Centre) 228, 231
Paraguay 155
parasitism rates 91
Pareto improvement 341
Pareto optimality 136
Parmesan, C. 13
Participatory Environment Management Programme 35
Participatory Rural Analysis (PRA) 338
pastoral lifestyle, causes and consequences of changes in 333–4
pastoral paradigm 298–9, 301, 302–3
pastoralism, definition of 332
pastoralists' policy stance on burning 178, 181, 182
 discourses adopted in support of 182, 183–4, 185, 186, 187, 189–90, 192
Paterson, M. 200, 201, 202, 203, 204, 216
Patzek, T.W. 421
Pauly, D. 259
Pautsch, G.R. 336

Pavasovic, A. 231
Payments for Environmental Services 43, 55
payoff, expected 312–28 *passim*
Pearce, D.W. 99
peasant cooperatives 155
peat bogs 101
Peberdy, J.R. 284
pedunculate oak (*Quercus robur*) 89–90
Pelletier-Fleury, N. 138
Pemba 246, 247, 248, 249, 250, 251, 255, 257, 260, 264, 265, 267, 270, 271, 275, 277
Penn World Tables 368, 371, 373, 374, 377, 379, 381, 386, 396, 397, 398
perennial herbaceous species 85
Perkin, S. 31
Perlez, J. 427
Perman, R. 95, 98, 99
Permit to Light Fire system 178, 179
Perrings, C. 95, 103, 366
Peru 337
pervasive power 193, 194
pesticides 104, 418, 419, 420
Peterken, G.F. 88
Peters, C.M. 335, 337, 338
Pezzey, J. 95
Pfaff, A.S.P. 392–3
pH 389
Phalaris aquatica 85
Philippines 14, 149–50, 375
Phillips, O. 337
phosphorous 387
phylogenetic diversity 21
Picea abies 88
pied flycatcher (*Ficedula hypoleuca*) 83
Pignatti, S. 82, 83
pigs 88
Pimbert, M.C. 30, 31
Pimentel, D. 421
Pimm, S.L. 8, 11, 13, 80
Pimmental, D. 75, 82
pine 92
 see also scots pine (*Pinus silvestris*)
Pinedo-Vasquez, M. 337, 338
pink ear emperor fish (*Lethrinus lentjan*) 247
pinna shells (*Pinna muricata*) 248, 252, 267, 268, 270
pioneer species 78, 92

plant products, valuation of 335, 337, 339
plant species, conservation of 9, 10, 12, 17, 18, 20, 35, 48, 55, 63, 82–3, 87, 92
Plassmann, F. 393–4
Platteau, J.P. 121, 128, 150, 151, 154, 155, 157, 161, 316
Poa trivialis 86
poaching 31
Pócs, T. 48
Poitrat, E. 417, 419, 420, 421
Poland 88
Polasky, S. 75
policy failure 97, 402
Polity III database 383
polluter pays principle 221
pollution
 economic growth and 6, 359–403, 408
 international agreements on 199, 211, 215
  see also Mediterranean Action Plan (MAP)
 measuring of 212
 in open access regimes 129
 and rainfall 284
 transaction costs associated with 148
 see also under specific types of pollution, e.g. carbon dioxide (CO_2) emissions; water pollution
pollution activity 379, 380
pollution dumping 221–3, 229, 379, 380, 384, 387–8, 400
pollution haven hypothesis (PHH) 387
pollution intensity 375–6, 379, 380, 384–5
Polunin, N.V.C. 242
Pomeroy, C. 269
Pope, R.D. 103
population density, effect on income–pollution relationship 364, 370, 371, 377, 381, 388–91
Populus 88
Portney, P.R. 101
Portugal 394
Posey, D.A. 34
Posner, R.A. 120
Possingham, H.P. 12

Poundstone, W. 329
poverty 42, 47, 244, 332, 333, 359
    alleviation of 22, 32, 34, 65, 67, 158, 282, 303–4, 360, 409
Power, M.E. 78
power
    discourse and 169–70, 172–5, 193, 194
    inequality and pollution 381–2
    realism and 200, 219
Powlson, D.S. 421
Pratt, D.J. 285
Pratt, J. 287
precautionary principle 107, 221
preferred strategy 312
pregnancy 260–61
Prendergast, J.R. 12
Pressey, R.L. 10, 11, 12, 14, 16
pressure groups 215
Pretty, J.M. 31
Price, A.R.G. 21
Price, G. 329
Priority Actions Programme (PAP) 217, 230, 231
Priority Actions Programme Regional Activity Centre (PAP/RAC) 228, 231
prisoner's dilemma game 125, 160, 201, 317, 319–20, 322–3
private cost 96, 129
private property 96, 121, 126–7, 129, 133
privatization 124, 138–9, 333, 334
problem-solving discourse 175, 185, 187
problem trees 40, 60
production function approach to valuing ecosystems 341–5
    direct use values measured by 339–40, 342–3, 345–50
    indirect use values measured by 102–4, 340, 343–4, 351–5
    limitations to use of 344–5
productivity, ecosystem 80–81, 84–7, 89–90, 93–4, 97, 103–4, 106
    valuation of 331–56
profit function 348, 349, 350, 353, 354
profit maximization 97, 317, 332, 341, 345, 361
'property', meaning of 126

property rights 2, 4, 119, 371, 383
    Coase theorem and 120
    collective action approach focusing on 119–20
    definition of 120
    empirical studies of transactions costs 146–50, 160–61
        problems associated with 140–41
    four different types of property regimes for common pool resources 126
    group heterogeneity and emergence of local management institutions 150–58, 161
    motivations for contracting 121, 123–4
    open access versus common property regimes 124–7, 131
    property rights transformation and resource management 121–4
    theoretical aspects of common property regimes 130–34, 160
    theoretical aspects of open access regimes 127–30, 159–60
    transactions costs and selection of appropriate governance structure 135, 138–9, 141–6, 159
protected area coverage 13, 75
Prunus africana trees 57, 59, 61
public goods 96–7, 127, 130, 134, 362
public relevance of transactions 137, 144, 159
public research and development (PRD) 394
public sector governance, transactions costs of 142–6, 149–50
pufferfish and tobies (*Tetraodontidae*) 260
Punjab, Pakistan 102
Puppan, D. 418, 419
purchasing power parity (PPP) 363, 367, 374, 377, 381, 382, 386, 394, 396, 397
pure Nash equilibrium 316
pure strategy 316
Putnam, R. 162
Putuhena, W.M. 336

QPWS (Queensland Parks and Wildlife Service) 178, 180, 184, 190

quasi-option value 98
Queensland Department of Natural Resources, Mines and Energy (DNR) 178, 179, 180
Queensland Environmental Protection Agency (EPA) 178, 179–80
Queensland Parks and Wildlife Service (QPWS) 178, 180, 184, 190
Queensland Rural Fire Service (RFS) 178, 179
Queensland State Government 178, 180
Quercus 88
questionnaire surveys 346–8, 352–3
Quinn, C.H. 287, 289, 290, 301, 302, 321, 332
Quirimba Island, Mozambique *see* seagrass fishery on Quirimba Island, socioeconomic aspects of
Quissanga 261

r-strategist species 91, 92
Rada, F. 336
radical uncertainty 171, 193
Raftopoulos, E. 220, 226, 227, 230, 231
rainfall
  decision problem dependent on pattern of 311–14
  in semi-arid regions 90, 91, 283–8, 292, 294, 297–8, 301, 321, 322, 332, 336
    calculating indirect use value of 351–5
Ramanathan, V. 284–5
Rambal, S. 85
Randall, A. 101
random coefficients method 398
random effects model 398
Random Roadside Test 392
rapeseed 414, 415, 416, 418, 419, 420, 421, 423, 425, 426
rapeseed methyl ester (RME) 418, 419
Ratcliffe, P.R. 92
rational choice theory 201–2
rationality, economic 98, 169–70
  bounded 135
Ravallion, M. 374
Raven, R.H. 75
Rawat, J.S. 336

Rawat, M.S. 336
realism 199–200, 219, 234
red deer (*Cervus elaphus*) 89–90
Redford, K.H. 10, 338
redstart 83
Rees, J. 171
'regime', meaning of 126, 204
regime theory 199–205, 219–20, 234
Regional Action Plans
  components of 220
  see also Mediterranean Action Plan (MAP)
Regional Activity Centres (RACs) 227, 228–9
Regional Marine Pollution Emergency Response Centre for the Mediterranean Sea (REMPEC) 224, 228
regulatory design 138
Reid, W.V. 21, 75
Reinhammar, L.-G. 83
Reinhardt, G. 418, 419, 420
religious rhetoric 190
Relva, M.A. 89
REMPEC (Regional Marine Pollution Emergency Response Centre for the Mediterranean Sea) 224, 228
Renewable Energy Road Map 417
repeated games 125, 201, 320
Repetto, R. 97
Republic of Mozambique State Secretariat of Fisheries 263, 264, 272
reputation 170
resilience, ecosystem 80, 87, 90, 94, 97, 103
resource context, description of 138
respondent bias 101
RFS (Queensland Rural Fire Service) 178, 179
rhetorical devices 176
  in Cape York seminar transcript 187–92 *passim*, 193
Richards, M. 120, 146–8, 149
Richardson, T. 173
Ricketts, T.H. 16, 23
Righelato, R. 410
rights, emphasis on 65, 301
rinderpest 292–3, 300
Ringia, O. 48, 52

Rio Declaration on Environment and Development (Agenda 21) 218, 229
rivalry in consumption 128, 131, 133, 134, 160, 317
  see also non-rivalry in consumption
river pollution 368, 369
  see also water pollution
RME (rapeseed methyl ester) 418, 419
roads 383
Roberts, C.M. 21
Roberts, J.T. 364, 373
Robinson, J. 338
Roca, J. 361
Rock, M. 385
rodents 61
Roderick, M.L. 285
Rodgers, W.A. 31, 37, 290, 293, 299, 300, 301
Rodrigues, A.S.L. 17, 20
Roe, D. 32, 43
roe deer (*Capreolus capreolus*) 88, 89–90
Rohde, R. 297, 303
Roland, M.A. 91
Romania 409
Room, J. 139
Rose, D.B. 176
Rosenzweig, M.L. 75
Rothman, D.S. 363, 386–7
Rotty, R.M. 374
Rousseau, J.J. 329
routines 174
Royal Society for the Protection of Birds (RSPB) 48, 69
rubber-vine 179
Rucker, R. 148
Ruddle, K.R. 241, 260, 262
rule of law 377, 383
rules for decision-making 120, 204
Runge, C.F. 125
Russell-Smith, J. 179
Russia 14, 423
Ryan, L. 419
Rydin, Y. 2, 168, 169, 170, 172, 173, 174, 175, 193, 194, 283, 304

safe minimum standards approach 107–8
Salafsky, N. 31, 40, 43, 56

San Salvador Island, Philippines 149–50
sanctions for non-compliance 211–12, 225, 236
Sanderson, E.W. 9, 12, 13, 16, 20
Sands, P. 199, 217
sanitation 368, 369, 382, 383, 390
Sanjayan, M.A. 31, 33
Santos, T. 82
Sarkar, S. 22
savannas 42, 86, 87, 90, 97, 176, 285, 288
  economic valuation of 331, 334–6, 338–40
scale effect 388, 396
scarids 248
Schemske, D.W. 92
Schindler, D.W. 78, 79
Schlager, E. 150
Schmalensee, R. 373–4
Schmida, A. 92
Schmidt, K. 10
Schmidt, L. 419
Scholes, R.J. 87, 94
Schrijver, N. 31
Schroeder, M. 91
Schubert, C. 414
Science 3
science assessment 208
Scoones, I. 173, 296, 297, 299, 334, 338–9, 343, 350
scorpionfish (*Scorpaenidae*) 260
Scotland 83, 90, 93, 101
scots pine (*Pinus silvestris*) 88–9, 90, 93
Scott, A. 124–5
Scrase, I. 169
SCUBA equipment 249, 251
sea cucumber (*Holothuriidae*) 241, 249, 250–51, 274
Sea of Marmara 221
Seabright, P. 151
seagrass fishery on Quirimba Island, socioeconomic aspects of 5, 241–79
  history of the Quirimbas 242–4
  reasons for study 241–2
  study area 242
  study conclusions 278–9
  study methods 244–5

study results and discussion 245–78
  alternative sources of income 268–70
  changes in fishing patterns and effect on economy 278
  division of fisheries between the sexes 266–8, 269
  economic gain from fishing 257–9, 278
  employment opportunities 264–6
  fish prices 250, 251–3, 257, 259, 263–4, 270
  fishing traditions 260–63
  integrated coastal zone management in northern Mozambique 271
  investment by fishers 254–7, 278
  itinerant fishers 271–3, 274, 276–7
  management implications and problems 275–7
  provincial and recent historical context 263–4
  Quirimba during the study period (1996–97) 245–6
  salaries 255, 256, 257, 258, 263, 264, 278–9
  social structure of fishing fleet 266
  socio-cultural aspects of fish use 259–60
  threats to sustainable resource use 273–5
  total value of fishery 263
  types of marine resource use 246–53
seagrass parrot fish (*Leptoscarus vaigiensis*) 247, 250, 251
seahorses 251
seamoth fish (*Pegasidae*) 260
search costs 135, 138
Sechrest, W. 21
sectoral composition of the economy 366, 375–6, 378–80, 385, 386, 387, 388, 396
seed dispersal 78
seine net fishing 246, 247, 254, 255–8, 260, 261, 263–4, 266–7, 269, 270, 272, 278–9
Selden, T.M. 360, 364, 366, 371–3
Seligman, N.G. 284, 287
Selous Game Reserve 293–4

Semesi, A.K. 267
semi-arid regions
  climatic and vegetative characteristics of 90–91, 283–8, 294, 321–2, 332
  definitions of 287–8
  ecological and social paradigms and management of 5–6, 295–304
  economic valuation of different forms of land-use 331–56
  game theory applied to CPR management in Tanzania 321–8
  role of people and livelihoods in 90–91, 288–95
Sen, A. 329
Serbia and Montenegro 221
Sere, C. 287
Serengeti 86, 90, 285
Serra King, H.A. 269
Seychelles 266
Shafik, N. 360, 363, 368–70, 385, 393
shark fin fisheries 274
Sharma, M.P. 421, 422
sheep 86–7, 89, 94, 97, 106
Shelanski, H.A. 135
Sherman, P.B. 332
Shetland 97, 106
Shi, H. 13
Shields, D. 158
Shorter, A. 289, 290, 291
shrub species 89
Sillitoe, P. 242
Silva, J. 90
Silva-Pando, F.J. 89
silver birch (*Betula pendula*) 89–90, 92
Simberloff, D. 94
Simpson, I.A. 97, 106
Sisk, T.D. 13, 17
SITES 41
sites of special scientific interest (SSSI) 107–8
Skjaerseth, J.B. 216–17, 232–3, 235
Slatter, S.M. 178
slave trade 243
Sloan, R.J. 87
Slovenia 221
Smale, M. 103
small animal husbandry 39
Smith, A. 401
Smith, C.D. 294, 303

Smith, G. 170
Smith, J.M. 329
Smith, R.J. 13, 121
Smith, T.B. 21
Smith, V.L. 125
smoke 367–8, 382, 391
snappers (*Lutjanidae*) 251
Snidal, D. 310
Snow, C.P. 1
social capital 143, 144–6
social constructivist approach 173–4, 175
social cost 96, 129
social norms 120, 122, 131, 204
social paradigms 298–9
socially optimal level 317
Sofala Province 271
soil erosion 298, 301, 321–3, 336, 344, 380, 390
soil type 343
Solow, A. 77
Somalia-Maasai region 285, 288, 291, 298
Song, D. 360, 364, 366, 371–2
Songorwa, A.N. 32, 34
soot 418
Soulé, M.E. 77
Sousa, A.G. 243
South Africa 14
South Asia 157–8
South East Asia 269, 270
South Pacific 260, 269
Southwood, T.R.E. 17
soya beans 410, 418, 419, 421, 423, 425, 426
SPA/RAC (Specially Protected Areas Regional Activity Centre) 228
Spain 82, 89, 221, 228, 394
SPAMI List 225
Spanish Pyrenees 84
Spash, C. 101, 105
spatial decision support system (SDSS) software 41
spear-fishing 248, 249, 265
Spears, J. 152, 157
specialist opinion, irreplaceability measures derived from 12, 17
Specially Protected Areas Regional Activity Centre (SPA/RAC) 228
species extinctions 107, 137

Species of European Conservation Concern (SPECS) 88
Spector, S. 21
Spinage, C.A. 30
Spracklen, D.V. 410
Sprinz, D.F. 209, 210
spruce 92
spruce bark beetle (*I. typographus*) 91
spruce budworm 93, 104
squid 252
Sri Lanka 375
stability, ecosystem 79–80, 87, 90–91, 94–5, 102–3
stag hunt game 201, 329
see also assurance game
Staines, B.W. 90
stakeholder involvement in ICDP design 36–7, 40, 41, 50, 52–3
standards, environmental 209, 214, 217, 218, 362, 376, 415, 427
safe minimum standards approach 107–8
Stanhill, G. 285
start-up costs 139, 154–5
state and transition models (STM) 91
state capability 144–6
state property 124, 126, 127
see also nationalization; public sector governance, transaction costs of
state security 200, 215–16, 234, 409
Stattersfield, A.J. 9, 12, 13, 14, 16
Stavins, R. 214, 362
Steenblik, R. 410
Steenkamp, J. 331
Stern, D.I. 398, 400
Sternberg, M. 93
Stevenson, G.G. 124, 128, 129, 130, 132–4, 160
Stevis, D. 202
Stockholm Ministerial Conference 217
Stocking, G.C. 176
Stocking, M. 31, 301
stocking rates 97, 106, 298–9, 301, 339
Stoll, J.R. 102
story-lines 173, 174, 176
in Cape York seminar transcript 181–5, 186, 188, 192
Stott, P. 301
Strict Nature Reserves 33, 34, 68

strong sustainability 402
structural adjustment programmes 401–2
structural diversity 92
structural realism 200
Stuart, S.N. 17
Sudan 85
sugarcane 414, 418, 423, 425, 426
Sugden, R. 125
Sullivan, S. 297, 301, 303
sulphur dioxide (SO_2) emissions 211, 363, 367–8, 369, 370, 372, 376, 377–9, 382, 384, 385, 387, 388–9, 390–91, 393, 395, 398, 399, 418
Summers, R. 367–8, 373, 379, 381, 382, 383, 390, 398
Sumner, D.R. 103
Sun, D. 93
sunflower oil 419, 421
supply gap 171–2
Supriatna, J. 22
Suri, V. 365, 385–6, 393
suri traps 247
survivalism discourse 175, 185, 186
suspended particulate matter (SPM) 363, 367–8, 369, 370, 372, 376, 382, 387, 389, 390, 391, 393, 395, 396, 399
sustainability discourse 174, 175, 185, 186
sustainable development
  definition of 282
  environmental management for 5–6, 282–304
  poverty reduction and 282, 303–4
  principles of 408
sustainable end points for ICDPs 42–3, 55–6, 61–2, 66
sustainable extraction 39
Svendsen, J.O. 48
Swanson, T. 334
Swatuk, L.A. 215
Sweden
  agri-environmental programme in 148
  biodiversity in 88, 91, 92
  biofuel usage in 412–13, 414, 422
  economic growth and pollution in 394, 397
Swift, D.M. 298

swimming crab (*Portunus pelagicus*) 249
Switzerland 397
Sykes, J.M. 90
symmetric equilibrium 316
Syria 221

Taffs, K.H. 11, 12, 16
Taghi Farvar, M. 126
Taiwan 375
Takekawa, J.Y. 87, 104
TAMARIN 41
Tang, S.Y. 153
TANU (Tanzania African National Union) 300
Tanzania
  climatic and vegetative characteristics of 284, 285–6, 288, 294, 332
  environmental management in colonial-influenced approaches to 6, 293–4, 295, 299–300
  ecological and social paradigms and 5–6, 295–304
  economic valuation of different forms of land use 331–56
  game theory applied to CPR management 321–8
  Integrated Conservation and Development Project (ICDP) in Uluguru Mountains 45–56, 67, 68–9
  Participatory Environment Management Programme 35
  tax system and 43
  fish consumption in 270
  historical development of societies in 289–95
  income sources in 270
  itinerant fishers from 272
Tanzania African National Union (TANU) 300
taxes 43, 105, 401, 413
Taylor, P.D. 91
Taylor, P.J. 107
technology shifts, effect on income–pollution relationship 366, 369–70, 385, 394–5
technology uptake 63–4, 227
Telleria, J.L. 82, 83, 92

temperate forests 14
Terborgh, J. 92
terminal nodes 312
termites 17–20
terrestrial vertebrate species 12, 17, 20, 48
Thailand 380
theft 261–2
Thomas, C.D. 13
Thomas, R.P. 131
threats to natural resources 35–6, 38–41, 44, 45, 47, 56, 66, 143, 200, 273–5
three Cs approach to measuring effectiveness of international institutions 206–7
Thucydides
History of the Peloponnesian War 199
tiger beetles 17–20
tigers 41
Tilia 88
Tilman, D. 80, 81
timber production/extraction 42, 53, 54–5, 100, 101, 102, 103–4, 125, 148, 275, 331, 335
Tinbergen, L. 87
Tole, L. 371
Toledo, V.M. 337
Torrano, L. 84
Torras, M. 365, 381–2
total coliform 368, 384
total economic value (TEV) 98, 99
tourism 39, 42, 43, 45, 184, 244, 269, 270, 276, 303, 333
toxic emissions 375
traffic volume 372
tragedy of the commons 124, 125–6, 128, 132, 160
transaction cost economics (TCE) 135
transactions costs
    and collective action 138–41
    definitions of 135, 140
    and effectiveness of international environmental regimes 213
    empirical studies of 146–50, 160–61
    problems associated with 140–41
    group heterogeneity and 150
    insights from economic literature 134–8

and selection of appropriate governance structure 135, 138–9, 141–6, 159
transactions costs approach to institutional analysis 119
transboundary effects 362
transparency 146, 208, 214
transport
    biofuel use in 409, 411–27
    livestock used for 335, 339, 340
Trapnell, C.G. 285
travel cost method (TCM) 100
tree planting 39, 171, 301
tree species 79, 88, 89
trespassing game 324–8
triage 11
tropes 192
trophic-level analysis 78
tropical rain forests
    biofuel crop production and 7, 410, 415, 423–4, 425, 426, 427
    burning of 176–7
    see also Amazonian rain forests
trust 150, 151, 152, 261, 303
trust funds 43, 55, 62, 69, 231
Tscharntke, T. 91
tsetse fly 286, 288, 293–4, 295
Tsongwain, D.V. 64
Tuan, N.A. 380–81, 389
Tubbs, C.R. 88
Tucker, A.W. 329
Tucker, G.M. 83, 88
tulip shells (*Pleuroploca trapezium*) 248, 268
Tumucumaque National Park 16
Tunis 225, 228
Tunisia 221, 228
Turkey 221
Turley, D. 419, 420, 421
Turner, R.K. 99
turtle (*Chelonidae*) 20, 249

Uganda 35
Ujamaa villages 294
UK Department for International Development (DFID) 242, 287, 288
Ukraine 423
Uluguru Mountains, Tanzania 45–56, 67, 68–9

uncertainty 107, 108, 136–7, 138, 139, 141, 159, 171, 174, 184, 188, 193, 220, 303, 312, 425
  radical uncertainty 171, 193
Underdal, A. 206, 209
underinvestment 130, 160
UNDP (United Nations Development Programme) 32, 64, 244, 382
UNEP (United Nations Environment Programme) 75, 76, 217, 218, 224, 227, 228, 229, 230, 231, 232, 287, 371, 381
UNEP/MAP 221, 230
  see also Mediterranean Action Plan (MAP)
UNEP Regional Seas Programme 217
UNESCO (United Nations Educational, Scientific and Cultural Organization) 227
UNIDO (United Nations Industrial Development Organization) 227
unique Nash equilibrium 316, 317, 319
United Kingdom
  biodiversity in managed landscapes in 83, 87, 88
  economic growth and pollution in 394, 395, 397
  ecosystem disturbance and productivity in 90
  Environmentally Sensitive Area (ESA) scheme in 105–6
  feedstock production for biofuels in 415, 420
  sites of special scientific interest (SSSI) in 107–8
  valuing biodiversity in 100, 101
United Nations 203, 374, 376, 384
United Nations Conference on Environment and Development 'Earth Summit' (Rio, 1992) 218, 221
United Nations Conference on the Human Environment (Stockholm, 1972) 199, 359
United Nations Development Programme (UNDP) 32, 64, 244, 382
United Nations Educational, Scientific and Cultural Organization (UNESCO) 227
United Nations Environment Programme (UNEP) 75, 76, 217, 218, 224, 227, 228, 229, 230, 231, 232, 287, 371, 381
United Nations Framework Convention on Climate Change 202, 408
United Nations Industrial Development Organization (UNIDO) 227
United Nations International Comparison Programme 386
United Nations Millennium Development Goals 7, 282, 409
United Nations Statistical Yearbook 390
United States
  biodiversity conservation priorities in 14
  biofuel use in 418, 421
  economic growth and pollution in 375, 376, 380, 393, 394–5, 397
  Kyoto Protocol not ratified by 408–9
  valuing biodiversity in 100, 102, 104
Unruh, G.C. 396–8
Uphoff, N. 150
Uri, N.D. 336
US Environmental Protection Agency (EPA) 376, 395
US–Mexico Free Trade Agreement 367
use values 98, 100
  direct use value 99–100, 335
  measurement of 101–2, 336–40, 342–3, 345–50, 355
  indirect use value 100, 331–2, 335–6
  measurement of 102–4, 340, 343–4, 351–5
utility maximization 119

Valderrábano, J. 84
Vallega, A. 231
valuation of biodiversity 98–100
  indirect use values and ecosystem function 102–4
  measuring biodiversity values 100–101
  value of species and habitats 101–2

valuation of different forms of land use 331–56
  ecosystem values 334–6
  importance of valuation of ecosystems 331–2
  past valuation studies 336–41
  production function approach to 341–5
    direct use values 339–40, 342–3, 345–50
    indirect use values 340, 343–4, 351–5
    limitations to use of 344–5
  Tanzanian land-use issues 332–4
Van der Elst, R. 242
van Dorp, D. 88
Van Hees, A.F.M. 89–90
Van Keulen, H. 284, 287
van Rensburg, T.M. 101
Vane-Wright, R.I. 77
variegated emperor fish (*Lethrinus variegatus*) 247, 250, 251
Varughese, J. 130, 150, 156, 160
Vätn, A. 146
Veblen, T.T. 89
Veech, J.A. 13, 14
vehicle emissions 379, 383, 392, 409, 417–18, 419, 420, 426, 427
Velded, T. 150, 156–7
verification of compliance 213–14, 229–30
vertical integration 138
Victor, D.G. 213–14
Vienna Convention (1985) 209, 376
Vietnam 380
village elders 262
Village Forest Reserves 49, 50, 53, 65
villagization 293–4, 295
Vincent, J.R. 364, 389–90
volatile organic compounds (VOCs) 387, 395
von Moltke, K. 207–8, 214
Von Neumann, J. 310
vulnerability
  measures of 12–13, 17
  relative to irreplaceability 10–11, 14–16, 18–19

Wagenaar, H. 171, 172
Wainwright, C. 31
Wakeford, R.C. 266
Walker, B.H. 78, 87, 90, 94, 103
Walker, T.S. 103
Wallace, M.B. 129, 130, 160
Walter, A. 284
Waltz, K.N. 200
Wamer, N.O. 92
Wang, N. 122, 123
Warming, J. 124
Warren, A. 297, 298
Wasser, S.K. 48
water payments 43, 55
water pollution
  biofuel feedstock production and 419, 420
  economic growth and 368, 369, 375–6, 381–2, 383–4, 387, 389, 390, 391
  see also river pollution
water supplies 42, 48, 52, 54, 55, 59, 64, 65, 295, 336
  clean 42, 360, 368, 369, 380, 382, 383
  see also rainfall
Watt, P. 172
Wattle Hills residents' policy stance on burning 180, 181, 182
  discourses adopted in support of 182, 184, 185, 186, 187, 190–91
WCED (World Commission on Environment and Development) 95, 104, 282
Weale, A. 214
Weber, G.E. 89
Wehrmeyer, W. 31
Weisbrod, B. 98
Weitzman, M.L. 77
Welch, D. 90
Wells, M. 31, 32, 43, 67
wells 148–9
Wenner, C.G. 301
Werner, U. 14
Weslien, J. 91
Wesseler, J. 421
West, H.G. 243, 262
Westoby, M. 91
Wettestad, J. 206
Wheeler, W.B. 171
Whelan, C.J. 87
Whigham, D.F. 88, 92
White, A.T. 241

White, F. 285, 286, 292
white stork 83
Whiteman, A. 100
Whittaker, R.H. 82
Whittaker, R.J. 14, 22
Whitten, T. 8
Whittington, M.W. 242
whooping crane (*Grus americana*) 102
Wikramanayake, E.D. 22
Wilcox, B.A. 77
Wildlife Conservation Society of Tanzania 48, 49, 51, 69
Williamson, O.E. 135, 136, 141, 147, 161–2
willingness to accept (WTA) compensation 98, 100
willingness to pay (WTP) 98, 100, 101, 151–2
Willis, K.G. 100, 101, 105
Willson, M.F. 92
Wilson, E.O. 75, 91, 92, 331
Wilson, J.D.K. 272
Wilson, K.A. 12, 13
Wilson, M.V. 92
Wilson, W.L. 88
Wily, L.A. 42, 68
wind 94
witch-doctors (*curandeiros*) 243, 262
Wittmer, H. 136, 137, 140, 141, 142, 143, 144, 145, 146, 147
women, fishing by 245, 248, 252, 254, 255, 266–8, 269, 270, 278
wood warbler (*Phylloscopus sibilatrix*) 83
World Bank 55, 373, 401
  World Bank Tables 378, 396
  World Development Report 381, 385

World Commission on Environment and Development (WCED) 95, 104, 282
World Health Organization (WHO) 227, 367, 383
World Petroleum Institute 377
World Resources Institute (WRI) 75, 372, 378, 381, 383
World Summit on Sustainable Development (WSSD) 32
World Trade Organization 426–7
World Wildlife Fund (WWF) 9, 11, 13, 16, 55
WRI (World Resources Institute) 75, 372, 378, 381, 383
WSSD (World Summit on Sustainable Development) 32
WWF (World Wildlife Fund) 9, 11, 13, 16, 55
WWF-World Bank Management Effectiveness Tracking Tool 56
Wyckoff-Baird, B. 31
Wynter, P. 270

Xai-Xai Integrated Coastal Area Management project 271

Yellowstone National Park 95
Yohe, G. 13
Young, O.R. 202, 204, 207, 210–11

Zak, P.J. 150
Zambesian region 285–6, 288, 298
Zambezi river 336
Zanzibar (Unguja) Island 270
Zimbabwe 43, 87, 339–40
Zohary, M. 82